ENCYCLOPEDIA OF MILITARY AIRCRAFT

ENCYCLOPEDIA OF MILITARY AIRCRAFT

FROM WORLD WAR I TO THE PRESENT

CHRIS McNAB

This book first published in 2024 by

Amber Books Ltd
United House
North Road
London N7 9DP
United Kingdom
www.amberbooks.co.uk
Facebook: amberbooks
YouTube: amberbooksltd
Instagram: amberbooksltd
X(Twitter): @amberbooks

Copyright © 2024 Amber Books Ltd

All rights reserved. With the exception of quoting brief passages for the purpose of review no part of this publication may be reproduced without prior written permission from the publisher. The information in this book is true and complete to the best of our knowledge. All recommendations are made without any guarantee on the part of the author or publisher, who also disclaim any liability incurred in connection with the use of this data or specific details.

ISBN: 978-1-83886-452-1

Project Editor: Michael Spilling
Picture Research: Terry Forshaw
Design: Mark Batley, Lewis Hughes-Batley, Keren Harragan

Printed in China

Contents

Introduction	8
WORLD WAR I	**10**
Blériot XI (1909)	12
Jeannin Taube (1910)	13
Royal Aircraft Factory B.E.2c (1912)	14
Albatros B.II (1913)	15
Vickers F.B.5 (1914)	16
Voisin III (1914)	17
LVG C.I & C.II (1915)	18
Fokker Eindecker (1915)	19
Caudron R.11 A.3 (1915)	20
Aviatik C.I (1915)	21
Albatros C.I & C.III (1915)	22
Nieuport 10 (1915)	23
Morane-Saulnier Type N (1915)	24
Rumpler C.I (1915)	25
Sopwith Pup (1916)	26
Sopwith Triplane (1916)	27
Nieuport 11 (1916)	28
Royal Aircraft Factory R.E.8 (1916)	29
Nieuport 17 (1916)	30
Airco DH.2 (1916)	31
LFG Roland C.II (1916)	32
Halberstadt D.V (1916)	33
Albatros D.I & D.II (1916)	34
Albatros D.III (1916)	35
Bristol F.2A & F.2B (1916)	36
Royal Aircraft Factory F.E.2b (1916)	37
Royal Aircraft Factory S.E.5a (1916)	38
Sopwith Camel (1916)	39
DFW C.V (1916)	40
Hanriot HD.1 (1916)	41
SPAD S.VII (1916)	42
Handley Page O/400 (1916)	43
Nieuport 27 (1917)	44
Fokker Dr.1	14 45
SPAD S.XIII (1917)	46
Pfalz D.III (1917)	47
Siemens-Schuckert D.III (1917)	48
Gotha G.IV (1917)	49
Albatros D.V (1917)	50
Airco DH.4 (1917)	51
Zeppelin-Staaken R.IV and R.VI (1917)	52
Airco DH.9 (1917)	53
Bréguet 14 (1917)	54
Felixstowe F.2A (1917)	55
Caproni Ca.3 (1917)	56
Fokker D.VII (1917)	57
Hansa-Brandenburg W.12 (1917)	58
Sopwith Dolphin (1918)	59
INTERWAR PERIOD	**60**
Bréguet 19 (1921)	62
Armstrong Whitworth Siskin (1923)	63
Martin T3M and T4M (1928)	64
Tupolev ANT-9 (1931)	65
Hawker Fury (1931)	66
Tupolev TB-3 (1931)	67
Grumman FF and F2F (1931)	68
Bristol Bulldog (1932)	69
Boeing P-26 Peashooter (1932)	70
Blackburn Shark (1933)	71
Avia B.534 (1933)	72
Curtiss P-6 Hawk (1933)	73
Polikarpov I-16 (1934)	74
PZL P.11c (1934)	75
Fiat CR.32 (1934)	76
Caproni Ca.133 (1935)	77
Douglas C-47 Skytrain (1935)	78
Heinkel He 111 (1935)	79
Heinkel He 51 (1935)	80
Supermarine Walrus (1935)	81
Messerschmitt Bf 109 (1935)	82
Hawker Hurricane (1935)	83
Junkers Ju 87 Stuka (1935)	84
Boeing B-17 Flying Fortress (1935)	85
Mitsubishi A5M4 'Claude' (1936)	86
Supermarine Spitfire (1936)	87
Junkers Ju 88 (1936)	88
Messerschmitt Bf 110 (1936)	89
Consolidated PBY Catalina (1936)	90
Henschel Hs 123 (1936)	91
Savoia-Marchetti SM.79 Sparviero (1936)	92
Tupolev SB-2 (1936)	93
Fairey Swordfish (1936)	94
Morane-Saulnier MS.406 (1937)	95
Nakajima B5N Tenzan (1937)	96
Bristol Blenheim Mk IV (1937)	97
Gloster Gladiator (1937)	98
Fieseler Fi 156 Storch (1937)	99
Ilyushin DB-3 (1937)	100
Douglas TBD Devastator (1937)	101
Seversky P-35 (1937)	102
Dornier Do 24 (1937)	103
Henschel Hs 126 (1937)	104
Blohm & Voss BV 138 (1938)	105
Vought SB2U Vindicator (1938)	106
Short Sunderland Mk III (1938)	107
Handley Page Hampden (1938)	108
Brewster F2A Buffalo (1938)	109
Vickers Wellington (1938)	110
Mitsubishi A6M Reisen 'Zeke' (Zero) (1938)	111
Curtiss SBC Helldiver (1938)	112
Vought OS2U Kingfisher (1938)	113
Blackburn Skua (1938)	114
Curtiss P-36 & Hawk Model 75 (1938)	115
WORLD WAR II	**116**
Ilyushin Il-2 Shturmovik (1939)	118
North American B-25 Mitchell (1939)	119
Dewoitine D. 520 (1939)	120
Arado Ar 196 (1939)	121
Dornier Do 17 (1939)	122
Aichi D3A (1939)	123
Nakajima Ki.43 Hayabusa (1939)	124
Polikarpov I-153 (1939)	125
Curtiss P-40 Kittyhawk (1939)	126
Lockheed Hudson Mk.1 (1939)	127
Ilyushin Il-4 (1939)	128
Focke-Wulf Fw 189 Uhu (1939)	129
Boulton Paul Defiant (1939)	130
Douglas A-20 Havoc (1939)	131
Fairey Albacore (1940)	132
Martin PBM Mariner (1940)	133
North American P-51 Mustang (1940)	134
Yakovlev Yak-1 (1940)	135
SBD-3 Dauntless (1940)	136
Focke-Wulf Fw 200 Condor (1940)	137
Mitsubishi Ki.46 Dinah (1940)	138
Petlyakov Pe-8 (1940)	139
Avro Manchester (1940)	140
Bristol Beaufighter TF.Mk X (1940)	141
Handley Page Halifax (1940)	142
Short Stirling (1940)	143
Douglas DB-7/A-20 Havoc/Boston (1940)	144
Grumman F4F Wildcat (1940)	145
Lavochkin-Gorbunov-Gudkov LaGG-3 (1940)	146
Fairey Fulmar (1940)	147
Vought F4U Corsair (1941)	148
Lockheed P-38 Lightning (1941)	149
Focke-Wulf Fw 190 (1941)	150

Republic P-47 Thunderbolt (1941)	151	**COLD WAR**	**202**	De Havilland Sea Vixen (1959)	254
De Havilland Mosquito (1941)	152	De Havilland Vampire (1946)	204	Bell UH-1 Iroquois (Huey) (1959)	255
Avro Lancaster (1941)	153	Douglas A-1 Skyraider (1946)	205	Saab J-35 Draken (1960)	256
Consolidated B-24 Liberator (1941)	154	De Havilland Hornet (1946)	206	English Electric Lightning (1960)	257
Mikoyan-Gurevich MiG-3 (1941)	155	Mikoyan-Gurevich MiG-9 (1946)	207	Convair B-58 Hustler (1960)	258
Petlyakov Pe-2 (1941)	156	McDonnell FH-1 Phantom (1947)	208	Westland Wessex (1960)	259
Supermarine Seafire (1941)	157	Lockheed P2V-1 Neptune (1947)	209	Nanchang CJ-6 (1960)	260
Martin B-26 Marauder (1941)	158	Hawker Sea Fury (1947)	210	Dassault Mirage III (1961)	261
Piper L-4 Grasshopper (1941)	159	Mikoyan-Gurevich MiG-15 (1948)	211	North American A-5A Vigilante (1961)	262
Messerschmitt Me 210 (1941)	160	McDonnell F2H Banshee (1948)	212	Boeing RC-135 (1961)	263
Bell P-39 Airacobra (1941)	161	Convair B-36 Peacemaker (1948)	213	Boeing CH-47 Chinook (1962)	264
Yermolaev Yer-2 (1941)	162	North American F-86 Sabre (1949)	214	Lockheed SR-71 'Blackbird' (1962)	265
Northrop P-61 Black Widow (1941)	163	Ilyushin Il-28 Beagle (1949)	215	Lockheed P-3 Orion (1962)	266
Yakovlev Yak-9 (1941)	164	Mikoyan-Gurevich MiG-17 (1950)	216	Tupolev Tu-22 Blinder (1962)	267
Grumman F6F Hellcat (1942)	165	Sikorsky H-19 Chickasaw/S-55/HRS-1 (1950)	217	Shenyang J-6/F-6 (1962)	268
Hawker Typhoon (1942)	166	Douglas C-124C Globemaster II (1950)	218	Grumman A-6 Intruder (1963)	269
Kawasaki Ki.45 Toryu (1942)	167	English Electric Canberra (1951)	219	General Dynamics F-111 (1964)	270
Kawasaki Ki.61 Hien (1942)	168	Boeing B-52 Stratofortress (1952)	220	Grumman E-2 Hawkeye (1964)	271
Curtiss SB2C Helldiver (1942)	169	Mil Mi-4 (1952)	221	Tupolev Tu-128 (1964)	272
Grumman TBF Avenger (1942)	170	Avro Vulcan (1952)	222	Lockheed C-141 StarLifter (1965)	273
Dornier Do 217 (1942)	171	Grumman F9F Panther (1952)	223	Sukhoi Su-15 (1965)	274
Heinkel He 177 (1942)	172	Westland Whirlwind (1952)	224	Transall C-160 (1965)	275
Lavochkin La-5 (1942)	173	Avro Canada CF-100 Canuck (1952)	225	Chengdu J-7/F-7 (1965)	276
Fairey Barracuda (1942)	174	Lockheed C-130 Hercules (1954)	226	Sikorsky S-65/CH-53 Sea Stallion (1966)	277
Messerschmitt Me 323 Gigant (1943)	175	Republic F-84F Thunderstreak (1954)	227	Bell OH-58 Kiowa (1966)	278
Fiat G.55 Centauro (1943)	176	Gloster Meteor NF.14 (1954)	228	Cessna O-2 Skymaster (1966)	279
Bell P-63 Kingcobra (1943)	177	Hawker Hunter (1954)	229	Cessna A-37B Dragonfly (1966)	280
Messerschmitt Me 410 Hornisse (1943)	178	Tupolev Tu-16 Badger (1954)	230	Antonov An-22 (1967)	281
Fairey Firefly (1943)	179	North American F-100 Super Sabre (1954)	231	Mil Mi-8 (1967)	282
Messerschmitt Me 262 (1944)	180	Lockheed U-2 (1955)	232	Bell AH-1 HueyCobra (1967)	283
Boeing B-29 Superfortress (1944)	181	Mikoyan-Gurevich MiG-21 (1955)	233	Lockheed AC-130 Spectre (1967)	284
Arado Ar 234 (1944)	182	Vickers Valiant (1955)	234	Vought A-7 Corsair II (1967)	285
Messerschmitt Me 163 Komet (1944)	183	Tupolev Tu-95 (1956)	235	Agusta-Bell 212 (1968)	286
Mitsubishi Ki.67 Hiryu (1944)	184	Boeing KC-135 Stratotanker (1956)	236	Aérospatiale Puma (1968)	287
Yokosuka MXY7 Ohka (1944)	185	Gloster Javelin (1956)	237	Lockheed C-5 Galaxy (1968)	288
Lavochkin La-7 (1944)	186	Douglas A-4 Skyhawk (1956)	238	BAC 167 Strikemaster (1968)	289
Tupolev Tu-2 (1944)	187	Vought F-8 Crusader (1956)	239	Sikorsky MH-53 Pave Low (1968)	290
Gloster Meteor III (1944)	188	Convair F-102 Delta Dagger (1956)	240	Ilyushin Il-38 (1968)	291
Hawker Tempest (1944)	189	Grumman OV-1 Mohawk (1956)	241	North American Rockwell OV-10 Bronco (1969)	292
Supermarine Spitfire FR. Mk XIVE (1944)	190	Douglas C-133 Cargomaster (1957)	242	Antonov An-30 (1969)	293
Dornier Do 335 Pfeil (1944)	191	Lockheed T-2V1 SeaStar (1957)	243	Ilyushin Il-20 and Il-22 (1969)	294
Lockheed F-80 Shooting Star (1944)	192	McDonnell F-101 Voodoo (1957)	244	McDonnell Douglas F-4 Phantom (1969)	295
Gloster Meteor F.4 (1944)	193	Supermarine Scimitar (1957)	245	Hawker Siddeley/BAE Nimrod (1969)	296
Douglas A-26 Invader (1944)	194	McDonnell Douglas F-4 Phantom II (1958)	246	Xi'an H-6 (1969)	297
Ryan FR-1 Fireball (1945)	195	Fiat/Aeritalia G.91 (1958)	247	Grumman F-14 Tomcat (1970)	298
Grumman F7F Tigercat (1945)	196	Beriev Be-10 'Mallow' (1958)	248	Antonov An-26 (1970)	299
Bachem Ba 349 Natter (1945)	197	Lockheed F-104 Starfighter (1958)	249	Hawker Siddeley/BAe Harrier (1970)	300
Heinkel He 162 (1945)	198	Republic F-105 Thunderchief (1958)	250	Mikoyan-Gurevich MiG-23 (1970)	301
Focke-Wulf Ta 152 (1945)	199	Handley Page Victor (1958)	251	Saab SF-37 Viggen (1971)	302
Junkers Ju 88 Mistel (1945)	200	Convair F-106 Delta Dart (1959)	252	Tupolev Tu-134 (1971)	303
Grumman F8F Bearcat (1945)	201	Sukhoi Su-7 (1959)	253	Grumman EA-6B Prowler (1971)	304
				Bell AH-1J SeaCobra (1971)	305

Tupolev Tu-22M (1972)	306	Beriev A-50 (1985)	354	Chengdu J-10 (2004)	404
Fairchild Republic A-10 Thunderbolt II (1972)	307	Rockwell B-1B Lancer (1985)	355	Antonov An-148 (2004)	405
		Bell AH-1W SuperCobra (1986)	356	Shaanxi Y-8 and Y-9 Special Mission (2004)	406
Aero L-39 Albatros (1972)	308	Changhe Z-8 (1986)	357		
SEPECAT Jaguar (1972)	309	Antonov An-124 (1986)	358	Lockheed Martin F-22 Raptor (2005)	407
Mi-24 'Hind' (1972)	310	Boeing AH-64 Apache (1986)	359	T-50 Golden Eagle (2005)	408
Tupolev Tu-142 (1972)	311	Atlas Cheetah (1987)	360	Bombardier CRJ (2005)	409
Lockheed Martin F-16 Fighting Falcon A/B (1974)	312	Tupolev Tu-160 (1987)	361	Shaanxi KJ-200 (2005)	410
				Xi'an KJ-2000 (2005)	411
Panavia Tornado (1974)	313	**MODERN ERA**	**362**	Leonardo C-27J Spartan (2006)	412
Dassault Mirage F.1 (1974)	314	Kamov Ka-29 (1987)	364	Hongdu JL-10 (2006)	413
Lockheed S-3 Viking (1974)	315	Boeing/Grumman E-8 J-Stars (1989)	365	General Dynamics F-16E/F (2007)	414
Ilyushin Il-76 and Il-78 (1974)	316	Northrop B-2 Spirit (1989)	366	PAC JF-17 Thunder (2007)	415
Northrop F-5E/F Tiger II (1974)	317	British Aerospace Harrier GR.7 (1989)	367	Mil Mi-28 (2007)	416
ShinMaywa SS-2 (1975)	318	Boeing F-15E Strike Eagle (1989)	368	MQ-8B Fire Scout (2009)	417
McDonnell Douglas F-15 Eagle (1975)	319	Mikoyan MiG-29K (1989)	369	Boeing 737 AEW&C (2010)	418
Mil Mi-14 (1975)	320	AMX International AMX (1990)	370	E-2D Advanced Hawkeye (2010)	419
Sukhoi Su-24 (1975)	321	Ilyushin Il-80 & Ilyushin Il-82 (1990)	371	Yakovlev Yak-130 (2010)	420
Mikoyan-Gurevich MiG-27 (1975)	322	MQ-1 Predator (1991)	372	Wing Loong II/Chengdu GJ-2 (2011)	421
FMA Pucará (1975)	323	Sukhoi Su-30 (1992)	373	Shaanxi Y-9 (2012)	422
Israeli Aircraft Industries (IAI) Kfir (1976)	324	AIDC F-CK-1 Ching-Kuo (1992)	374	Tupolev Tu-214 (2012)	423
Yakovlev Yak-38 (1976)	325	Xi'an JH-7 (1992)	375	Changhe Z-10 (2012)	424
British Aerospace Hawk (1976)	326	Saab JAS 39 Gripen (1993)	376	Kazan Ansat (2013)	425
Sukhoi Su-27 'Flanker' (1977)	327	McDonnell Douglas/Boeing C-17 Globemaster (1993)	377	Shenyang J-15 (2013)	426
Mikoyan MiG-29 'Fulcrum' (1977)	328			HAL LCA Tejas (2013)	427
Dassault/Dornier Alpha Jet (1977)	329	Mikoyan MiG-29M and MiG-35 (1994)	378	Sukhoi Su-34 (2014)	428
Mitsubishi F-1 (1978)	330	Harbin Z-9 (1994)	379	Sukhoi Su-35S (2014)	429
Boeing E-3 Sentry (1978)	331	Kamov Ka-31 (1995)	380	Airbus A400M Atlas (2014)	430
Dassault Mirage 2000 (1978)	332	Kamov Ka-52 (1995)	381	Shaanxi KJ-500 (2014)	431
McDonnell Douglas F/A-18 Hornet (1978)	333	Mil Mi-171 (1995)	382	Bayraktar TB2 (2015)	432
McDonnell Douglas AV-8B Harrier II (1978)	334	Saab Erieye (1996)	383	Lockheed Martin F-35 Lightning II (2015)	433
		AH-64D Longbow/AH-64E Guardian (1997)	384	Shenyang J-16 (2015)	434
Westland Lynx (1978)	335			Chengdu J-20 (2016)	435
Dassault Super Etendard (1978)	336	Bell V-22 Osprey (1997)	385	Kawasaki C-2 (2016)	436
Eurocopter AS332 Super Puma (1978)	337	RQ-4 Global Hawk (1998)	386	Xi'an Y-20 (2016)	437
British Aerospace Sea Harrier (1978)	338	Shenyang J-11 and J-11A (1999)	387	CH-5 Rainbow (2017)	438
Aermacchi M.B.339A (1979)	339	AgustaWestland EH101/AW101 Merlin (1999)	388	Harbin Z-20 (2017)	439
Sikorsky UH-60 Black Hawk/ SH-60 Sea Hawk (1979)	340			Embraer C-390 Millennium (2019)	440
		Sukhoi Su-27K and Su-33 (1999)	389	Sukhoi Su-57 (2021)	441
Mil Mi-26 (1980)	341	Boeing F/A-18E Super Hornet (1999)	390		
Shenyang J-8 (1980)	342	Lockheed Martin C-130J Hercules (1999)	391	**Glossary**	**442**
Lockheed F-117 Nighthawk (1981)	343	Mitsubishi F-2 (2000)	392	**Index**	**443**
Mikoyan MiG-31 (1981)	344	Aero Vodochody L-159 ALCA (2000)	393	**Picture Credits**	**448**
Sukhoi Su-25 (1981)	345	Boeing 737 (2001)	394		
Shaanxi Y-8 (1981)	346	Dassault Rafale (2001)	395		
Mil Mi-17 (1981)	347	Airbus C295 (2001)	396		
Kamov Ka-27 (1982)	348	MQ-9 Reaper (2001)	397		
Sikorsky HH-60 (1982)	349	Antonov An-140 (2002)	398		
General Dynamics EF-111A Raven (1983)	350	Kamov Ka-226 (2002)	399		
Xi'an Y-7 (1984)	351	Guizhou JL-9/FTC-2000 (2003)	400		
Vickers VC10 (1984)	352	Eurofighter Typhoon (2003)	401		
Lockheed Martin F-16 Fighting Falcon C/D (1984)	353	Shenyang J-11B (2003)	402		
		Beriev Be-200 (2003)	403		

Introduction

Military aviation literally added a new dimension to the history of warfare. The earliest military aircraft were pure observation types, aerial platforms for reconnaissance. But it did not take long before weaponry took to the air in the world's first fighters and bombers.

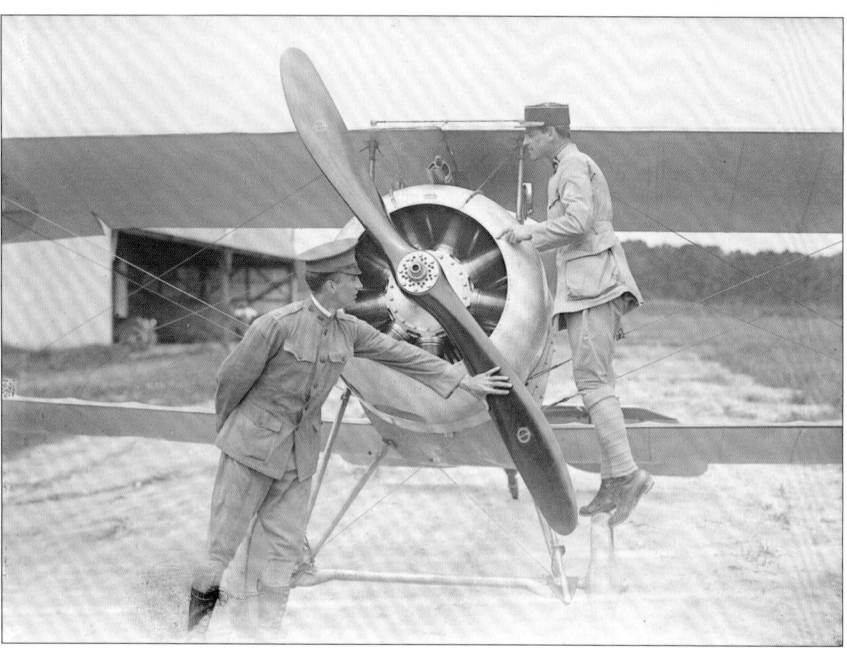

ABOVE:
A French Nieuport 17 fighter is examined at Langley Field, Virginia, by US Army Air Corps Captain JC Bartolf and French Air Service Lt. LeMaitre, 1917.

Aviation has become so quotidian that it is easy to forget just what a profound and rapid transformation it wrought on this planet. The first powered, controlled flight occurred, famously, on 17 December 1903, when Orville Wright piloted the 'Wright Flyer' for a momentous and linear 12 seconds at Kitty Hawk, North Carolina, travelling 36m (120ft) in the process. While this flight was distinctly civil in nature, the possible military applications of aircraft were almost immediately seized upon. Just six years later, in 1909, the Wright Flyer was repurposed as the Wright Military Flyer, history's first military aircraft, designed to meet a 1908 Signal Corps competitive bid for a new two-seat observation aircraft. The age of military aviation had begun.

INNOVATION AND ACCELERATION
The period between the early 20th century and our present age is unrivalled in terms of the sheer rapidity of technogical progress. The arena of military aviation is no exception. We can clarify this by a simple benchmarking exercise across roughly the first half century of combat aircraft. Our variables for this exercise are based on a common scenario: two opposing fighter aircraft approach each other at 32km (20 miles) distance, engaging each other when it is ready and possible to do so.

If we go back to 1915, for example, a British Vickers FB.5 biplane and a German Fokker E.IV monoplane, enemies over the World War I battlefields of the Western Front, would have a combined closing speed of approximately 274km/h (170mph). At this speed, it would take about six full minutes before they come into visual range, and 30 seconds later they would be able to engage each other with their machine guns, dogfighting their aircraft to align guns with opponent – this was the age of direct fire.

Now add just 50 years to our start date. It's 1965 and the aerial combat zone is over North Vietnam. Our two aircraft are a US F-8 Crusader and a North Vietnamese MiG-17. This time, the closing speed of these monoplane jets going flat out is around 2896km/h (1800mph), much of that accounted for by the supersonic Crusader. Instead of the time measured in minutes, the 32km (20 miles) separation will be closed entirely in about 40 seconds. And while the MiG will still rely on its cannon, the Crusader will engage the opponent with Sidewinder AIM-9C air-to-air missiles, a semi-active radar homing missile with a maximum range of 18km (11 miles), well beyond visual range.

AIR SUPERIORITY

These two examples, confortably within a human lifetime, give just a hint of the incredible evolution in air power outlined in this book. The aircraft of the present age are themselves several technological generations ahead of those 1960s, in terms of their performance, avionics, weaponry, material construction, digital intelligence and much more. This book brings together in historical sequence more than 400 different aircraft types, from rattling subsonic biplanes to stratospheric, hemispheric and supersonic bombers. The advances in aircraft capability are a testimony both to the profound ingenuity of engineers and designers, but also to the fact of a world locked in the race to achieve dominance of the skies.

LEFT:
The German Dornier Do 17 had a bombload of about 1000kg (2200lg); the major strategic bombers of World War II could carry more than six times that volume of ordnance.

BELOW:
A US Marine aviator exits a Marine Corps F-35B aircraft at Eglin Air Force Base. Standards of combat pilot training have in many ways evolved as much as the technologies around the pilots.

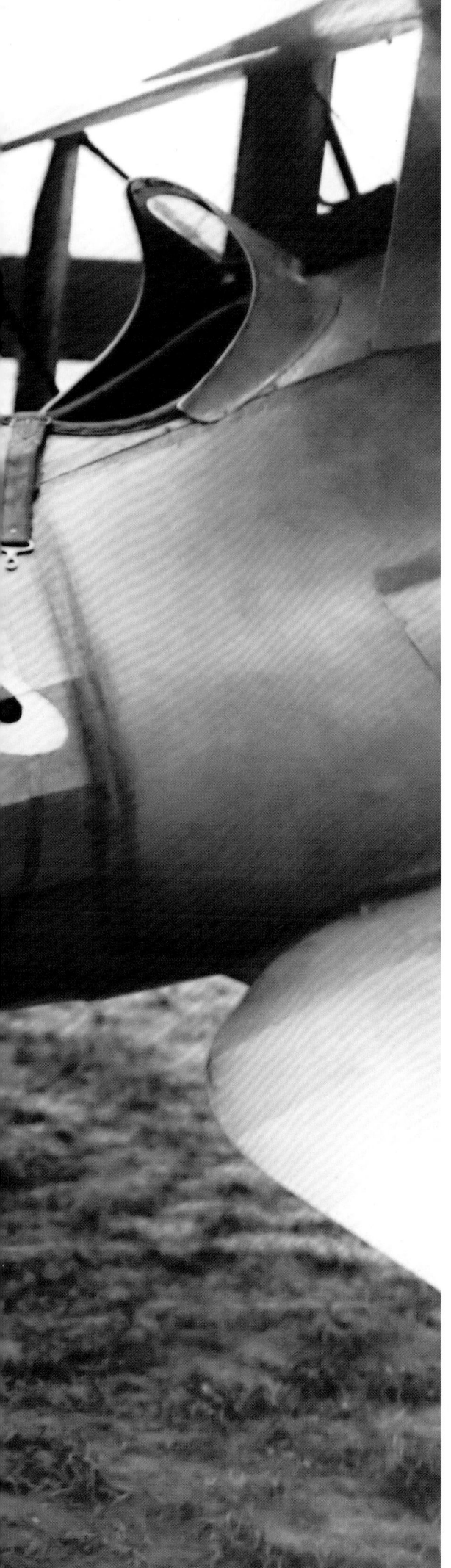

World War I

World War I was the proving ground for combat aviation. In 1914–15, reconnaissance aircraft usefully demonstrated their value in roles such as observation and artillery spotting. Then came the fighters, built for performance and aerial firepower. At first they were purposed mainly with downing enemy reconnaissance types, but in the resulting arms race the focus switched to fighter vs fighter dogfighting, facilitated by the invention of machine guns synchronized to fire safely through the arc of the propeller. The war also saw the emergence of the first strategic bomber and ground-attack aircraft, the latter laying the foundations of close air support Although many of the aircraft of World War I were crude, uncomfortable and mechanically dangerous, the conflict ended with aviation as a formal, established and effective branch of service amongst all the combatants.

SPAD ACE
The American combat pilot Eddie Rickenbacker stands aside his SPAD XIII aircraft. Rickenbacker was the most successful American fighter ace of World War I, with 26 kills. In 1930 he was awarded the Medal of Honor for his service.

Blériot XI (1909)

TYPE • *Reconnaissance* COUNTRY • *France*

A 1910 design, the Blériot XI was in reconnaissance service with France, Britain, Russia and Italy at the start of the war. Possessing modest performance and reliability, it was replaced by newer, more effective types as soon as practical.

SPECIFICATIONS

DIMENSIONS:	Length: 7.65m (25ft 1in); Wingspan: 8.9m (29ft 3.5in); Height: 2.5m (8ft 2in)
WEIGHT:	800kg (1764lb) loaded
POWERPLANT:	1 x 37kW (50hp) or 60kW (80hp) Gnome rotary engine
MAX SPEED:	120km/h (75mph)
ENDURANCE:	3 hours 30 minutes
CEILING:	2000m (6560ft)
CREW:	1
ARMAMENT:	None

Left: The Blériot XI came to fame on 25 July 1909, when the great Louis Blériot used the aircraft to make the first heavier-than-air crossing of the English Channel.

PROPELLER
The Chauvière Intégrale two-bladed scimitar propeller was made from walnut. It was a particularly efficient design for the time.

WING WARPING
The Blériot XI featured wing warping for lateral control. The system used pulleys and cables to twist the trailing edges of the wings, in much the same way as a bird controls its flight.

FUSELAGE
The fuselage was a partially covered box-girder type constructed from ash with wire cross bracing, a design that provided a good mix of rigidity and flexibility.

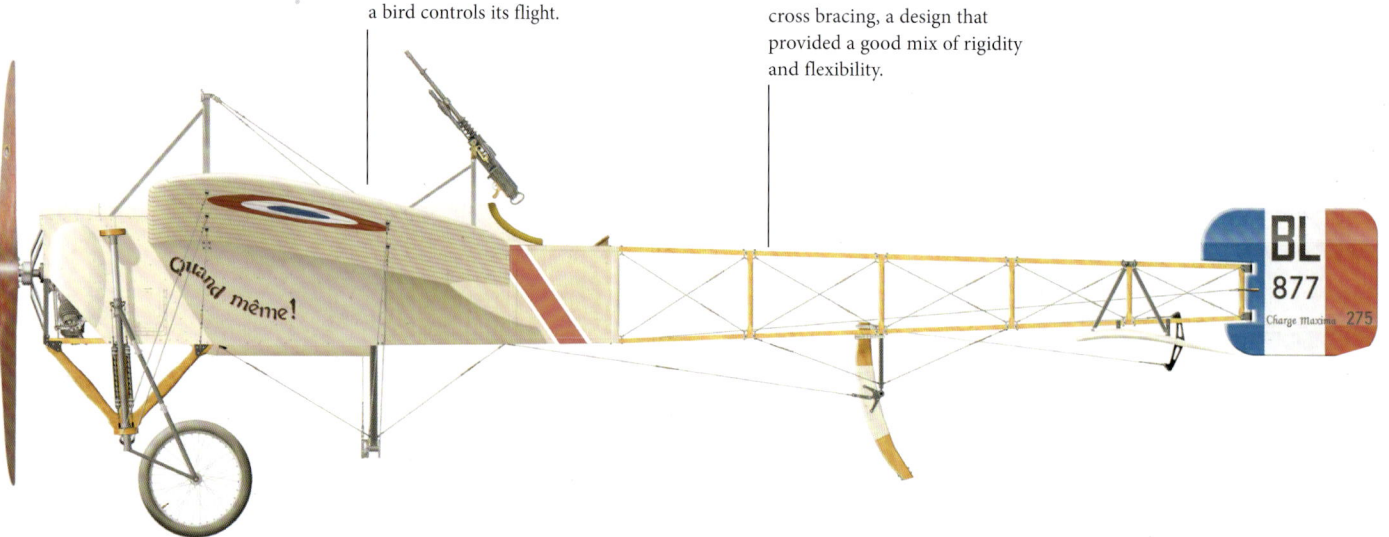

MILITARY USE
The Blériot XI was soon recognized for its military potential. Italy and France were the first military adopters of the aircraft in 1910, with the RFC following two years later. Typically, the aircraft were used as unarmed observer or trainer types, but during the war some aircraft were modified to carry a light 25kg (55lb) bombload.

RFC BLERIOT XI
Although designed as civilian aircraft, the Blériot XI found several military users, including the British Royal Flying Corps (RFC), which received its first XIs in 1912.

Jeannin Taube (1910)

TYPE • *Fighter* **COUNTRY** • *Germany*

SPECIFICATIONS

DIMENSIONS:	Length: 9.69m (31ft 10in); Wingspan: 13.87m (45ft 6in); Height: 2.97m (9ft 9in)
WEIGHT:	(gross) 1035kg (2020lb)
POWERPLANT:	One 89kW (120hp) Argus As.II 6-cylinder water-cooled inline piston engine
MAX SPEED:	100km/h (62mph)
ENDURANCE:	4 hours
CEILING:	2000m (6600ft)
CREW:	2
ARMAMENT:	None

Germany's most iconic aircraft of the pre-war period, the Taube (Dove) had already made history as the first aircraft to drop a bomb in combat. Its operational career during World War I was brief but eventful.

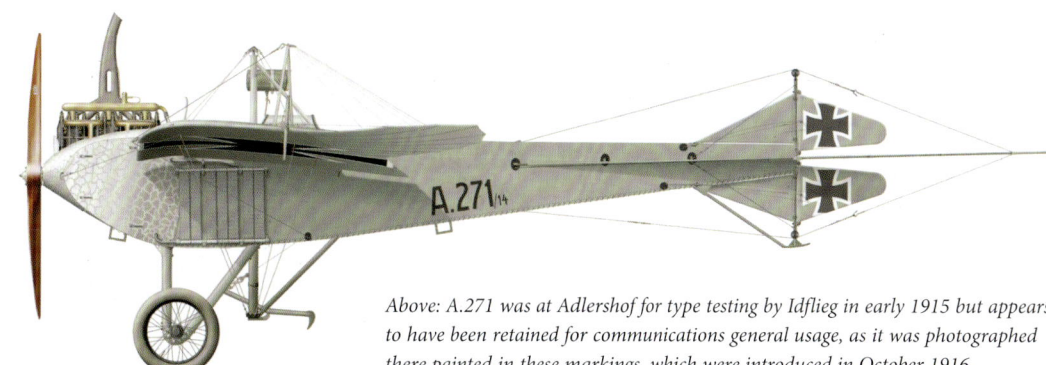

Above: A.271 was at Adlershof for type testing by Idflieg in early 1915 but appears to have been retained for communications general usage, as it was photographed there painted in these markings, which were introduced in October 1916.

ENGINES
Taubes were fitted with a variety of engines, mostly inline units, initially the 4-cylinder Mercedes E4F, but Taubes from some manufacturers also utilized rotary engines.

CONTROLS
Roll control was by wing warping, specifically controlling the rearward projecting wingtips; there was no elevator as such.

STEEL TUBE STRUCTURE
Jeannin utilized a steel tube internal structure, hence the 'Stahltaube' name, and this may have contributed to the longevity of A.180, the world's sole surviving Taube.

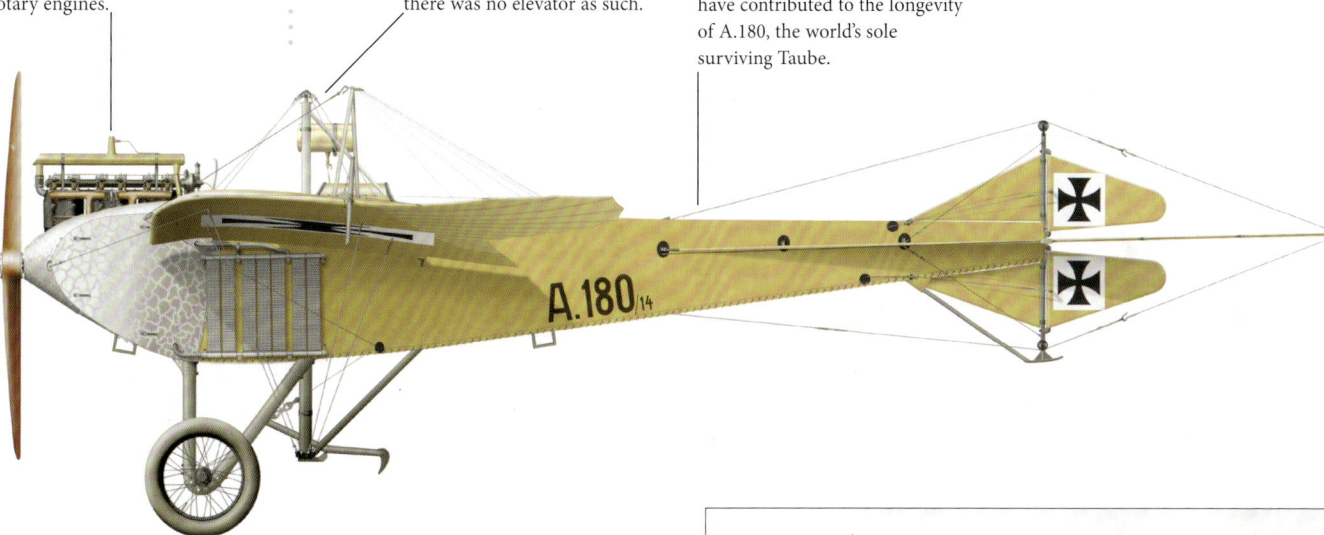

TEST CRASH
Designed in 1909 by Austro-Hungarian Igo Etrich, the Taube's distinctive wing planform possessed inherent stability in both pitch and roll. The Taube flew for the first time in 1910 with Etrich at the controls, although he handed testing duties to pilot Karl Illner after he nearly broke his back when the aircraft crashed on an early test flight.

ETRICH'S TAUBE
Etrich tailored the designs to meet the specific criteria of the military, which included such 'useful' attributes as the ability to land on a freshly ploughed field. Although Rumpler was the largest producer of Taubes, Jeannin was responsible for one of the next most common Taube clones.

Royal Aircraft Factory B.E.2c (1912)

TYPE • *Reconnaissance* **COUNTRY** • *United Kingdom*

In the early war years the B.E.2 was the most widely used British reconnaissance airplane. It was designed before the war, when great stability was thought to be important for reconnaissance.

SPECIFICATIONS

Dimensions:	Length: 8.31m (27ft 3in); Wingspan: 12.42m (40ft 9in); Height: 3.66m (12ft)
Weight:	971kg (2142lb) maximum take-off
Powerplant:	1 x 67kW (90hp) RAF 1a inline engine
Max Speed:	120km/h (75mph)
Endurance:	3 hours 15 minutes
Ceiling:	3048m (10,000ft)
Crew:	2
Armament:	1 x 7.62mm (0.303in) Vickers MG in front cockpit

Right: Royal Aircraft Factory Engine B.E.2c 100HP. The Royal Aircraft Factory BE 2e 100H, was a British aircraft of World War I.

Powerplant
The B.E.2c had a 67kW (90hp) RAF 1a inline engine generating enough power to take the aircraft to a maximum speed of 120km/h (75mph).

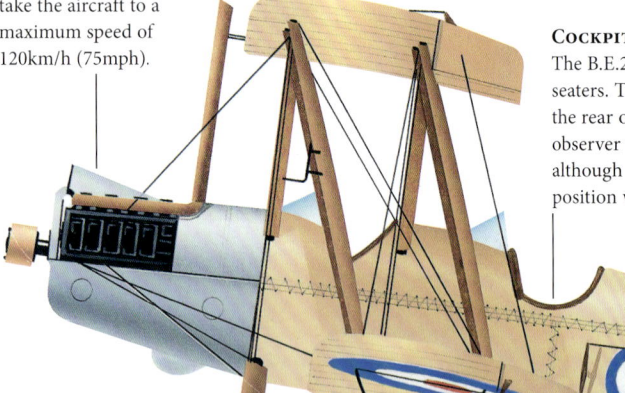

Cockpit
The B.E.2 aircraft were two-seaters. The pilot actually sat in the rear of the two seats. The observer was in the front seat, although the view from this position was not optimal.

New design
The B.E.2c was essentially the fifth variant in the B.E.2 series, but it was a significant redesign. The tailplane was one new element, with an improved rudder configuration.

Fokker Scourge
His Majesty's Balloon Factory at Farnborough diversified into heavier-than-air machines in 1909. The B.E.2c (Blériot Experimental) introduced the 67kW (90hp) RAF 1a engine and was the first to be armed with a machine gun. In wartime service, the B.E.2 was a fine reconnaissance platform, but its stability proved lethal in aerial combat and many were lost during the 'Fokker Scourge' of 1915–16.

RFC B.E.2 WITH CAMERA
The B.E.2 could be fitted with a side-mounted camera for aerial photography, this duty adding to the already formidable manual demands upon the pilot.

Albatros B.II (1913)

TYPE • *Reconnaissance* **COUNTRY** • *Germany*

SPECIFICATIONS

DIMENSIONS:	Length: 7.63m (25ft 0.38in); Wingspan: 12.8m (42ft); Height: 3.15m (10ft 4in)
WEIGHT:	(gross) 1071kg (2361lb)
POWERPLANT:	One 74.5kW (100hp) Mercedes D.I, or 89.5kW (120hp) Mercedes D.II, or 74.5kW (100hp) Argus As.I, or 89.5kW (120hp) Argus As.II, or 82kW (110hp) Benz Bz.II 6-cylinder water-cooled inline piston engine
MAX SPEED:	120km/h (75mph)
ENDURANCE:	4 hours
CEILING:	3000m (9840ft)
CREW:	2
ARMAMENT:	None

The Albatros B-types were very successful in the early war period, and the B.II subsequently enjoyed an astonishingly long career as a trainer. However, it also achieved a certain notoriety by becoming the first German aircraft to attack Britain.

Left: Remarkably, the Albatros B.II would be manufactured by seven different companies in large numbers, although exact figures are unknown.

ENGINE MOUNTING
As built, it would have mounted a radiator above the engine but this example has been retrofitted with fuselage-mounted Hazet radiators. The engine here is the Mercedes D.II.

TAIL
This early production B.II had a new rudder fitted, replacing the original smaller unit and obliterating half the tailplane *Eisernkreuz* (Iron Cross) in the process.

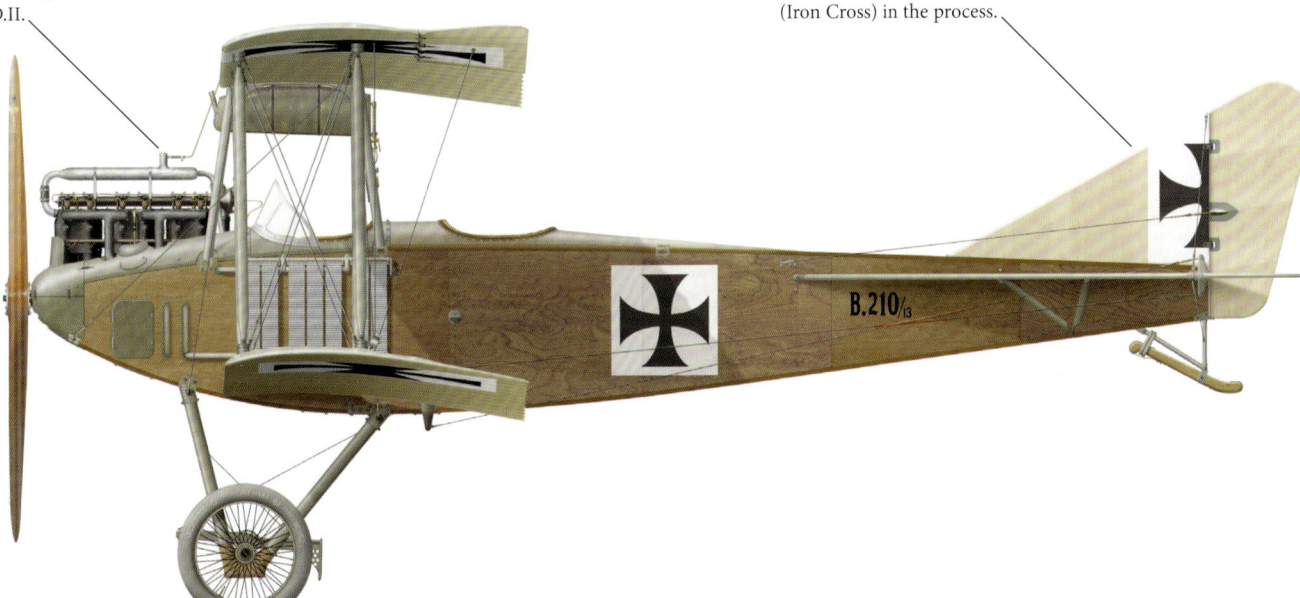

ATTACK ON BRITAIN
On 16 April 1915, the observer of a B.II from Feldflieger Abteilung 41 dropped 10 bombs, by hand, in the area around Sittingbourne in Kent, causing virtually no damage. From around mid-1915, the B.IIs were gradually replaced by armed C-type aircraft and the unarmed and slow B-types were relegated to training.

ALBATROS BII (PK)
An Albatros B.II (PK) produced by Phönix for the Austro-Hungarian Imperial and Royal Aviation Troops. The aircraft was operated by 10 nations in total, including, surprisingly, a single evaluation aircraft in British service.

Vickers F.B.5 (1914)

TYPE • Fighter COUNTRY • United Kingdom

SPECIFICATIONS

Dimensions:	Length: 8.28m (27ft 2in); Wingspan: 11.13m (36ft 6in); Height: 2.60m (8ft 6in)
Weight:	930kg (2050lb) maximum take-off
Powerplant:	1 x 82kW (110hp) Gnôme engine
Max Speed:	112.65km/h (70mph)
Endurance:	4hrs 30 minutes
Ceiling:	2743m (9000ft)
Crew:	2
Armament:	1 x 7.7mm (.303in) Lewis MG

The Vickers F.B.5 was based on the pre-war concept of the Gunbus and served reliably on the Western Front from 1915 to 1916. It was one of the first such types to see combat, although others would soon follow.

Above: The original belt-fed Vickers gun was later replaced by the drum-fed Lewis, which was more convenient to operate in the aircraft.

Structure
The F.B.5's airframe was constructed from light steel tubing, with only the wings, tail and cockpit area covered in fabric.

Pusher aircraft
The F.B.5 was a pusher-type aircraft with the 82kW (110hp) Gnôme engine mounted behind the crew compartment unlike tractor-type planes.

Two-man crew
The F.B.5 was designed with two crew members, a pilot to fly the craft and a gunner to shoot at enemy airplanes with a Lewis or Vickers machine gun.

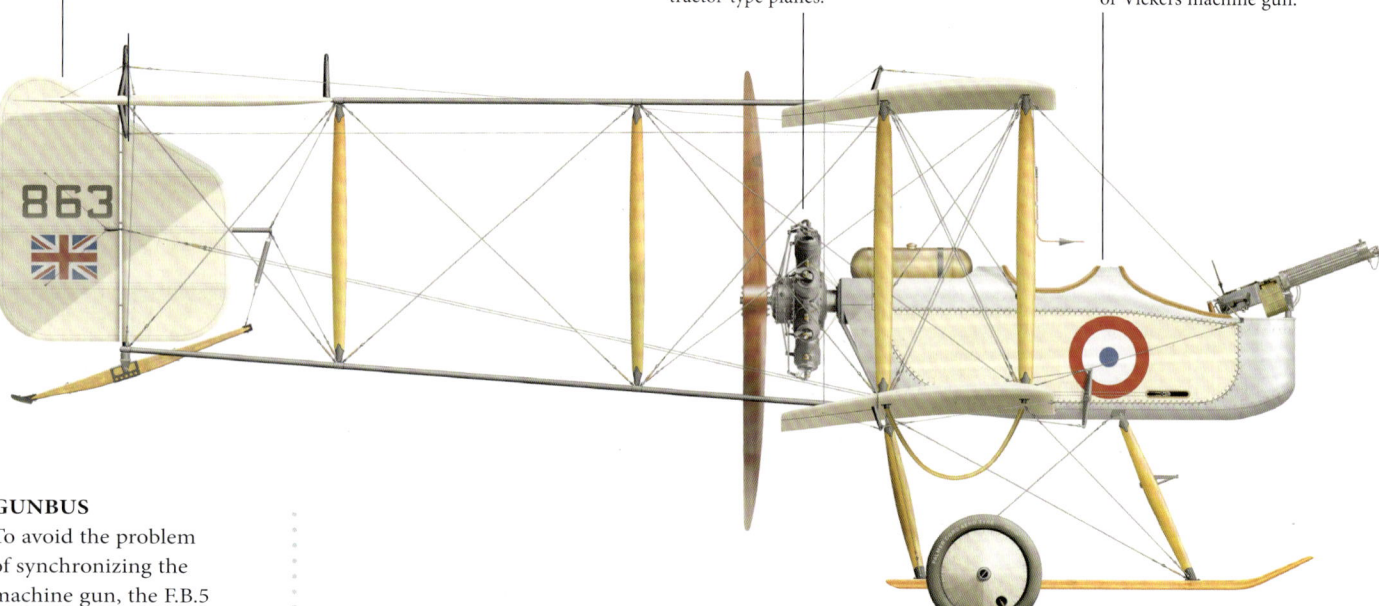

Gunbus
To avoid the problem of synchronizing the machine gun, the F.B.5 was designed as a pusher, with the propeller behind the fuselage nacelle, the pilot in the middle and the gunner in front, where he had the maximum field of fire. The Gunbus was reasonably effective against the slow, unarmed two-seat observation airplanes it met over the front, but the main difficulty it had was catching them because the F.B.5 was so slow.

Air Victories
It was too slow to be very effective intercepting German reconnaissance airplanes, but the F.B.5 scored some victories and chased many intruders back behind their own lines. Lt G.S.M. Insall was awarded the Victoria Cross for an action in a Gunbus on 7 November 1915.

Voisin III (1914)

TYPE • Bomber COUNTRY • France

SPECIFICATIONS

DIMENSIONS:	Length: 9.50m (31ft 2in); Wingspan: 14.74m (48ft 4.3in); Height: 2.95m (9ft 8in)
WEIGHT:	1350kg (2976lb) maximum take-off
POWERPLANT:	1 x 89.5kW (120hp) Salmson M9 engine
MAX SPEED:	98km/h (60.5mph)
ENDURANCE:	4 hrs 30 minutes
CEILING:	4000m (13,123ft)
CREW:	2
ARMAMENT:	1 x 8mm (0.315in) Hotchkiss MG

The Voisin III was a two-seat, pusher-type biplane developed from the earlier Voisin I. Its claim to fame was secured by one of the first air-to-air combat kills of the war, achieved in October 1914 near Reims in France.

Right: The Voisin III was a successful export aircraft for France. International operators during the war included Belgium, Italy, Romania, Russia (as seen in this photograph) and the United Kingdom.

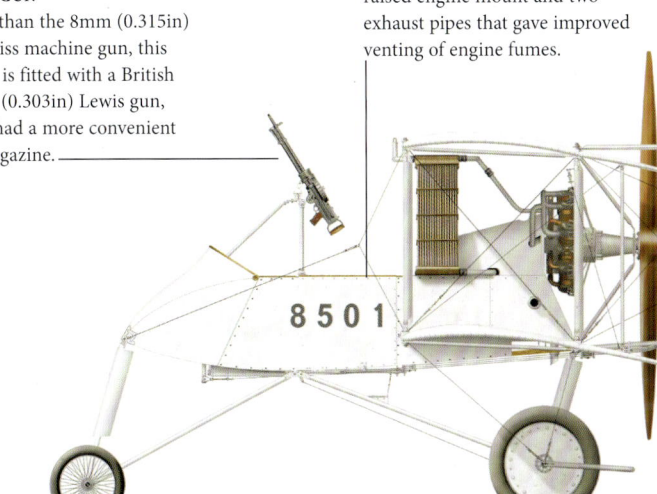

LEWIS GUN
Rather than the 8mm (0.315in) Hotchkiss machine gun, this aircraft is fitted with a British 7.7mm (0.303in) Lewis gun, which had a more convenient pan magazine.

VOISIN LAS
The Voisin LAS variant had a raised engine mount and two exhaust pipes that gave improved venting of engine fumes.

STEEL FRAME
The open steel frame structure made the aircraft extremely light, which combined with the tricycle undercarriage meant it was well-suited to taking off from rough fields.

BOMBER AND FIGHTER
The Voisin III took both bomber and fighter roles over its career. As a fighter, it utilized the observer-operated Hotchkiss M1909 machine gun, flexibly mounted. The aircraft's first fighter kill came on 5 October 1914, when a Voisin III of Escadrille V.24 shot down a German Aviatik B.I. As the LA bomber variant, it carried a bombload of approximately 150kg (330lb).

OBSERVER WEAPON
A view of a Mr Sauzay and Mr Roussillon posing in a Voisin biplane in a hangar near Amiens (Somme) May–June 1915. Note how the observer is using a rifle mounted on a tripod as a primary weapon, rather than the usual Hotchkiss MG.

LVG C.I & C.II (1915)

TYPE • *Reconnaissance* COUNTRY • *Germany*

The first German aircraft to be armed with a machine gun for the observer, the LVG C.I and its near identical derivative, the C.II, gave excellent service over the combat fronts as reconnaissance and light bombing machines.

SPECIFICATIONS (C.II)

DIMENSIONS:	Length: 8.10m (26ft 7in); Wingspan: 12.85m (42ft 2in); Height: 2.93m (9ft 7in)
WEIGHT:	(gross) 1405kg (3097lb)
POWERPLANT:	One 120kW (160hp) Mercedes D.III 6-cylinder water-cooled inline piston engine
MAX SPEED:	130km/h (81mph)
ENDURANCE:	4 hours
CEILING:	4000m (13,125ft)
CREW:	2
ARMAMENT:	One 7.92mm (0.312in) Parabellum MG14 machine gun flexibly mounted in rear cockpit; up to 60kg (132lb) bombload; later production aircraft added one 7.92mm LMG 08/15 'Spandau' fixed-forward firing

Right: This photograph of an LVG C.II on the Western Front illustrates how the pilot's forward view was obscured by the large 120kW (160hp) Mercedes D.III 6-cylinder water-cooled inline piston engine.

RING MOUNT
Schneider's ring mount, here fitted with a Parabellum MG14, would become a standard feature of all German C-types.

UPGRADES
The later C.III inexplicably featured the positions of the pilot and observer reversed, but did not enter production, unlike the slightly enlarged 8-cylinder Mercedes D.IV powered LVG C.IV, the first fixed-wing aircraft to bomb London.

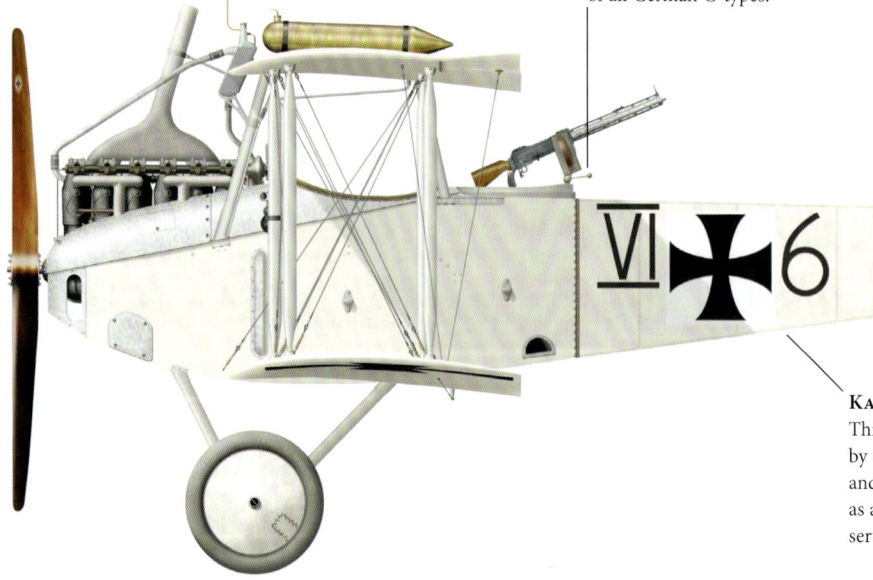

KASTA 6
This LVG C.II of Kasta 6 was apparently flown by a pilot and observer known only as Jureck and Christensen. The C.II was never regarded as a particularly impressive aircraft but was serviceable enough for its intended role.

ARMED AIRCRAFT
Luftverkehrsgesellschaft built Farman aircraft under licence. In September 1914, their chief designer, Franz Schneider, developed a machine gun ring mounting, allowing the weapon to be fired in any direction above the fuselage and downwards to either side. This was mounted on a strengthened airframe and resulted in it becoming the first C-type armed aircraft.

LVG C.1
The C.I was in other regards a very typical German two-seater for its era. Limited production ensued before switching to the much more common C.II that appeared towards the end of 1915 and was built in numbers.

Fokker Eindecker (1915)

TYPE • *Fighter* COUNTRY • *Germany*

The first aircraft to enter service equipped with an interrupter gear allowing the machine gun to fire through the propeller, Fokker's E (*Eindecker*) series of monoplanes was the world's first effective fighter and one of the most significant combat aircraft in history.

SPECIFICATIONS (E.I)

DIMENSIONS:	Length: 6.95m (22ft 8in); Wingspan: 8.95m (29ft 4in); Height: 2.9m (9ft 6in)
WEIGHT:	(gross) 550kg (1212 lb)
POWERPLANT:	One 60kW (80hp) Oberursel U.O 9-cylinder air-cooled rotary piston engine
SPEED:	130km/h (81mph)
ENDURANCE:	1 hour 30 minutes
CEILING:	3000m (9800ft)
CREW:	1
ARMAMENT:	One 7.92mm (0.312in) Parabellum LMG 14 or one 7.92mm lMG 08 'Spandau' machine gun

Above: Baron Kurt von Crailsheim transferred to aviation after being wounded in August 1914. Crailsheim flew this E.II with FAA 53 in the autumn of 1915, replacing an earlier Eindecker *that he had written off in a crash landing after engine failure, which happened often at this stage of military aviation development.*

INTERRUPTER GEAR
The interrupter gear meant that the machine gun could be fired safely through the arc of the propeller without hitting the spinning blades.

WINGS
Fitted with the same engine as the E.II, the E.III differed only in that it was fitted with new wings of a slightly narrower chord and possessed a larger fuel tank allowing for a two-and-a-half-hour flight endurance.

FUSELAGE
The E.III was larger than the E.I and E.II, with an overall length of 7.25m (23ft 9in) and a wingspan of 10.05m (32ft 11in).

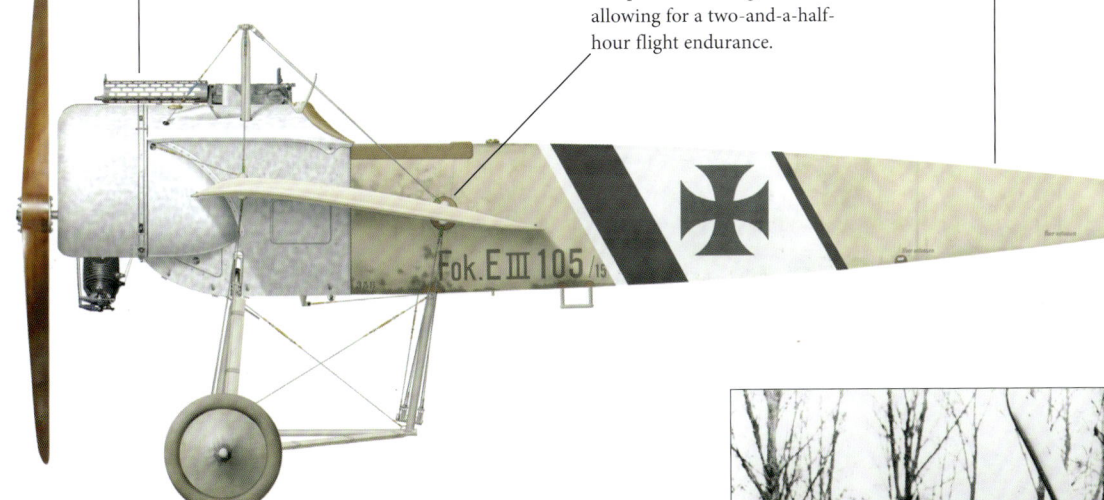

MAX IMMELMANN
The most famous of the *Eindecker* pilots, Max Immelmann scored all 15 of his victories with Fokker monoplanes. As well as giving his name to an aerial manoeuvre known as the 'Immelmann Turn,' Prussia's highest military honour, the 'Pour le Mérite,' was allegedly nicknamed the 'Blue Max' in his honour – Immelmann being one of the first two recipients of this medal in the Great War.

DOWNED *EINDECKER*
A downed German *Eindecker* is inspected. In April 1915, Roland Garros's aircraft, with his self-designed bullet deflector gear, fell into German hands, prompting them to develop a more effective interrupter gear. This was fitted to a short-span M.5k scout to produce the E.I, and from April until the end of December 1915 the Fokker monoplane was the scourge of Allied pilots on the Western Front. The E.III was the definitive model.

Caudron R.11 A.3 (1915)

TYPE • *Escort fighter* **COUNTRY** • *France*

SPECIFICATIONS

DIMENSIONS:	Length: 11.22m (36ft 9.7in); Wingspan: 17.92m (58ft 9.5in); Height: 2.8m (9ft 2in)
WEIGHT:	2165kg (4773lb) loaded
POWERPLANT:	2 x 160kW (215hp) Hispano-Suiza 8Bda engines
MAX SPEED:	183km/h (114mph)
ENDURANCE:	3–4 hours
CEILING:	5950m (19,521ft)
CREW:	3
ARMAMENT:	5 x 7.7mm (0.303in) flexible Lewis MGs (2 in nose, 2 dorsal, 1 firing down)

During 1918 the three-seat Caudron R.11 flew along with Bréguet 14 day bombers as close escort fighters, a role in which they were highly effective. The R.11 was the most successful – perhaps the only successful – multi-seat escort fighter.

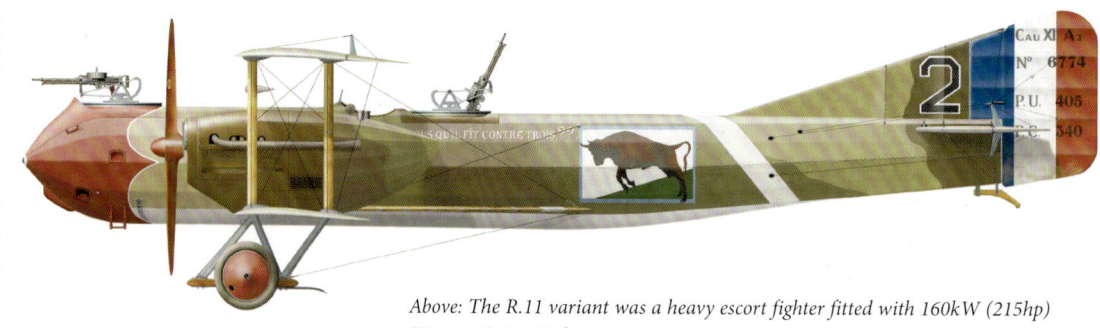

Above: The R.11 variant was a heavy escort fighter fitted with 160kW (215hp) Hispano-Suiza 8Bda engines.

TWIN MACHINE GUNS
In the front and rear gun positions, the R.11 was fitted with twin 7.7mm (0.303in) Lewis guns; two Lewis guns firing together had a combined cyclical rate of about 1000rpm.

CREW
The R.11 had a three-man crew, with the pilot in the middle under the wings between the two gunners.

CAUDRON R.11
This R.11 served with Escadrille C46 on the Western Front in the summer of 1918. The white trident is the Escadrille insignia; the tactical number is on the fin.

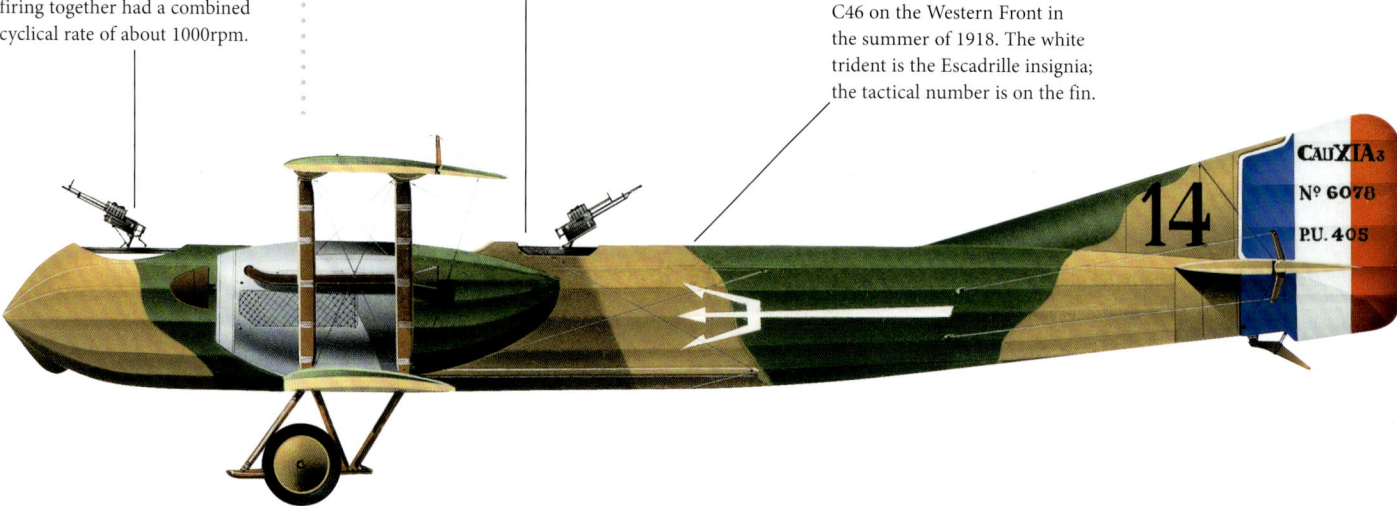

CAUDRON BROTHERS
The Caudron brothers, René and Gaston, began to design and manufacture aircraft in 1909. Former Aviation Militaire pilot Paul Delville took over the job of chief designer when Gaston was killed testing the R.4 twin-engine reconnaissance/bomber aircraft. Delville improved on the R.4 design with the R.5 and R.10 prototypes, but it was the R.11 that proved most successful.

IMPROVED DESIGN
The R.11 owed much to the R.4, having a full-length fuselage, single fin and rudder, unequal-span wings and twin tractor engines, but the R.11 differed by having a more streamlined nose, no nose wheel and engines mounted in nacelles. Production began in 1917, although the first aircraft were not delivered until February 1918.

Aviatik C.I (1915)

TYPE • Reconnaissance COUNTRY • Germany

SPECIFICATIONS

DIMENSIONS:	Length: 7.93m (26ft); Wingspan: 12.5m (41ft); Height: 2.95m (9ft 8in)
WEIGHT:	(gross) 1340kg (2954lb)
POWERPLANT:	One 119kW (160hp) Mercedes D.III 6-cylinder water-cooled inline piston engine
SPEED:	142km/h (88mph)
ENDURANCE:	3 hours
CEILING:	3500m (11,500ft)
CREW:	2
ARMAMENT:	One 7.92mm (0.312in) Parabellum MG14 machine gun flexibly mounted in forward cockpit

The first Aviatik aircraft designed from the outset for military use was the C.I. Aviatik's armed two-seaters were typical of their kind and delivered reliable, if unspectacular, service until being replaced by higher performance machines during 1917.

Right: The C.I aircraft bore a marked resemblance to the earlier B series as seen in this Aviatik B.II. The B.II appeared in 1915 and had a lighter and stronger rudder and elevator structure compared to the B.I, as well as a more powerful Mercedes engine.

FIREPOWER
This Aviatik C.I of Kampfstaffel 7, KG2, was operating during 1915 and has a second Parabellum machine gun fitted to the observer's cockpit.

LAYOUT
A totally conventional aircraft, the C.I was arranged, like the equivalent British BE.2, with the observer in the front cockpit and pilot at the rear.

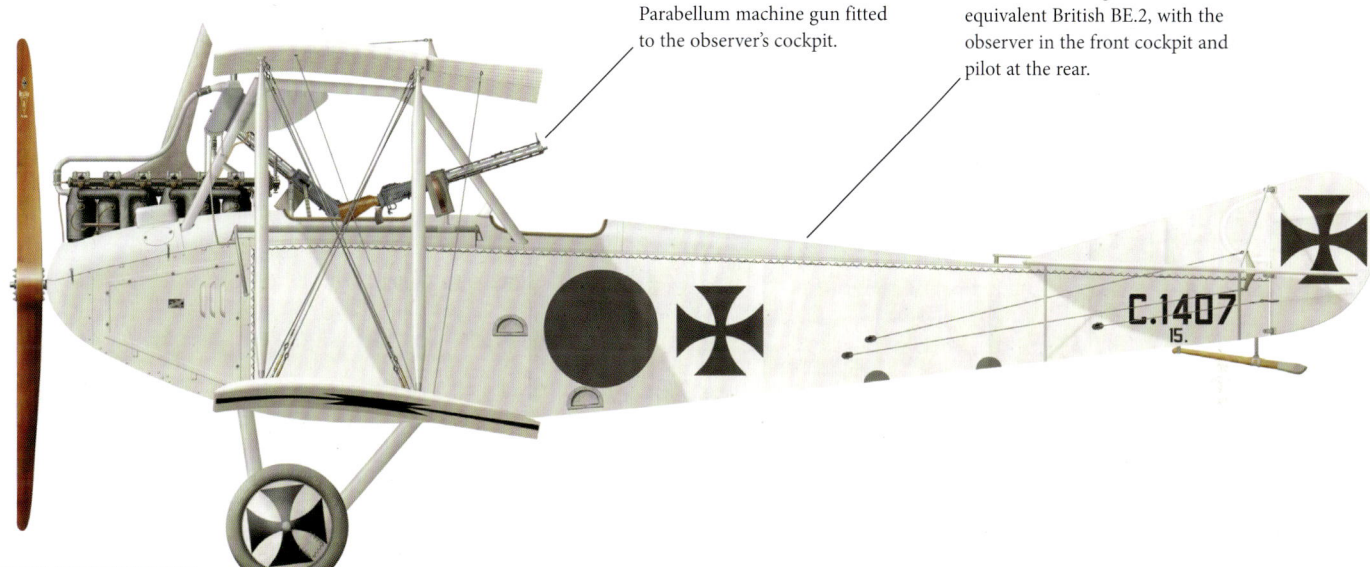

MILITARY PURPOSE
In the C.Ia the personnel layout of the C.I was reversed, a change that resulted in a much more potent aircraft. Only a small number of C.Ia aircraft were built before production switched to the C.II with 149kW (200hp) Benz Bz.IV power and significantly revised tail surfaces. Most widely produced was the C.III, which had reduced span wings, a streamlined nose, an improved exhaust system, and two machine guns.

AVIATIK C.I
This Aviatik C.I has yet to be fitted with its defensive armament. The flexibly mounted Parabellum machine gun was clipped to a sliding rail on either side of the cockpit that featured a quick release mechanism allowing the observer to transfer the gun to whichever side it was required. This system was unwieldy and the gun's position resulted in a badly restricted field of fire.

Albatros C.I & C.III (1915)

TYPE • *General purpose* **COUNTRY** • *Germany*

Albatros produced their first armed aircraft in the form of the C.I, an aircraft that proved highly successful in reconnaissance, bombing and artillery observation roles. A developed version, the C.III, became the most produced of the Albatros two-seaters.

SPECIFICATIONS (C.III)

DIMENSIONS:	Length: 8m (26ft 3in); Wingspan: 11.69m (38ft 4in); Height 3.1m (10ft 2in)
WEIGHT:	(gross) 1353kg (2977lb)
POWERPLANT:	One 110kW (150hp) Benz Bz.III or 120kW (160hp) Mercedes D.III or 130kW (180hp) Argus As.III 6-cylinder water-cooled inline piston engine
MAX SPEED:	132km/h (82mph)
ENDURANCE:	2.5 hours
CEILING:	4900m (11,000ft)
CREW:	2
ARMAMENT:	One 7.92mm (0.312in) LMG 08/15 MG fixed and one 7.92mm Parabellum MG14 machine gun flexibly mounted in rear cockpit; up to 90kg (200lb) bombload

Below: The C.III (the C.II designation being used for a pusher aircraft of which only one example was built) appeared on the Western Front during December 1915. It was easily identified by its elegantly rounded tail surfaces that bestowed greater responsiveness of control.

EASTERN FRONT
Pictured is the C.III of Lieutenant Bruno Maas of Feldflieger Abteilung 14, flying on the Eastern Front in January 1917.

FORWARD ARMAMENT
The C.III introduced a synchronized forward-firing machine gun for the pilot and is believed to be the first German two-seater to feature machine gun armament for both its crew members.

C.III

The C.III, which first entered service in late 1916, was Albatros's most prolific two-seater. It followed a generally similar configuration to the C.I, with the observer aft of the pilot, but had a redesigned tail. Later aircraft were equipped with a synchronized forward-firing machine gun, and had a bay between the two crew for the stowage of small bombs.

GROUND CREW
Two ground crew pose in front of their Albatros C.III. Eventually surpassed by later aircraft in operational units, the C.III had largely disappeared from the front by mid-1917, but like the C.I before it production continued for training purposes.

Nieuport 10 (1915)

TYPE • *Fighter* **COUNTRY** • *France*

SPECIFICATIONS

DIMENSIONS:	Length: 7.05m (23ft 2in); Wingspan: 7.92m (25ft 11.8in); Height: 2.67m (8ft 9in)
WEIGHT:	660kg (1452lb) maximum take-off
POWERPLANT:	1 x 59.65kW (80hp) Le Rhône rotary engine
MAX SPEED:	146km/h (91.3mph)
ENDURANCE:	3 hours
CEILING:	About 4000m (13,123ft)
CREW:	1 or 2
ARMAMENT:	1-2 x 7.7mm (.303in) Lewis MGs

The famous Nieuport fighters designed by Gustave Délage had the lower wing much narrower than the upper wing, thus making them 'sesquiplanes'. This, and a slight wing sweepback on most models, gave a good view downwards, but structural strength suffered.

Right: A Nieuport 10 single-seat fighter armed with a Lewis machine gun over the top wing photographed at Rosnay, Marne on 21 August 1915.

POWERPLANT
The standard powerplant of the Nieuport 10 was the 9.65kW (80hp) Le Rhône rotary engine.

V STRUTS
The Nieuports were used by French, British, Italian and American units and were the mounts of many aces. The distinctive 'V' struts were replaced in the totally revised Nieuport 28 with parallel struts and a streamlined fuselage.

GEORGES GUYNEMER
This aircraft was flown by Corporal Georges Guynemer, of Escadrille N-3, based at Breuil-le-Sec in December 1915.

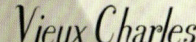

SESQUIPLANE
Gustave Délage's Nieuport 10 was a small two-seat reconnaissance sesquiplane (with the lower wing much smaller than the upper). However, it proved to be underpowered, and most of the two-seaters were converted to single-seat scouts. Délage also designed the Nieuport Bébé and developed this into the Type 11 scout.

RFC NIEUPORT 10
The Nieuport 10 was exported to many countries, including Britain, where it was used by both the Royal Flying Corps (RFC) and the Royal Naval Air Service (RNAS).

Morane-Saulnier Type N (1915)

TYPE • Fighter COUNTRY • France

SPECIFICATIONS

DIMENSIONS:	Length: 5.83m (19ft 2in); Wingspan: 8.15m (26ft 9in); Height: 2.25m (7ft 5in)
WEIGHT:	444kg (979lb) maximum take-off
POWERPLANT:	1 x 60kW (80hp) Le Rhône 9C rotary piston engine
MAX SPEED:	144km/h (89mph)
ENDURANCE:	1hr 30 minutes
CEILING:	4000m (13,123ft)
CREW:	1
ARMAMENT:	1 x 7.9mm (.31in) Hotchkiss MG

While the majority of the French fighters during World War I were biplane types, the Morane-Saulnier company built a series of successful monoplane designs, including the streamlined Type N single-seat fighter.

Right: A Morane-Saulnier Type N sits alongside a more archaic-looking Henry Farman biplane, in the background. The monoplane seen is in Royal Flying Corps (RFC) service.

HOTCHKISS GUN
The armament was a single unsynchronized Hotchkiss gun firing through the propeller. Note the bullet deflectors on the propeller blades.

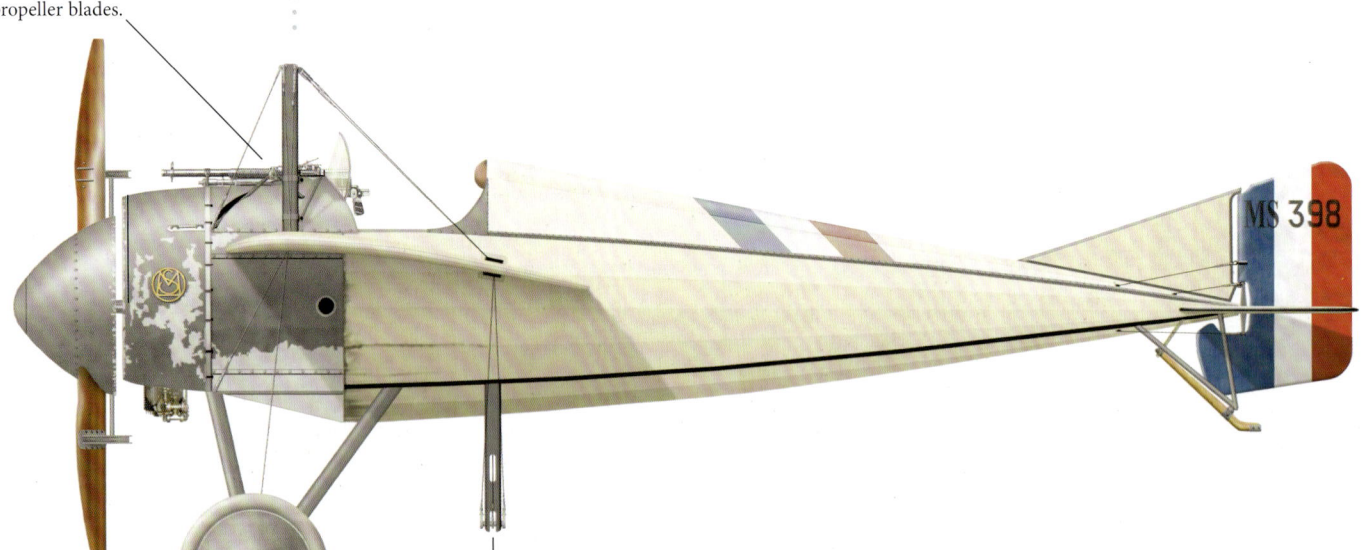

DEFLECTORS
The Morane-Saulnier Type N did not have a synchronized gun, hence relied on deflector plates, which literally bounced the bullets out of harm's way if they struck the propeller blades. The gun mount that most biplane fighter crews preferred and continued in use to the end of the war, was mounting the gun above the top wing to fire over the propeller arc, but being a monoplane this option was not available for the Type N.

HANDLING
The Type N had smooth lines visually, but practically it was not the easiest aircraft to fly, being very 'twitchy' under the controls. Its lateral control was via wing warping, aided by the pulleys beneath the fuselage.

TYPE N
Comparing aircraft with pilot, this photograph gives an immediate insight into the compact dimensions of the Morane-Saulnier Type N, which measured 5.83m (19ft 2in) from nose to tail.

Rumpler C.I (1915)

TYPE • *Reconnaissance* COUNTRY • *Germany*

The Rumpler C.I was the fastest of the armed two-seaters that began to appear during 1915. Popular due to its excellent performance and sturdy construction, the Rumpler enjoyed a long and effective career.

SPECIFICATIONS

Dimensions:	Length: 7.85m (25ft 9in); Wingspan: 12.15m (39ft 10in); Height: 3.06m (10ft)
Weight:	(gross) 1333kg (2939lb)
Powerplant:	One 120kW (160hp) Mercedes D.III 6-cylinder water-cooled inline piston engine
Max Speed:	152km/h (94mph)
Endurance:	4 hours
Ceiling:	5050m (16,570ft)
Crew:	2
Armament:	One 7.92mm (0.312in) Parabellum MG14 machine gun flexibly mounted in rear cockpit; up to 100kg (220lb) bombload; later production aircraft added one 7.92mm (0.312in) LMG 08/15 fixed-forward firing

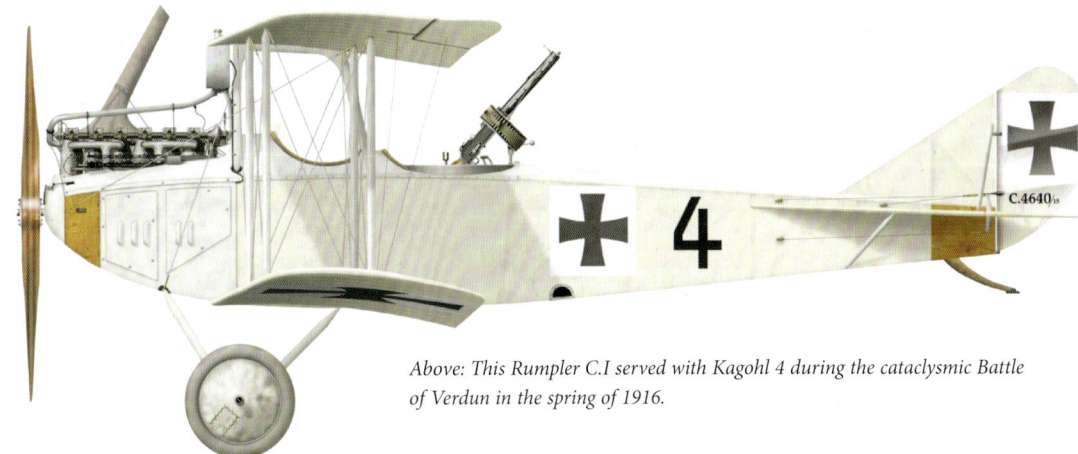

Above: This Rumpler C.I served with Kagohl 4 during the cataclysmic Battle of Verdun in the spring of 1916.

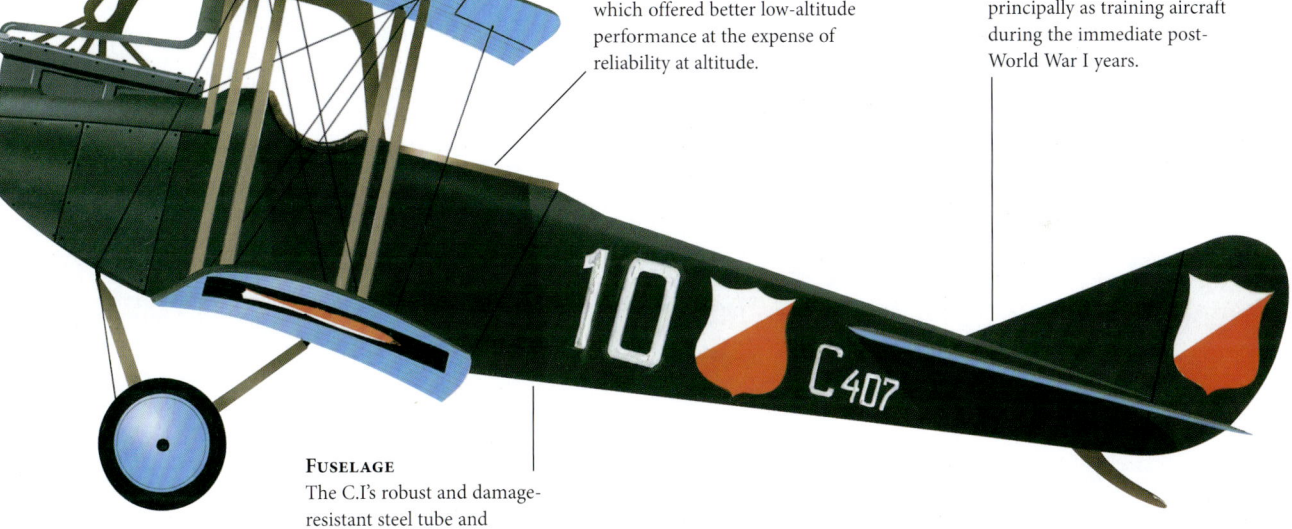

Handling
The aircraft offered better performance and easier handling than its contemporaries. A 134kW (180hp) Argus As.III-powered version was also built under licence by Hannover, which offered better low-altitude performance at the expense of reliability at altitude.

Polish C.I
Some 84 Rumpler C.Is were used by the Polish Air Force, principally as training aircraft during the immediate post-World War I years.

Fuselage
The C.I's robust and damage-resistant steel tube and plywood fuselage conferred great survivability in the event of a crash.

DEPLOYMENT
Notable for being faster than the contemporary Fokker *Eindecker* fighter and possessing a similar rate of climb, the C.I first took to the air during the summer of 1915. Acceptance testing was completed by the end of October, the aircraft having already been ordered into production in July, and the first C.Is started to arrive at the front during November and December 1915.

C.I IN FLIGHT
The C.I was in service throughout 1916, but the advent of higher-performance aircraft saw it gradually reassigned from the Western Front to the training role. Its continued combat usefulness in less demanding areas is demonstrated by the fact that new C.Is were being supplied to the front in the Middle East as late as the spring of 1917.

Sopwith Pup (1916)

TYPE • Fighter COUNTRY • United Kingdom

Famed for its delightful handling, the Pup was the first of Sopwith's 'flying zoo' of aircraft named after animals. Arguably the first unequivocally successful British single-seat fighter, the Pup was built in large numbers and cemented Sopwith's reputation as a combat aircraft producer.

SPECIFICATIONS

DIMENSIONS:	Length: 5.89m (19ft 4in); Wingspan: 8.08m (26ft 6in); Height: 2.87m (9ft 5in)
WEIGHT:	Maximum take-off: 556kg (1225lb)
POWERPLANT:	One 60kW (80hp) Le Rhône 9C 9-cylinder air-cooled rotary piston engine
MAX SPEED:	180km/h (112mph)
ENDURANCE:	3 hours
CEILING:	5300m (17,500ft)
CREW:	1
ARMAMENT:	One 7.7mm (0.303in) Vickers machine gun fixed-forward firing on upper forward fuselage

Above: Sopwith built 170 Pups for the RNAS, and another 1600 were built for the RFC. Because of their exceptional time-to-height performance, many Pups were assigned to home defence units to counter the German bombing threat.

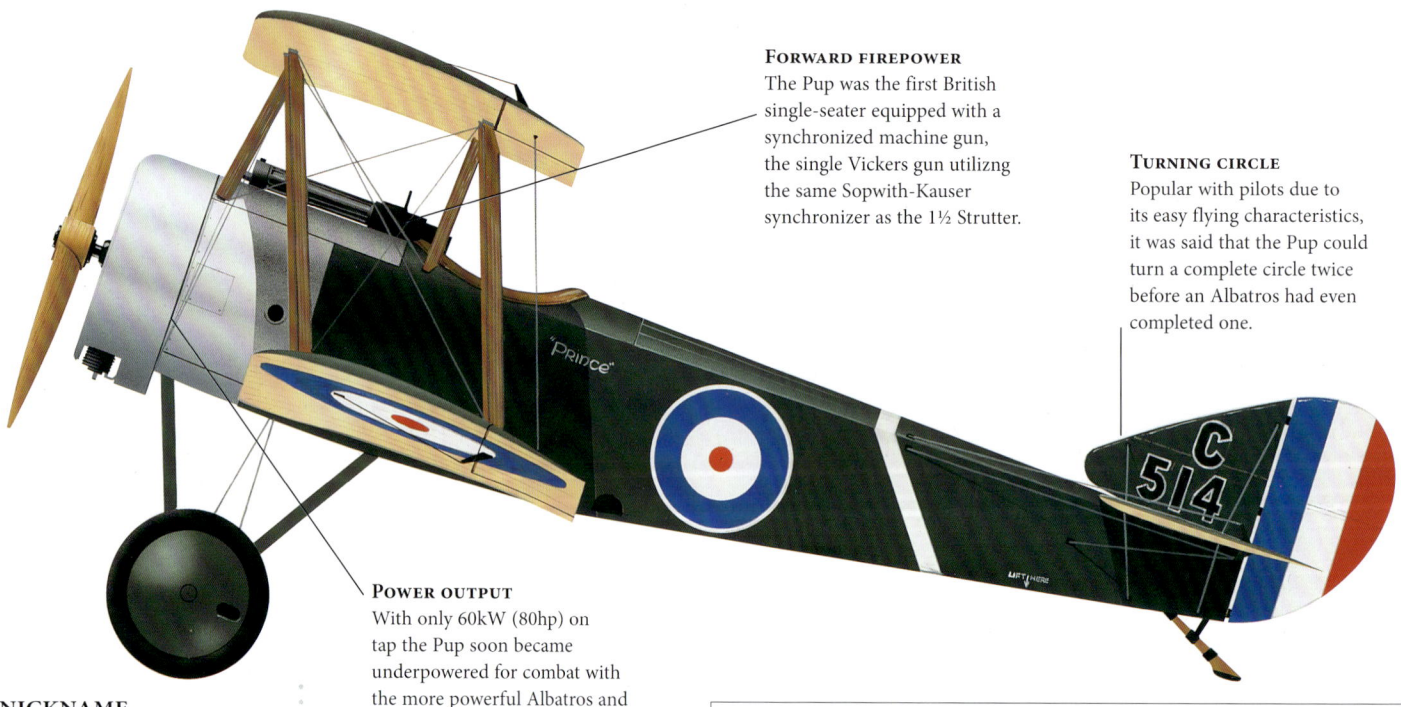

FORWARD FIREPOWER
The Pup was the first British single-seater equipped with a synchronized machine gun, the single Vickers gun utilizng the same Sopwith-Kauser synchronizer as the 1½ Strutter.

TURNING CIRCLE
Popular with pilots due to its easy flying characteristics, it was said that the Pup could turn a complete circle twice before an Albatros had even completed one.

NICKNAME
The Pup got its nickname from its likeness to a scaled-down 1½ Strutter. It first flew in February 1916 with an 60kW (80hp) Le Rhône rotary engine. Given the relatively small power output of this engine, the fact that the Pup was such a pleasure to fly speaks volumes for its design and construction. It was small, simple and reliable, but its generous wing area gave it excellent performance.

POWER OUTPUT
With only 60kW (80hp) on tap the Pup soon became underpowered for combat with the more powerful Albatros and Halberstadt scouts, although it was more manoeuvrable.

TAKE-OFF
The Pup also enjoyed a secondary career as a naval aircraft. On 2 August 1917, a Pup became the first aircraft to land on a moving ship at sea. The Pup pioneered operations aboard the first extemporized aircraft carriers as well as flying off platforms mounted on gun turrets on cruisers and battleships.

Sopwith Triplane (1916)

TYPE • *Fighter* **COUNTRY** • *United Kingdom*

SPECIFICATIONS

DIMENSIONS:	Length: 5.74m (18ft 10in); Wingspan: 8.08m (26ft 6in); Height: 3.2m (10ft 6in)
WEIGHT:	699kg (1541lb) maximum take-off
POWERPLANT:	1 x 97kW (130hp) Clerget 9B 9-cylinder rotary piston engine
MAX SPEED:	188km/h (117mph)
ENDURANCE:	2 hours 45 minutes
CEILING:	6250m (20,500ft)
CREW:	1
ARMAMENT:	1–2 x 7.62mm (0.303in) fixed forward-firing Vickers MGs

In the spring of 1917 the new British Sopwith Triplane was the only Allied fighter that could beat the dominant German Albatros fighter, creating in response a year-long German craze for developing triplanes.

Right: Despite the trend for triplanes started by the Sopwith Triplane, the biplane configuration is actually a better compromise between strength, weight and aerodynamic drag than the triplane, which has too much drag.

POWERPLANT
The Sopwith Triplane retained the single gun and general lines of the Pup, but possessed a more powerful 97kW (130hp) rotary engine.

ARMAMENT
The Sopwith Triplane was one of the few Allied fighters able to best the Albatros D.III in early 1917. It had good climb and manoeuvrability, but with only a single gun was under-armed.

ACE RAYMOND COLLISHAW
'Black Maria' was one of 62-victory ace Lt Col Raymond Collishaw's airplanes and one of a handful with two guns. The black nose, fin and wheel covers are the markings of 'B' Flight.

PROTOTYPE
The lessons from early aerial combats over the Western Front emphasized the need for the highest possible rate of climb and manoeuvrability and, by fitting three slender planes to an airframe derived from the Pup, Sopwith designer Herbert Smith sought to exceed even this aircraft in these respects. The prototype flew in May 1916, and demonstrated an exceptional rate of climb, at only a small cost to manoeuvrability.

TRIPLANE FIGHTER
The first aircraft entered service in November 1916 and over the following six months the Triplane gained almost complete ascendancy over enemy fighters. The German aircraft industry was launched into frenetic activity and by early 1917 almost every company was designing a triplane that could match it.

Nieuport 11 (1916)

TYPE • Fighter **COUNTRY** • France

SPECIFICATIONS

DIMENSIONS:	Length: 5.8m (19ft); Wingspan: 7.55m (24ft 9in); Height 2.45m (8ft)
WEIGHT:	Empty 350kg (772lb); maximum take-off weight 480kg (1058lb)
POWERPLANT:	One 60kW (80hp) Le Rhône 9C 9-cylinder rotary engine
MAX SPEED:	155km/h (97mph)
ENDURANCE:	2 hours 30 minutes
CEILING:	4500m (14,765ft)
CREW:	1
ARMAMENT:	1 x 7.7mm (.303in) Lewis machine gun

French designer Gustave Délage designed and built the civilian Nieuport Bébé for the Gordon Bennett aviation race, and developed this into the Type 11 scout. Hundreds were built for the RFC, RNAS, French and Belgian L'Aviation Militaire and the Imperial Russian Air Service.

Above: This Nieuport 11 was flown by Raoul Lufberry, the leading ace of N124, the Lafayette Escadrille, who eventually achieved 16 victories before being killed in action on 19 May 1918. The intials on the fuselage were Lufberry's personal marking.

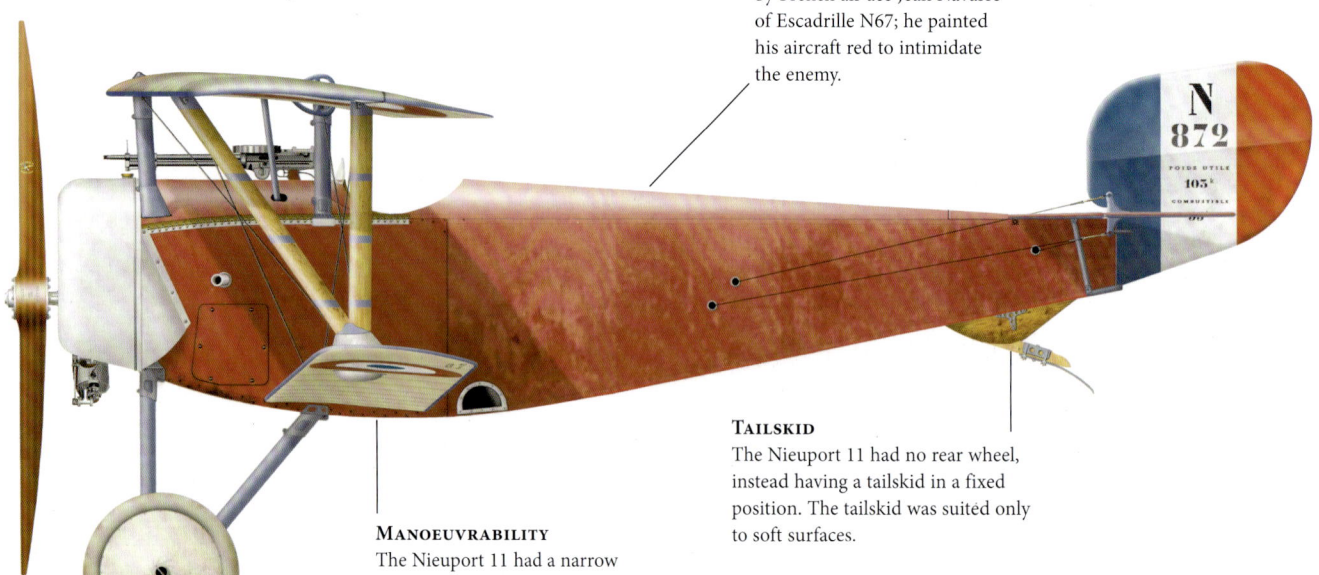

ESCADRILLE N67, VERDUN
This Nieuport 11 was flown by French air ace Jean Navarre of Escadrille N67; he painted his aircraft red to intimidate the enemy.

TAILSKID
The Nieuport 11 had no rear wheel, instead having a tailskid in a fixed position. The tailskid was suited only to soft surfaces.

MANOEUVRABILITY
The Nieuport 11 had a narrow lower wing for good downward visibility and was faster and more manoeuvrable than the Fokker *Eindeckers*.

FIGHTER ACE
Flying a distinctively marked Nieuport 11, French ace Jean Navarre became renowned as the 'Sentinel of Verdun' for flying in constant patrols over that battlefield. Navarre scored 12 victories before being severely wounded in aerial combat on 17 June 1916, which ended his illustrious wartime flying.

NIEUPORT 11S
Nieuport 11 aircraft line up ready for operations. The position of the wing-mounted Lewis guns, firing over the propellers, illustrates how much the pilot had to compensate for the elevation of the weapon.

Royal Aircraft Factory R.E.8 (1916)

TYPE • *Reconnaissance* COUNTRY • *United Kingdom*

SPECIFICATIONS

DIMENSIONS:	Length: 6.38m (20ft 11in); Wingspan: 12.98m (42ft 7in); Height: 2.9m (9ft 6in)
WEIGHT:	1301kg (2869lb) maximum take-off
POWERPLANT:	1 x 112kW (150hp) RAF 4a 12-cylinder V-piston engine
MAX SPEED:	164km/h (102mph)
ENDURANCE:	4 hours 15 minutes
CEILING:	4115m (13,500ft)
CREW:	2
ARMAMENT:	1 x 7.62mm (.303in) forward-firing Vickers MG, 1 x 7.62mm (.303in) Lewis MG over rear cockpit; 102kg (224lb) bombload

The 'Harry Tate', as it was dubbed by the Royal Flying Corp's cockney contingent, was designed to meet an RFC requirement for a two-seat reconnaissance/artillery-spotting aircraft. Despite promising early performance, it was not a convincing combat machine.

Right: An Australian R.E.8 of No. 3 Sqn AFC. The observer/gunner is equipped with the Lewis gun Mk 2, distinguished by its slender barrel shroud and projecting muzzle.

TWIN GUNS
As well as the rear-mounted defensive Lewis gun, the R.E.8 also had a fixed forward-firing Vickers machine gun on the side of the fuselage.

STABILITY
Like the earlier B.E.2, the R.E.8's inherent stability proved to be a major handicap in aerial combat. Among the rank and file it was never a really popular aircraft.

FIRST DELIVERIES

The R.E.8 resembled a scaled-up version of the B.E.2 and shared the same staggered biplane wing configuration, but it had a far sturdier fuselage and more substantial armament. Early tests revealed good all-round handling, encouraging the RFC to place a large order. The first aircraft were delivered in autumn 1916, but these were grounded after a series of mysterious accidents. As a result the tail was redesigned and the mass production of an eventual 4077 aircraft was resumed.

BOMB-LOADING AN R.E.8
Ground crew of the 69th Australian Squadron fix incendiary bombs to an R.E.8 on 22 October 1917. The aircraft could carry an external bombload of up to 102kg (224lb).

Nieuport 17 (1916)

TYPE • Fighter COUNTRY • France

SPECIFICATIONS

DIMENSIONS:	Length: 5.96m (19ft 7in); Wingspan: 8.2m (26ft 11in); Height: 2.44m (8ft)
WEIGHT:	560kg (1235lb) maximum take-off
POWERPLANT:	1 x 82kW (110hp) Le Rhône 9J rotary piston engine
MAX SPEED:	170km/h (106mph)
ENDURANCE:	1.5–2 hours
CEILING:	1980m (6500ft)
CREW:	1
ARMAMENT:	1 x 7.7mm (0.303in) fixed forward-firing Vickers MG

The Nieuport 17 was unquestionably one of the finest Allied combat aircraft of World War I. The aircraft first flew in January 1916 and the first deliveries were made in May, helping to end the 'Fokker Scourge' of previous months.

Above: A Nieuport 17 fitted with a bulbous propeller hub cap. French aces who flew the aircraft include Georges Guynemer, René Fonck, and Charles Nungesser.

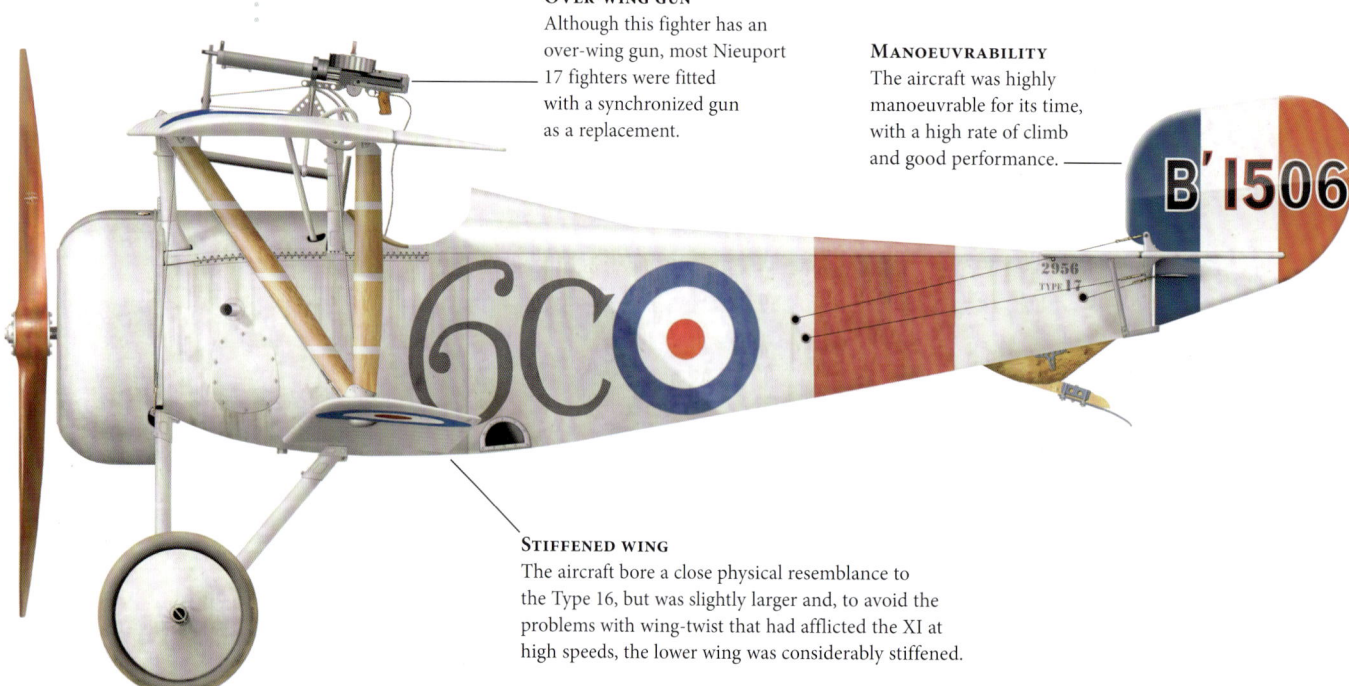

OVER-WING GUN
Although this fighter has an over-wing gun, most Nieuport 17 fighters were fitted with a synchronized gun as a replacement.

MANOEUVRABILITY
The aircraft was highly manoeuvrable for its time, with a high rate of climb and good performance.

STIFFENED WING
The aircraft bore a close physical resemblance to the Type 16, but was slightly larger and, to avoid the problems with wing-twist that had afflicted the XI at high speeds, the lower wing was considerably stiffened.

EVOLUTION
While the Germans introduced the superior Albatros and other new biplane fighters, the Allies also continued developing new fighters. The 60kW (80hp) Nieuport 11 was given a more powerful 82kW (110hp) engine to create the nose-heavy Nieuport 16; an enlarged, more balanced airframe resulted in the Nieuport 17 that arrived in June.

WIDESPREAD SERVICE
This Nieuport 17 is seen on the Italian Front in World War I. Many hundreds were built for service with the RFC and RNAS, L'Aviation Militaire in France and Belgium, Russia, Holland, Italy, Finland and the USAAF.

Airco DH.2 (1916)

TYPE • *Fighter* **COUNTRY** • *United Kingdom*

SPECIFICATIONS

DIMENSIONS:	Length: 7.68m (25ft 3in); Wingspan: 8.61m (28ft 3in); Height: 3m (9ft 7in)
WEIGHT:	Maximum take-off: 654kg (1441lb)
POWERPLANT:	One 75kW (100hp) Gnome Monosoupape 9-cylinder air-cooled rotary piston engine
MAX SPEED:	150km/h (93mph)
ENDURANCE:	2 hours 45 minutes
CEILING:	4300m (14000ft)
CREW:	1
ARMAMENT:	One 7.7mm (0.303in) Lewis machine gun fixed-forward firing in nose

The world's first purpose-designed, single-seat fighter, the DH.2 overcame an unfortunate early reputation to emerge as an excellent fighting scout and the machine of choice for most early British aces.

Right: The first DH.2s were accepted by the Flying Corps in December 1915. The first RFC unit to be wholly equipped with single seaters, 24 Squadron, flew its DH.2s to France in February 1916.

PUSHER LAYOUT
A pusher layout was adopted by designer Geoffrey de Havilland for the DH.2 to allow a machine gun to fire directly forwards.

GNOME ENGINE
The DH.2's Gnome engine was known to occasionally shed cylinders that could then smash into the tail booms with catastrophic results.

LAYOUT
The DH.2 was essentially a scaled-down development of the earlier two-seat DH.1, with a wire-braced wooden airframe. It flew for the first time on 1 July 1915.

DANGEROUS SPIN
Often described as a response to the appearance of the Fokker *Eindecker*, development of the DH.2 had in fact started before the *Eindecker* appeared over the front. Although possessing an excellent rate of climb for its day and outclassing the *Eindecker* in both speed and manoeuvrability, the DH.2 was initially regarded with some trepidation by its pilots. The controls of the aircraft were very sensitive and the aircraft gained a reputation for falling into an unheralded spin.

DH.2 UNIT, FRANCE
Experience proved that the evil reputation of the DH.2 as a flying machine was unfounded. In combat, the DH.2 actually proved highly successful and most of the first RFC aces would score some or all of their victories with the type. Furthermore, the DH.2 units quickly gained a reputation for almost willfully aggressive flying.

LFG Roland C.II (1916)

TYPE • *Reconnaissance* **COUNTRY** • *Germany*

The Roland C.II and C.IV were aerodynamically ahead of their time and boasted an unusually good performance, although at the cost of somewhat capricious handling characteristics.

SPECIFICATIONS

DIMENSIONS:	Length: 7.7m (25ft 3in); Wingspan: 10.3m (33ft 10in); Height 2.9m (9ft 6in)
WEIGHT:	(gross) 1284kg (2831lb)
POWERPLANT:	One 120kW (160hp) Mercedes D.III 6-cylinder water-cooled inline piston engine
MAX SPEED:	165km/h (103mph)
ENDURANCE:	4 hours
CEILING:	4000m (13,000ft)
CREW:	2
ARMAMENT:	One 7.92mm (0.312in) MG14 in rear cockpit; up to 50kg (110lb) bombload; later production aircraft added one 7.92mm LMG 08/15

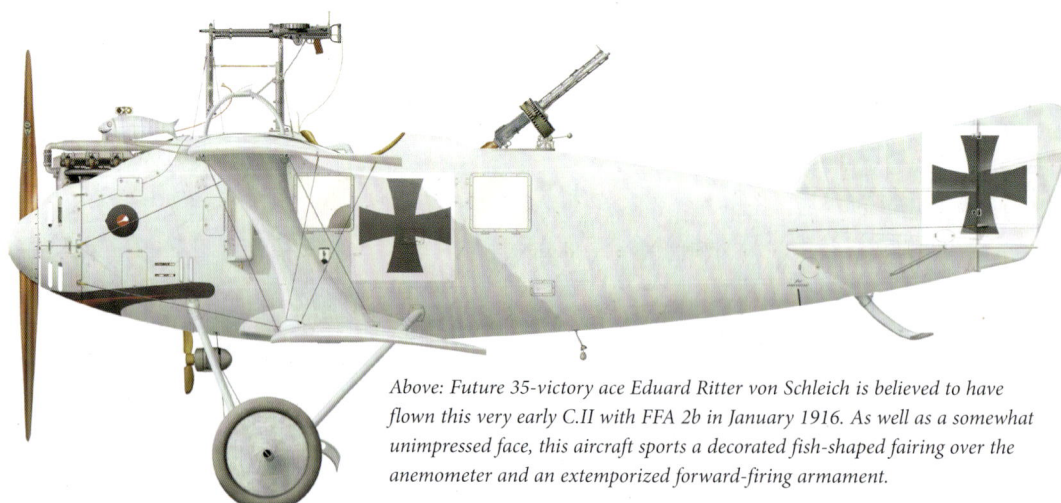

Above: Future 35-victory ace Eduard Ritter von Schleich is believed to have flown this very early C.II with FFA 2b in January 1916. As well as a somewhat unimpressed face, this aircraft sports a decorated fish-shaped fairing over the anemometer and an extemporized forward-firing armament.

SYNCHRONIZED GUN
On the second production batch a synchronized MG was introduced for the pilot, which necessitated a redesign of the roll-over pylon above the cockpit.

CREW POSITIONS
The crew positions gave them a commanding view upwards and sideways with an excellent field of fire for the observer but visibility downwards was poor, ameliorated somewhat by large windows cut into the fuselage.

AIRFRAME
The C.II featured LFG's patented Wickelrumpf fuselage that was created in halves from two layers of thin plywood strips, each layer applied at an opposing angle of around 60 degrees formed over a mould.

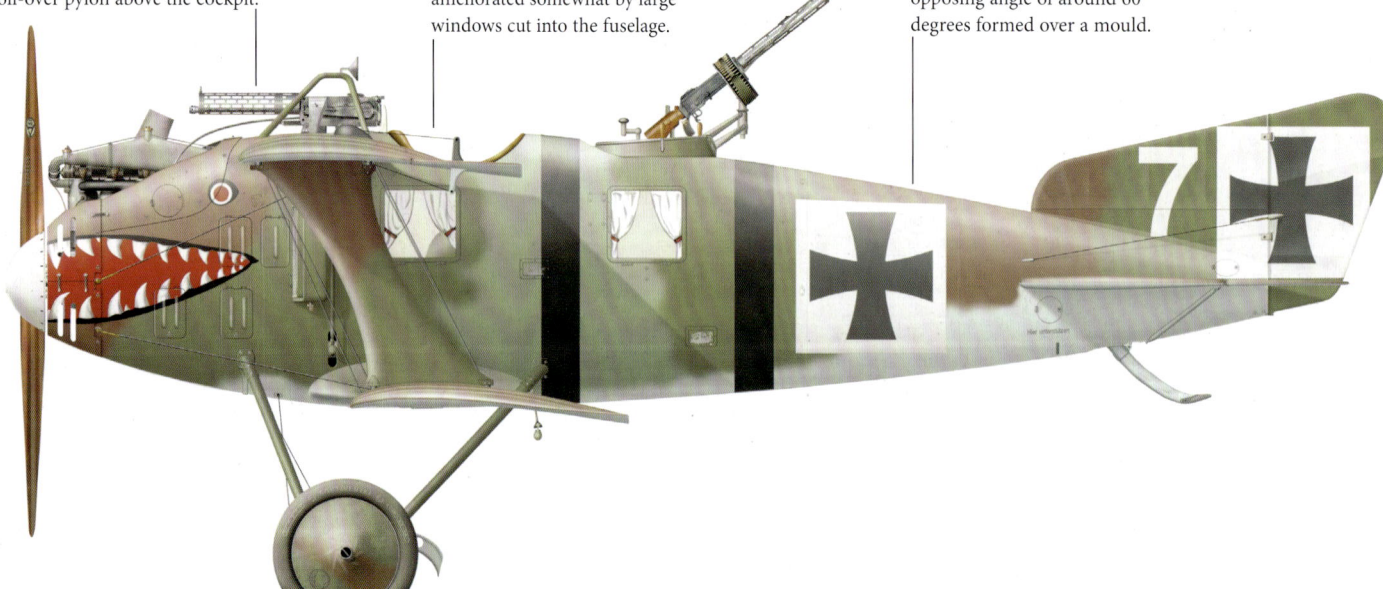

STREAMLINING
The C.II was the first and most successful aircraft design of LFG (Luftfahrzeug-Gesellschaft), working under the trade name Roland. Great care was taken to streamline the aircraft as far as possible. Wire bracing was kept to a minimum and the aircraft featured broad chord 'I' struts between the wings that caused less drag than conventional struts.

ROLAND C.II
Nicknamed *Walfisch* (Whale) due to its corpulent yet streamlined shape, the C.II was actually as fast as it looked, boasting a top speed some 30km/h (19mph) faster than contemporary two-seaters and faster than many fighters.

Halberstadt D.V (1916)

TYPE • Fighter COUNTRY • Germany

SPECIFICATIONS	
DIMENSIONS:	Length: 7.3m (23ft 11in); Wingspan: 8.7m (28ft 6in); Height: 2.50m (8ft 2in)
WEIGHT:	(gross) 728kg (1606lb)
POWERPLANT:	One 90kW (120hp) Argus As.II 6-cylinder water-cooled inline piston engine or 90kW (120hp) Mercedes D.II 6-cylinder water-cooled inline piston engine
MAX SPEED:	160km/h (99mph)
ENDURANCE:	2 hours
CEILING:	4500m (14,764ft)
CREW:	1
ARMAMENT:	One 7.92mm (0.312in) LMG 08/15 'Spandau' machine gun

For a brief period, the finest fighter available to the Jagdstaffeln (Jastas) was the Halberstadt single-seater that supplemented Fokker's biplane fighters, which bridged the gap between the *Eindeckers* and the arrival of the Albatros scouts.

Left: Halberstadt developed the D.III (seen here), of which 50 were produced. It was basically the same aircraft as the D.II, but with an Argus As.II engine of the same 6-cylinder layout and power output in place of the original Mercedes D.II.

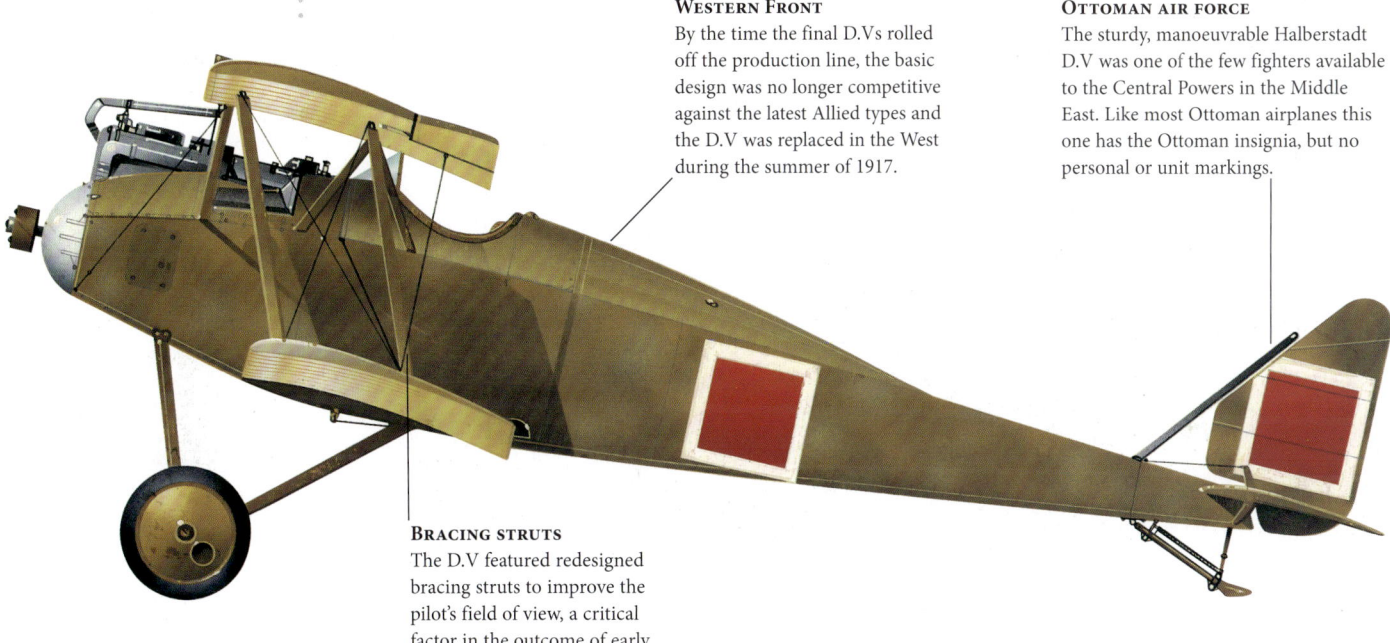

WESTERN FRONT
By the time the final D.Vs rolled off the production line, the basic design was no longer competitive against the latest Allied types and the D.V was replaced in the West during the summer of 1917.

OTTOMAN AIR FORCE
The sturdy, manoeuvrable Halberstadt D.V was one of the few fighters available to the Central Powers in the Middle East. Like most Ottoman airplanes this one has the Ottoman insignia, but no personal or unit markings.

BRACING STRUTS
The D.V featured redesigned bracing struts to improve the pilot's field of view, a critical factor in the outcome of early aerial combat.

IMPROVED VARIANT
The Halberstadt D.V adopted the basic design of the series, with the single machine gun and Argus engine of the D.III combined with improved ailerons and a different cabane and upper wing centre section design, resulting in better handling and improved pilot view. Around 90 examples of the D.V were constructed in two batches from late 1916 until June 1917, and developed a reputation as a delight to fly.

HALBERSTADT D.II
During 1915, Halberstadt developed the world's first fighter powered by a water-cooled inline engine, the D.I. Although production of this aircraft did not go ahead, Halberstadt developed an improved version with the radiator moved from the nose to the wing and a built-up 'turtledeck' surrounding a raised pilot's seat to improve the view from the cockpit. In this form the aircraft was redesignated D.II

33

Albatros D.I & D.II (1916)

TYPE • *Fighter* COUNTRY • *Germany*

SPECIFICATIONS

DIMENSIONS:	Length: 7.4m (24ft 3in); Wingspan: 8.5m (27ft 11in); Height: 2.64m (8ft 6in)
WEIGHT:	(gross) 888kg (1954lb)
POWERPLANT:	One 120kW (160hp) Mercedes D.III 6-cylinder water-cooled inline piston engine
MAX SPEED:	175km/h (109mph)
ENDURANCE:	1 hour 30 minutes
CEILING:	5200m (17,000ft)
CREW:	1
ARMAMENT:	Two 7.92mm (0.312in) LMG 08/15 'Spandau' machine guns

Although produced in comparatively small numbers, on its debut the D.I was arguably the world's finest fighter aircraft and heralded the vast swarms of Albatros fighters that would dominate the Jastas during 1917.

Above: The Albatros D.I was the fastest, most powerful fighter at the front when it first arrived, and immediately established superiority over Allied fighters. This is the D.I of Ltn Karl-Heinrich Büttner of Jasta Boelcke, who was downed on 16 November 1916 by Capt Parker and Lt Harvey of No. 8 Sqdn RFC in a BE.2C.

RADIATOR
Late-production D.IIs had an airfoil radiator instead of this ear radiator. The airfoil radiator reduced drag and retained water in the engine in event of a leak, enabling the engine to run longer.

UPPER WING
The Albatros D.II differed from the earlier D.I by its upper-wing position, which was lowered to improve the pilot's field of view.

JOSEF JACOBS
This D.II was flown by Ltn Josef Jacobs in March 1917. Jacobs went on to score 48 victories and was awarded the Pour le Mérite. Jacobs survived the war; 'Kobes' was his personal marking.

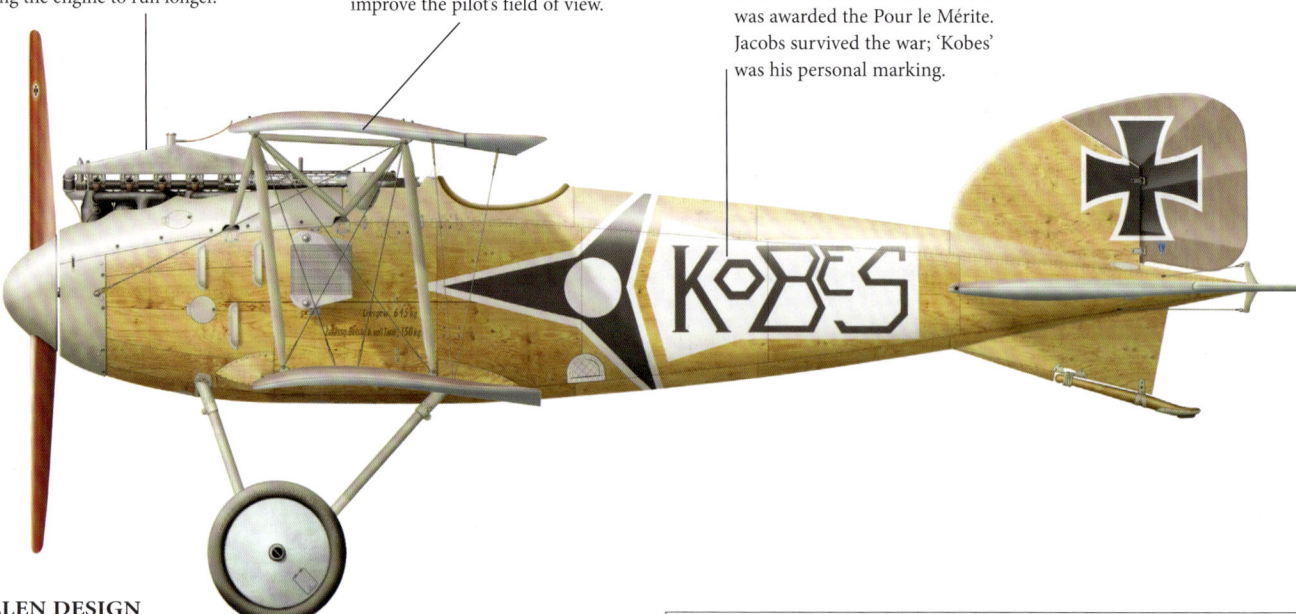

THELEN DESIGN
Designed by Robert Thelen as a replacement for the successful Fokker *Eindecker*, the Albatros D.I was an advanced design that set a precedent by favouring speed and climb performance over outright agility. It was also significant for effectively doubling the installed armament of German fighters with its twin synchronized LMG 08/15 machine guns.

ALBATROS D.I
The D.I was viewed as quite unorthodox on its appearance in the summer of 1916, its semi-monocoque plywood fuselage was lighter and stronger than the conventional fabric-skinned structure of the time, and could more readily be constructed in an aerodynamic shape, contrasting sharply with that of the *Eindeckers* that preceded it.

Albatros D.III (1916)

TYPE • *Fighter* **COUNTRY** • *Germany*

SPECIFICATIONS	
DIMENSIONS:	Length: 7.33m (24ft); Wingspan: 9.05m (29ft 8in); Height: 2.98m (9ft 10in)
WEIGHT:	(gross) 886kg (1949lb)
POWERPLANT:	One 120kW (160hp) Mercedes D.IIIa 6-cylinder water-cooled inline piston engine
MAX SPEED:	175km/h (109mph)
ENDURANCE:	2 hours
CEILING:	4877m (18,000ft)
CREW:	1
ARMAMENT:	Two 7.92mm (0.312in) LMG 08/15 'Spandau' machine guns

The first German fighter to be truly mass-produced, the Albatros D.III was largely responsible for the Jastas' air supremacy during early 1917. Despite its success, the design contained a fatal structural design flaw.

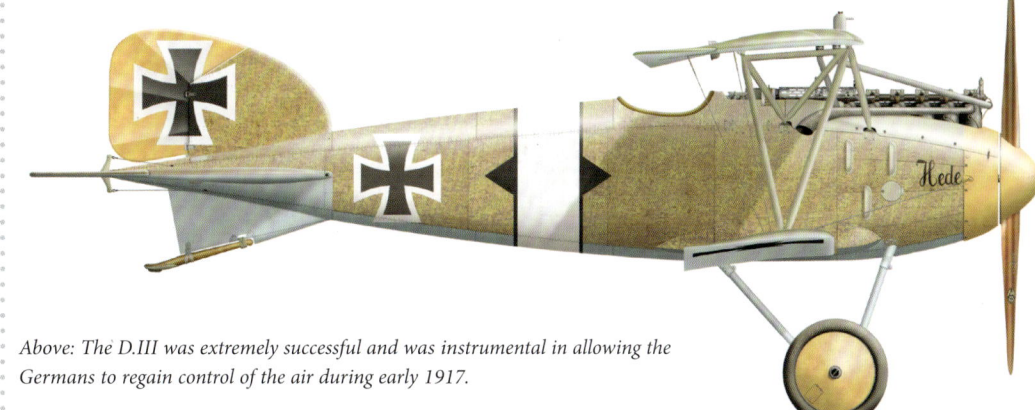

Above: The D.III was extremely successful and was instrumental in allowing the Germans to regain control of the air during early 1917.

POWERPLANT
The Albatros D.III boasted an engine that delivered just under twice the power of the Nieuport 17's Le Rhône rotary.

WING STRESS
Because of the problems with wing stress, pilots were encouraged to 'avoid' prolonged dives in the D.III and the problem, although never eradicated, was largely brought under control.

COLOUR SCHEME
Jasta 18's standard colour scheme was this striking red and blue fuselage with each pilot identifiable by a white personal marking, in this case the axe of Paul Strähle. Strähle would survive the war with 15 victories and found a successful interwar aerial photography business in 1921.

DESIGN FLAW
The D.III adapted the Nieuport 17 sesquiplane (one and a half wing) wing cellule to the basic Albatros design, improving its downward visibility, speed and climb rate. However, the single-spar lower wing of the Nieuport was not robust, a weakness exacerbated by the greater weight and speed of the Albatros D.III, which subjected the wing to problematic stresses.

HEINRICH GONTERMANN
The German fighter ace Heinrich Gontermann and his dog pose in front of his Albatros D.III in 1917. He was credited with a total of 39 victories before his death in a flying accident in a Fokker Dr.I on 30 October 1917.

Bristol F.2A & F.2B (1916)

TYPE • Fighter   COUNTRY • United Kingdom

The Bristol F.2A and F.2B were among the great warplanes of World War I. Initially, however, these types were misused, being flown as a typical two-seater with painful losses against nimble German fighters.

SPECIFICATIONS (F.2B)

Dimensions:	Length: 7.87m (25ft 10in); Wingspan: 11.96m (39ft 3in); Height: 2.97m (9ft 9in)
Weight:	650kg (2779lb)
Powerplant:	1 x 205kW (275hp) Rolls-Royce Falcon III V-12
Max Speed:	182km/h (113mph)
Endurance:	3 hours
Ceiling:	6096m (20,000ft)
Crew:	2
Armament:	1 x 7.7mm (.303in) fixed Vickers MG and 1–2 x 7.7mm (.303in) flexible Lewis MGs

Above: Bristol F.2B (D8084) was flown by Australian pilot ace Captain Sydney Dalrymple of No. 139 Squadron, which was based in northern Italy in 1918.

Engine fittings
The wartime scarcity of Rolls-Royce engines meant that the F.2B was fitted with a variety of powerplants, including 150kW (200hp) Sunbeam Arab and Hispano-Suiza types.

Lewis Gun
The Lewis gun, minus the cooling jacket often seen on infantry versions, was a very common flexible-mount weapon on British and French aircraft during the war.

Imperial gift
This particular aircraft, F.2B H.1557, was actually sent from the UK to Christchurch, New Zealand, in 1919 as an imperial gift aircraft.

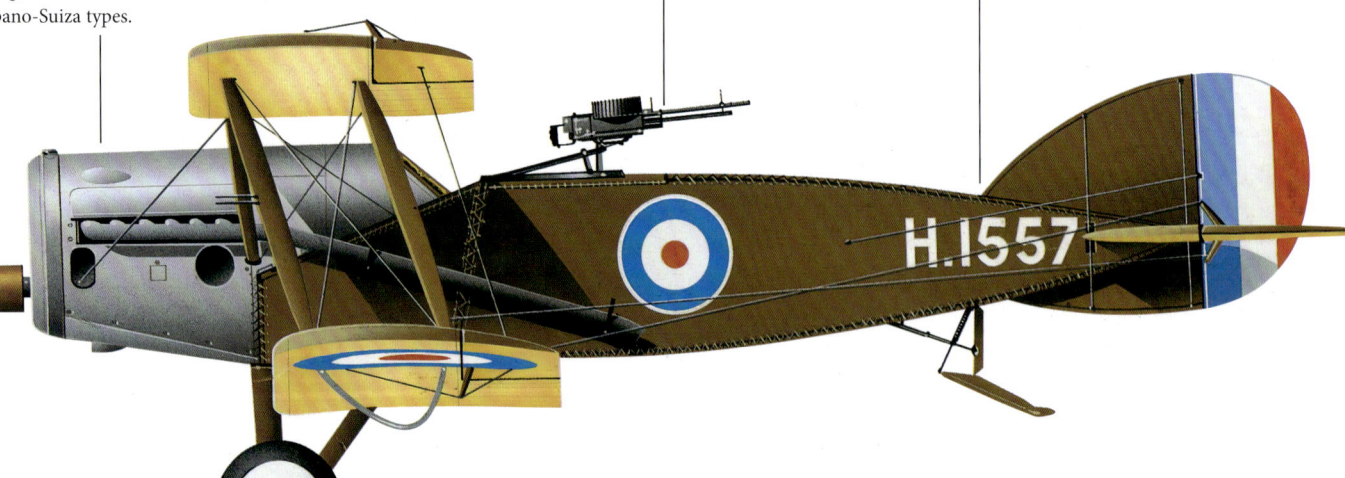

Tactics
Frank Barnwell designed the Type 9 R.2A as a two-seat reconnaissance aircraft, but by August 1916 this had been re-engined and redesignated as the Type 12 F.2A, to denote its new fighter role. The first production F.2As were delivered in February 1917 and, although their early operational tactical employment as a gun platform was naive and resulted in heavy casualties, pilots soon adopted single-seat fighter tactics with great success.

BRISTOL F.2B
The F.2B was the main production variant and incorporated modified upper longerons for improved pilot visibility, enlarged fuel tanks and a variety of engines.

Royal Aircraft Factory F.E.2b (1916)

TYPE • Bomber   COUNTRY • United Kingdom

The F.E.2 series served in the RFC and RAF from early 1916 to the end of the war, serving longer at the front than any other type of either side. Initially used for day fighting and reconnaissance, it was also effective as a night bomber despite its obsolete configuration.

SPECIFICATIONS

DIMENSIONS:	Length: 9.83m (32ft 3in); Wingspan: 14.55m (47ft 9in); Height: 3.85m (12ft 7.5in)
WEIGHT:	1378kg (3037lb) maximum take-off
POWERPLANT:	1 x 119kW (160hp) Beardmore inline piston engine
MAX SPEED:	147km/h (91.5mph)
ENDURANCE:	4–5 hours
CEILING:	3353m (11,000ft)
CREW:	2
ARMAMENT:	1 x 7.7mm (0.303in) Lewis MG, plus up to 136kg (300lb) of bombs

Left: Nearly a year elapsed after the first flight of the F.E.2a before a production order was even placed, and by this time the whole concept of a pusher biplane was rapidly becoming obsolete.

CREW
The two-man crew of the F.E.2b consisted of pilot and observer. For day-fighter operations the pilot occupied the rear cockpit, although for night operations the order was reversed.

LAYOUT
The layout of the F.E.2b was dictated by the need to fire a machine gun dead ahead. In 1913 there was no way of safely firing ahead through a tractor propeller.

FRAMEWORK
The F.E.2 aircraft series had an open strut-and-wire framework with the crew seated in a wood and canvas nacelle.

RAF DESIGN
The first successful Fighter Experimental design from the Royal Aircraft Factory was ready to go into production in January 1914, and had it done so the RFC would have been able to match the Fokker monoplanes of the German Air Service on better terms. Instead, nearly a year elapsed after the first flight before a production order was even placed.

RFC RAF F.E.2B
The first order for 12 F.E.2as was placed in August 1914, followed by the progressively more powerful F.E.2b and F.E.2c. Altogether some 1939 of these were built, as well as 386 long-span F.E.2d models.

Royal Aircraft Factory S.E.5a (1916)

TYPE • *Fighter* COUNTRY • *United Kingdom*

Entering service at the same time as the tricky Sopwith Camel, the S.E.5a is today less well known, but was one of the exceptional fighting scouts of its day. It was the mount of famous Allied aces including William Bishop, James McCudden and Edward Mannock.

SPECIFICATIONS

DIMENSIONS:	Length: 6.38m (20ft 11in); Wingspan: 8.11m (26ft 7in); Height 2.89m (9ft 6in)
WEIGHT:	Empty 639kg (1410lb); maximum take-off weight 902kg (1988lb)
POWERPLANT:	One 150hp (112kW) Hispano-Suiza 8a 8-cylinder inline piston engine
MAX SPEED:	212km/h (132mph) at 1980m (6500ft)
ENDURANCE:	3 hours
CEILING:	5182m (17,000ft)
CREW:	1
ARMAMENT:	1 × fixed 7.7mm (0.303in) forward-firing machine gun, 1 × trainable 7.7mm (0.303in) machine gun

This page: This S.E.5a was flown by the highest-scoring British pilot, Captain Edward Mannock VC DSO MC, while serving with No. 74 Squadron. Despite having sight in only one eye, Mannock scored more than 70 kills before his death.

ARMAMENT
The S.E.5a usually had one 7.7mm (0.303in) Vickers machine gun fixed in the top of the fuselage to the left of centre, firing through the propeller disc, with a 400-round belt, plus one 7.7mm (0.303in) Lewis machine gun on a Foster mount, with four 97-round drum magazines.

LANDING GEAR
Towards the end of 1917, a stronger landing gear became standard with substantially tapered forward legs.

WINDSCREEN
Early production aircraft had an inconveniently large windscreen fitted in front of the cockpit. This actually obscured the pilot's view so was quickly removed.

SCOUT PILOTS
The most celebrated of British scout pilots flew the SE.5 and SE.5a, of whom the greatest exponent was Major James McCudden, whose total score of 57 aerial victories included 50 while serving with No. 56 Squadron. Other Allied aces to have flown the SE.5 and/or SE.5a included Mannock, Bishop, Beauchamp-Proctor and Ball.

PATROL
Undoubtedly the best warplane to come from the Royal Aircraft Factory at Farnborough, the SE.5 (Scout Experimental) was one of the great combat aircraft of World War I. Here a squadron of SE.5as prepare to head out on patrol in 1918.

Sopwith Camel (1916)

TYPE • Fighter   COUNTRY • United Kingdom

SPECIFICATIONS	
DIMENSIONS:	Length: 5.64m (18ft 6in); Wingspan: 8.53m (28ft); Height: 2.59m (8ft 6in)
WEIGHT:	667kg (1471lb) maximum take-off
POWERPLANT:	Bentley B.R.1 9-cylinder rotary piston engine, 112kW (150hp)
MAX SPEED:	187km/h (117mph)
ENDURANCE:	2 hrs 30 minutes
CEILING:	6095m (20,000ft)
CREW:	1
ARMAMENT:	2 × 7.7mm (0.303in) fixed forward-firing Vickers machine guns

One of the most famous aircraft of World War I, the Camel was so-called because of its distinctive 'humped' back. Between June 1917 and November 1918 it destroyed at least 3000 enemy aircraft – more than any other plane.

Above: The Camel F6314 was on strength with No. 120 Squadron, part of the newly established Royal Air Force in 1918. The unit saw much action during the German offensive on the Western Front in spring 1918.

CONSTRUCTION
The Camel was built on the basis of a conventional wire-braced wooden box girder structure, with aluminium covering immediately aft of the engine.

REAR FUSELAGE
Further aft, the covering was plywood as far as the rear of the cockpit, while the rear fuselage was covered in fabric.

ARMAMENT
The two synchronized forward-firing Vickers machine guns could be supplemented by up to four 11kg (25lb) bombs carried externally.

IN FLIGHT
The Camel's combat performance was achieved at some cost to the peerless handling of the earlier types. In inexperienced hands the Camel could bite, and the engine's torque was such that it had a nasty tendency to flip suddenly to the left on take-off. Casualties among trainee pilots were high, but once mastered it was a superb dogfighter.

CARRIER STRIKE
The first aircraft carrier strike in history was flown on 19 July 1918 when six Sopwith navalized 2F1 Camels, seen here, launched from HMS *Furious* and attacked the Zeppelin sheds at Tondern. Two Zeppelins, L.54 and L.60, were destroyed in their sheds.

DFW C.V (1916)

TYPE • Reconnaissance COUNTRY • Germany

Introduced in late 1916, the DFW C.V was the quintessential German two-seater. More DFW C.Vs were built than any other German warplane, and it served with distinction until the end of the war.

SPECIFICATIONS

DIMENSIONS:	Length: 7.88m (25ft 10in); Wingspan: 13.27m (43ft 6in); Height: 3.25m (10ft 8in)
WEIGHT:	(gross) 1430kg (3153lb)
POWERPLANT:	One 150kW (200hp) Benz Bz.IV 6-cylinder water-cooled inline piston engine
MAX SPEED:	155km/h (96mph)
ENDURANCE:	3 hours 30 minutes
CEILING:	5000m (16,400ft)
CREW:	2
ARMAMENT:	One 7.92mm (0.312in) LMG 08/15 'Spandau' fixed forward-firing and one 7.92mm (0.312in) Parabellum MG14 machine gun in rear cockpit; up to 100kg (220lb) bombload

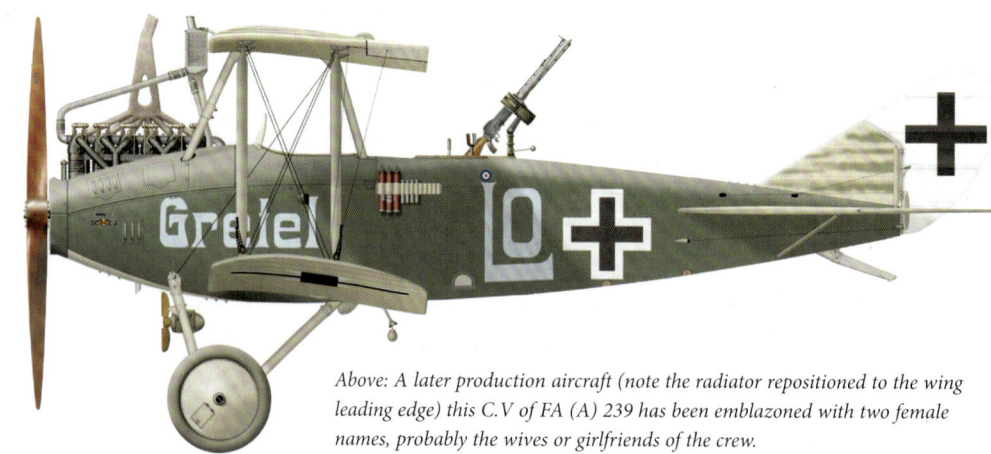

Above: A later production aircraft (note the radiator repositioned to the wing leading edge) this C.V of FA (A) 239 has been emblazoned with two female names, probably the wives or girlfriends of the crew.

NOSE CAP
Later aircraft benefitted from an aerodynamically improved nose with a streamlined spinner.

WINGS
The wings were of conventional wood, wire and fabric construction and the tail featured a steel tube structure.

FUSELAGE
The fuselage sides and underside were constructed of plywood sheets with the smooth curved upper surface formed from mouldings of strip ply, then covered with doped fabric, which proved to be light but sturdy.

FRONTLINE C.V
Ordered into production in August 1916, the first C.Vs reached the front in late September 1916. Although the C.V was intended as a reconnaissance and observation platform, it proved exceptionally versatile and over the next two years would be used for bombing, as a fighter and ground-attack aircraft and as a trainer.

IN FLIGHT
Tough, manoeuvrable and with good handling characteristics, the DFW C.V was a dangerous opponent of even the best fighters. For its original reconnaissance/observation role it could be fitted with a variety of cameras and wireless equipment.

Hanriot HD.1 (1916)

TYPE • Fighter COUNTRY • France

SPECIFICATIONS

DIMENSIONS:	Length: 5.84m (19ft 2in); Wingspan: 8.69m (28ft 6in); Height: 2.94m (9ft 7.7in)
WEIGHT:	605kg (1334lb) maximum take-off
POWERPLANT:	1 x 89kW (120hp) Le Rhône 9Jb 9-cylinder rotary piston engine
SPEED:	186km/h (115.6mph)
ENDURANCE:	2 hours 30 minutes
CEILING:	6000m (19,700ft)
CREW:	1
ARMAMENT:	1 x 7.7mm (.303in) Vickers machine gun with 110 rounds

The Hanriot HD.1 was a French design, although it was not actually used by France, which preferred the sturdy SPAD fighters. The manoeuvrable HD.1 was both imported from France and built under licence in Italy.

Left: This Hanriot HD.1 belonged to the Italian 71st Fighter Squadron.

POWERPLANT
Standard powerplant on the Italian HD.1 was a 82kW (120hp) Le Rhône 9Jb 9-cylinder rotary piston engine.

ARMAMENT
This Swiss aircraft is unarmed, but the armed versions of the HD.1 had a single fixed 7.7mm (0.303in) Vickers. Attempting to boost firepower with a second machine gun gave too much weight to the aircraft.

SWISS HD.1
The HD.1 saw service with Belgian and Italian forces in World War I, but it was also exported to neutral Switzerland, to which this aircraft from circa 1920 belonged.

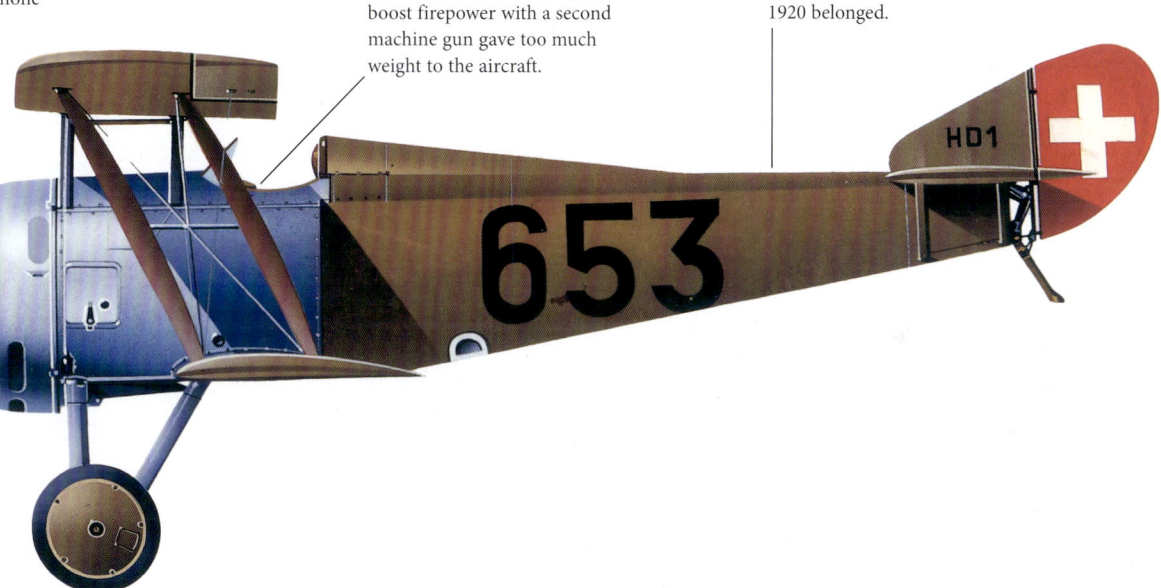

ITALIAN SERVICE

As air combat developed, Italy purchased Nieuport, then Hanriot and SPAD fighters from France and also produced them under licence in Italy. The Hanriot HD.1, not used by France itself, which preferred the SPAD, was the fighter of choice in Italian service. The HD.1s had some notable kills. With 27 victories, fourth-ranking Austrian ace Frank Linke-Crawford was killed in a combat with two Hanriots on 31 July 1918.

SINGLE MACHINE GUN
The Hanriot was a stronger airplane than the similarly powered Nieuport and was preferred by Italian pilots. It was the most popular fighter used by Italy during the war, despite mounting only a single machine gun, as seen in this photograph.

SPAD S.VII (1916)

TYPE • *Fighter* **COUNTRY** • *France*

SPECIFICATIONS

DIMENSIONS:	Length: 6.08m (19ft 11in); Wingspan: 7.82m (25ft 8in); Height: 2.20m (7ft 3in)
WEIGHT:	705kg (1554lb) maximum take-off
POWERPLANT:	1 x 134kW (180hp) Hispano-Suiza 8Ab V8 in-line piston engine
MAX SPEED:	212km/h (132mph)
ENDURANCE:	1 hour 30 minutes
CEILING:	6553m (21,499ft)
CREW:	1
ARMAMENT:	1 x 7.7mm (0.303in) Vickers MG

Powered by the innovative Hispano-Suiza V-8 engine, SPAD fighters were fast and strong with good manoeuvrability and handling qualities. In French service, the SPAD S.VII proved to be a popular and effective fighter.

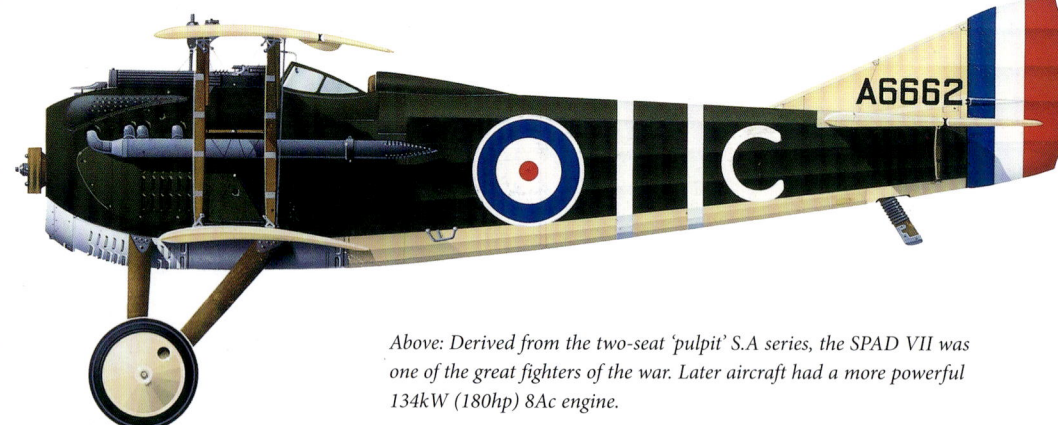

Above: Derived from the two-seat 'pulpit' S.A series, the SPAD VII was one of the great fighters of the war. Later aircraft had a more powerful 134kW (180hp) 8Ac engine.

POWERPLANT
The key to the success of the aircraft was the new V-engine designed in 1915 by Marc Birkigt, co-founder of Hispano-Suiza.

INTERNATIONAL USE
The SPAD VII was used by Britain, Belgium, Italy and Russia in addition to France. This is an Italian aircraft, flown by XXIII Gruppo in the 1920s.

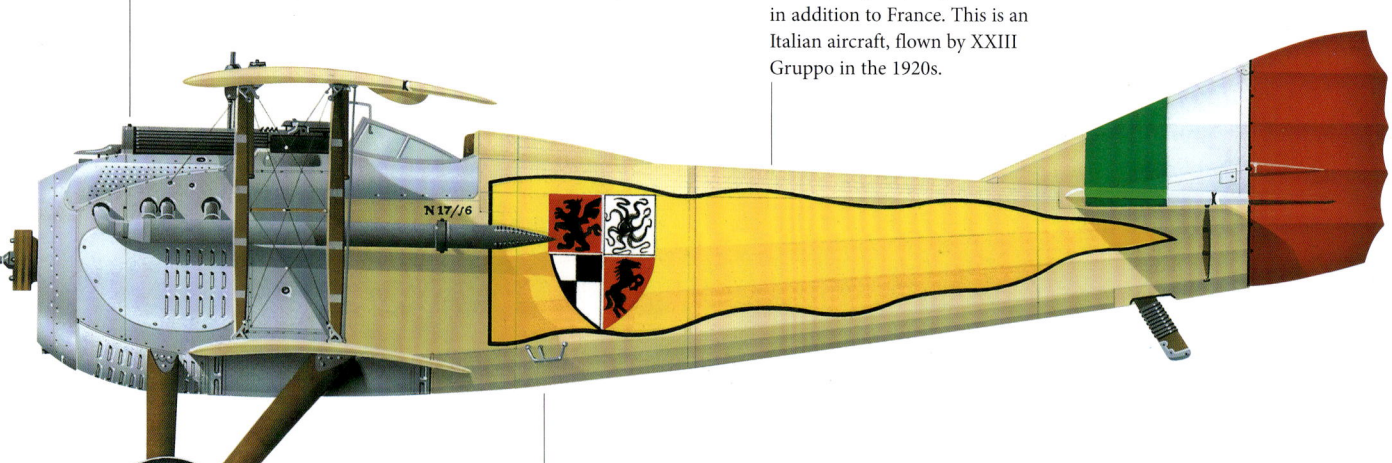

COAT OF ARMS
The coat of arms on the side of the aircraft features the emblems of the four squadrons that composed the Italian Air Force group.

SPAD DESIGN
Just prior to World War I, SPAD undertook the design and manufacture of one of the unremarkable 'A' series of two-seat fighters, but from early 1915 moved on to the development of the SPAD S.V tractor biplane. This was the company's first truly successful military aircraft. The first of these began to arrive at the squadrons in September, with the service designation S.VII.

COMBAT AIRCRAFT
The first combat success with the new type was scored by Lieutenant Armand Pinsard of Escadrille N26 on 23 August 1916. By the summer of 1917 the S.VII was in service with 50 *Escadrilles de chasse* (fighter squadrons) on the Western Front.

Handley Page O/400 (1916)

TYPE • Bomber COUNTRY • United Kingdom

One of the larger aircraft in service in World War I, the O/400 was a more powerful development of the O/100 bomber. Replacing the O/100 from 1916, the O/400 served successfully as a night bomber until the end of the war.

SPECIFICATIONS

DIMENSIONS:	Length: 19.16m (62ft 10.5in); Wingspan: 30.48m (100ft); Height: 6.7m (22ft)
WEIGHT:	6060kg (13,360lb) maximum take-off
POWERPLANT:	2 x 268kW (360hp) Rolls-Royce Eagle II engines
MAX SPEED:	122km/h (76mph)
ENDURANCE:	8 hours
CEILING:	2590m (8500ft)
CREW:	4
ARMAMENT:	1–2 x 7.7mm (0.303in) flexible Lewis guns in nose; 1–2 x 7.7mm (0.303in) flexible Lewis guns in dorsal positions, 1 x 7.7mm (0.303in) flexible Lewis gun in ventral position; 16 x 51kg (112lb) bombs

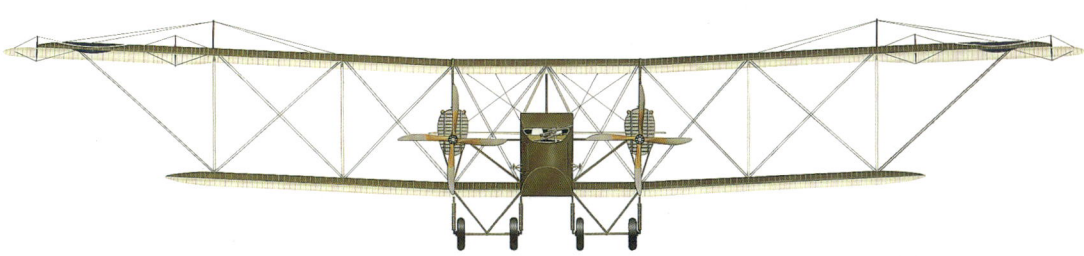

Above: Both the O/100 and the O/400 had a wingspan of 30.48m (100ft), necessitating a four-wheel front undercarriage.

FOLDING WINGS
Despite the impressive wingspan, the wings were actually designed to fold back along the fuselage so the aircraft could be stored conveniently in existing hangars.

US VARIANT
Pictured here is a British O/400, but the US Air Service was also eager to acquire the type and a production licence was negotiated for Standard Aircraft Corporation to build the aircraft with Packard Liberty 12 engines.

BOMBLOAD
By the end of the conflict, O/400s were carrying 748kg (1650lb) bombs. A variant used by the United States later dropped a 1815kg (4000lb) bomb.

UPGRADE
In 1916 Handley Page designer George Volkert had modified the O/100 into the O/400 by moving the fuel tanks from the nacelles into the fuselage and fitting Rolls-Royce Eagle VIII engines. The other minor modification was the introduction of a compressed air engine starting system. Some 554 O/400s were built by British contractors with any one of four different engines fitted.

CAPTURED O/100
German personnel stand around a captured British O/100, the precursor to the O/400. The O/100 was developed to meet the challenge presented by the Admiralty's O/100 Specification of December 1914, which called for a large bombing aircraft at a time when few people could conceive of such a machine.

Nieuport 27 (1917)

TYPE • Fighter COUNTRY • France

SPECIFICATIONS

DIMENSIONS:	Length: 5.87m (19ft 2in); Wingspan: 8.21m (26ft 11in); Height: 2.40m (7ft 11in)
WEIGHT:	535kg (1179lb) loaded
POWERPLANT:	1 x 97kW (130hp) Le Rhône rotary piston engine
MAX SPEED:	172km/h (107mph)
ENDURANCE:	2 hours 15 minutes
CEILING:	5550m (18,210ft)
CREW:	1
ARMAMENT:	1 x 7.7mm (0.303in) Vickers MG

Improvements to the aerodynamics of the Nieuport Scout resulted in a more streamlined fuselage and distinctive rounded tail surfaces, although negligible increases in engine power meant the Type 27 possessed a similar performance to its predecessor.

Left: Lt Destainville of Escadrille N87 preparing for take-off in his Nieuport 27 fighter at Lunéville airfield, France.

WING-MOUNTED LEWIS
To amplify the volume of firepower beyond the single fixed Vickers, this aircraft also had a wing-mounted 7.7mm (0.303in) Lewis gun.

FUSELAGE
Compared to the Nieuport 17 from which it evolved, the Type 27 had a circular section fuselage that improved the streamlining of the aircraft.

PERFORMANCE
The 97kW (130hp) Le Rhône rotary piston engine gave the aircraft a top speed of 172km/h (107mph).

OFFSHOOTS
The were many offshoots from the original Nieuport 17. The Type 24, for example, had improved streamlining, a fixed fin and a circular section fuselage. The Type 24-bis trainer had the original Type 17 tail unit, and the Type 25 was the Type 24 prototype fitted with the tailplane and skid seen on the Type 27. This had a 89kW (120hp) Le Rhône engine and was used by Sweden, the RFC and RNAS, and the USAAF.

US TYPE 27
Seen here in September 1918, just weeks before the end of the war, a French Nieuport 27 is used to instruct air cadets in Mineola, Long Island, in the United States.

Fokker Dr.I 14 (1917)

TYPE • *Fighter* COUNTRY • *Germany*

Thanks to the exploits of the legendary 'Red Baron', the Fokker Dr.I has become the archetypal fighter of World War I in the public imagination. Aside from its fame, the Fokker Triplane suffered from a number of operational shortcomings.

SPECIFICATIONS

DIMENSIONS:	Length: 5.77m (18ft 11in); Wingspan: 7.19m (23ft 7in); Height: 2.95m (9ft 8in)
WEIGHT:	(gross) 586kg (1291lb)
POWERPLANT:	One 82kW (110hp) Oberursel Ur.II 9-cylinder air-cooled rotary piston engine
MAX SPEED:	180km/h (110mph)
ENDURANCE:	1 hour 30 minutes
CEILING:	6100m (20,000ft)
CREW:	1
ARMAMENT:	Two 7.92mm (0.312in) LMG 08/15 'Spandau' machine guns

Above: In its three-wing layout, the Dr.I was among several contemporary fighting scouts to adopt the proven configuration of the British Sopwith Triplane.

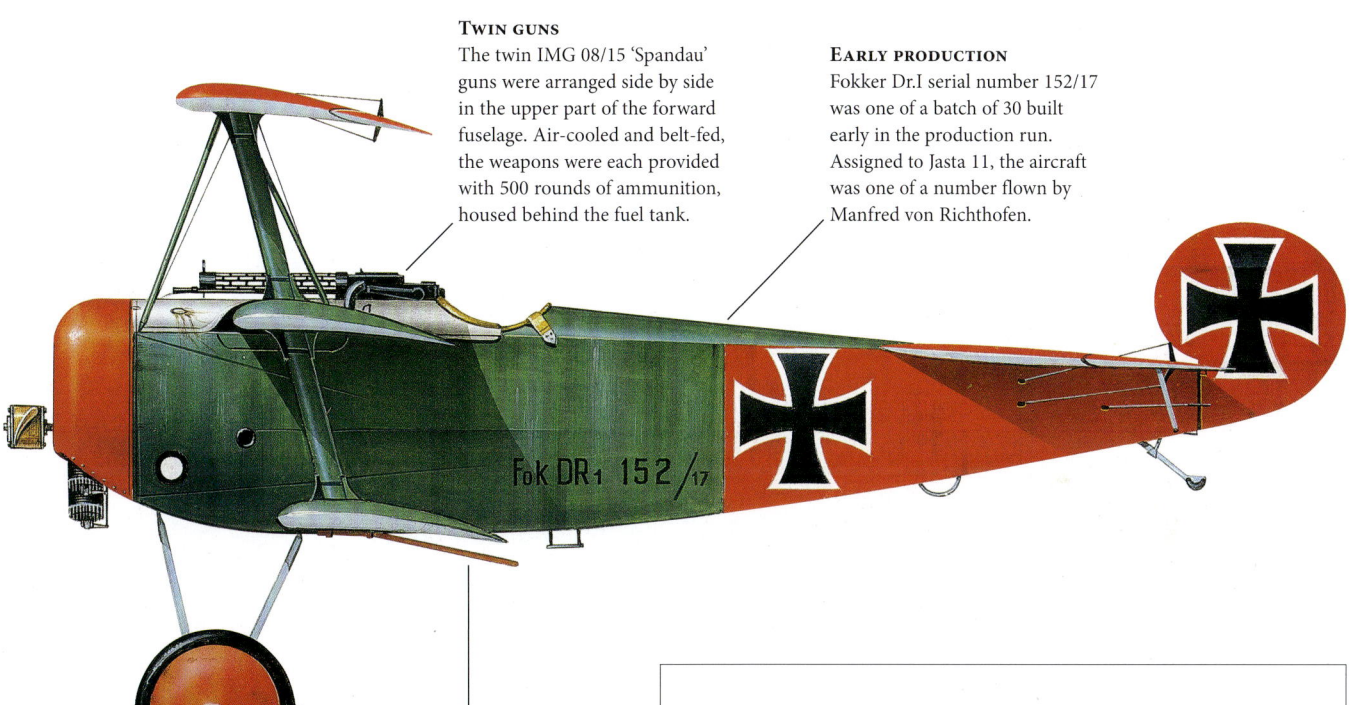

TWIN GUNS
The twin IMG 08/15 'Spandau' guns were arranged side by side in the upper part of the forward fuselage. Air-cooled and belt-fed, the weapons were each provided with 500 rounds of ammunition, housed behind the fuel tank.

EARLY PRODUCTION
Fokker Dr.I serial number 152/17 was one of a batch of 30 built early in the production run. Assigned to Jasta 11, the aircraft was one of a number flown by Manfred von Richthofen.

LAYOUT
A side view of Dr.I 152/17 reveals the characteristic slab-sided fuselage, strut-braced tailplane and fixed landing gear. A key attribute of the Dreidecker was its simple, low-cost construction.

LIMITED NUMBERS

The reputation of the Dr.I was such that certain German leading exponents continued to fly the type even after the arrival of more modern equipment. Although 320 examples were eventually built by the time production ended in May 1918, at no time were there more than 171 Dr.Is in service, making it all the more impressive that it became such a feared foe.

JG 1 DR. 1
A German Fokker Dr.I of Jagdgeschwader 3 (JG 3). Alongside its undoubted prowess in aerial combat, the notoriety of the Dr.I among Allied pilots was certainly sealed by the activities of fighter ace Manfred von Richthofen's 'Flying Circus' (JG 1).

SPAD S.XIII (1917)

TYPE • *Fighter* COUNTRY • *France*

Famed as the colourful mount of the American Expeditionary Forces' 94th Aero Squadron, the French SPAD S.XIII was one of the finest Allied fighting scouts of the war. It was also flown by renowned aces Georges Guynemer and René Fonck.

SPECIFICATIONS

DIMENSIONS:	Length: 6.15m (20ft 2in); Wingspan: 7.8m (25ft 8in); Height: 2.12m (6ft 11in)
WEIGHT:	740kg (1631lb) maximum take-off
POWERPLANT:	Hispano-Suiza 8Aa V-8 inline piston engine, 112kW (150hp)
MAX SPEED:	192km/h (119mph)
ENDURANCE:	2 hours 15 minutes
CEILING:	5300m (17,390ft)
CREW:	1
ARMAMENT:	1 or 2 × 7.7mm (0.303in) fixed forward-firing Vickers machine guns

Above: A French SPAD S.XIII that served with Escadrille SPA.48 in the spring of 1918 displays the cockerel insignia of the squadron.

CONSTRUCTION
Typical for its day, the SPAD S.XIII was fabric-covered from behind the cockpit, with aluminium panels for the nose and cowling. The fuselage was constructed of four longerons with spruce struts and stringers.

INSIGNIA
The insignia of the Lafayette Escadrille (manned largely by US volunteers) is worn on the fuselage of SPAD S.XIII C.1, serial number S7714. Many members of Lafayette joined the 103rd Aero Squadron after the United States entered the war.

POWERPLANT
Advanced for its day, the water-cooled Hispano-Suiza 8B inline engine was a refined version of the 8A engine that powered the SPAD S.VII. This V-8 unit was designed by the Swiss engineer Marc Birkigt in 1915 and featured overhead camshafts and single-piece aluminium cylinders.

SCOUT AIRCRAFT
Boosted by the lobbying of air ace Georges Guynemer and by the evolution of new German scouts, aviation designer Louis Béchereau made strident efforts to improve the performance of the SPAD S.VII. The resulting SPAD S.XII prototype first flew in April 1917 and series aircraft began entering service at the end of May. The S.XIII soon replaced the S.VII in service.

SPAD X.III, FRANCE, 1918
One of the most famous SPAD X.III pilots was American aviator Eddie Rickenbacker, who emerged as the leading US ace of World War I, with 26 confirmed victories.

Pfalz D.III (1917)

TYPE • *Fighter* COUNTRY • *Germany*

SPECIFICATIONS

DIMENSIONS:	Length: 6.95m (22ft 10in); Wingspan: 9.4m (30ft 10in); Height: 2.67m (8ft 9in)
WEIGHT:	933kg (2056lb) maximum take-off
POWERPLANT:	One 130kW (180hp) Mercedes D.IIIa 6-cylinder water-cooled inline piston engine
MAX SPEED:	165km/h (102mph)
ENDURANCE:	2 hours
CEILING:	5180m (16,995ft)
CREW:	1
ARMAMENT:	Two 7.92mm (0.312in) LMG 08/15 'Spandau' machine guns

Powered by the same engine as the Albatros and featuring the same armament and airframe technology, the Pfalz D.III had essentially the same performance. However, the Pfalz was much stronger and proved to be a useful supplement to the Albatros.

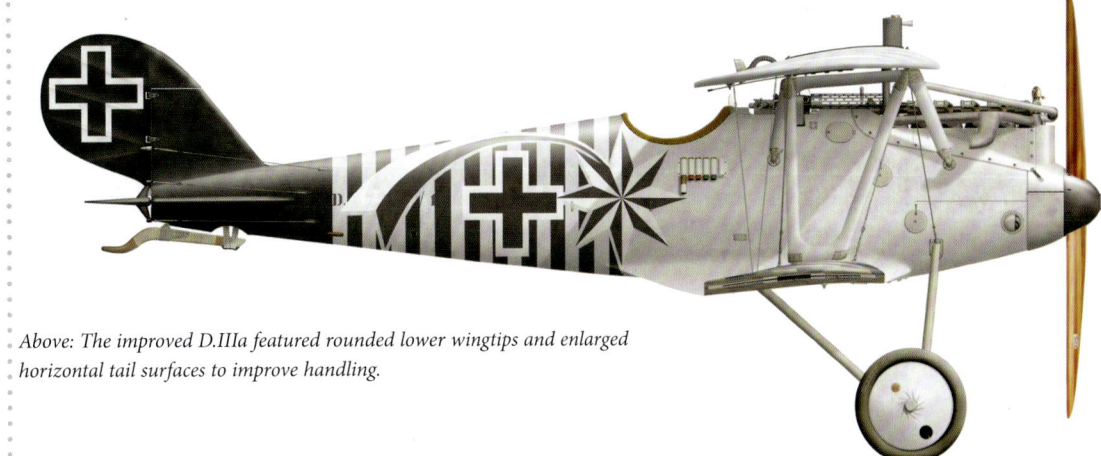

Above: The improved D.IIIa featured rounded lower wingtips and enlarged horizontal tail surfaces to improve handling.

GUN LOCATION
The most serious operational problem was the location of its machine guns in the fuselage, inaccessible in the event of the guns jamming. In response, in November 1917 Pfalz flew the improved D.IIIa in which the guns were relocated directly ahead of the pilot.

WING DESIGN
Although the Pfalz engineers adapted the Nieuport's sesquiplane wing cellule, they used two spars in the lower wing, which minimized the twisting of the wing during combat manoeuvres.

PERFORMANCE
The Pfalz was criticized for its comparatively poor manoeuvrability, slow rate of climb, heavy controls and vicious stall that led to a dangerous flat spin.

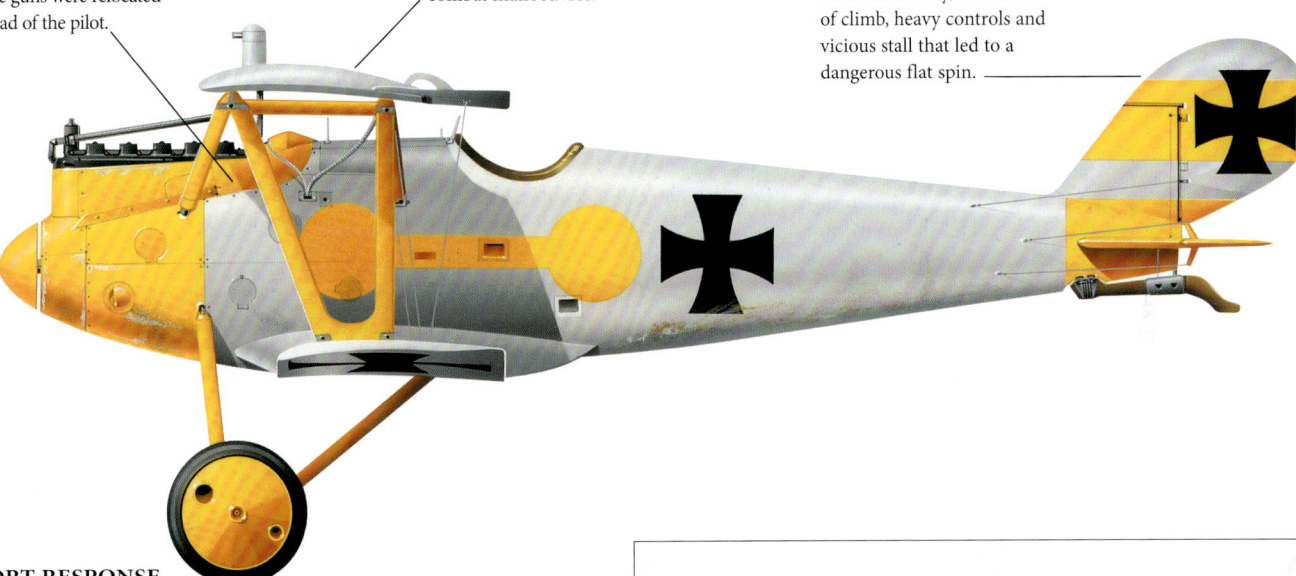

NIEUPORT RESPONSE
Similar to the Albatros D.III and D.V, the Pfalz D.III was designed as a response to the Nieuport 17, which had proved superior to any fighter the Central Powers then possessed. The Pfalz D.III did not ape the Nieuport's sesquiplane layout to the same extent as the Albatros, and as a result never suffered from the structural concerns that bedevilled its more numerous contemporary.

PFALZ D.III
In use, the Pfalz D.III and D.IIIa gave good service although they were always overshadowed by Albatros fighters and somewhat outperformed by contemporary Allied fighters. Most pilots preferred the Albatros with its better handling and climb performance.

Siemens-Schuckert D.III (1917)

TYPE • *Fighter* COUNTRY • *Germany*

SPECIFICATIONS

DIMENSIONS:	Length: 5.7m (18ft 8in); Wingspan: 8.43m (27ft 8in); Height: 2.8m (9ft 2in)
WEIGHT:	(gross) 725kg (1598lb)
POWERPLANT:	One 120kW (160hp) Siemens-Halske Sh.III 11-cylinder air-cooled geared rotary piston engine
MAX SPEED:	177km/h (110mph)
ENDURANCE:	2 hours
CEILING:	8000m (26,000ft)
CREW:	1
ARMAMENT:	Two 7.92mm (0.312in) LMG 08/15 'Spandau' machine guns

Despite its rotund appearance, the Siemens-Schuckert D.III was one of the most potent fighters of the entire conflict. Possessed of an outstanding rate of climb and high manoeuvrability, the D.III was likely the finest interceptor to see widespread service before the armistice.

Above: Joachim von Ziegesar flew this early D.III with Jasta 15. Ziegesar scored three confirmed victories and went on to survive the war.

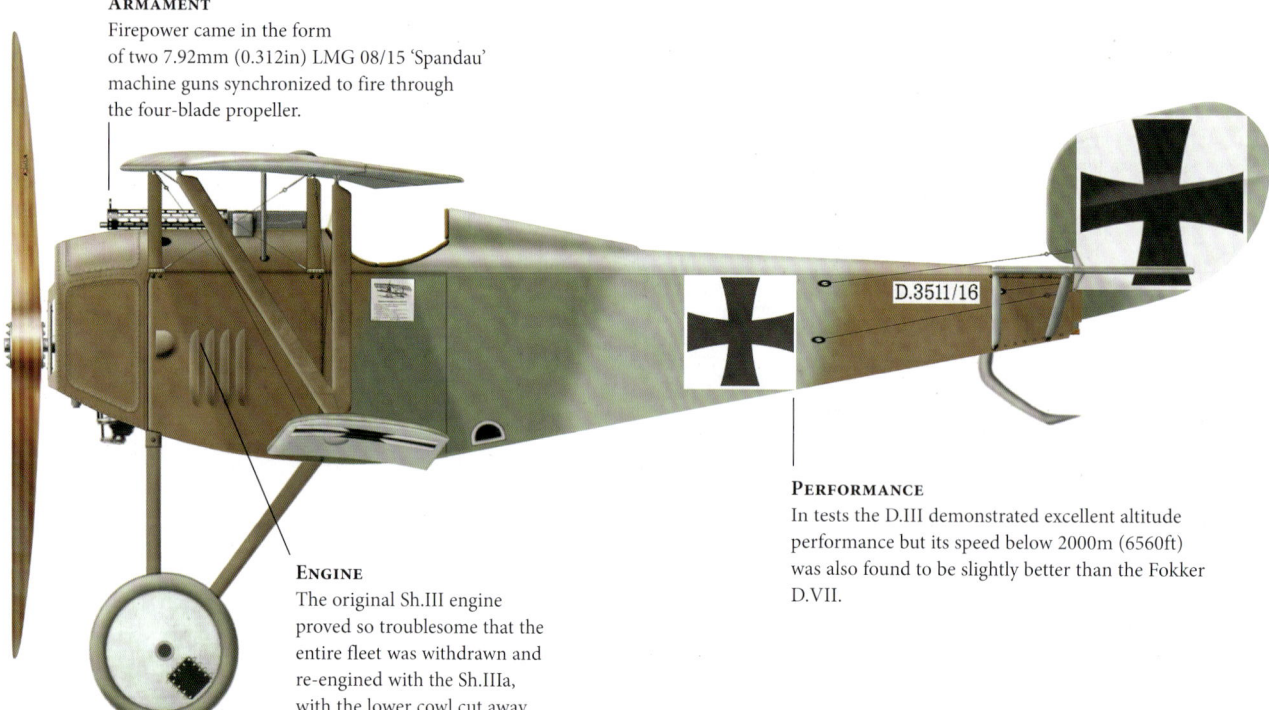

ARMAMENT
Firepower came in the form of two 7.92mm (0.312in) LMG 08/15 'Spandau' machine guns synchronized to fire through the four-blade propeller.

ENGINE
The original Sh.III engine proved so troublesome that the entire fleet was withdrawn and re-engined with the Sh.IIIa, with the lower cowl cut away to improve cooling.

PERFORMANCE
In tests the D.III demonstrated excellent altitude performance but its speed below 2000m (6560ft) was also found to be slightly better than the Fokker D.VII.

PILOT SKILL
It was noted that the D.III required a higher degree of piloting skill than the docile Fokker D.VII. As a result, two Siemens-Schuckert test pilots, Hans Muller and Bruno Rodschinka toured frontline units to instruct service pilots and perform flying demonstrations. Muller was present on 21 August 1917 when Naval ace Theo Osterkamp intercepted a DH.4 at 6000m (19,685ft).

JASTA 2 D.III, 1917
In total, 41 D.IIIs were delivered to combat units by May 1918, most going to Jasta 2 under Rudolf Berthold where they proved immediately effective, Berthold noting the D.III's 'brilliant' rate of climb.

Gotha G.IV (1917)

TYPE • Bomber COUNTRY • Germany

The Gotha bombers were the world's first mass-produced large aircraft. Their strategic bombing campaign against London had a psychological effect out of all proportion to the actual damage inflicted upon the British capital.

SPECIFICATIONS

DIMENSIONS:	Length: 12.2m (40ft); Wingspan: 23.7m (77ft 9in); Height: 3.9m (12ft 10in)
WEIGHT:	(gross) 3648kg (8042lb)
POWERPLANT:	Two 190kW (260hp) Mercedes D.IVa 6-cylinder water-cooled inline piston engines
MAX SPEED:	135km/h (84mph)
ENDURANCE:	6 hours
CEILING:	5000m (16,400ft)
CREW:	3
ARMAMENT:	One 7.92mm (0.312in) Parabellum MG14 machine gun in nose position, one 7.92mm (0.312in) Parabellum MG14 machine gun in dorsal/ventral position; up to 500kg (1100lb) bombload

Right: From August 1917, the G.IV had been joined in the London raids by the similar G.V (seen here), which differed primarily in having the fuel tanks relocated from the engine nacelles to a less vulnerable position in the fuselage.

ENGINES
The engines were Mercedes D.IVa 6-cylinder units with pusher propellers rather than the tractor arrangement of the original G.I.

'GUN TUNNEL'
The G.IV was basically the same as the G.III, but with ailerons on all four wingtips and the Gotha 'gun tunnel', a trough-shaped cut-out in the rear fuselage that allowed the rear gunner to fire downwards.

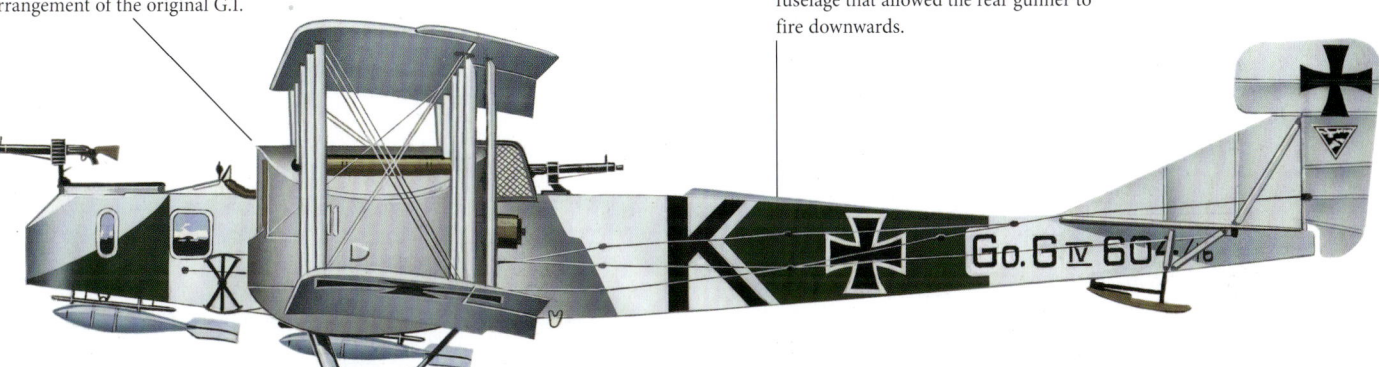

FIRST RAID
Entering service in the autumn of 1916, the Gotha G.IV allowed the first strategic raids on London to be mounted by fixed-wing aircraft. The raid flown on 13 June 1917 was the first time London had been bombed by day. It proved spectacularly effective, all 20 aircraft dispatched returned safely and the 14 that bombed London caused an estimated £125,953 of damage and killed 162 people.

BOMBLOAD
The G-series of aircraft was one of the early efforts to create a true strategic bomber. The G.IV could carry up to 500kg (1100lb) of bombs under the fuselage and wings.

ARMING A GOTHA
A German ground crew arm a Gotha G.V with bombs ready for a raid. Both the G.IV and G.V became tail heavy and difficult to control once the bombs had been released and 36 were lost in landing accidents, a third more than to enemy action.

Albatros D.V (1917)

TYPE • *Fighter* **COUNTRY** • *Germany*

SPECIFICATIONS (D.Va)

DIMENSIONS:	Length: 7.33m (24ft); Wingspan: 9.05m (29ft 8in); Height: 2.7m (8ft 10in)
WEIGHT:	(gross) 937kg (2066lb)
POWERPLANT:	One 127kW (170hp) Mercedes D.IIIaü 6-cylinder water-cooled inline piston engine
MAX SPEED:	172km/h (107mph)
ENDURANCE:	2 hours
CEILING:	5800m (19,029ft)
CREW:	1
ARMAMENT:	Two 7.92mm (0.312in) LMG 08/15 'Spandau' machine guns

The Albatros D.V's performance was little better than the D.III it replaced, and it suffered from the same major structural problem. Despite this, the D.V and improved D.Va were produced in large numbers and served until the armistice.

Above: Despite its shortcomings, the D.V was produced in large numbers for lack of a better alternative.

POWERPLANT
The D.Va had 127kW (170hp) Mercedes D.IIIaü 6-cylinder water-cooled inline piston engine, an upgraded unit compared to the D.V.

WING STABILITY
Issues with wing integrity were never entirely resolved in the D.Va, with pilots being most concerned about losing a wing in a steep dive.

SCHEMATICS
German fighters during World War I often had highly elaborate paint schemes that expressed the personality of their pilots and squadrons rather than seeking uniformity.

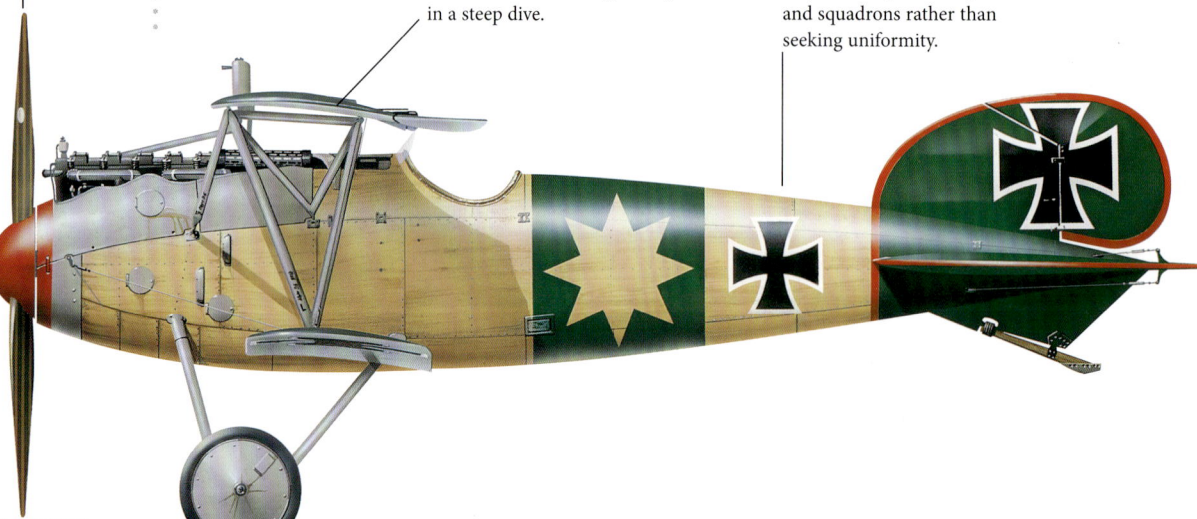

STRUCTURAL ISSUES
The first D.Vs reached the Jastas in June 1917 but proved a disappointment for pilots, initially due to the very modest performance gain delivered by the new aircraft. However, this was compounded by the realization that the wing-twisting structural issues plaguing the D.III were not improved upon, but had worsened. To combat the problems, Albatros introduced the D.Va.

ALBATROS D.VA
The D.Va featured a strengthened structure and a return to the D.III's wing design, a process resulting in an aircraft heavier than the D.III, although the use of the high compression Mercedes D.IIIaü compensated for the weight increase.

Airco DH.4 (1917)

TYPE • Bomber   **COUNTRY** • United Kingdom

SPECIFICATIONS

DIMENSIONS:	Length: 9.35m (30ft 8in); Wingspan: 12.92m (42ft 4in); Height: 3.35m (11ft)
WEIGHT:	1575kg (3742lb) maximum take-off
POWERPLANT:	1x 280kW (375hp) Rolls-Royce Eagle VIII inline piston engine
MAX SPEED:	230km/h (143mph)
ENDURANCE:	3 hours 45 minutes
CEILING:	6705m (22,000ft)
CREW:	2
ARMAMENT:	4 x 7.7mm (0.303in) Vickers MGs (two forward-firing, two in rear cockpit); 209kg (460lb) of bombs

De Havilland designed the Airco DH.4 around the 200 BHP (Beardmore-Halford-Pullinger) engine in response to an Air Ministry request for a new day bomber. In this role, the DH.4 was the best aircraft in its class during the war.

Left: The DH.4 excelled as a long-range reconnaissance airplane, day bomber and even as a Zeppelin interceptor.

ENGINES
A variety of engines were used in the DH.4, but the 280kW (375hp) Rolls-Royce Eagle made the DH.4 the fastest airplane on the Western Front – remarkable for such a large two-seater.

CREW
The wide separation between pilot and observer was a controversial and potentially dangerous feature as it hampered communication in the air.

INLINE LAYOUT
The DH.4 was a classic design. Using an inline piston engine, de Havilland employed a clean tractor layout. The DH.4 was also built under licence in the United States.

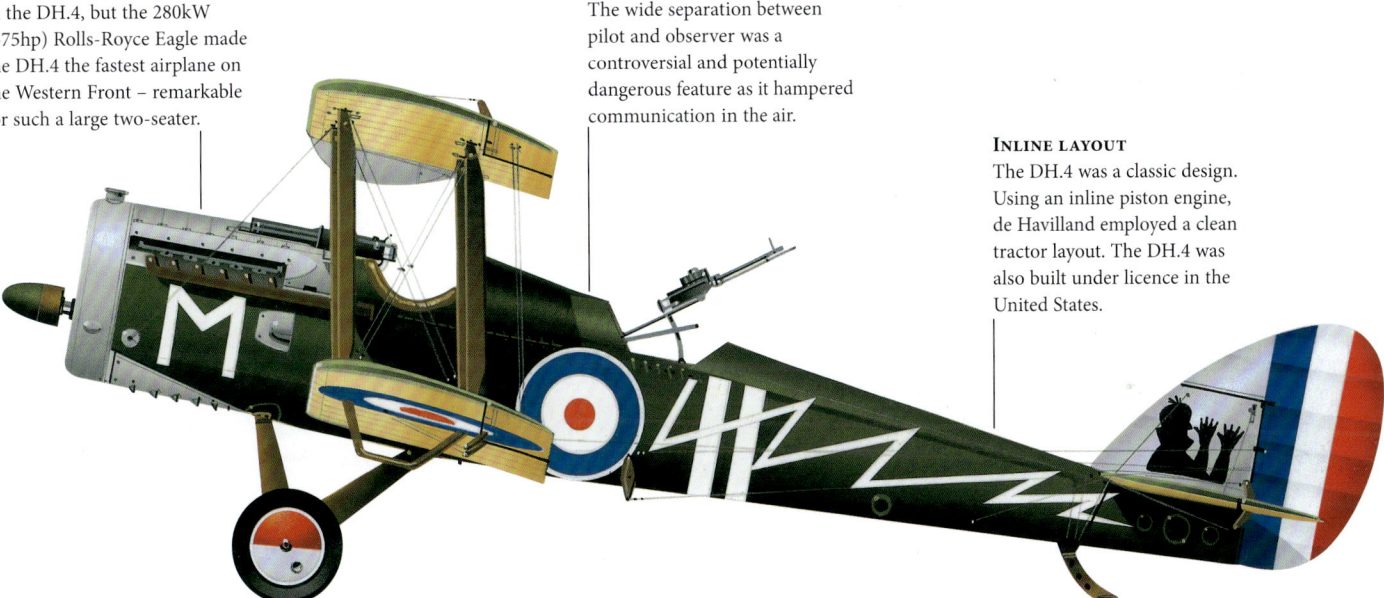

ANTI-ZEPPELIN
Britain pressed many high-performance airplanes into the anti-Zeppelin role and this DH.4 was particularly successful. On 5 August 1918, Eagle-powered A8032 flown by Major E. Cadbury and Capt R. Leckie, downed *L.70*, which was carrying Peter Strasser, commander of the German naval airships. They were experienced: Cadbury had downed *L.21* in 1916 while flying a BE.2c and Leckie had flown a Curtiss H-12 when he downed *L.22* in 1917.

DH.4 PROFILE
The 1449 British-built DH.4s were manufactured by various subcontractors, although delayed production of the BHP engine meant that other engines were employed on production aircraft. By spring 1918 the DH.4 equipped nine RAF squadrons and was also in use with the Royal Naval Air Service.

Zeppelin-Staaken R.IV and R.VI (1917)

TYPE • Bomber   COUNTRY • Germany

As well its famous line of massive airships, the Zeppelin company also built equally huge aeroplanes. The impressive R.VI was the largest heavier-than-air aircraft of World War I to enter serial production.

SPECIFICATIONS (R.VI)

Dimensions:	Length: 22.1m (72ft 6in); Wingspan: 42.2m (138ft 5in); Height: 6.3m (20ft 8in)
Weight:	(gross) 11,848kg (26,120lb)
Powerplant:	Four 183kW (245hp) Maybach Mb.IVa high-compression, 6-cylinder water-cooled inline piston engines
Max Speed:	135km/h (84mph)
Endurance:	7–10 hours
Ceiling:	4320m (14,170ft)
Crew:	10
Armament:	One 7.92mm (0.312in) Parabellum MG14 machine gun flexibly mounted in nose, dorsal and ventral positions; up to 2000kg (4409lb) bombload

Left: The Zeppelin-Staaken R.IV was the only one of the early Zeppelin Giants to survive the war.

DEFENCES
Featuring a gun position in each nacelle as well as one on the trailing edge of the upper wing, the R.IV was an unusually well defended aircraft.

LAYOUT
A development of the VGO aircraft, the R.IV featured six engines, four Benz Bz.IVs in pairs, each pair driving a propeller at the rear of the engine nacelle and two Mercedes D.IIIs mounted side by side in the nose geared to a single large propeller.

DESIGN CHANGES
In contrast to the R.IV seen here, the R.VI dispensed with the nose engine layout and featured a fully enclosed cabin for the sizeable 10-man crew.

DESIGN TEAM
The largest aircraft deployed in World War I was the sluggish but capable *Riesenflugzeug* (giant aeroplane) series produced by the Zeppelin Werke Staaken (originally Gotha). Via several other one-off bombers, with three, four or five engines and different schemes of defensive armament, the design team of Baumann, Hirth and Klein eventually produced R.VI.

ZEPPELIN-STAAKEN R.VI
The R.VI was visually distinguished from the R.IV by its four 183kW (245hp) Maybach Mb.IVa tractor props, rather than the R.IV's pusher props.

Airco DH.9 (1917)

TYPE • Bomber **COUNTRY** • United Kingdom

SPECIFICATIONS (DH.9A)

- **DIMENSIONS:** Length: 9.22m (30ft 3in); Wingspan: 14.01m (45ft 11in); Height 3.45m (11ft 4in)
- **WEIGHT:** 1575kg (3742lb) maximum take-off
- **POWERPLANT:** 1 x 313kW (420hp) Packard Liberty 12 V-12 piston engine
- **MAX SPEED:** 198km/h (123mph)
- **ENDURANCE:** 5 hours 15 minutes
- **CEILING:** 5105m (16,750ft)
- **CREW:** 2
- **ARMAMENT:** 1 x 7.7mm (0.303in) Vickers MG and 1–2 x 7.7mm (0.303in) Lewis MGs in rear cockpit; external pylons with provision for 299kg (660lb) of bombs

Derived from the excellent DH.4 by moving the pilot's cockpit aft and substituting the low-powered and unreliable Puma engine, the DH.9 suffered greatly from the attentions of German fighters during its daylight bombing raids.

Right: Airco DH.9s, belonging to the RAF No. 120 Squadron, line up on a wintry airfield at Hawkinge in Kent, England.

ENGINE
Substituting the powerful Rolls-Royce Eagle or American Liberty engine for the wretched Siddeley Puma of the DH.9, resulted in the DH.9A's excellent speed.

IMPROVEMENTS
Other improvements of the DH.9A compared to the DH.9 included a larger wing, which combined with the better engine gave the aircraft a much improved service ceiling.

GUN ARRANGEMENT
The rear observer/gunner was armed with either a single 7.7mm (0.303in) Lewis MG or twin guns to increase the volume of defensive firepower.

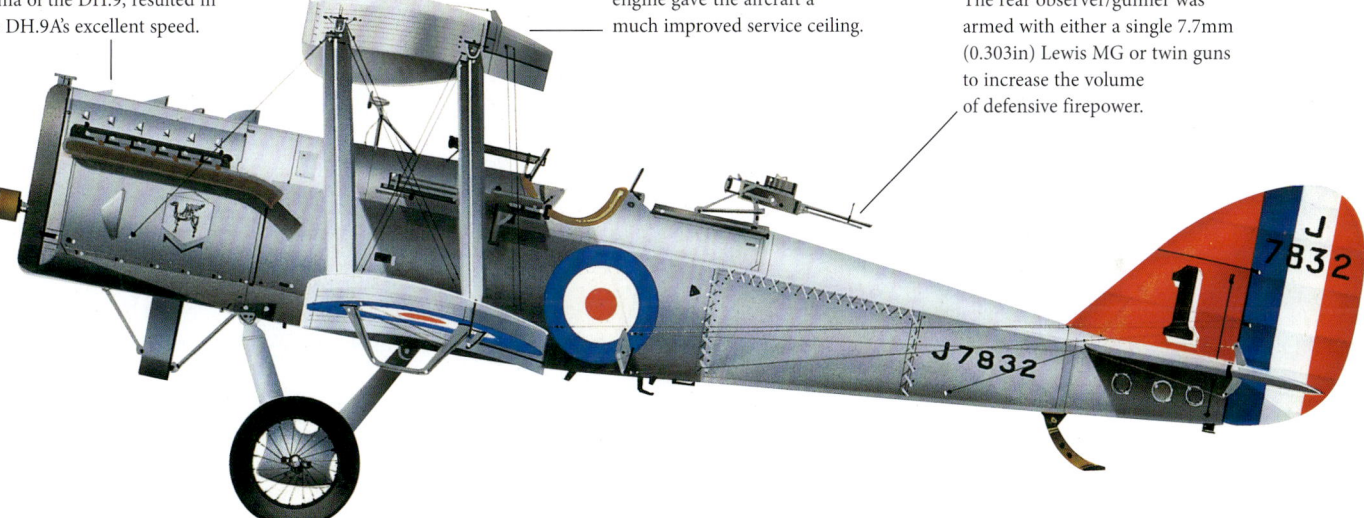

REVISED SEATING
Persistent German raids on Britain during World War I prompted a doubling in the size of the Royal Flying Corps, with most of the new squadrons equipped with day bombers. The DH.4 was the expected type, but de Havilland had already attempted to rectify a glaring weakness of this aircraft by designing a modified version designated DH.9, with the pilot and observer accommodated in back-to-back seating.

DH.9As IN FLIGHT,
Two British DH.9A biplanes in flight in 1917; the observer in the nearest aircraft is looking somewhat precarious. The DH.9A could be distinguished from the DH.9 by the latter's rounded nose, as opposed to the flat engine profile seen here.

Bréguet 14 (1917)

TYPE • Bomber COUNTRY • France

The Bréguet 14 was another of the great warplanes of World War I. Produced as the Bréguet 14B2 day bomber, in 1918 it equipped all French day-bomber units. It was also widely used as the Bréguet 14A2 reconnaissance plane.

SPECIFICATIONS

DIMENSIONS:	Length: 8.87m (29ft 1in); Wingspan: 14.36m (47ft 2in); Height: 3.03m (9ft 11in)
WEIGHT:	1765kg (3883lb) maximum take-off
POWERPLANT:	1 x 224kW (300hp) Renault 12 Fe liquid-cooled V-12 engine
MAX SPEED:	177km/h (110mph)
ENDURANCE:	2 hours 45 minutes
CEILING:	5750m (18,850ft)
CREW:	2
ARMAMENT:	1 x fixed and 2 x flexible 7.7mm (0.303in) Lewis MGs, up to 320kg (705lb) of bombs

Left: A Bréguet 14B2, operating with Escadrille BR 117. The 14B2 was the best day bomber of the war and was used in large formations during 1918. It was fast, reliable and carried a good bombload.

POWERPLANT
The engine on the Bréguet 14 was a single Renault 12 Fe liquid-cooled V-12, generating 224kW (300hp) and helping the aircraft achieve a respectable maximum speed of 177km/h (110mph)

METAL FRAMEWORK
The Bréguet 14 was especially appreciated for its modern metal framework, which included duralumin components. This structure was much safer than wood in a crash.

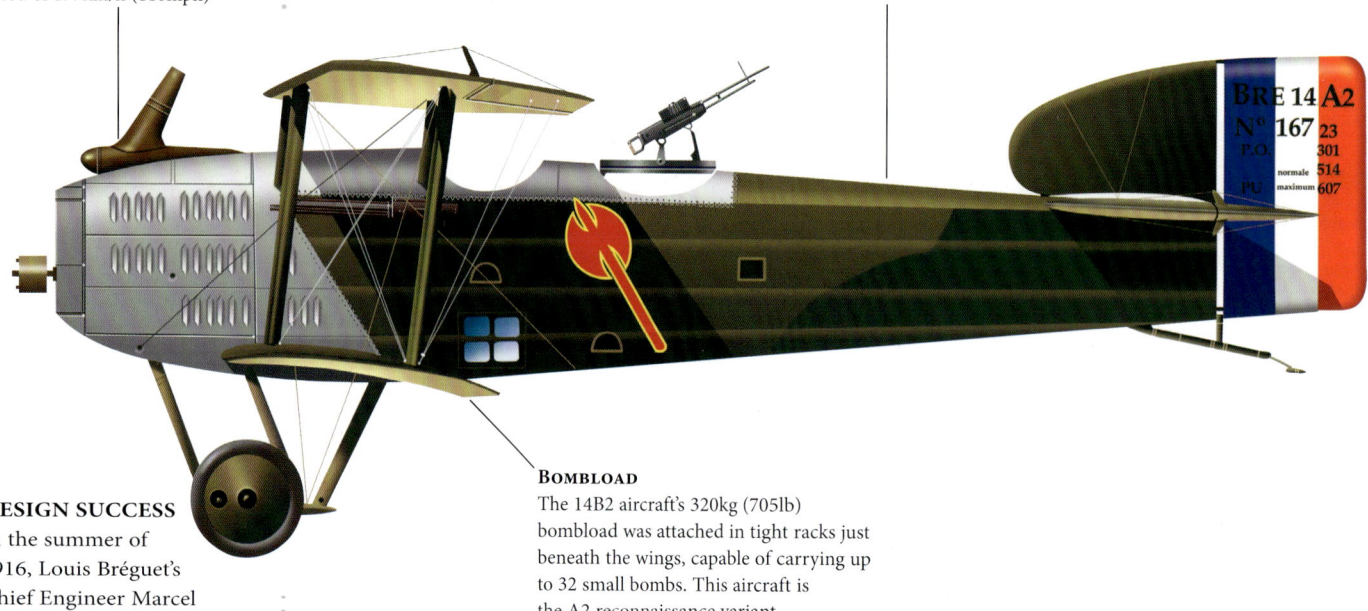

DESIGN SUCCESS
In the summer of 1916, Louis Bréguet's Chief Engineer Marcel Vullierme began the design of Bréguet's most successful wartime product, the Bre.14. The prototype of this two-seat reconnaissance/light bomber aircraft made its first flight barely five months later, and the first Bre.14 A2 production aircraft entered service with the Aéronautique Militaire the following spring.

BOMBLOAD
The 14B2 aircraft's 320kg (705lb) bombload was attached in tight racks just beneath the wings, capable of carrying up to 32 small bombs. This aircraft is the A2 reconnaissance variant.

DAY BOMBER
This 14B2, seen here in March 1918, was operated by Escadrille BR 128 at Champien (Somme). The French mounted large, heavily escorted day-bombing raids with the 14B2 and the Germans could do little to stop the high-performance aircraft.

Felixstowe F.2A (1917)

TYPE • *Flying boat* COUNTRY • *United Kingdom*

The Felixstowe F.2A was active on anti-submarine patrols over the North Sea. Many were painted flamboyantly to increase their visibility should they need rescue after ditching, which resulted in some outlandish colour schemes.

SPECIFICATIONS

DIMENSIONS:	Length: 14.10m (46ft 3in); Wingspan: 29.15m (95ft 8in); Height: 5.33m (17ft 6in)
WEIGHT:	4980kg (10,978lb) maximum take-off
POWERPLANT:	2 x 257kW (345hp) Rolls-Royce Eagle VIII V-piston engines
MAX SPEED:	153km/h (95.5mph)
ENDURANCE:	6 hours
CEILING:	2925m (9,600ft)
CREW:	2
ARMAMENT:	3–6 x 7.7mm (.303in) Lewis MGs (1–2 forward-firing; 1–2 rearward-firing; 1–2 lateral-firing); 209kg (460lb) bombload

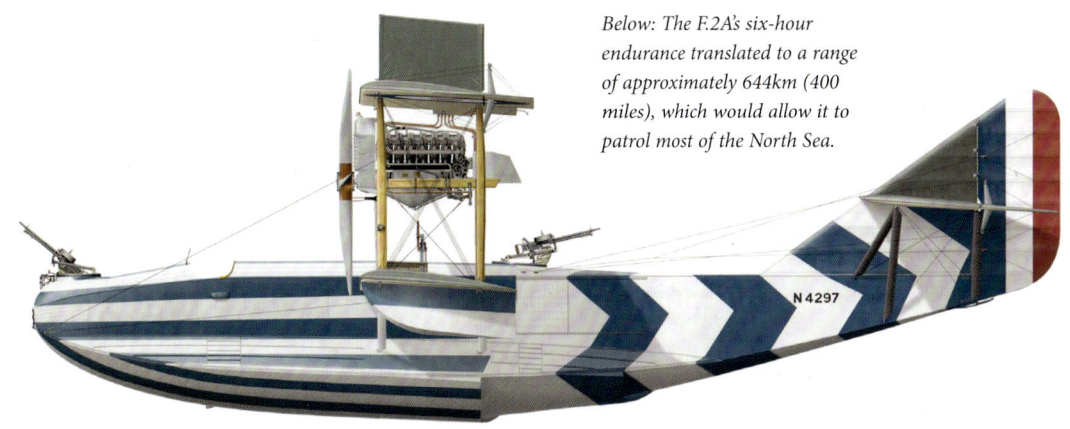

Below: The F.2A's six-hour endurance translated to a range of approximately 644km (400 miles), which would allow it to patrol most of the North Sea.

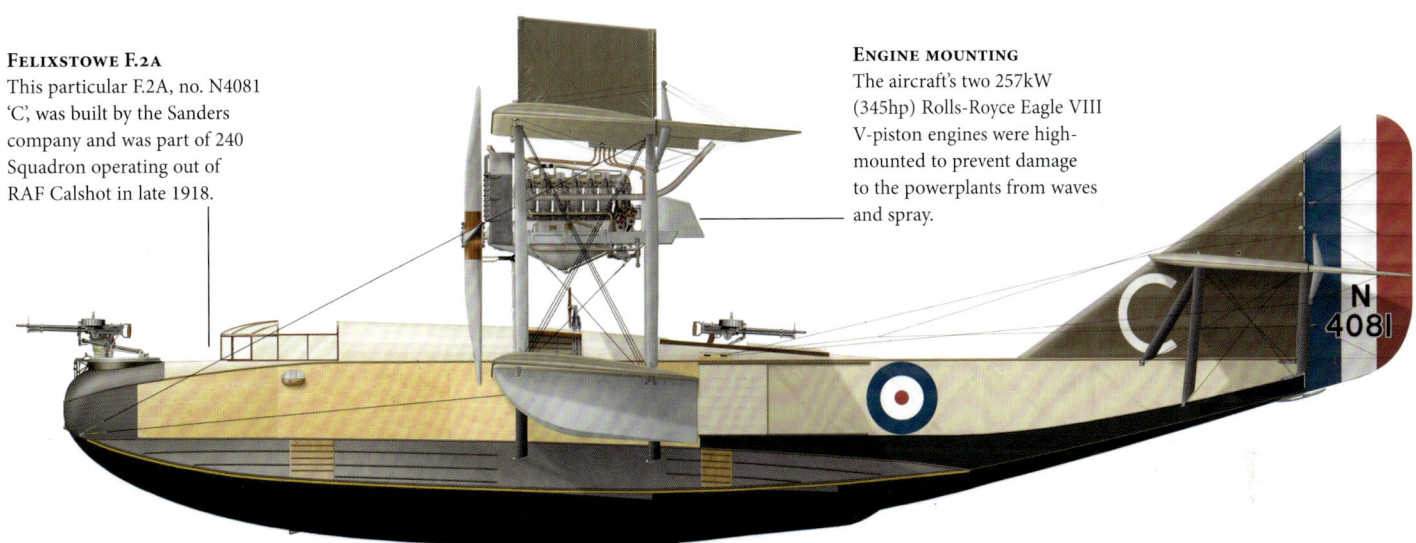

FELIXSTOWE F.2A
This particular F.2A, no. N4081 'C', was built by the Sanders company and was part of 240 Squadron operating out of RAF Calshot in late 1918.

ENGINE MOUNTING
The aircraft's two 257kW (345hp) Rolls-Royce Eagle VIII V-piston engines were high-mounted to prevent damage to the powerplants from waves and spray.

HULL
The F.2A's hull was known for its excellent sea-handling characteristics, and the profile had a significant impact on post-war flying boat design.

ANTI-SUBMARINE
The Felixstowe F.2A fulfilled a requirement that had not existed half a decade earlier. As German U-boats preyed on Allied merchant shipping in the Irish Sea, North Sea and around British coastal waters, the F.2A had the endurance (6 hours) and the bombload to mount extended anti-submarine operations. It also had heavy defensive armament to deal with enemy fighters.

F.2A FLIGHT LINE
The Felixstowe F.2A flying boats were designed and developed by Lieutenant Commander John Cyril Porte RN, who worked at the Seaplane Experimental Station, Felixstowe, seen here during World War I.

Caproni Ca.3 (1917)

TYPE • Bomber COUNTRY • Italy

SPECIFICATIONS

DIMENSIONS:	Length: 11.05m (36ft 3in); Wingspan: 22.2m (72ft 10in); Height: 3.84m (12ft 7in)
WEIGHT:	3312kg (7302lb) maximum take-off
POWERPLANT:	3 x 112kW (150hp) Isotta-Fraschini V.4B inline piston engines
MAX SPEED:	138km/h (85.7mph)
ENDURANCE:	4 hours
CEILING:	4500m (14,764ft)
CREW:	4
ARMAMENT:	2 or 4 x 6.5mm (0.256in) Fiat-Revelli MGs on flexible mounts in cockpit; plus a maximum bombload of 240kg (529lb)

When first entering service, the Caproni Ca.3 series of biplanes were the best bombers available to the Western Allies. The heavy bombers were sold to France and even built under licence in France.

Left: Some 83 Ca.3s were built under licence in France by Robert Esnault-Pelterie, and equipped two units of the Aéronautique Militaire.

CREW POSITIONS
The crew positions were completely open to the elements. The crew consisted of a front gunner, two pilots and a rear-gunner/flight engineer.

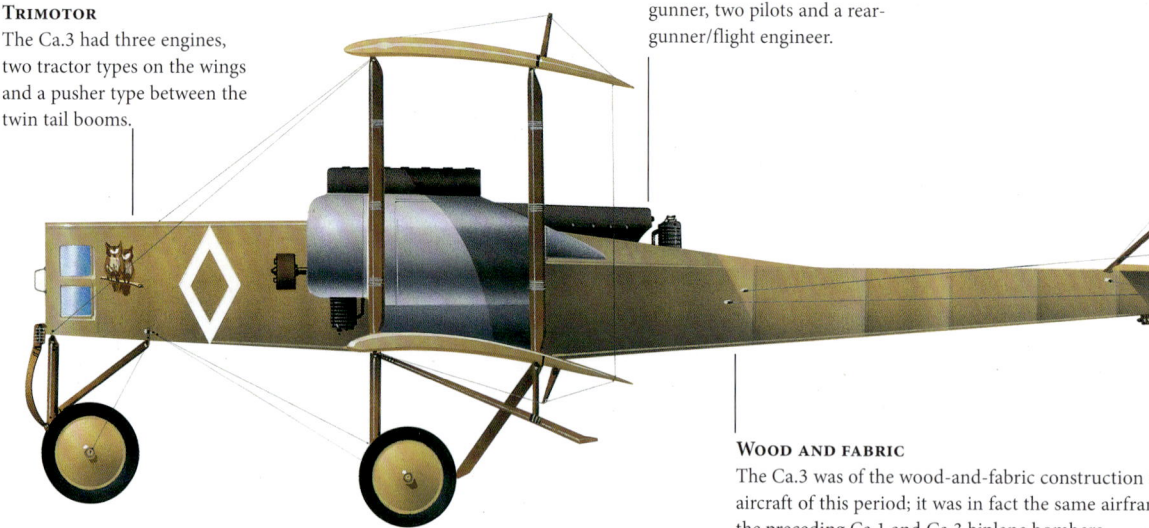

TRIMOTOR
The Ca.3 had three engines, two tractor types on the wings and a pusher type between the twin tail booms.

WOOD AND FABRIC
The Ca.3 was of the wood-and-fabric construction typical of aircraft of this period; it was in fact the same airframe as the preceding Ca.1 and Ca.3 biplane bombers.

PROUD HISTORY

After entering service with the Corpo Aeronautica, the Ca.2 carried out the first Italian bombing raids of the war on 25 August 1915, and soon established a proud tradition of arduous missions on the Austro–Hungarian front over difficult mountainous terrain for what were basically extremely flimsy aircraft. They were followed into service in 1917 by the Ca.3. This aircraft had more powerful engines and greater bombload, and was undoubtedly the most successful Allied bomber of the war.

ITALO BALBO
Italo Balbo (centre, in dark jacket), the Italian fascist politician and Air Minister under Mussolini, here poses for a photograph with a group of officers in front of an Italian Air Force Caproni Ca.3 bomber at Mellaha airfield, near Tripoli, Libya, in the 1930s.

Fokker D.VII (1917)

TYPE • *Fighter* COUNTRY • *Germany*

By consensus one of the most formidable fighters to see widespread service during the conflict, the Fokker D.VII transformed the fortunes of both the Jagdflieger and the Fokker company. Over 3000 were built in total.

SPECIFICATIONS
(Early production)

DIMENSIONS:	Length 6.95m (22ft 10in); Wingspan: 8.9m (29ft 2in); Height: 2.75m (9ft)
WEIGHT:	(gross) 906kg (1997lb)
POWERPLANT:	One 120kW (160hp) Mercedes D.III 6-cylinder water-cooled inline piston engine
MAX SPEED:	188km/h (117mph)
ENDURANCE:	1 hour 30 minutes
CEILING:	6000m (19,685ft)
CREW:	1
ARMAMENT:	Two 7.92mm (0.312in) LMG 08/15 'Spandau' machine guns

Left: This Fokker D.VII has been captured and placed under British markings.

AMMUNITION
Early examples placed the ammunition supply too close to the engine and there were instances in which bullets 'cooked off' in flight.

HANDLING
The single greatest attribute that the D.VII possessed was its benign handling characteristics courtesy of the thick wing. The D.VII was almost laughably easy to fly, particularly at high altitude and near the stall.

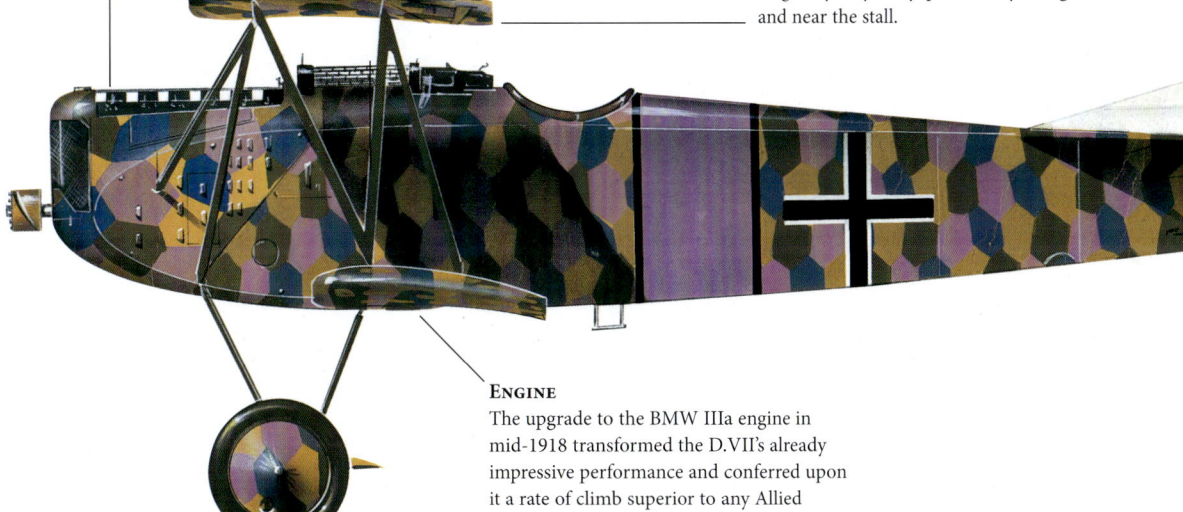

ENGINE
The upgrade to the BMW IIIa engine in mid-1918 transformed the D.VII's already impressive performance and conferred upon it a rate of climb superior to any Allied aircraft then in service.

FIGHTER UNIT
The D.VII proved vastly superior to any of the other submissions in the German standard fighter competition of January 1918. The first unit to receive the type was Manfred von Richthofen's unit JG I, which was commanded by Hermann Göring after the death of the Red Baron in April 1918. Approximately 1000 of this extremely capable aircraft had been completed by the time of the armistice.

COMBAT CAPABILITY
Unsure of what to make of the new boxy aircraft – lacking the sleek lines of the Albatros fighters it replaced – Allied aircrew initially underestimated the capabilities of the D.VII. In combat from May 1918 onwards, the D.VII soon dazzled its opponents.

Hansa-Brandenburg W.12 (1917)

TYPE • *Seaplane* COUNTRY • *Germany*

A versatile and formidable aircraft, the W.12 proved to be a particularly successful maritime fighter in long-range skirmishes with RNAS flying boats. The prototype flew in early 1917 and was followed by 146 production aircraft.

SPECIFICATIONS (late production)

DIMENSIONS:	Length: 9.6m (31ft 6in); Wingspan: 11.2m (36ft 9in); Height: 3.3m (10ft 10in)
WEIGHT:	(gross) 1454kg (320lb)
POWERPLANT:	One 110kW (150hp) Benz Bz.III 6-cylinder water-cooled inline piston engine or one 120kW (160hp) Mercedes D.III 6-cylinder water-cooled inline piston engine
MAX SPEED:	161km/h (100mph)
ENDURANCE:	3 hours 30 minutes
CEILING:	5000m (16,400ft)
CREW:	2
ARMAMENT:	Two 7.92mm (0.312in) LMG 08/15 'Spandau' MG guns, one 7.92mm Parabellum MG14 MG in rear cockpit

Left: A W.12 on dry land awaits movement to water. The aircraft's appearance attracted many early criticisms, but most of the critical voices were shut down by the aircraft's operational performance.

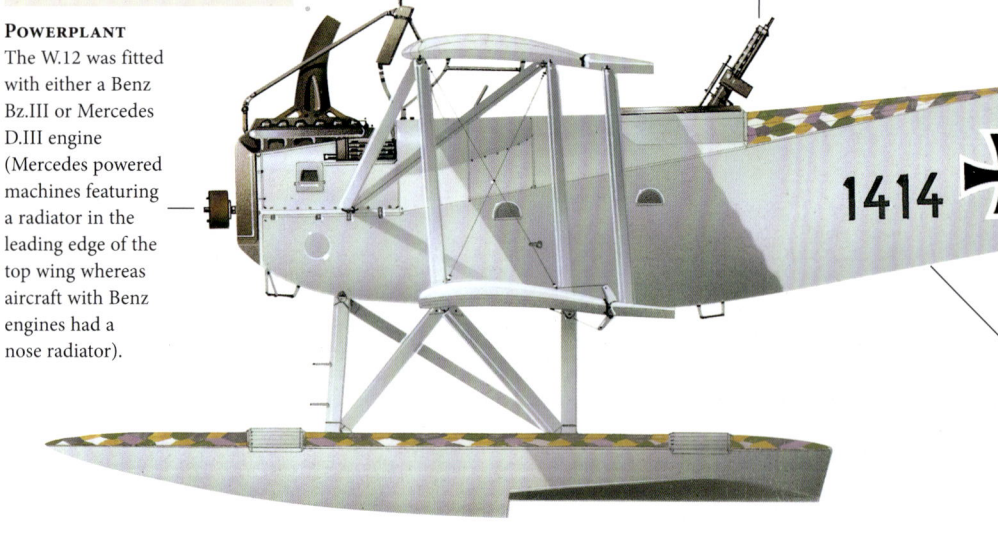

REAR GUNNER
In addition to the downturned rudder, the rear gunner's field of view was improved still further by the cantilever horizontal tail surfaces that required no struts or bracing.

POWERPLANT
The W.12 was fitted with either a Benz Bz.III or Mercedes D.III engine (Mercedes powered machines featuring a radiator in the leading edge of the top wing whereas aircraft with Benz engines had a nose radiator).

FUSELAGE
The W.12 featured a distinctive deep, slab-sided, upswept fuselage that compensated for the side area of the floats.

REAR GUNNER

The Hansa-Brandenburg W.12 was designed by Ernst Heinkel in late 1916 and made its first flight in February the following year. The most striking aspect of the design was the rudder that protruded below the aircraft, rather than above as is more conventional. The reason for this was to give the rear gunner an unparalleled field of fire. The wings also featured no wire bracing and the gunner could fire between, or even through, them without fear of breaking a bracing cable.

W.12 TAKE-OFF
Operations proved that the new W.12 aircraft was a great success. The two-man crew allowed for a much more useful patrol aircraft and later W.12s were fitted with radio allowing for great operational flexibility.

Sopwith Dolphin (1918)

TYPE • Fighter COUNTRY • United Kingdom

SPECIFICATIONS

DIMENSIONS:	Length: 6.785m (22ft 3in); Wingspan: 9.9m (32ft 6in); Height: 2.59m (8ft 6in)
WEIGHT:	1959kg (4319lb) maximum take-off
POWERPLANT:	1 x 149kW (200hp) Hispano-Suiza engine
MAX SPEED:	195.5km/h (121.5mph)
ENDURANCE:	n/a
CEILING:	6096m (20,000ft)
CREW:	1
ARMAMENT:	2 x 7.7mm (0.303in) fixed Vickers and 1 x 7.7mm (0.303in) Lewis MGs

At high altitude the Dolphin outperformed the SE.5a and was probably the best British fighter of the war, but the pilots' lack of enthusiasm for the unusual configuration, which looked dangerous in case of a nose-over on landing, limited it to a small number of squadrons.

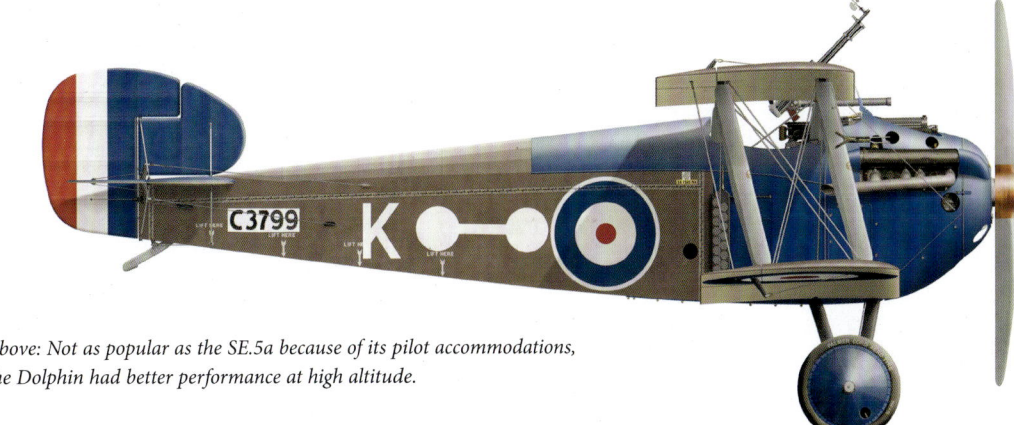

Above: Not as popular as the SE.5a because of its pilot accommodations, the Dolphin had better performance at high altitude.

ARMAMENT
Like the earlier Camel, the Dolphin had two fixed, synchronized Vickers machine guns, firing through the propeller arc.

UPPER WING
In an attempt to maximize the pilot's field of view, the upper wing was staggered back so the pilot's head was in the middle of the upper wing middle section.

SQUADRON MARKINGS
The white squares behind the cockpit were the squadron marking: the 'T' indicates a spare aircraft not assigned to a specific flight.

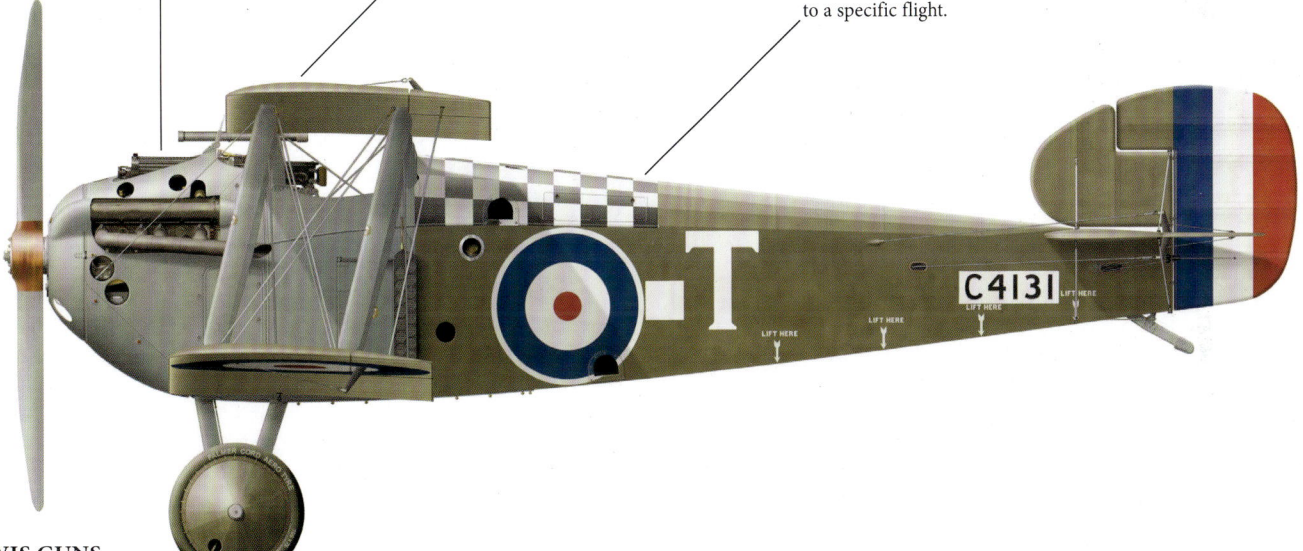

LEWIS GUNS
In early 1918 the Sopwith Dolphin appeared. Like the SE.5a, it was powered by the 149kW (200hp) engine. Conceived as a multi-gun fighter, the Dolphin had two fixed Vickers like the Camel, but also had one or two fixed Lewis guns firing upwards over the propeller arc. To save weight and improve performance, normally only one Lewis was fitted in combat, and some pilots flew with no Lewis guns at all.

UNGAINLY AIRCRAFT
The Sopwith Dolphin had a curiously ungainly appearance that belied a solid performance in the air. At the end of the war Sopwith was also planning the Dolphin II with the 224-kW (300-hp) Hispano-Suiza.

Interwar Period

The interwar period was a time when military aviation was pursuing an uneven and often uncertain transition. The past was represented by the biplanes and triplanes of World War I. The future lay in the emerging generation of monoplanes, which promised greater performance and combat power. But the economic constraints of the 1920–30s, and the unequal pace of military rearmament amongst the former combatants, resulted in a patchwork of new types. Some of these aircraft, especially those that clung to traditional biplane layouts, quickly reached obsolescence. Others, meanwhile, began lines of development that led to some of the greatest combat aircraft of the 20th century, proven in the global conflict that lay on the horizon of the 1930s.

HAWKER FURY
Aircraft such as the Hawker Fury, seen here, represented the summit of the biplane fighter, being relatively fast (359km/h, 223mph), manoeuvrable and robust. But by the end of the 1930s, such aircraft were utterly outclassed by new monoplane fighter types.

Bréguet 19 (1921)

TYPE • Bomber COUNTRY • France

SPECIFICATIONS
(Bre.19A2)

DIMENSIONS:	Length: 9.51m (31ft 2in); Wingspan: 14.83m (48ft 7in); Height: 3.69m (12ft 1in)
WEIGHT:	Empty 1722kg (3796lb); maximum take-off 3110kg (6856lb)
POWERPLANT:	One 383kW (513hp) Renault 12Kd water-cooled V-engine
MAX SPEED:	235km/h (146mph)
RANGE:	1200km (746 miles)
CEILING:	7000m (22,970ft)
CREW:	2
ARMAMENT:	One 7.7mm (.303in) Vickers MG; three 7.7mm (.303in) Lewis MGs, plus maximum internal bombload of 400kg (882lb) with a further 400kg (882lb) carried underwing

The Bréguet 19 was a sesquiplane bomber and reconnaissance aircraft developed in the years immediately after World War I ended in 1918. It was produced with a wide range of engine types and in many specialized variants.

Left: The Bre.19B2, seen here in Polish service, was a two-seat light bomber.

POWERPLANT
The Super-Bidon *Point d'Interrogation* was initially powered by a Hispano-Suiza 12Lb engine, but was subsequently re-engined with a 650hp (485kW) Hispano-Suiza 12Nb.

ARMAMENT
Military variants were heavily armed with up to four machine guns and internal and external bombloads totalling 800kg (1760lb).

VOYAGE OF FRIENDSHIP
In 1930 the aircraft embarked on a *Tour d'Amitié* (Voyage of Friendship) of the USA. Its landing points are recorded on the rear fuselage tricolour chevron.

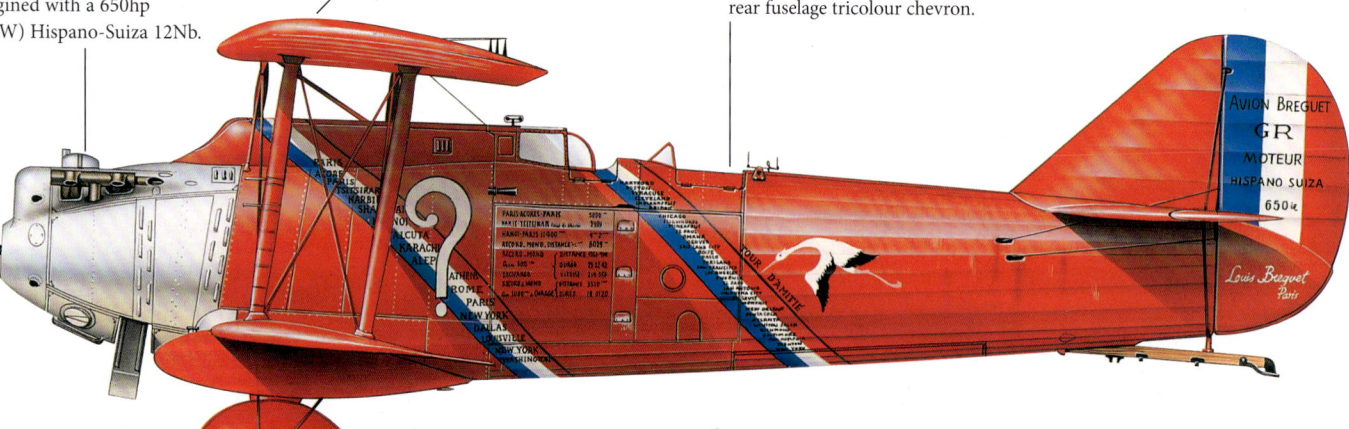

SUPER-BIDON
The most illustrious of the Bréguet 19 family was the French-built Super-Bidon *Point d'Interrogation* (see main picture). The fuel capacity was raised to 5170 litres (1137 imp gal) to enable a series of record-breaking flights, establishing a new distance record of 7905km (4912 miles) when it was flown by Dieudonné Costes and Maurice Bellonte from Le Bourget in France to Tsitsihar (Qiqihar) in Manchuria, China, between 27 and 29 September 1929.

BRÉGUET 19, 1930
This Polish Bre.19 is in flight over the Tatra Mountains. The service ceiling of these aircraft varied according to the type and engine, but was typically in the region of 7000m (22,970ft).

Armstrong Whitworth Siskin (1923)

TYPE • Fighter COUNTRY • United Kingdom

SPECIFICATIONS (Siskin IIIA)

DIMENSIONS:	Length: 7.72m (25ft 4in); Wingspan: 10.11m (33ft 2in); Height: 3.10m (10ft 2in)
WEIGHT:	Empty 935kg (2061lb); maximum take-off weight 1366kg (3012lb)
POWERPLANT:	One 313kW (420hp) Armstrong Siddeley Jaguar IV radial engine
MAX SPEED:	251km/h (156mph)
ENDURANCE:	3 hours
CEILING:	8230m (27,000ft)
CREW:	1
ARMAMENT:	Two fixed forward-firing 7.7mm (.303in) Vickers machine guns; underwing racks with provision for up to four 9kg (20lb) Cooper practice bombs

This superbly aerobatic aircraft formed the vanguard of Britain's home defence squadrons from March 1927, but faded swiftly in the 1930s as technology developed with newer types such as the Bristol Bulldog.

Left: In total, 11 RAF squadrons would fly the Siskin during the interwar years.

POWERPLANT
The IIIA had a single 313kW (420hp) Armstrong Siddeley Jaguar IV radial engine, which was fitted to more than 40 aircraft types.

METAL FRAME
The Siskin IIIA was the first all-metal framed aircraft to be adopted in significant numbers for the RAF. The internal metal frame was made from aluminium alloy.

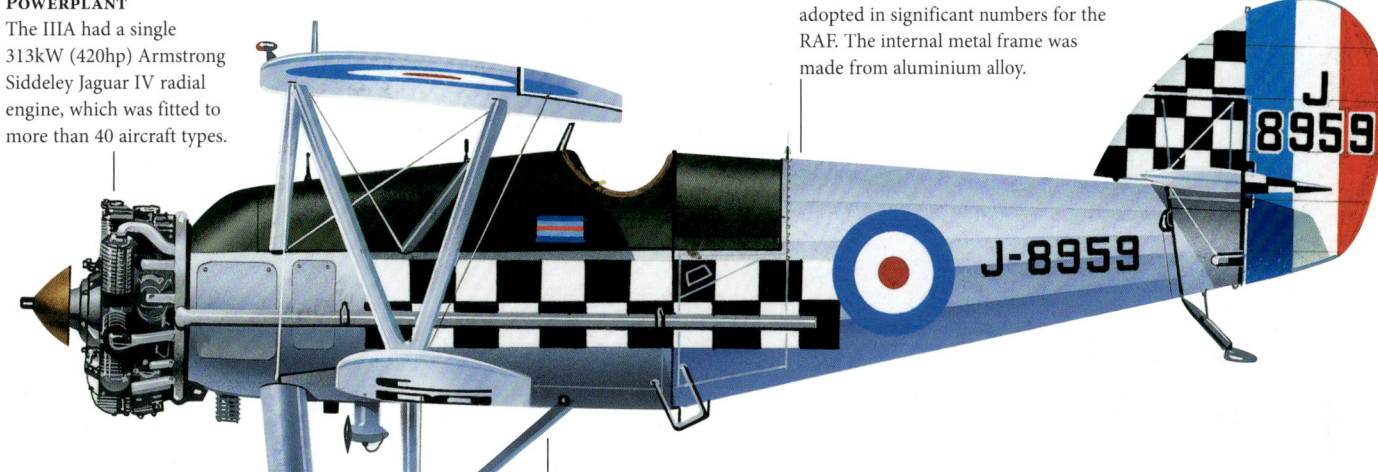

SISKIN IIIA
Pictured is a Siskin IIIA of No. 43 Squadron from 1929. It had a maximum speed of 251km/h (156mph), at this time comparable to many emerging monoplanes.

ORIGINS
The aircraft has its origins in the Siddeley S.R.2 Siskin, produced by the Siddeley-Deasy Motor Car Company in 1918 for the 224kW (300hp) Royal Aircraft Factory R.A.F.8 engine, but which in fact first flew with the 239kW (320hp) ABC Dragonfly. The poor performance of this latter engine prompted Armstrong Siddeley to equip the aircraft with its own 242kW (325hp) Jaguar engine.

SISKIN MK IIIA
After the S.R.2 had received the Jaguar engine and been redesigned according to Air Ministry policy with an all-metal structure, it was ordered for the RAF in 1923 as the Siskin Mk IIIA. The aircraft seen here is in Canadian service.

Martin T3M and T4M (1928)

TYPE • *Torpedo bomber* **COUNTRY** • *United States*

The Martin T3M and T4M biplanes were a new breed of interwar biplane torpedo bombers, both of which operated from US carriers during the 1930s. Indeed, the T4M was the last of the biplane torpedo-bombers in US Navy service.

SPECIFICATIONS (T4M-1)

DIMENSIONS:	Length: 10.85m (35ft 7in); Wingspan: 16.16m (53ft); Height: 4.5m (14ft 9in)
WEIGHT:	Empty 264 kg (5814lb)
POWERPLANT:	1 × 392kW (525hp) Pratt & Whitney R-1690 Hornet 9-cylinder air-cooled radial engine
MAX SPEED:	184km/h (114mph)
RANGE:	584km (363 miles)
CEILING:	3095m (10,105ft)
CREW:	3
ARMAMENT:	1 × flexibly mounted 7.62mm (0.3in) machine gun in rear cockpit; 1 × torpedo or bomb under fuselage

Above: The Martin T3M-2 was the principal variant of the T3M, with 100 produced, powered by a Packard 3A-2500 engine.

POWERPLANT
The T4M-1 production version had a Pratt & Whitney R-1690 Hornet 9-cylinder air-cooled radial engine.

SPACIOUS INTERIOR
The T4M was known by its three-man crew for having a surprisingly spacious fuselage interior, in which crew members were able to switch positions during flights.

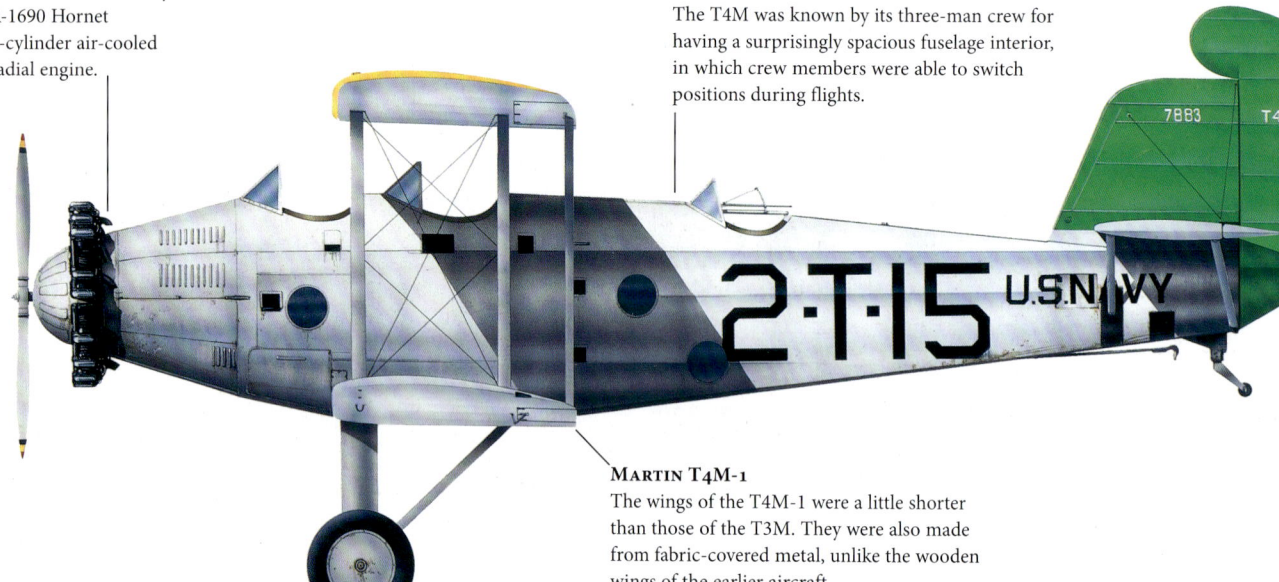

MARTIN T4M-1
The wings of the T4M-1 were a little shorter than those of the T3M. They were also made from fabric-covered metal, unlike the wooden wings of the earlier aircraft.

CARRIER AGE
The T3M and T4M, introduced into service in 1926 and 1928 respectively, arrived as the navies of the world were steadily waking up to the strategic and tactical potential of aircraft carriers. While the T3M had limited use, the T4M participated in fleet exercises in 1929, in which squadrons of the biplanes flew from the carrier USS *Saratoga* and successfully demonstrated a mock attack on the Panama Canal and installations around it.

MARTIN T3M-1
Both the T3M and T4M aircraft had the option of being fitted with either wheels or floats. Operationally, the T3M was mostly used from its floats, whereas the T4M relied more on its wheels, to suit its increased use on carrier platforms.

Tupolev ANT-9 (1931)

TYPE • Transport COUNTRY • Soviet Union

SPECIFICATIONS	
DIMENSIONS:	Length: 16.8m (55ft 1in); Wingspan: 23.7m (78ft 9in); Height: 4.86m (15ft 11in)
WEIGHT:	Maximum take-off: 5043kg (11,118lb)
POWERPLANT:	Three 172kW (230hp) Gnome-Rhône 5K Titan 5-cylinder air-cooled radial piston engines
MAX SPEED:	215km/h (134mph)
RANGE:	1800km (1118 miles)
CEILING:	3750m (12,300ft)
CREW:	2
ARMAMENT:	NONE

The ANT-9 was the most produced large Soviet airliner of the interwar period. Like its more famous contemporary, the Junkers Ju 52/3m, the ANT-9 was designed to be primarily an airliner but with the secondary ability to operate as a bomber.

Above: Of 16 ANT-9s listed as operational in September 1941, ten were listed as serviceable in December 1943, but all had disappeared by August 1944.

RECONFIGURATION
The all-metal ANT-9 initially suffered from the lack of a good engine. Early examples such as this were trimotors, but by June 1941 all had been re-engined with two Mikulin M-17s.

ENGINES
Production aircraft were initially fitted with Wright J-5s, but they were expensive and the Bessonov M-26 was used instead. That engine proved extremely unreliable and in 1933 was swapped for the M-17 engine.

CARGO
The ANT-9 had the capacity to carry nine passengers and more than a metric tonne of cargo. In 1941, a pair of ANT-9s transported 655 people and 68 tonnes (67 tons) of cargo, flying six sorties per night.

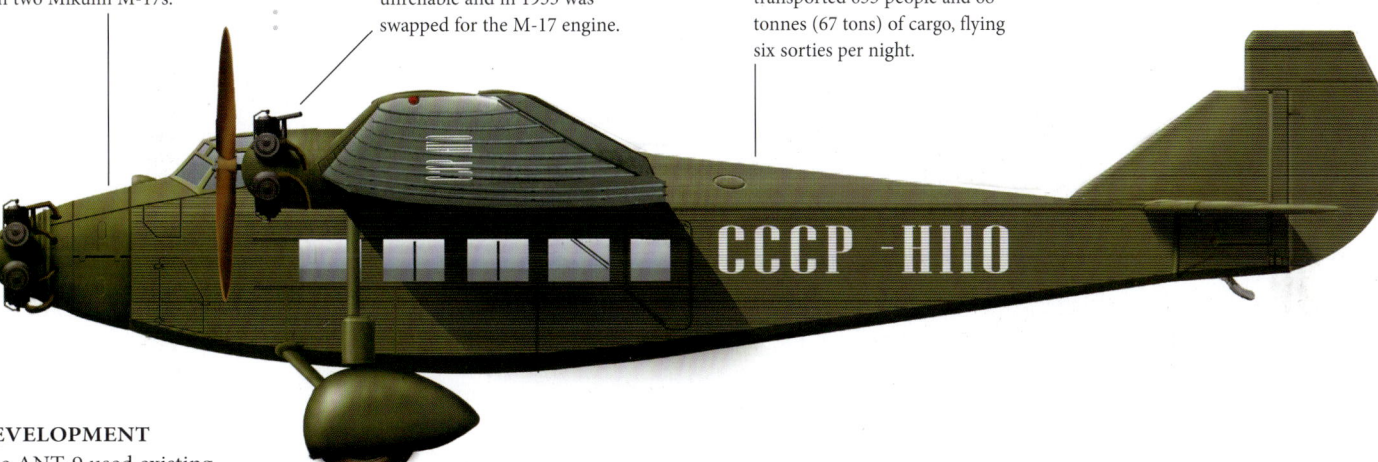

DEVELOPMENT
The ANT-9 used existing components to speed up the development process, the wings and tail being taken from the Tupolev R-6. The ANT-9 was first flown on 5 May 1929 powered by three Gnome-Rhône Titans, and after successful testing, entered production in early 1930. From 1935 the aircraft was known as the PS-9 and the last were withdrawn from Aeroflot use in early 1941. By this time, they had seen military service as medical transports during the 1940 Winter War in Finland.

NATIONAL AIRCRAFT
Here the distinctive shape of the ANT-9 features on a Soviet postage stamp.

Hawker Fury (1931)

TYPE • Fighter COUNTRY • United States

SPECIFICATIONS (Mk II)

DIMENSIONS:	Length: 8.13m (26ft 8in); Wingspan: 9.14m (30ft); Height: 3.1m (10ft 2in)
WEIGHT:	Maximum take-off 1637kg (3609lb)
POWERPLANT:	One 447kW (700hp) Rolls-Royce Kestrel VI 12-cylinder V-piston engine
MAX SPEED:	359km/h (223mph)
RANGE:	435km (270 miles)
CEILING:	8990m (29,500ft)
CREW:	1
ARMAMENT:	Two fixed forward-firing 7.7mm (.303in) Vickers Mk III machine guns

In 1931 the fastest fighter in the RAF was the Hawker Fury, which had a maximum speed of 333 km/h (207 mph) – about half as fast as the Supermarine S.6B.

Above: Only three Hispano-engined Hawker Spanish Furies reached the Republican government before the Spanish Civil War. At least one was captured and used by the Nationalists and later recaptured by the Republicans.

POWERPLANT
The Fury II was fitted with the Kestrel VI engine. The aircraft was accepted on the basis of being an interim fighter needed to serve during the Hawker Hurricane's development period.

ARMAMENT
Both the Mk I and the Mk II Fury were equipped with the same armament, two fixed forward-firing 7.7mm (.303in) Vickers Mk III machine guns atop the engine cowling.

HAWKER FURY MK II
This Hawker Fury Mk II belonged to No. 25 Squadron, RAF Fighter Command, RAF Hawkinge, Kent. The squadron operated Fury aircraft from 1931 until replacement by Gladiators and Demons in the late 1930s.

INTERCEPTOR
One of a number of successful types derived from the Hart bomber of 1928, the Fury was selected as the RAF's first dedicated interceptor fighter in 1930. Powered by a water-cooled V-12 Kestrel engine, Furies were more streamlined and faster than contemporary RAF fighters, but also more expensive. Only three squadrons were equipped with Fury Is and four with the improved Fury II. The aircraft served as the basis for the Nimrod naval fighter.

HAWKER FURY, 1938
The Hawker Fury fighter had its roots in the radial-engined Hoopoe prototype. This design exceeded the requirements of Air Ministry specification and was developed further. Due to its higher cost than the Bristol Bulldog (£700 more per aircraft), the Fury I was issued to only a handful of 'elite' fighter squadrons.

Tupolev TB-3 (1931)

TYPE • Bomber **COUNTRY** • Soviet Union

SPECIFICATIONS

DIMENSIONS:	Length: 24.4m (80ft 1in); Wingspan: 41.8m (137ft 2in); Height: 8.5m (27ft 11in)
WEIGHT:	19,300kg (42,549lb)
POWERPLANT:	Four 526kW (705hp) Mikulin M-17FV-12 liquid-cooled piston engines
MAX SPEED:	212km/h (132mph)
RANGE:	2000km (1200 miles)
CEILING:	4800m (15,700ft)
CREW:	8–10
ARMAMENT:	Two paired 7.62mm (0.3in) Degtyaryov DA machine guns flexibly mounted in nose, two paired 7.62mm (0.3in) DA machine guns flexibly mounted in port and starboard dorsal gun positions, one 7.62mm (0.3in) DA machine gun in ventral 'dustbin' turret; up to 5000kg (11,000lb) bombload

The world's first four-engined cantilever monoplane heavy bomber, the TB-3 was effectively obsolete at the outbreak of hostilities, yet performed valuable work during the initial phases of the conflict.

Left: Finnish troops examine a heavily damaged Tupolev TB-3 bomber that has been captured following an emergency landing in March 1940.

PERFORMANCE
About half the production aircraft were fitted with Mikulin M-17 engines, but later production TB-3s utilized the more powerful Mikulin AM-34.

ARMAMENT
Defensive armament was comprehensive, featuring twin gun positions in the nose and two amidships, as well as two underwing 'dustbin' turrets.

CAPACITY
The TB-3 had a large crew of up to 10 people. Bombload throughout its life remained a maximum of 5000kg (11,000lb).

DESIGN
Essentially a stretched TB-1, the TB-3 was first flown in December 1930 and entered service during 1932. The TB-3 performed impressively, setting several payload-to-altitude records and achieving a record endurance of over 18 hours. Given the stately performance of the TB-3 – early M-17 aircraft struggled to exceed 200km/h (124mph) – it was fitted with heavy defensive armament.

OPERATIONS
Despite having been officially retired in 1939, the Soviet Air Force still had more than 500 operational TB-3s at the time of the German invasion in June 1941. These escaped the widespread destruction and were pressed back into frontline service. When daylight raids proved near suicidal, a switch was made to nocturnal operations. The aircraft was also used as a paratroop transport (pictured here).

Grumman FF and F2F (1931)

TYPE • Naval fighter   COUNTRY • United States

Encouraged by the success of the FF/SF-1 aircraft, Grumman designers drew up a proposal for a single-seat version and in June 1932 offered it to the US Navy, which ordered a prototype in November. This aircraft became the F2F-1.

SPECIFICATIONS (F2F-1)

DIMENSIONS:	Length: 6.53m (21ft 5in); Wingspan: 8.69m (28ft 6in); Height: 2.77m (9ft 1in)
WEIGHT:	Maximum take-off 1745kg (3847lb)
POWERPLANT:	One 480kW (650hp) Pratt & Whitney R-1535-72 Twin Wasp Junior radial piston engine
MAX SPEED:	383km/h (238mph)
RANGE:	1585km (985 miles)
CEILING:	8380m (27,500ft)
CREW:	1
ARMAMENT:	Two fixed forward-firing 7.62mm (0.3in) Browning machine guns; underwing racks for two 53kg (116lb) bombs

Above: Grumman SF-1 was a two-seat carrier scout variant of the single-seat FF-1 fighter aircraft.

ENGINE
Power was provided by a 466kW (625hp) XR-1535-44 Twin Wasp Junior engine that could deliver a maximum speed of 383km/h (238mph).

ARMAMENT
The F2F-1 was slightly smaller than its predecessor and had ailerons on the upper wing only. The aircraft had twin forward-firing Brownings and underwing racks for two small bombs.

GRUMMAN F2F-1
The Navy placed an order for 54 F2F-1 fighters in May 1934 and the first was delivered the following January. The aircraft served with VF-2B on USS *Lexington* until September 1940.

F2F-1 DEVELOPMENT
The F2F-1 was slightly smaller than its predecessor, and had ailerons on the upper wing only. Power was provided by a 466kW (625hp) XR-1535-44 Twin Wasp Junior engine. The aircraft had twin forward-firing Brownings and underwing racks for two bombs. This XF2F-1 first flew in October 1933 and after a six-month evaluation at Anacostia, Washington DC, returned to Grumman for minor modifications to resolve problems with directional instability.

GRUMMAN FF-1
The Grumman FF-1 is most distinguished by being the first carrier aircraft to have retractable landing gear. The aircraft type began operational service aboard USS *Lexington* in 1933.

Bristol Bulldog (1932)

TYPE • *Fighter* **COUNTRY** • *United Kingdom*

Together with the Hawker Fury, the Bristol Bulldog epitomized RAF Fighter Command in the 1930s. Immediately prior to WWII, it was still the principal aircraft of the interceptor squadrons.

SPECIFICATIONS (Mk IIA)

DIMENSIONS:	Length: 7.7m (25ft 2in); Wingspan: 10.3m (33ft 10in); Height: 2.7m (8ft 9in)
WEIGHT:	Maximum take-off 1583kg (3490lb)
POWERPLANT:	One 365kW (490hp) Bristol Jupiter VIIF radial piston engine
MAX SPEED:	280km/h (174mph)
RANGE:	482km (300 miles)
CEILING:	8940m (29,300ft)
CREW:	1
ARMAMENT:	Two fixed forward-firing 7.7mm (303in) Vickers machine guns; underwing racks with provision for up to four 9kg (20lb) bombs

Above: Sweden took a total of 11 of the Bulldog Mk. IIA variant. Other export customers for the aircraft included Denmark, Latvia, Estonia and Siam (present-day Thailand).

BRISTOL BULLDOG II
A Bristol Bulldog Mk II of No. 23 Squadron, RAF Fighter Command, RAF Biggin Hill, Kent, in 1932.

PILOT FEATURES
Although possessing better handling than its predecessor, the Sopwith Snipe, the Bulldog had similar armament and performance. The Bulldog did, however, introduce a radio and oxygen supply for the first time in an RAF fighter.

FLY-OFF
Along with the Hawker Fury, the Bristol Bulldog typified RAF Fighter Command in the 1930s. The prototype Bristol Type 105 Bulldog flew as a contender for an RAF fighter specification in May 1927. A 'fly-off' was arranged between the Bristol and the slightly superior Hawker Hawfinch, and the modified Type 105A was selected as the Bulldog II in 1928.

ARMAMENT
As well as two fixed forward-firing 7.7mm (.303in) Vickers machine guns, some Mk IIA types had underwing racks with provision for up to four 9kg (20lb) bombs.

BULLDOG MK IV
This Bristol Bulldog Mk IV fighter biplane, serving for the Royal Air Force from 1933, was equipped with a Bristol Mercury engine plus ring cowling. Some 17 of the Mk IVA variant were built for Finland with the Mercury VIS2, and delivered in January 1935. These aircraft were flown in combat during the Winter War of 1939–40, and were often fitted with ski-landing gear.

Boeing P-26 Peashooter (1932)

TYPE • Fighter COUNTRY • United States

A groundbreaking design, the P-26 was both the first all-metal fighter aircraft produced in America and the first monoplane fighter adopted by the US Army Air Corps, in 1933. By 1941, however, it was obsolete.

SPECIFICATIONS

DIMENSIONS:	Length: 7.19m (23ft 7in); Wingspan: 8.5m (28ft); Height: 3m (10ft)
WEIGHT:	Maximum take-off 1524kg (3360lb)
POWERPLANT:	One 450kW (600hp) Pratt & Whitney R-1340-27 Wasp 9-cylinder air-cooled radial piston engine
MAX SPEED:	377km/h (234mph)
RANGE:	580km (360 miles)
CEILING:	8400m (27,400ft)
CREW:	1
ARMAMENT:	Two 7.62mm (0.3in) M1919 Browning MGs or one 7.62mm (0.3in) M1919 MG and one 12.7mm (0.5in) M2 Browning MG

Left: USAAC Boeing P-26 pursuit fighters stand on station at an American airbase. By December 1941, the only P-26s in operational US service on the American continent were nine examples based at Albrook Field in the Panama Canal zone.

LAYOUT
Despite representing a clear advance over the biplane fighters it replaced, the P-26 marked a transitional moment in fighter design, with its fixed undercarriage, open cockpit, and wire-braced wings.

WARTIME PAINT
This Peashooter is in the blue/yellow standard colour scheme prior to the introduction of the familiar olive drab/neutral grey wartime camouflage. It is aircraft 33-125 from the 34th Pursuit Squadron of 1st Pursuit Group, based at Selfridge Field, Michigan, in 1936.

POWERPLANT
The engine was a 450kW (600hp) Pratt & Whitney R-1340-27 Wasp 9-cylinder air-cooled radial.

P-12 REPLACEMENT
First flown in 1932, the P-26, with its art deco spats and flamboyant colour schemes, epitomized the US Army Air Corps in the last years of peace. Yet this rotund little fighter saw brief combat service in World War II. Initial deliveries to the army began in December 1933, and the P-26 was unusual among military aircraft in that it was cheaper than the aircraft it replaced: Boeing's own P-12 biplane fighter.

P-26 FORMATION
Throughout December 1941, the P-26s fought an overwhelming force of more modern Japanese aircraft, and were credited with shooting down one Mitsubishi G3M bomber and three A6M Zero fighters before the surviving Boeings were burned to prevent their capture.

Blackburn Shark (1933)

TYPE • *Torpedo bomber* **COUNTRY** • *United Kingdom*

SPECIFICATIONS

DIMENSIONS:	Length: 10.74m (35ft 3in); Wingspan: 14m (456ft); Height: 3.68m (12ft 1in)
WEIGHT:	Maximum take-off 3679kg (8111lb)
POWERPLANT:	One 570kW (760hp) Armstrong Siddeley Tiger VI 14-cylinder air-cooled radial piston engine
MAX SPEED:	240km/h (150mph)
RANGE:	1006km (625 miles)
CEILING:	4800m (15,600ft)
CREW:	3
ARMAMENT:	One 7.7mm (0.303in) Vickers MG in nose, one 7.7mm Vickers K MG mounted in dorsal position; up to 730kg (1600lb) bombload or one 460mm (18in) torpedo

The Blackburn Shark was a contemporary of the Fairey Swordfish and, in some regards, a superior aircraft. However, an unfortunate choice of engine effectively condemned the Shark to obscurity in service.

Left: A Royal Navy Fleet Air Arm Shark II overflies an Allied convoy during World War II. Blackburn had supplied several successful carrier-based torpedo bombers to the Fleet Air Arm – indeed they had been the sole supplier of this type of aircraft from 1921 to 1936.

ENGINE PROBLEMS
The Tiger VI was prone to severe vibration, leading to severed oil pipes, engine seizure and metal fatigue in the mounts.

COCKPIT
The Shark at first utilized an open cockpit design, exposing the pilot to the harsh elements, but a fully enclosed cockpit was introduced in later versions.

BLACKBURN SHARK MK I
In many ways a superior aircraft to the Fairey Swordfish, the failure to develop the Shark remains a mystery. This example was serving with 820 Squadron aboard HMS *Courageous* in 1937 during the Shark's brief service as a frontline carrier aircraft.

ENGINE ISSUES
When the Shark first flew on 24 August 1933, it was a sound design that was widely expected to emulate the success experienced by its progenitors. Although Blackburn wanted the Bristol Pegasus to power the Shark, the Air Ministry insisted that they utilize the Armstrong Siddeley Tiger VI instead. This engine, however, was critically unreliable and severely affected the operational reliability of the aircraft.

CARRIER LANDING
A Blackburn Shark catches an arrester cable on an aircraft carrier. Although Blackburn worked hard to eradicate the engine problems and largely succeeded, the Fairey Swordfish was available by 1937 and the Air Ministry ordered that aircraft instead, relegating the Sharks to target tugging duties.

Avia B.534 (1933)

TYPE • Fighter   COUNTRY • *Czechoslovakia*

SPECIFICATIONS (B.534-IV)

DIMENSIONS:	Length: 8.2m (26ft 10in); Wingspan: 9.4m (30ft 10in); Height: 3.1m (10ft 2in);
WEIGHT:	Maximum take-off 2120kg (4674lb)
POWERPLANT:	One 634kW (850hp) Hispano-Suiza HS 12Ydrs inline piston engine
MAX SPEED:	394km/h (245mph)
RANGE:	580km (360 miles)
CEILING:	10,600m (34,775ft)
CREW:	1
ARMAMENT:	Four fixed forward-firing 7.7mm (0.303in) Model 30 machine guns in forward fuselage; underwing Pantof racks with provision for up to six 20kg (44lb) bombs

The B.534 proved to be the outstanding fighter of the 1937 International Flying Meet in Zürich. It was widely regarded as the finest fighter of its day, until being inevitably outclassed by the new breed of low-wing monoplanes.

Below: Pictured is an Avia B.534-IV of 3.Staffel, Jagdgeschwader 71, based at Eutlingen near Stuttgart in late 1939.

UNDERCARRIAGE
In German service the B.534-IV aircraft were seen with spatted mainwheels on the undercarriage, as pictured here.

SLOVAKIAN AVIA B.534-IV
In contrast to the early models of the B.534, the definitive B.534-IV variant had an enclosed cockpit and revised aft fuselage decking.

MG BLISTERS
The blisters on the sides of the fuselage, just beneath and forward of the cockpit, served to cover the large breeches of the two machine guns.

GERMAN SERVICE
At the time of the Munich crisis in September 1938 the B.534 had been in large-scale production for three years, and around 300 of the eventual total of 445 were in service with 21 Czech fighter squadrons. After the occupation of Czechoslovakia in March 1939 a large number of aircraft passed to the Luftwaffe, and also to Hungary and Bulgaria, the latter's only combat outing being against the Consolidated B-24 Liberators that bombed the Ploesti oilfields in 1943.

STRIPPED-DOWN AVIA B.534-I
The first production model of this outstanding fighter was the B.534-I, which had a wooden screw to replace the metal unit of the prototype, an open cockpit and twin fuselage machine guns with two more carried in lower-wing fairings.

Curtiss P-6 Hawk (1933)

TYPE • Fighter COUNTRY • United States

The Curtiss P-6 Hawk was introduced into US service as a fighter aircraft in 1927. It proved to be a successful, if progressively obsolescent, design, with nearly 2000 of the aircraft produced. In its defining variant, the P-6E, the aircraft served until the beginning of the 1940s.

SPECIFICATIONS (P-6E)

DIMENSIONS:	Length: 7.06m (23ft 2in); Wingspan: 9.6m (31ft 6in); Height: 2.72m (8ft 11in)
WEIGHT:	Maximum take-off 1559kg (3436lb)
POWERPLANT:	One 448kW (600hp) Curtiss V-1570-23 Conqueror inline piston
MAX SPEED:	319km/h (198mph)
RANGE:	459km (285 miles)
CEILING:	7530m (24,700ft)
CREW:	1
ARMAMENT:	Two fixed forward-firing 7.7mm (0.303in) MGs

Below: This P-6E Hawk is of the 8th Pursuit Group, United States Army Air Corps, Langley, Virginia. The aircraft is as seen in 1933.

PERFORMANCE
The aircraft demonstrated exceptional manoeuvrability and useful top speed, serving with the 1st and 8th Pursuit Groups. One P-6E was powered by an unsupercharged V-1570F engine and designated XP-6G.

OPEN COCKPIT
The open cockpit of the P-6E was beginning to look like something from a bygone age. It was also deeply uncomfortable for the pilot, especially as the aircraft had a service ceiling of 7530m (24,700ft).

POWERPLANT
The engine on the P-6E was a 448kW (600hp) Curtiss V-1570-23 Conqueror inline piston engine.

P-6E HAWK
Most prolific and impressive of the P-6 Hawk family was the P-6E, the last biplane fighter to be delivered to the United States Army Air Corps. Generally similar to the P-6D, it had a slimmer forward fuselage, with the engine radiator mounted slightly forward of the landing gear, which comprised single-strut main legs with spat-type wheel fairings. Forty-six were ordered in July 1931.

CURTISS P-6A HAWK
To produce the prototype XP-6 Hawk, Curtiss took the airframe of a P-1 and installed a Curtiss V-1570 Conqueror engine. A second (XP-6A) conversion had the same Conqueror engine, untapered wings and drag-reducing wing radiators. It took first place at the US National Races with the-then remarkable speed of 323km/h (201mph).

Polikarpov I-16 (1934)

TYPE • Fighter COUNTRY • Soviet Union

Given the pace of fighter development in the 1930s, it is all the more remarkable that the Soviet-designed I-16 was essentially the best fighter in the world for a period of six years. It only met its match once pitted against the Luftwaffe's Messerschmitt Bf 109E.

SPECIFICATIONS

DIMENSIONS:	Length: 6.04m (19ft 10in); Wingspan: 8.88m (29ft 1in); Height: 3.25m (10ft 8in)
WEIGHT:	1475kg (3252lb)
POWERPLANT:	M-62 radial piston engine, 820kW (1100hp)
MAX SPEED:	490km/h (304mph) at 3000m (9845ft)
RANGE:	700km (435 miles)
CEILING:	9700m (31,825ft)
CREW:	1
ARMAMENT:	2 × 7.62mm (0.3in) ShKAS machine guns; 2 × 20mm (0.79 in) ShVAK cannons; 6 × unguided RS-82 rockets or up to 500kg (1102lb) of bombs

Above: An I-16 Type 28 of an unknown unit pictured during the summer of 1942. The cannon-armed Type 28 was built in relatively small numbers.

TAIL SURFACE
The tail surfaces were necessarily large to counter the lack of stability caused by the short rear fuselage, although this instability brought great manoeuvrability.

COCKPIT
The cramped cockpit was equipped with only rudimentary instruments. No radio or oxygen equipment was fitted, and there was no indicator for the undercarriage.

ARMAMENT
Gun armament comprised four 7.62mm (0.3in) ShKAS machine guns, two synchronized in the forward fuselage and two in the wings; the wing machine guns were replaced on some aircraft by two 20mm (0.78in) ShVAK cannon.

WING STRUCTURE
The I-16 had a metal two-spar wing structure, with trussed KhMA chrome molybdenum steel alloy centre-section spars and tubular outer spars. Wing ribs were made of dural and skinning was aluminium inboard and fabric outboard. The long ailerons were operated by rods and bell cranks. They could be drooped to act as flaps on landing.

PILOT HANDLING
The abbreviated fuselage and massive ailerons rendered the I-16 extremely sensitive to control movements on all three axes and it was susceptible to stalling and spinning with little provocation. However, the performance of the new aircraft, even with the low-powered engine, was at the forefront of international fighter design, with rate of climb and ceiling being considered the best in its class.

PZL P.11c (1934)

TYPE • Fighter **COUNTRY** • Poland

SPECIFICATIONS

Dimensions:	Length: 7.55m (24ft 9in); Height 2.85m (9ft 4in); Wingspan 10.72m (35ft 2in)
Weight:	Loaded 1650kg (3638lb)
Powerplant:	418kW (560hp) Bristol Mercury VS2 radial
Max Speed:	375km/h (233mph)
Range:	550km (341 miles)
Ceiling:	8000m (26,246ft)
Crew:	1
Armament:	2–4 7.92mm (0.31in) machine guns, plus 50kg (110lb) bombload

The gull-winged monoplane fighters produced by the Polish *Panstwowe Zaklady Lotnicze* (National Aviation Establishments) during the interwar years were among the best in service with any air force, but they were outmoded by the time Germany invaded Poland in 1939.

Above: The P.11's pilot sat in an open cockpit, in front of which a simple ring-and-bead gunsight was mounted. The P.11's successor, the PZL P.24, featured a fully enclosed cockpit.

Engine
The P.11c was powered by the extremely reliable Bristol Mercury radial engine, which was also the powerplant chosen for aircraft like Britain's Gloster Gladiator biplane fighter.

Gull wing
The cut-out centre section of the P.11's gull wing gave the pilot a good view forward, further improved in the P.11c by lowering the engine mounting and re-positioning the pilot farther to the rear.

Identities
Polish pilots were fiercely proud of their squadron identities, as this colourful *Jedenastka* reveals. The squadron emblem, a turkey cock, is emblazoned on the fuselage side.

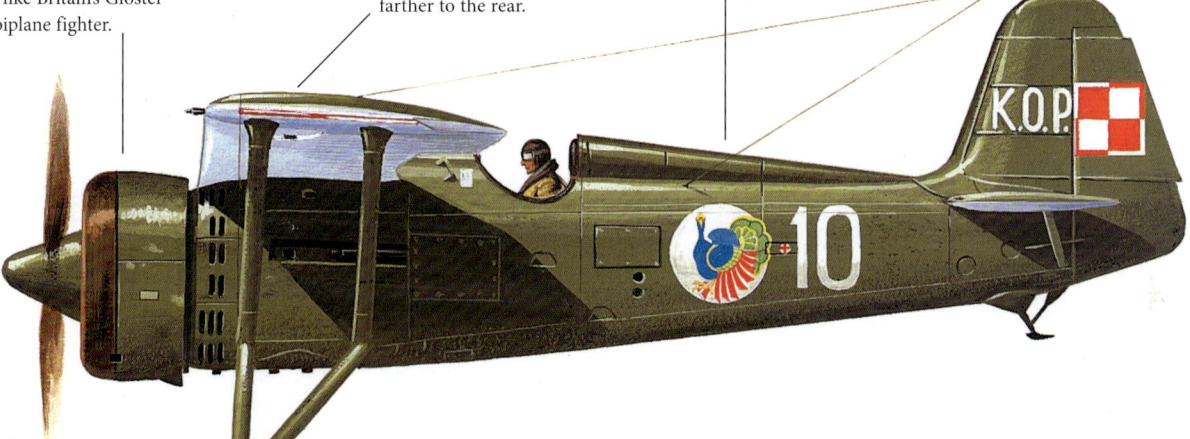

SUCCESSOR
The gull-winged PZL P.11 was a more powerful derivative of the PZL P.7, which equipped all first-line fighter squadrons of the Polish Air Force's 1st, 2nd, 3rd and 4th Air Regiments at the end of 1933 and was one of the best fighter aircraft of its day. Its successor, the PZL P.11, was basically a more powerful derivative, which first flew in September 1931, with deliveries beginning in 1934.

POLISH P-11S IN FLIGHT
Most P.11s were powered by Bristol Mercury engines built under licence by Skoda. The definitive version of the fighter was the P.11c, of which 175 were built. They suffered heavy losses during the German invasion of Poland in September 1939 against superior German Bf 109 fighters.

Fiat CR.32 (1934)

TYPE • Fighter COUNTRY • Italy

SPECIFICATIONS (CR.32quater)

- **DIMENSIONS:** Length: 7.45m (24ft 5in); Wingspan: 9.5m (31ft 2in); Height: 2.63m (8ft 7in)
- **WEIGHT:** Maximum take-off 1850kg (4079lb)
- **POWERPLANT:** One 447kW (600hp) Fiat A.30 RA bis 12-cylinder V-piston engine
- **MAX SPEED:** 375km/h (233mph)
- **RANGE:** 680km (422 miles)
- **CEILING:** 8800m (28,870ft)
- **CREW:** 1
- **ARMAMENT:** Two fixed forward-firing 7.7mm (0.303in) Breda-SAFAT machine guns

The CR.32 was the most important biplane fighter of the 1930s with 1712 built. It stemmed from the CR.30, designed by Chief Engineer Rosatelli in 1931 as a single-seat fighter and bearing many of his hallmarks, such as W-form interplane bracing.

Above: Over 400 CR.32s were sent to Spain during the civil war. Sgt Tarantella of the XVI Gruppo 'La cucaracha' (cockroach), Aviación del Tercio in Spain, flew this aircraft during 1937 in support of Nationalist forces.

ARMAMENT
The CR.32ter differed from previous variants mainly in terms of its armament, with a pair of 12.7mm (0.5in) guns installed, firing forward through the propeller.

LAYOUT
The CR.32 had a similar light alloy and steel structure to the previous CR.30 and was powered by the same 447kW (600hp) Fiat A.30 RA 12-cylinder inline engine.

CR.32TER OF 1939
By 1939, with war looming, the Regia Aeronautica's CR.32s were finished in this 'temperate zone' olive green with a darker green dapple. This machine of the 360a Squadriglia, 52° Stormo was based at Pontedera, Pisa in mid-1939.

CIVIL WAR
For five years the CR.32 was the mainstay of the Regia Aeronautica (RA) fighter element, with production continuing until May 1939. Entering service in 1934, the type soon equipped 1°, 3° and 4° Stormi and went on to serve with most of the RA's fighter units. CR.32s saw extensive service during the Spanish Civil War, the Aviación del Tercio and Aviazione Legionaria operating the aircraft in support of the Nationalist forces.

ITALIAN CR.32
The most numerous CR.32 variant was the CR.32quater of which 398 were completed. This was a lightweight version of the CR.32ter, with the same armament as the latter.

Caproni Ca.133 (1935)

TYPE • Bomber COUNTRY • Italy

SPECIFICATIONS

DIMENSIONS:	Length: 15.34m (50ft 4in); Wingspan: 21.23m (69ft 8in; Height: 4.0m (13ft 2in)
WEIGHT:	6700kg (14,771lb) loaded
POWERPLANT:	3 × Piaggio Stella P.VII C.16 7-cylinder air-cooled radial piston engines, 343kW (460hp) each
MAX SPEED:	230 km/h (140mph)
RANGE:	1350km (840 miles)
CEILING:	5500m (18,000ft)
CREW:	5 (bomber); 2 (transport)
ARMAMENT:	4 × 7.7mm (0.303 in) Breda-SAFAT machine guns; 1200kg (2646lb) of bombs

In military service, the Caproni Ca.133 was a capable transport, logistical support and paratrooper-deployment aircraft that saw operational use in many theatres of World War II, from North Africa to the Eastern Front.

Left: Italy was one of the European pioneers of paratrooper warfare. With its high-wing layout, the Ca.133 was ideally suited as a drop aircraft.

HIGH WING
The Ca.133 had a high-mounted wing made of wood and steel. Each wing had a single engine in a faired nacelle, with a further engine in the nose.

CA.133 AUSTRIAN AIR FORCE
The militarized Ca.133 was exported to several European air forces, including five aircraft going to the Austrians. During the war, even the UK put small numbers into service.

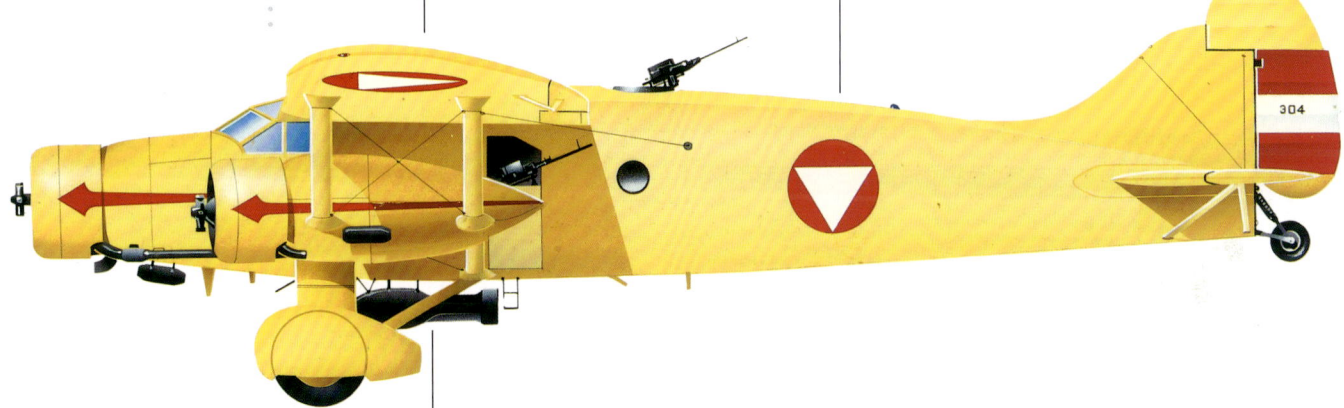

BOMB BAY
As well as fittings for external ordnance, as seen here, the Ca.133 as a bomber had two small internal bomb bays, for a total bombload of 1200kg (2646lb).

CIVIL TO MILITARY

The Ca.133 is one of several examples of interwar civil airliners being repurposed as military aircraft. In its civil format, which first flew in 1934, the three-engined aircraft was designed to carry a total of 16 passengers. The military adaptation involved several key modifications, naturally removing any passenger luxuries and instead fitting internal bomb bays, defensive armament and a militarized cabin.

CA.133, 1937
A Caproni Ca.133 aircraft in Novi Ligure, Italy, in June 1937. As the passengers of this aircraft clearly indicate, this particular aircraft is the civil version.

Douglas C-47 Skytrain (1935)

TYPE • *Transport*   COUNTRY • *United States*

SPECIFICATIONS	
DIMENSIONS:	Length: 19.43m (63ft 9in); Wingspan: 29.11m (95ft 6in); Height: 5.18m (17ft)
WEIGHT:	11,793kg (26,000lb) maximum take-off
POWERPLANT:	4 × Pratt & Whitney R-1830-92 14-cylinder radial piston engines, 895kW (1200hp)
MAX SPEED:	370km/h (230mph)
RANGE:	2575km (1600 miles)
CEILING:	7315m (24,000ft)
CREW:	3
ARMAMENT:	None

The fact that the C-47 remains in frontline military service in the second decade of the twenty-first century demonstrates the excellence of the basic 1930s design. In World War II, the C-47 played a leading role in air assault and logistical operations in European and Pacific theatres.

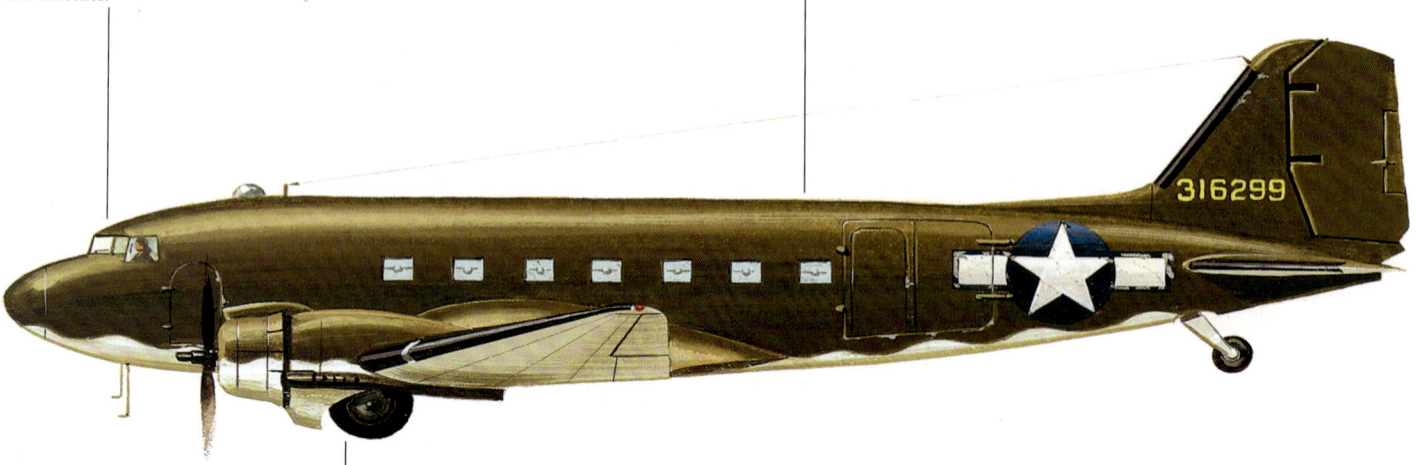

This page: The C-47 pictured here is camouflaged in standard Troop Carrier Command scheme of olive drab upper surfaces and grey under surfaces, the two separated by an undulating line.

COCKPIT
While the C-47 captain occupies the usual left-hand seat, the co-pilot sits on the right, and is also responsible for the radios and throttles.

CABIN
The typical interior layout for trooping operations comprises a row of utility bucket seats fitted along each cabin wall. These can be removed for freighting work.

FUEL
The main fuel tanks (containing 795 litres/210 Imp gals) are located in the centre section forward of the wing spar. These are supplemented by two auxiliary tanks (760 litres/201 Imp gals) aft of the spar.

TRANSPORTER
The first C-47s began to equip the USAAF in 1941, initially in only small numbers. From 1942, however, production accelerated, and the C-47 served as the USAAF's standard transport and glider tug and as such took part in every US airborne operation during the war. General Dwight D. Eisenhower ranked the C-47 alongside the bazooka, jeep and atomic bomb as weapons that contributed most to the Allied victory.

TROOP TRANSPORT
With longitudinal seats along the sides of the cabin for 28 troops and a port-side passenger door, the C-53 Skytrooper was a troop transport version of the C-47. A relatively small total of 404 C-53s was built for the USAAF.

Heinkel He 111 (1935)

TYPE • *Bomber* **COUNTRY** • *Germany*

SPECIFICATIONS (He 111H-6)

DIMENSIONS: Length: 16.40m (53ft 9.5in); Height: 3.40m (11ft 1.5in); Wingspan: 22.60m (74ft 1.33in)

WEIGHT: Maximum take-off 14,000kg (30,865lb)

POWERPLANT: Two 1007kW (1350hp) Junkers Jumo 211F inverted V-12 engines

MAX SPEED: 436km/h (271mph) at 6000m (19,685ft)

RANGE: 1950km (1212 miles)

CEILING: 6700m (21,980ft)

CREW: 5

ARMAMENT: One 20mm (0.79in) cannon, one 13mm (0.51in) machine gun and 4 x 7.92mm (0.31in) machine guns plus a max bombload of 4000kg (8818lb)

The He 111 was a major part of the Luftwaffe's expansion in the 1930s. It was often described as 'the wolf in sheep's clothing' because it first masqueraded as a transport aircraft, but its duty was to provide the Luftwaffe with a fast medium bomber.

Above: The He 111H-6 came to be the most widely used of all He 111s, entering production at the end of 1940.

NOSE SECTION
The standard crew of the He 111H was five: pilot, navigator/bombardier and three gunners, one of whom was also the radio operator. The pilot sat offset to port in the glazed nose section, the navigator/bombardier sitting beside him on a folding seat.

WING SHAPE
Early variants of the Heinkel He 111 had an elliptical wing; the He 111F was the first to feature a straight leading edge. The aircraft here is an He 111P-2, which conducted nocturnal raids against the United Kingdom in late 1940 and early 1941.

ARMAMENT
The He 111H-6 could carry a pair of 765kg (1687lb) LT F5b torpedoes, and was armed with six 7.9mm (0.31in) MG 15 machine guns and a forward-firing 20mm (0.79in) cannon. Some aircraft featured an MG 17 or remotely operated grenade-launcher in the extreme tail.

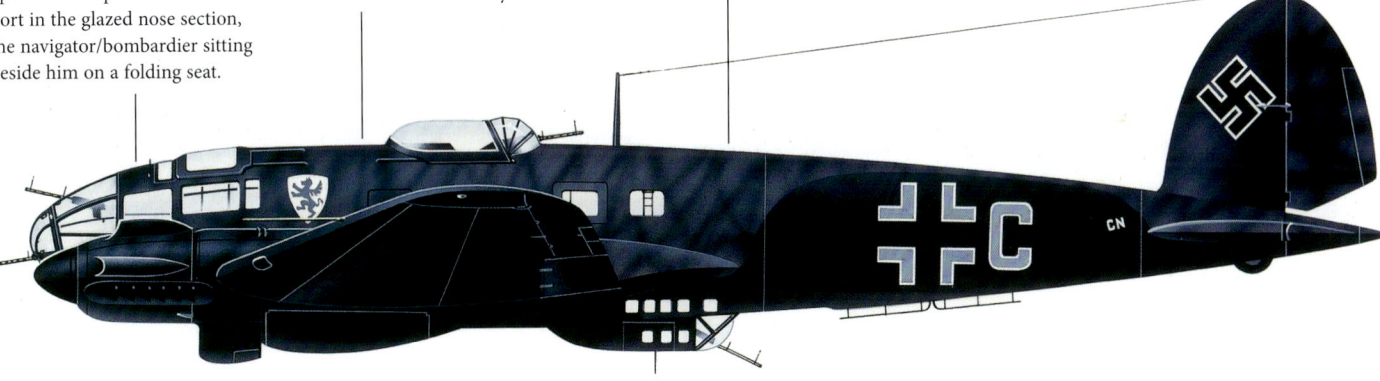

BOMBER BACKBONE
In mid-1939 the He 111P variant made its appearance, powered by two 857kW (1150hp) Daimler-Benz DB 601Aa engines, which incorporated a fully glazed asymmetric nose with an offset ball turret. Relatively few He 111Ps were completed before production switched to the He 111H, the variant that formed the backbone of the Luftwaffe's bomber force between 1940 and 1943, with about 6150 being built before production ended in 1944.

HEINKEL FLIGHT
Fitted with Jumo engines, the He 111H became the definitive version of the Luftwaffe's standard bomber. Here a squadron of He 111H-16s maintains a tight formation while returning from a sortie on the Russian Front. The H-16 was the third 'standard' production model (following the H-3 and H-6), and was powered by the Jumo 211F-2 engine.

Heinkel He 51 (1935)

TYPE • *Fighter* **COUNTRY** • *Germany*

SPECIFICATIONS

DIMENSIONS:	Length: 8.4m (27ft 7in); Wingspan: 11m (36ft 1in); Height: 3.2m (10ft 6in)
WEIGHT:	1900kg (4189lb)
POWERPLANT:	One 559kW (750hp) BMW VI 7.3 Z V-12 liquid-cooled piston engine
MAX SPEED:	330km/h (210mph)
RANGE:	570km (350 miles)
CEILING:	7700m (25,300ft)
CREW:	2
ARMAMENT:	Two 7.92mm (0.31in) MG 17 machine guns in nose; six 10kg (22lb) bombs (C-1)

The lineage of the He 51 – regarded as the Luftwaffe's first true fighter – stretches back to the He 49a, a single-seat biplane first flown in November 1932. In contrast to the He 51 that followed, the He 49a was, ostensibly at least, a civilian advanced trainer.

Above: Heinkel He 51 B-1. This aircraft also served as a fighter trainer with A/B Schule 71 at Prossnitz (Prostejov) in 1942.

POWERPLANT
The BMW VI 7.3 Z V-12 engine gave the aircraft a top speed of 330km/h (210mph), a respectable performance at the time of design but one that quickly became obsolete in the era of monoplane fighters.

WING CONFIGURATION
The He 51 had wings of an uneven span, the upper wings being longer than the lower wings, and the wing sets also staggered in front-to-back alignment. The aircraft was highly manoeuvrable.

GROUND ATTACK
As well as two 7.92mm (0.31in) MG 17 machine guns in the nose, the He 51 could carry six 10kg (22lb) bombs, enabling it to perform the ground-attack mission, formalized in the He 51C-1 variant.

PRODUCTION
Experience with the He 49 led directly to the He 51a, which first flew in summer 1933. Next came a pre-production batch of nine He51A-0 aircraft completed to the same standard and delivered from spring 1934, before the aircraft entered full-scale production as the He 51A-1. The first of these were delivered to the Luftwaffe in April 1935 and a total of 150 examples were built, by both Heinkel and Arado.

PRE-PRODUCTION FLIGHT
Three of the nine He 51A-0 pre-production machines – He 51A-01 D-IQ EE, A-04 D-IJAY and A-05 D-IDIE – were delivered from April 1934. The He 51A-0s were initially used for covert trials by the military, so wore civil markings and registrations, with no swastikas.

Supermarine Walrus (1935)

TYPE • *Seaplane* COUNTRY • *United Kingdom*

Designed for use on warship catapults, the elderly but versatile Walrus flying boat ultimately made its greatest contribution to the Allied war effort as an air-sea rescue aircraft, an aerial lifeline for downed pilots.

SPECIFICATIONS
(Walrus Mk I)

DIMENSIONS:	Length: 11.46m (37ft 7in); Wingspan: 13.97m (45ft 10in); Height: 4.65m (15ft 3in)
WEIGHT:	8050kg (3651lb) maximum take-off
POWERPLANT:	One 560kW (750hp) Bristol Pegasus VI nine-cylinder air-cooled radial piston engine
MAX SPEED:	217km/h (135mph)
RANGE:	970km (600 miles)
CEILING:	5600m (18,500ft)
CREW:	4
ARMAMENT:	One 7.7mm (0.303in) Vickers K machine gun in nose, one 7.7mm (0.303in) Vickers K machine gun in dorsal position; up to 450kg (600lb) bombload

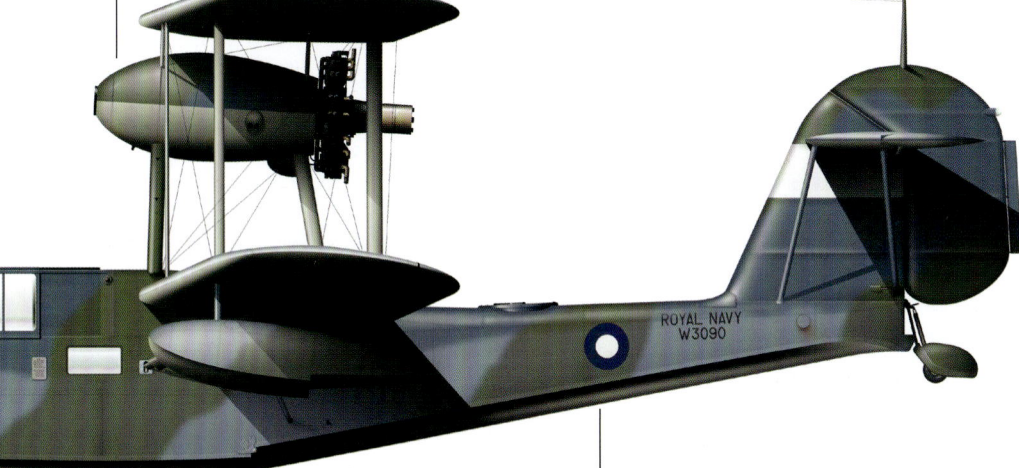

ENGINE NACELLE
A curious feature of the aircraft was that the entire engine nacelle assembly was angled three degrees to the right to counteract the torque of the propeller.

HULL
The hull was initially of anodized alloy construction although the Mk.II version substituted this for an all-wood hull due to wartime shortages of light alloys.

ADVANCED DESIGN
Despite its biplane configuration, the Walrus was quite an advanced aircraft for its era, being the first British military aircraft to combine the features of retractable undercarriage, a fully enclosed cockpit and a metal fuselage in the same airframe.

ORIGINS

The Walrus stemmed from an Australian requirement calling for a successor to their 1921 vintage Supermarine Seagull III flying boats. Supermarine's chief designer, R.J. Mitchell, had already designed an amphibian flying boat that fitted the Australian specification called the Seagull V and the RAAF duly ordered a prototype from the firm. Supermarine were extremely busy in the early 1930s, with the Seagull V considered a low priority and prototype construction, despite being started in 1930, was severely delayed. The aircraft eventually made its first flight on 21 June 1933, but only after pressure had been put on Supermarine from the Australian High Commission in London. The British variant became the Walrus.

Messerschmitt Bf 109 (1935)

TYPE • Fighter **COUNTRY** • Germany

The classic Luftwaffe fighter of World War II, the Bf 109 served throughout the conflict in a series of increasingly capable variants. It was the mount for Germany's most celebrated aces, including Erich Hartmann, Gerhard Barkhorn and Hans-Joachim Marseille.

SPECIFICATIONS (Bf 109E)

DIMENSIONS:	Length: 9.02m (29ft 7in); Wingspan: 9.92m (32ft 6in); Height: 3.4m (11ft 2in)
WEIGHT:	6600kg (14,551lb) maximum take-off
POWERPLANT:	Daimler-Benz DB 605AM inverted V-12 piston engine, 1342kW (1800hp)
MAX SPEED:	621km/h (386mph)
RANGE:	720km (447 miles)
CEILING:	11750m (38,550ft)
CREW:	1
ARMAMENT:	2 × 13mm (0.51in) MG 131 machine guns and 3 × 20mm (0.79in) MG 151 cannon

Above: Bf 109G-6/R6 'Red 13' was a 'Kanonenboote' flown by Feldwebel Heinrich Bartels of 11./JG 27, from Kalamaki, Greece in September 1943. The rudder displays 56 kills of Bartels's eventual tally of 99, most of which had been scored on the Russian Front. It also records his award of the Knight's Cross.

ARMAMENT
Armament of the K-4 consisted of a 30mm (1.2in) MK 108 cannon mounted through the engine and two 13mm (0.51in) MG 131s atop the engine cowling.

MESSERSCHMITT BF 109K-4
Flying with III./JG 27, this Gruppe was fully equipped with the K-4 model by the end of 1944. It was engaged in the fierce fighting over the Reich in the closing stages of the war.

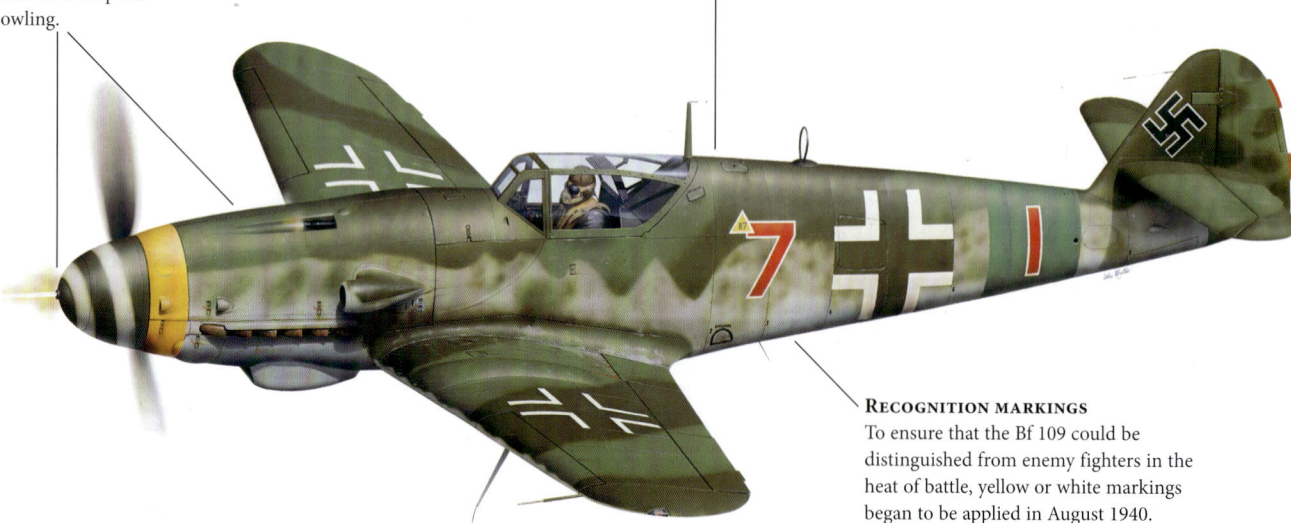

RECOGNITION MARKINGS
To ensure that the Bf 109 could be distinguished from enemy fighters in the heat of battle, yellow or white markings began to be applied in August 1940.

BF 109 ACES
By making continual improvements to the basic design, the Bf 109 remained viable right until the end of World War II and was the backbone of the Luftwaffe fighter arm. The Bf 109 is associated, therefore, with the legendary aces of the Jadgverband. The top-scoring ace of all time, Erich Hartmann, achieved his 352 victories in the space of three and a half years, all at the controls of a Bf 109.

MESSERSCHMITT BF 109E-4
Lieutenant Steindl, headquarters adjutant of Jagdgeschwader 54, overflies the Stalingrad sector in summer 1942 in a Bf 109E-4.

Hawker Hurricane (1935)

TYPE • *Fighter* **COUNTRY** • *United Kingdom*

The Royal Air Force's first monoplane fighter began a dynasty of Hawker warplanes. The most successful British fighter during the Battle of Britain subsequently excelled in the ground attack role in North Africa and the Far East.

SPECIFICATIONS

DIMENSIONS:	Length: 9.75m (32ft); Wingspan: 12.19m (40ft); Height: 4m (13ft 1in)
WEIGHT:	3583kg (7800lb) maximum take-off
POWERPLANT:	Rolls-Royce Merlin XX V-12 liquid-cooled piston engine, 954kW (1280hp)
MAX SPEED:	546km/h (339mph)
RANGE:	740km (460 miles)
CEILING:	10,850m (35,600ft)
CREW:	1
ARMAMENT:	4 x 20mm (0.79in) cannon plus (intruder) two 227kg (500lb) bombs

Above: This Hurricane Mk IIc, based in Egypt with No. 94 Squadron in 1942, carries the normal camouflage for RAF fighters in that theatre – dark earth, middle stone and azure blue.

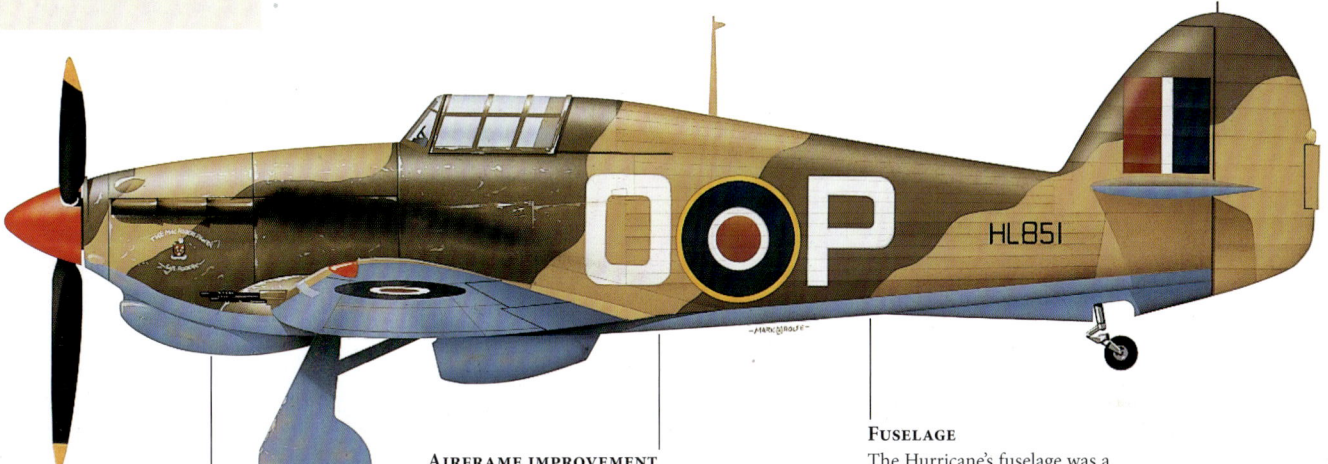

FILTERS
Tropicalized Hurricanes were fitted with Vokes filters to protect the ventral radiator and oil cooler from sand and dust. They proved equally useful on aircraft operating from primitive airstrips in Russia.

ANTI-TANK AIRCRAFT
A specialist anti-tank version was the Hurricane Mk IID, armed with 40mm (1.57in) cannon under the wings. The Mk IID appeared in service in 1942 and was primarily engaged in North Africa. A 'universal wing' was the primary feature of the Hurricane Mk IV; this could mount up to eight rocket projectiles or other external stores. Indeed, the Hurricane Mk IV was the first Allied aircraft to deploy air-to-ground rockets.

AIRFRAME IMPROVEMENT
During its production run, the Hurricane benefitted from changes including metal-skinned wings, an enlarged rudder and (on later Mk Is) a ventral underfin.

FUSELAGE
The Hurricane's fuselage was a box structure of round-section steel and duralumin type, wire-braced and connected by wooden stringers attached to 11 tapering metal frames. The structure was covered with doped Irish linen.

HURRICANE FORMATION
These uncoded aircraft were captured on film shortly after delivery, but not before a squadron badge had been applied to their tailfins and the CO's aircraft (second from left) had gained a squadron leader's pennant below its canopy.

Junkers Ju 87 Stuka (1935)

TYPE • Dive-bomber **COUNTRY** • Germany

SPECIFICATIONS

DIMENSIONS:	Length: 11.5m (37ft 9in); Wingspan: 13.8m (45ft 3in); Height: 3.9m (12ft 9in)
WEIGHT:	3205kg (7086lb)
POWERPLANT:	Junkers Jumo 211J-1 inverted-V piston engine, 1044kW (1400hp)
MAX SPEED:	410km/h (255mph)
RANGE:	1535km (954 miles)
CEILING:	6100m (20,015ft)
CREW:	2
ARMAMENT:	3 × 7.92mm (0.31in) machine guns plus up to 1800kg (3968lb) of external ordnance

The Stuka is remembered as a symbol of the success of Nazi Germany's Blitzkrieg operations in 1939 and 1940. Although rapidly outclassed in its original dive-bomber role, the Ju 87 saw service with the Luftwaffe until the end of the war.

Above: The robust 'spatted' main undercarriage was a key feature of the Ju 87, although the fairings were often removed to cope with winter conditions on the Eastern Front.

JUMO ENGINE
In Ju 87B form, the Stuka was powered by a 12-cylinder liquid-cooled Junkers Jumo 211Da engine. This was more powerful than the Jumo 210 in the A-series.

GULL WING
The very strong inverted gull wing was based on a two-spar structure with closely spaced ribs. The centre section was integral with the fuselage. Ailerons and flaps were provided.

TAILPLANE
To provide additional strength, the Ju 87B's two-spar tailplane was braced by two external struts, replaced on the improved Ju 87D by single aerodynamic struts. The elevators were used in conjunction with aerodynamic brakes to help pull out of a dive.

WING DESIGN
The Stuka's 'inverted' gull wing permitted the fixed undercarriage to be kept short, reducing drag as much as possible. The wing featured large dive brakes, and had underwing pylons for two 250kg (551lb) or four 50kg (110lb) bombs. The Stuka could carry either a single 500kg (1100lb) or 250kg (551lb) bomb on a cradle under the fuselage. The cradle ensured that the bomb swung clear of the propeller arc after it was released in a dive attack.

WINTER WARPLANES
Ju 87-Ds wear temporary winter camouflage for operations on the Eastern Front in 1942. The aircraft are armed with AB 500 cluster bomb containers and centreline SC 250 bombs. The Eastern Front also saw heavy use of the Ju 87G, a specialist anti-tank aircraft, armed with a powerful pair of 37mm (1.45in) cannon under the wings.

Boeing B-17 Flying Fortress (1935)

TYPE • Bomber COUNTRY • United States

SPECIFICATIONS

DIMENSIONS:	Length: 22.78m (74ft 9in); Wingspan: 31.62m (103ft 9in); Height: 5.82m (19ft 1in)
WEIGHT:	32,660kg (72,000lb) maximum take-off
POWERPLANT:	4 × Wright Cyclone R-1820-97 radial piston engines, 895kW (1200hp)
MAX SPEED:	462km/h (287mph)
RANGE:	3220km (2000 miles) with bombload
CEILING:	11,280m (37,000ft)
CREW:	10
ARMAMENT:	13 × 12.7mm (0.5in) machine guns plus a maximum bombload of 7800kg (17,160lb)

The B-17 was a mainstay of the US Eighth Air Force, ranging across occupied Europe as a daylight bomber and experiencing some of the hardest-fought air battles in history. The Flying Fortress also served widely in other theatres and in a variety of different roles.

Above: The B-17G 'Short Arm' represents the definitive standard for a late-production Flying Fortress, as delivered to the Eighth Air Force around the end of 1944.

COCKPIT
Well-planned and spacious, the cockpit of the B-17 benefitted from Boeing's experience in airliner design. The pilot (and aircraft commander) sat on the right, with the co-pilot on the left.

TOP TURRET
The power-operated Sperry top turret was operated by the flight engineer, who was also responsible for fuel management and basic in-flight repairs.

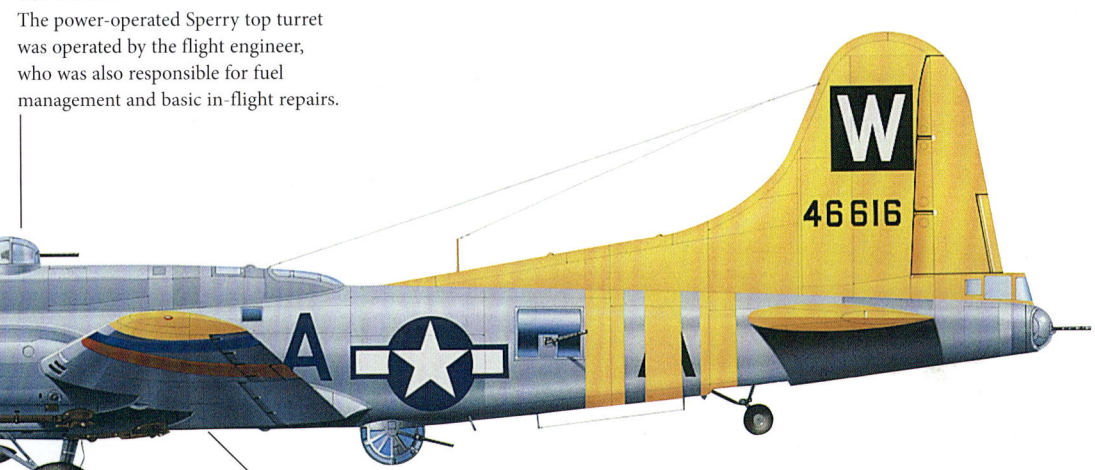

UPGRADES
The Flying Fortress was the iconic aircraft of the Eighth Air Force. As German defences improved, early models like the B-17E were found to lack defensive armament and armour plate. The B-17F and the B-17G were the main versions used in Europe by the Eighth Air Force from 1942 to 1945. The chin turret was introduced on late-model B-17Fs and was fitted to all production Gs. A new tail turret, provided better visibility.

BOMBLOAD
Although the B-17G could in theory carry a bombload of up to 7800kg (17,160lb), in practice a typical long-range load amounted to 2000kg (4400lb). Bombs included the smallest 0.9kg (2lb) incendiaries up to the heaviest 907kg (2000lb) demolition bombs.

SAFE LANDING
A B-17F is watched by ground crew as it comes in for a landing. Safe return was a blessing. In the last deep-penetration raid without 'all the way' fighter escort, 291 B-17s and B-24s were sent against the Schweinfurt ball-bearing factory in October 1943. The result was 60 bombers posted missing, 17 crashed or written off on return and 121 that suffered some kind of damage.

Mitsubishi A5M4 'Claude' (1936)

TYPE • *Naval fighter* **COUNTRY** • *Japan*

When it entered service at the beginning of 1937, the A5M 'Claude' represented a giant leap forwards in carrier-based fighters. Replacing antiquated biplane aircraft, it was the fastest naval fighter in the world and would remain so for almost two years.

SPECIFICATIONS

DIMENSIONS:	Length: 7.55m (24ft 9in); Wingspan: 11m (36ft 1in); Height: 3.2m (10ft 6in)
WEIGHT:	1216kg (2681lb)
POWERPLANT:	Nakajima Kotobuki 41 KAI nine-cylinder air-cooled radial engine, 585kW (785hp) at 3000m (9840ft)
MAX SPEED:	440km/h (273mph)
RANGE:	1200km (746 miles)
CEILING:	9800m (32,150ft)
CREW:	1
ARMAMENT:	2 × 7.7mm (0.303in) Type 89 machine guns

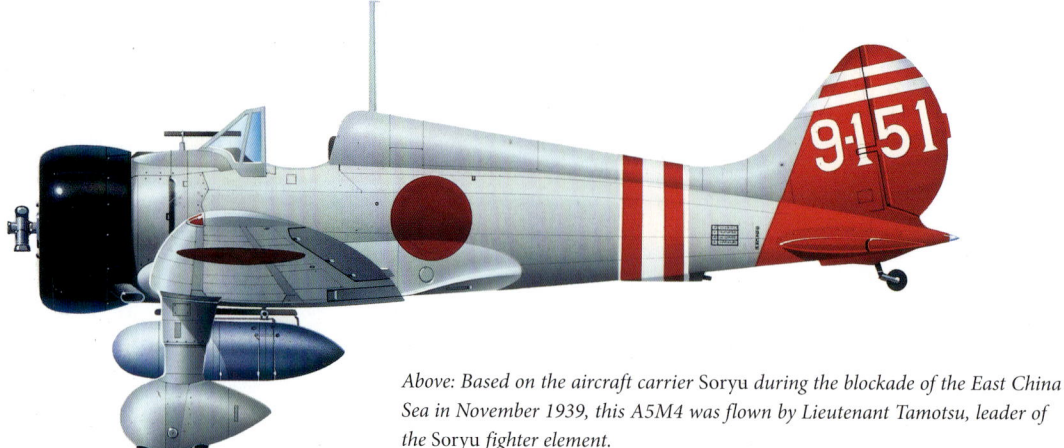

Above: Based on the aircraft carrier Soryu *during the blockade of the East China Sea in November 1939, this A5M4 was flown by Lieutenant Tamotsu, leader of the* Soryu *fighter element.*

ENGINE
The Nakajima Kotobuki 41 KAI nine-cylinder radial engine drove a three-bladed propeller. To improve forward visibility from the cockpit, a NACA (National Advisory Committee for Aeronautics) cowling with cooling flaps was fitted.

MARKINGS
Japanese fighters were generally colourfully decorated in their squadron markings. Many aircraft were funded by public subscription and carried the inscription 'Hokokugo' (patriotism).

IMPORTANCE
It is impossible to overstate the importance of the A5M carrierborne fighter in the development of Japanese industry and military capabilities in the mid-1930s. With this type, Japan moved from dependence on Western imports and thinking to an indigenous product that became its first carrierborne monoplane fighter – one that was comparable with the best of its Western equivalents.

OPEN COCKPIT
Pilots of the 'Claude' had to face the elements; this was the last Japanese naval fighter to be fitted with an open cockpit. It did, however, have an exceptional field of view. As seen here, the pilot often had a gunsight extending from the centre of the windshield.

Supermarine Spitfire (1936)

TYPE • Fighter COUNTRY • United Kingdom

SPECIFICATIONS (Spitfire Mk 1)

DIMENSIONS: Length: 9.12m (29ft 11in); Wingspan: 11.23m (36ft 10in); Height: 3.86m (12ft 8in)

WEIGHT: Empty 2049kg (4517 lb); loaded 2651kg (5844lb)

POWERPLANT: One Rolls-Royce Merlin II or III rated at 1030hp (768kW) at 4953m (16,250ft)

MAX SPEED: 557km/h (346mph)

RANGE: 1014km (630 miles)

CEILING: 9296m (30,500ft)

CREW: 1

ARMAMENT: 8 x Browning Mk II 7.7mm (0.303in) machine guns

The pre-eminent British fighter of World War II was a thoroughbred with a racing lineage. A masterpiece of design from R.J. Mitchell, the Spitfire is remembered as one of the classic fighters of all time, seeing service in every theatre of combat from 1939 until 1945.

Above: The grey-and-white colour scheme of this Mk II of the Czech-manned 312 Squadron, was adopted by RAF Fighter Command to better camouflage the aircraft over water, reflecting the offensive sweeps over occupied France regularly conducted after the Battle of Britain.

CANOPY
The straight-topped canopy of the first Mk Is was replaced by the familiar 'blown' hood. This alteration was more to accommodate taller pilots than to improve visibility.

COMBAT VETERAN
Flt Lt John Dundas flew Spitfire Mk I R6690/'PR-Q' in August 1940. By 9 October he had claimed nine kills and was No. 609 Sqn's top-scorer of the Battle of Britain. His aircraft was a standard machine-gun-armed Mk I.

WINGS
Much of the success of the Spitfire was due to its thin and gracefully tapered wing. There was little room for larger weapons, however, and later marks gained bulges on the wing to allow for the ammunition drums of two or four Hispano cannon.

MK V
The Mk V, which first appeared in March 1941, was the most significant production Spitfire, accounting for 6479 of the total 20,351 aircraft built. As RAF Fighter Command's standard fighter, the Mk V introduced the Merlin Mk 45. This was produced in a number of subvariants, including the Mk VB with cannon and machine-gun armament, and the Mk VC fighter-bomber with provision for external stores.

CLIPPED WINGS
Most Mk VCs were completed with 'clipped' wings (seen here) to improve performance at altitudes below 1525m (5000ft). From mid-1941 until mid-1942, the Spitfire Mk VB was the backbone of Fighter Command, until it was superseded by the Mk IX.

Junkers Ju 88 (1936)

TYPE • *Heavy fighter-bomber* **COUNTRY** • *Germany*

SPECIFICATIONS

DIMENSIONS:	Length: 15.58m (51ft 1in); Wingspan: 20m (65ft 7in); Height: 4.85m (15ft 11in)
WEIGHT:	13,100kg (28,880lb) maximum take-off
POWERPLANT:	2 × Junkers Jumo 213E liquid-cooled piston engines, 1342kW (1800hp)
MAX SPEED:	625km/h (388mph)
RANGE:	2250km (1398 miles)
CEILING:	10,000m (32,810ft)
CREW:	3
ARMAMENT:	(Ju 88 A-4) 5 × 7.92mm (0.31in) MG 18J machine guns on multiple mounts; up to 2000kg (4400lb) of ordnance

One of the most versatile warplanes to see service in World War II, the Ju 88 excelled in roles as diverse as medium bomber, anti-shipping strike, close support and night-fighter. It was a mainstay of the Luftwaffe throughout the conflict.

Above: This Ju 88G-7a of IV./NJG 6 has had its tail painted to represent the older and less potent Ju 88C.

REAR MG
The flight engineer had the secondary task of operating the rearward-firing 7.92mm (0.31in) MG 15 machine gun in the rear of the glazed cabin.

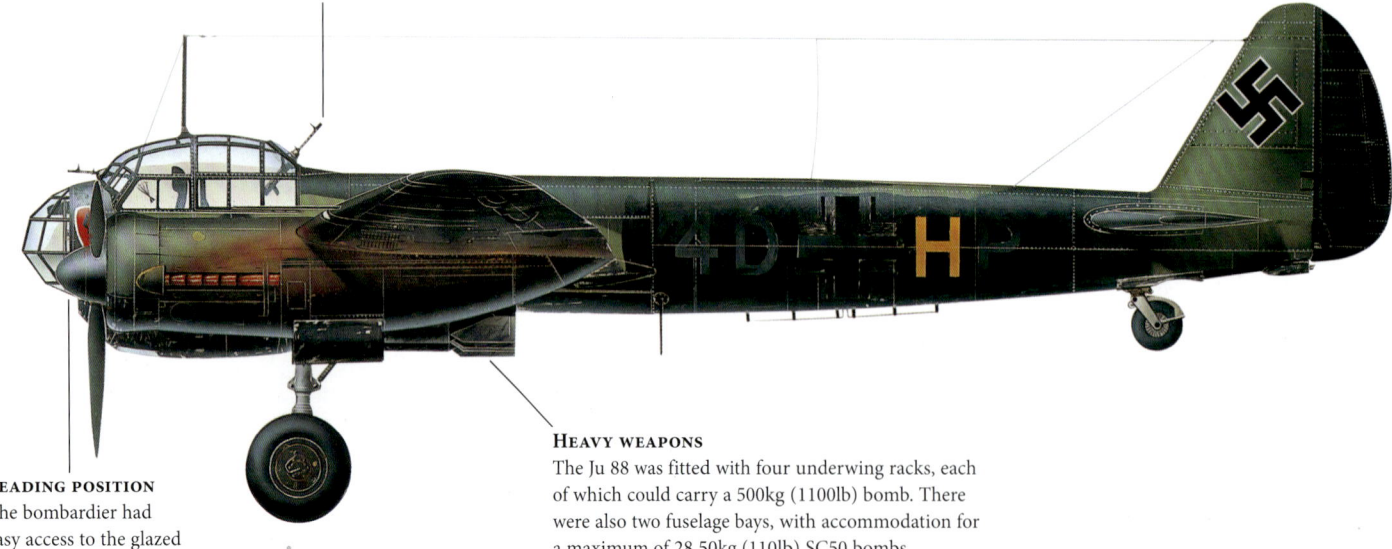

LEADING POSITION
The bombardier had easy access to the glazed nose section, where a bombsight was located for conventional bombing.

HEAVY WEAPONS
The Ju 88 was fitted with four underwing racks, each of which could carry a 500kg (1100lb) bomb. There were also two fuselage bays, with accommodation for a maximum of 28 50kg (110lb) SC50 bombs.

SERVICE
The prototype Ju 88 flew for the first time on 21 December 1936 and the first pre-series Ju 88A-0s were delivered to the Luftwaffe in August 1939. The Ju 88A was built in 17 different variants up to the Ju 88A-17, with progressively uprated engines and enhanced defensive armament. The most widely used variant was the Ju 88A-4.

DOWNED JU 88
British aircrew swarm around a downed Ju 88 during the Battle of Britain. The Ju 88A saw considerable action in the Balkans and the Mediterranean, and on the Eastern Front. Some of their most outstanding service, however, was in the Arctic, where aircraft of KG 26 and KG 30, based in northern Norway, carried out devastating attacks on Allied convoys to Russia.

Messerschmitt Bf 110 (1936)

TYPE • *Heavy fighter* COUNTRY • *Germany*

SPECIFICATIONS

DIMENSIONS:	Length: 12.3m (40ft 6in); Wingspan: 16.3m (53ft 4in); Height: 3.3m (10ft 9in)
WEIGHT:	4500kg (9921lb)
POWERPLANT:	2 × Daimler-Benz DB 601B-1 liquid-cooled inverted V-12, 809kW (1085hp) each
MAX SPEED:	560km/h (348mph)
RANGE:	2410km (1500 miles)
CEILING:	10,500m (35,000ft)
CREW:	2/3
ARMAMENT:	2 × 20mm (0.79in) MG FF/M cannon, 4 × 7.92mm (0.31in) MG 17 machine guns, 1 × 7.92mm (0.31in) MG 15 machine gun; up to 2000kg (4400lb) bombload

Messerschmitt's Bf 110 was one of the Luftwaffe's great hopes at the start of World War II, but the twin-engined Zerstörer (destroyer) proved hopelessly vulnerable to single-engined fighter opposition during the Battle of Britain.

Above: Messerschmitt Bf 110G-4b/R3. Successive series of aircraft were equipped with ever better radars, in particular the FuG 212 Lichtenstein C-1 and FuG 220 Lichtenstein SN-2.

ENGINES
All Bf 110s were fitted with the Daimler-Benz DB 601 engine. The prototype had used the DB 600, which was plagued by reliability problems.

CREW
The Bf 110 was designed to carry a crew of three, comprising pilot, radio operator and gunner. In practice a crew of two was usually carried, the radio operator also acting as the gunner. The Bf 110C-4 was the first variant to introduce armour protection for the crew.

STABILIZERS
The Me 110 had twin vertical stabilizers at the rear of the aircraft, designed to improve the aircraft's control and manoeuvrability.

NEW ROLES
The Bf 110's performance during the Battle of Britain did not consign the aircraft to obsolescence. As the war progressed, it proved to be a solid fighting machine as a long-range fighter and fighter-bomber, as a bomber-destroyer and particularly as a night-fighter. It had become outclassed on daylight operations by late 1941, but when equipped with radar the Bf 110 night-fighters proved deadly against the RAF's bombers and made up 75 per cent of the night-fighter force by late 1942.

MESSERSCHMITT BF 110C
With a combat weight of 5900kg (13,007lb), the Bf 110C attained 540 km/h (336mph) at a rated altitude of 6050m (19,850ft), faster than most contemporary Allied fighters, and only 32–43km/h (20–30mph) slower than its next opponents, the French Dewoitine D.520 and the British Supermarine Spitfire Mk I.

Consolidated PBY Catalina (1936)

TYPE • Seaplane COUNTRY • United States

SPECIFICATIONS

DIMENSIONS:	Length: 19.47m (63ft 10in); Wingspan: 31.7m (104ft); Height: 6.15m (25ft 1in)
WEIGHT:	16,066kg (35,420lb)
POWERPLANT:	2 × Pratt & Whitney R1830-92 twin Wasp radial piston engines, 895kW (1200hp)
MAX SPEED:	288km/h (175mph)
RANGE:	4030km (2520 miles)
CEILING:	4000m (15,800ft)
CREW:	8
ARMAMENT:	3 x 7.62mm (0.3in) machine guns, 2 x 12.7mm (0.5in) machine guns; up to 1814kg (4000lb) of bombs or depth charges

The exceptional flying boat of World War II, the 'Cat' was perhaps all the more remarkable since it had first been ordered for the US Navy back in 1933. It served with great success throughout World War II, and it became the most extensively built flying boat in aviation history.

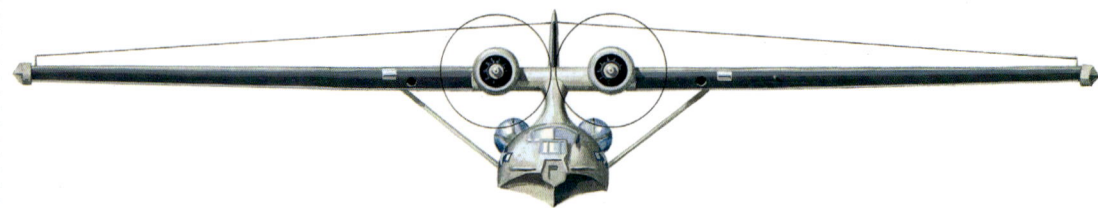

Above: The nose section provided accommodation for one crew member who acted as an observer. The panel below the station was blind, but would have been used as a bomb-aiming window.

ENGINE MOUNTS
The PBY's twin Pratt & Whitney Twin Wasp radial engines were mounted as high as possible to escape the effects of spray, which could be quite serious if the aircraft was taking off or alighting in a heavy swell.

BEAM GUNS
The Catalina carried two 12.7mm (0.5in) machine guns in cupolas on each side of the fuselage. The large observation blisters, first installed in the PBY-4, provided excellent visual coverage. The waist gunners stood on a semi-circular platform, allowing them to traverse their guns over a wide arc.

FLIGHT DECK
The pilot and co-pilot sat side by side on the flight deck and were provided with a roof escape hatch for emergencies.

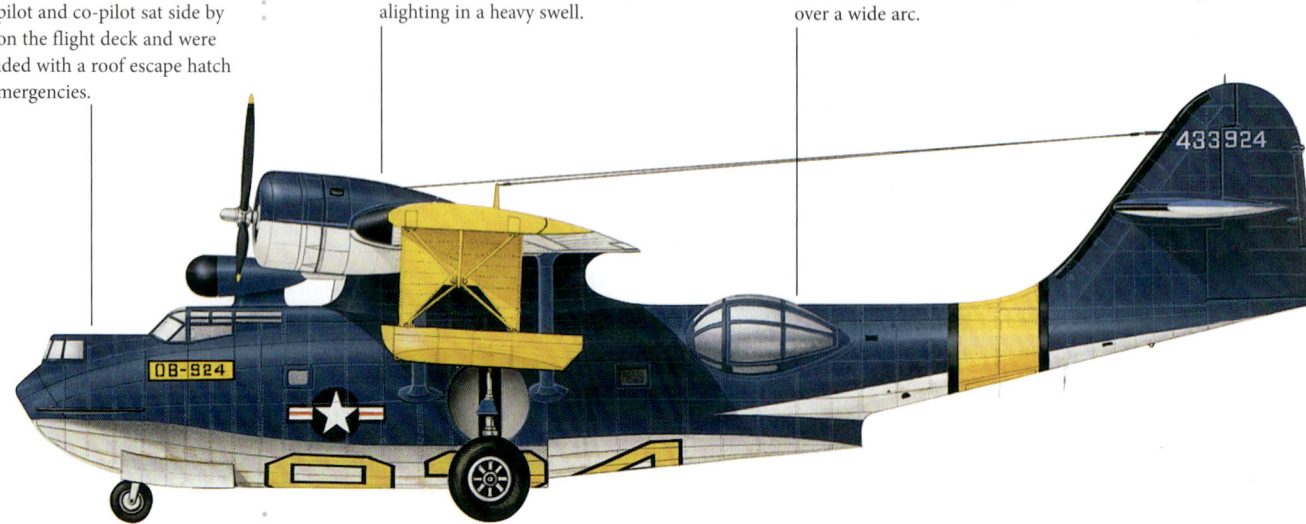

AMPHIBIOUS
Following tests with a retractable tricycle wheel landing gear in the last PBY-4, the final 33 US Navy PBY-5s were completed in the amphibian form, as were 761 PBY-5A aircraft. Following early successful use of the PBY-5 by the RAF's Coastal Command in 1941 as the Catalina Mk I, large orders continued to be placed for the US Navy, additional production being undertaken by Canadian Vickers and Boeing of Canada.

ON PATROL
Catalinas flew long, deep missions hundreds of miles from the nearest land. More than 800 versions of the amphibious version of the PBY-5 were built, mainly for the US Navy. The retractable tricycle landing gear meant it could be used for conventional runways.

Henschel Hs 123 (1936)

TYPE • Dive-bomber **COUNTRY** • Germany

Despite its apparent obsolescence and production stopping in 1937, the tough and accurate Hs 123 found plenty of employment on the steppes of Russia and Ukraine from 1941 to 1944.

SPECIFICATIONS

DIMENSIONS:	Length: 8.33m (27ft 4in); Wingspan: 10.50m (34ft 5.5in); Height: 3.20m (10ft 6in)
WEIGHT:	Empty 1500kg (3308lb); maximum take-off 2215kg (4883lb)
POWERPLANT:	656kW (880hp) BMW 132Dc radial
MAX SPEED:	340km/h (211mph) at 2000m (6562ft)
RANGE:	855km (531 miles)
CEILING:	9000m (29,530ft)
CREW:	1
ARMAMENT:	2 x 7.92mm (0.31in) MG17 machine guns; up to 450kg (992lb) of bombs

Above: After early crashes caused by the upper wing tearing away from the centre section struts, the Hs 123's structure was substantially strengthened, resulting in an extremely robust aircraft that could endure a great deal of punishment.

ENGINE NOISE
Although the Hs 123 was lightly armed with only two machine guns, the roar of its engine was a powerful psychological weapon against troops.

COCKPIT
The Hs 123 was never given an enclosed cockpit, so conditions for its pilots on the Russian Front in winter were not exactly comfortable – although the heat generated by the big radial helped somewhat.

WINTER CAMOUFLAGE
This appropriately finished Hs 123A equipped the 5.(Schl)/LG 2 in the Central Sector, Moscow front during 1941/42. For winter operations a white soluble distemper was applied over the permanent splinter camouflage. The yellow bands are theatre identification bands, while the black triangle was a ground attack unit marking, believed to date back to World War I.

POLISH OPERATIONS
Equipped with Hs 123 aircraft, II (Schlacht)./LG 2 was in the lead air assault against Poland on 1 September 1939 that opened World War II. Armed with 50kg (110lb) bombs on the wing racks and the MG 17 guns, the Hs 123s flew just feet above the heads of the Polish cavalry brigades for 10 days. More effective than the armament was the terrifying noise of the BMW radial, dispersing Polish mounted columns.

ATTACK RUN
An Hs 123 in typical pose, turning to dive on a target. In fact the Henschel biplane was more widely employed in the close support role, in which its ability to absorb damage from ground fire was greatly appreciated.

Savoia-Marchetti SM.79 Sparviero (1936)

TYPE • *Bomber* COUNTRY • *Italy*

SPECIFICATIONS

DIMENSIONS:	Length: 15.62m (51ft 3in); Wingspan: 21.20m (69ft 6.5in) Height: 4.40m (14ft 5.25in)
WEIGHT:	Empty 6800kg (14,991lb); maximum take-off 11,300kg (24,912lb)
POWERPLANT:	Three 746kW (1000hp) Piaggio P.XI RC 40 radial engines
MAX SPEED:	435km/h (270mph) at 3650m (11,975ft)
RANGE:	1900km (1181 miles)
CEILING:	6500m (21,325ft)
CREW:	5/6
ARMAMENT:	3 x 12.7mm (0.5in) machine guns and one 7.7mm (0.303in) machine gun; 2 x 450mm (17.7in) torpedoes or 1250kg (2756lb) of bombs

Of all the Italian bombers that served during World War II, the SM.79 *Sparviero* (Sparrowhawk) was probably the best known and most effective. It was one of the finest anti-shipping aircraft of the war, not least because it could carry two torpedoes.

Above: On its upper wings, this SM.79 carried the insignia of fascist Italy, the 'fasces' – a bundle of rods around an axe that was carried before a magistrate in ancient Rome as a symbol of authority.

REAR MG
Defence of the upper rear of the aircraft was the responsibility of the flight engineer or radio operator, who used a 12.7mm (0.50in) gun on a flexible mount. The gun could be retracted and a panel put in place to cover the hole in the fuselage.

SQUADRON NUMBER
This SM.79 carries its *squadriglia* (squadron) number on the rear fuselage, which was the usual practice. This indicates that the aircraft belonged to the 192nd Squadriglia, which was active over Malta in 1941.

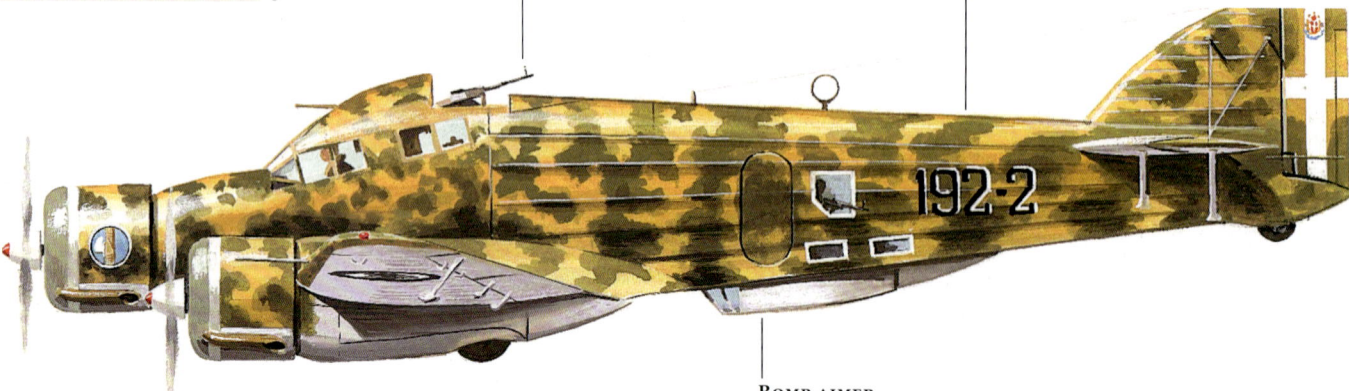

BOMB AIMER
The ventral gondola contained the bomb aimer, who was equipped with a small wheel enabling him to make corrections to the aircraft's course by adjusting the rudder. The Italians were renowned for their extremely accurate high-level bombing.

MARITIME PREDATOR
When Italy entered World War II in June 1940, SM.79s accounted for well over half the Italian Air Force's total bomber strength. SM.79s saw continual action in the air campaign against Malta and in North Africa, becoming renowned for their high-level precision bombing, while the torpedo-bomber version was active against British shipping in the Aegean and the central Mediterranean.

DESERT PATROL
Two SM.79 *Sparviero* aircraft hunt for targets over the coastal strip of the Mediterranean. Production of the aircraft, which was the military counterpart of an eight-seat civil airliner, began in October 1936 and was to have an uninterrupted run until June 1943, by which time 1217 aircraft had been built.

Tupolev SB-2 (1936)

TYPE • Bomber COUNTRY • Soviet Union

SPECIFICATIONS (SB-2bis)

DIMENSIONS:	Length: 12.57m (41ft 2.33in); Wingspan: 20.33m (66ft 8.5in); Height: 3.25m (10ft 8in)
WEIGHT:	Empty 4768kg (10,511lb); maximum take-off 7880kg (17,372lb)
POWERPLANT:	2 x 716kW (960hp) Klimov M-103 12-cylinder V-type
MAX SPEED:	450km/h (280mph) at 1000m (3281ft)
RANGE:	2300km (1429 miles)
CEILING:	9000m (29,530ft)
CREW:	3
ARMAMENT:	4 x 7.62mm (0.30in) machine guns; bombload of 600kg (1323lb)

The Tupolev SB-2 was almost certainly the most capable light bomber in service anywhere in the world in the mid-1930s. It was the first aircraft of modern stressed-skin construction to be produced in the USSR, and in numerical terms was also the most important bomber of its day.

Right: The SB-2 was used operationally in Spain, where its crews held it in great esteem and gave it the nickname 'Katushka'. It was considered to be invulnerable, as it was faster than most fighters then in service.

PERFORMANCE
The SB-2's broad, high aspect ratio wing gave it a good altitude performance of nearly 9150m (30,000ft). Russian crews nicknamed the bomber the 'Pterodactyl'.

ENGINES
Most production SB-2s were fitted with the Klimov M-103 engine. They drove new VISh variable-pitch propellers. In order to make full use of the engine's performance, a new engine cowling was produced without frontal radiators; these were replaced with new radiators placed under the engine nacelle.

CREW POSITIONS
The aircraft attracted complaints about the high noise level, cramped crew compartments, hard undercarriage suspension and in particular about the front gunner's position. This could only be reached through a hatch under the fuselage, so he had no means of escape in the event of a belly landing or a ditching.

DEVELOPMENT
The story of the SB-2 began in the early 1930s, when Andrei N. Tupolev embarked on design studies of a fast tactical bomber. He built three prototypes, designated ANT-40, ANT-40-1 and ANT-40-2. The latter proved the best variant. The type entered service in 1936, and 6967 aircraft were built before production ended in 1941. Among the principal variants were the SB-2bis of 1938, with uprated engines and greater fuel capacity, and the SB-2RK dive-bomber version of 1940.

FACING THE GERMANS
The crew of an SB-2 warm up the engines ready for a sortie. By the time of the German invasion of Russia in 1941 the SB-2 was obsolescent, and heavy losses sustained in daylight attacks led to the aircraft being switched to night bombing. Later, many SB-2s were used as target tugs, crew trainers and troop/cargo transports.

Fairey Swordfish (1936)

TYPE • Torpedo bomber COUNTRY • United Kingdom

Affectionately known as the 'Stringbag', the Fairey Swordfish made a decisive contribution to the war, especially in the Mediterranean. Archaic in appearance even when it first flew, it was the Fleet Air Arm's premier torpedo-bomber at the outbreak of hostilities.

SPECIFICATIONS

DIMENSIONS:	Length: 10.87m (35ft 8in); Wingspan: 12.97m (42ft 6in); Height: 3.76m (12ft 4in)
WEIGHT:	Empty 2132kg (4700lb); maximum take-off 4196kg (9250lb)
POWERPLANT:	611kW (820hp) Bristol Pegasus XXX radial engine
MAX SPEED:	222km/h (138mph)
RANGE:	879km (546 miles)
CEILING:	5867m (19,250ft)
CREW:	2/3
ARMAMENT:	Two 7.7mm (0.303in) machine guns, plus one 457mm (18in) torpedo or eight 27.2kg (60lb) rocket projectiles

Above: Primary armament of the Swordfish was the 457mm (18in) torpedo. In May 1941, this weapon crippled the mighty German battleship Bismarck, *enabling warships to close in and finish her off.*

POWERPLANT
Early production Mk IIs retained the Pegasus IIIM engine which delivered 514kW (690hp), but later examples were fitted with the more powerful 560kW (750hp) Pegasus XXX.

DINGHY
The Swordfish's rubber dinghy was stowed in the central portion of the upper wing, which was fitted with a quick-release handle. In the event of an emergency, the crew could be out of the aircraft and into the dinghy in seconds.

ARRESTER HOOK
The arrester hook, retracted here, was a vital part of the aircraft's fittings for service on carriers. The Swordfish's wings were also hinged to fold back against the fuselage to reduce the amount of stowage space needed by each aircraft.

TARANTO
In 1940 came the supreme triumph of the Swordfish – the memorable assault on the Italian fleet at anchor in Taranto harbour. The attack, made by 21 Swordfish on the night of 11 November 1940, was launched in two waves. The Italian Navy was dealt a shattering blow: three battleships were severely damaged; a cruiser and two destroyers had been hit; and two auxiliary vessels had been sunk.

CARRIER AIRCRAFT
Seen overflying the newly commissioned British aircraft carrier HMS *Ark Royal* in early 1939, this Swordfish Mk I is from No. 820 Squadron. The unit had become the first squadron to deploy aboard the carrier in January of that year.

Morane-Saulnier MS.406 (1937)

TYPE • *Fighter* COUNTRY • *France*

SPECIFICATIONS

DIMENSIONS:	Length: 8.17m (26ft 9in); Wingspan: 10.62m (34ft 5in); Height: 3.25m (10ft 8in)
WEIGHT:	Maximum take-off 2722kg (6000lb)
POWERPLANT:	641kW (860hp) Hispano-Suiza 12Y-31 12-cylinder V-type
MAX SPEED:	490km/h (304mph)
RANGE:	1500km (932 miles)
CEILING:	9400m (30,850ft)
CREW:	1
ARMAMENT:	One 20mm (0.79in) cannon or 7.5mm (0.295in) machine gun and two 7.5mm machine guns

The MS.406 was France's first 'modern' monoplane fighter, but was in all major respects – except firepower and its ability to withstand battle damage – an indifferent warplane with little to commend it except ease of manufacture and availability.

This page: The aircraft shown here is a Morane-Saulnier MS.406C-1 based in France in 1940, before the armistice with Germany. The number of the aircraft's Groupe de Chasse, in this case GC 11, is shown on the aircraft's tail.

NOSE CANNON
In common with other fighter aircraft of European design, the MS.406 mounted a 20mm (0.79in) cannon between the engine blocks, firing through the propeller hub.

ANTENNAE
The MS.406 had a rather curious antenna arrangement, with an aerial under the fuselage as well as one above. The one under the fuselage retracted automatically when the main undercarriage was lowered.

RADIATOR AND COOLANT
A prominent radiator was mounted under the fuselage for cooling the engine. The coolant reservoir was mounted beneath the engine in the fuselage.

OUTFOUGHT
In terms of numbers, the MS.406 was the most important fighter in French service in September 1939. The MS.406 equipped 16 Groupes de Chasse and three Escadrilles in France and overseas, and 12 of the Groupes saw action against the Luftwaffe. The aircraft was highly manoeuvrable and could withstand heavy battle damage, but it was outclassed by the Bf 109 and losses were heavy.

POLISH AIRCRAFT
The number of MS.406 aircraft eventually built reached 1080, some of which were exported to Switzerland and Turkey. Poland placed an order for 50 MS.406s, but they were taken back by the French at the port of Gdynia. These Armée de l'Air aircraft were flown by pilots who had escaped the German onslaught on Poland in September 1939.

Nakajima B5N Tenzan (1937)

TYPE • *Torpedo-bomber* COUNTRY • *Japan*

Used to devastating effect at Pearl Harbor, the B5N *Tenzan* (called 'Kate' by the Allies) was an important weapon in the initial stages of the Pacific War. However, by the time its replacement, the B6N, had entered service the tide had turned against the Japanese.

SPECIFICATIONS

DIMENSIONS:	Length: 10.30m (33ft 9.5in); Wingspan: 15.51m (50ft 11in); Height: 3.70m (12ft 1.33in)
WEIGHT:	Empty 2279kg (5025lb); maximum take-off 4108kg (9056lb)
POWERPLANT:	746kW (1000hp) Nakajima NK1B Sakae 11 14-cylinder rad
MAX SPEED:	378km/h (235mph) at 3600m (11,811ft)
RANGE:	2000km (1243 miles)
CEILING:	8260m (27,100ft)
CREW:	3
ARMAMENT:	One 7.7mm (0.303in) machine gun; one 800kg (1764lb) torpedo, or up to 800kg (1764lb) of bombs

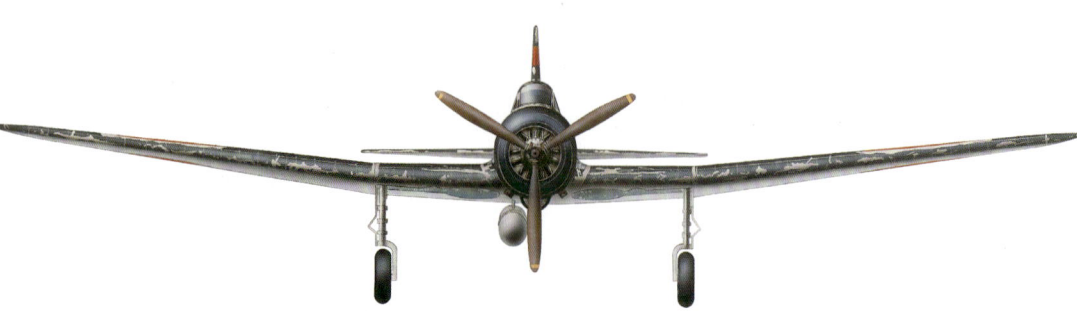

Above: The B5N was tested in two versions, one with Fowler-type flaps, hydraulic flaps and hydraulic wing folding, and the other with plain flaps and manual wing folding. It was the latter version, seen here, that was ordered into production.

VISIBILITY
The adoption of the 14-cylinder Sakae 11 engine in the B5N2, which had a relatively small diameter, meant that a smaller cowling could be fitted, improving the pilot's forward visibility and also reducing drag.

COCKPIT
The B5N2 had a crew of three in a fully enclosed cockpit, comprising the pilot, an observer/navigator who also acted as bomb aimer, and the radio operator, who also manned the trainable 7.7mm (0.303in) machine gun that was the aircraft's sole means of defence.

TAIL LETTERS
The 'AI' code letters on the tail of the B5N2 seen here indicate that it belongs to the *Akagi* Air Group, the *Akagi* being one of the carriers assigned to what the Japanese termed the 'Hawaiian Operation' – in other words, the attack on Pearl Harbor.

B5N2
Designed in 1936, the prototype B5N carrier attack bomber first flew in January 1937 and became operational as the B5N1 light bomber during the Sino–Japanese war. After its usefulness in that conflict was assessed, no major modifications were found necessary, but the need to improve the aircraft's performance was apparent. This led to a more powerful, definitive version, the B5N2.

CARRIER LAUNCH
Armed with its standard anti-ship weapon load of one 800kg (1764lb) torpedo, a B5N2 Tenzan rolls down the deck of a Japanese Imperial Navy carrier for a Nakajima raid on Allied shipping in late 1944.

Bristol Blenheim Mk IV (1937)

TYPE • Bomber COUNTRY • United Kingdom

SPECIFICATIONS

DIMENSIONS:	Length: 12.98m (42ft 7in); Wingspan: 17.70m (58ft 1in); Height: 2.99m (9ft 10in)
WEIGHT:	Maximum take-off 6537kg (14,400lb)
POWERPLANT:	Two 674kW (905hp) Bristol Mercury XV radial engines
MAX SPEED:	428km/h (266mph) at 3595m (11,800ft)
RANGE:	2340km (1460 miles)
CEILING:	6705m (22,000ft))
CREW:	2
ARMAMENT:	Five 7.7mm (0.303in) machine guns; internal bombload of 454kg (1000lb).

Hailed at the time as the fastest light bomber in the world, the Blenheim was a fundamental part of the RAF's attack force, only to achieve disappointing results and heavy losses in combat during the early years of the war.

This page: The navigator had a chart table and instrument panel on the port side of his station in the glazed nose. Below this were two flatpane windows for bomb aimer.

ESCAPE HATCHES
The gunner had an emergency exit hatch behind him in the roof of the fuselage. The pilot and navigator had sliding panels in the cockpit roof, and an escape hatch in the floor of the nose section.

TAIL SURFACES
Only a small proportion of the Blenheim's vertical tail surfaces was fixed, the aircraft having a large full-height rudder controlled by cables running the length of the fuselage.

ENGINES
The Blenheim's Bristol Mercury engines had a prominent intake projecting forward from the engine cowling. These were ram air intakes for the oil cooler.

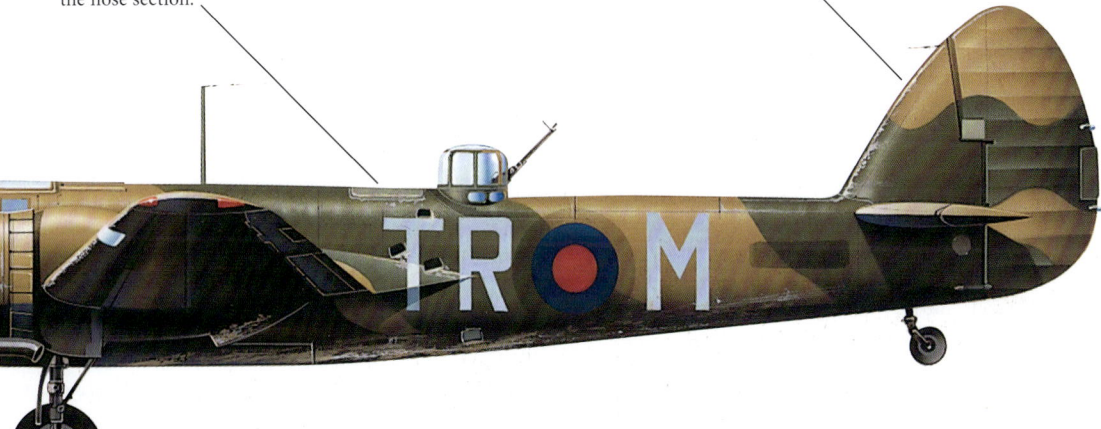

COMBAT PERFORMANCE
On the second day of the war, Blenheims of Nos. 107 and 110 Squadrons from Marham, Norfolk, carried out the RAF's first offensive operation when they unsuccessfully attacked units of the German Navy in the Elbe Estuary. The total inadequacy of the Blenheim's defensive armament became apparent in the battles of Norway and France, when the Blenheim squadrons engaged in anti-shipping operations in the North Sea and those deployed to France suffered appalling losses.

DEFENSIVE ARMAMENT
The single turret-mounted Lewis or Vickers gun, as carried by these pre-war Blenheim Mk IVs, provided insufficient defence and was replaced by twin-gun Browning or Vickers installations. Initially, the Blenheim IV also had a single forward-firing Vickers K in the nose, with a single gun in the dorsal turret. This proved inadequate, and a new BI.Mk IV turret housing two belt-fed Brownings was added.

Gloster Gladiator (1937)

TYPE • Fighter **COUNTRY** • United Kingdom

Somewhat overrated and largely sentimentalized as a result of myth rather than actual achievement, the Gladiator stands in history as the RAF's last biplane interceptor. It served just prior to World War II, as the era of the classic monoplane fighter was dawning.

SPECIFICATIONS

DIMENSIONS:	Length: 8.36m (27ft 5in); Wingspan 9.83m (32ft 3in); Height 3.53m (11ft 7in)
WEIGHT:	Empty 1562kg (3444lb); maximum take-off 2206kg (4864lb)
POWERPLANT:	619kW (830hp) Bristol Mercury VIIIA 9-cylinder radial
MAX SPEED:	414km/h (257mph)
RANGE:	708km (440 miles)
CEILING:	10,210m (33,500ft)
CREW:	1
ARMAMENT:	4 x 7.7mm (0.303in) machine guns

This page: This Gladiator Mk I, K7986, bears the pre-war colours of No. 79 Squadron, which was based at Biggin Hill, Kent, in 1937. No. 79 Squadron replaced its Gladiators with Hawker Hurricane monoplane fighters in November 1938.

ENGINE MOUNT
The Gladiator's Bristol Mercury engine was secured by eight bolts to a hexagonal engine ring on the front of the engine mounting bay structure.

COCKPIT
For the first time, RAF fighter pilots had the benefit of a fully enclosed cockpit. Some diehards complained that it reduced visibility to an unacceptable level.

WINGS
The Gladiator's wings had a distinctive stagger. Each wing was built up on two spars, braced by drag struts, ribs and stringers. Flaps and Frise-type ailerons were fitted to all four wings, the ailerons on the bottom wings actuating those above by means of tie rods.

FIGHTER COMMAND
Designed as a more advanced successor to the open-cockpit Gauntlet fighter, the prototype Gladiator was flown in September 1934 and evaluated by the Air Ministry in the following year, the trials resulting in a production order for 23 machines, followed by further orders for 128 aircraft. First deliveries were made in February 1937, and the aircraft went on to equip eight squadrons of Fighter Command.

BEF GLADIATORS
Cheerful pilots and their No. 615 Sqn Gladiator Mk IIs, part of the Air Component of the British Expeditionary Force, are seen in France in 1939/40. In fact, the BEF's Gladiators had an unhappy time, suffering heavy losses in the face of experienced Luftwaffe pilots flying modern fighter aircraft.

Fieseler Fi 156 Storch (1937)

TYPE • *Reconnaissance* COUNTRY • *Germany*

SPECIFICATIONS

DIMENSIONS:	Length: 9.9m (32ft 6in), Wingspan: 14.3m (46ft 11in), Height: 3.1m (10ft 2in)
WEIGHT:	Maximum take-off 1260kg (2778lb)
POWERPLANT:	One 180kW (240hp) Argus As 10 V-8 inverted air-cooled piston engine
MAX SPEED:	175km/h (109mph)
RANGE:	380km (240 miles)
CEILING:	4600m (15,090ft)
CREW:	2
ARMAMENT:	One 7.92mm (0.31in) MG 15 machine gun

Widely used by the Luftwaffe throughout the conflict, the Fieseler Fi 156 Storch ('Stork') set the trend for subsequent aircraft in its class, with a combination of excellent short take-off and landing (STOL) performance and an extensively glazed cockpit.

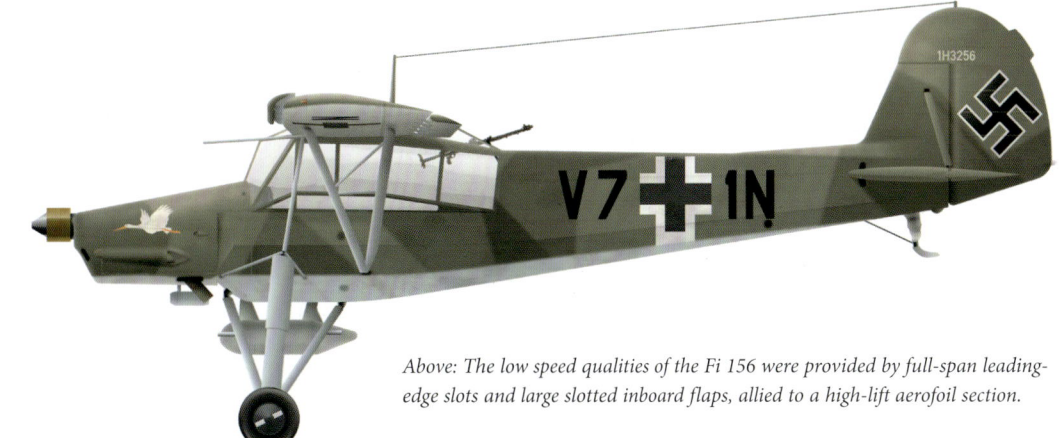

Above: The low speed qualities of the Fi 156 were provided by full-span leading-edge slots and large slotted inboard flaps, allied to a high-lift aerofoil section.

PROPELLER
The second prototype Fi 156 was tested with a variable-pitch metal propeller, but all other Storch aircraft had a wooden, fixed-pitch Schwarz propeller.

DEFENCES
Against enemy fighters, the Storch's incredible low speed and agility rendered it one of the most difficult of aerial targets – if flown well. Nevertheless, it was given a single 7.9mm MG 15, firing through the rear canopy to fend off any fighters attacking from the rear.

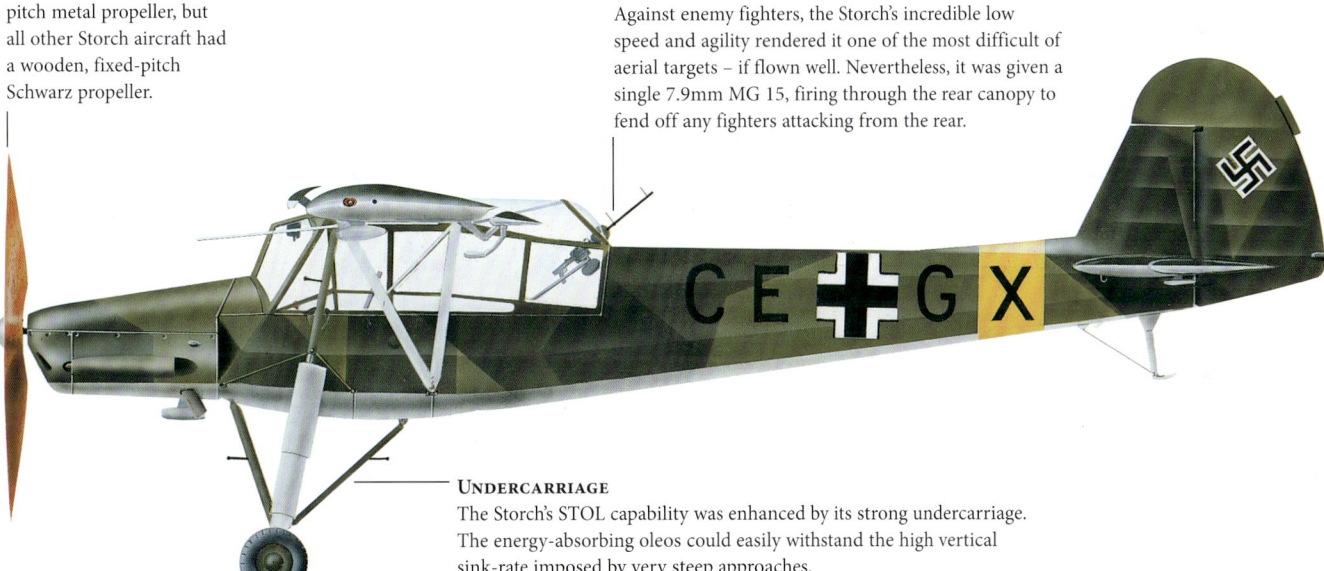

UNDERCARRIAGE
The Storch's STOL capability was enhanced by its strong undercarriage. The energy-absorbing oleos could easily withstand the high vertical sink-rate imposed by very steep approaches.

FAMOUS EXPLOITS
The Storch's wartime exploits in the hands of the Luftwaffe included the rescue of Italian leader Benito Mussolini from his imprisonment in a hotel high in the Apennine Mountains on 12 September 1943. Meanwhile, test pilot Hanna Reitsch flew into the ruins of Berlin on 26 April 1945 carrying General Ritter von Greim, who Hitler would appoint as the new – and final – commander of the Nazi-era Luftwaffe.

PARIS LANDING
Shortly after the fall of Paris to the German Blitzkrieg in 1940, this Storch landed in the Place de la Concorde, demonstrating its excellent STOL characteristics. The aircraft's STOL capability was the result of a wing equipped with high-lift devices, including a fixed slat that extended over the full span of the leading edge, while the trailing edge was equipped with slotted ailerons and, across its full length, slotted camber-changing flaps.

Ilyushin DB-3 (1937)

TYPE • Bomber COUNTRY • Soviet Union

After demonstrating spectacular long-range capability in propaganda flights in the late 1930s, on 7 August 1941 the DB-3 set an important landmark in Soviet wartime history – it became the first Soviet aircraft to bomb Berlin.

SPECIFICATIONS

DIMENSIONS:	Length: 14.22m (46ft 8in); Wingspan: 21.44m (70ft 4in); Height: 4.19m (13ft 9in)
WEIGHT:	Maximum take-off 9450kg (20,834lb)
POWERPLANT:	Two 709kW (951hp) Nazarov 9-cylinder air-cooled radial engines
MAX SPEED:	439km/h (273mph)
RANGE:	3800km (2400 miles)
CEILING:	9600m (31,500ft)
CREW:	3
ARMAMENT:	One 7.62mm (0.3in) ShKAS machine gun flexibly mounted in nose, one 7.62mm (0.3in) ShKAS machine gun in dorsal turret, one 7.62mm (0.3in) ShKAS flexibly mounted in ventral hatch; up to 2500kg (5511lb) bombload

Above: The DB-3TP floatplane was converted from a standard DB-3 with the addition of a pair of British-built floats. Tested at Sevastopol during 1938, the aircraft was found unsatisfactory and no production ensued.

CREW
The crew of three consisted of the pilot, navigator/front gunner with a 7.62mm (0.3in) ShKAS and the radio operator/rear gunner who was required to divide his time between the dorsal turret and a ventral gun position.

AIRFRAME
The aircraft featured new construction techniques that allowed for a light structure but was difficult to build and production was slow – only 45 aircraft being built by two factories during 1937.

BLACK SEA FLEET AIRCRAFT
An M-86 engined DB-3T of 2 MTAP serving with the Black Sea Fleet in the spring of 1941. Black Sea Fleet DB-3Ts were the first Soviet aircraft to bomb Berlin in August 1941.

MODERN FEATURES
Developed from the all-wood TsKB-26, the first Soviet twin engine to be looped, the DB-3 was an all-metal long-range bomber that, although it could carry five times the bombload of Tupolev's SB-2, was only 24km/h (15mph) slower. It also boasted modern features like self-sealing fuel tanks and a crew intercom system, both included as a result of Soviet aircraft experience over Spain.

GERMAN INVASION
After seeing combat over China and Finland, the DB-3 was heavily engaged in fighting the German invaders from June 1941. Initially used by day, losses of the lightly defended DB-3 were considerable. In a single raid on 26 June 1941, for example, 43 DB-3s were lost. By 1942 the units operating the aircraft had wisely switched to nocturnal operations.

Douglas TBD Devastator (1937)

TYPE • Dive-bomber **COUNTRY** • United States

SPECIFICATIONS

DIMENSIONS:	Length: 10.67m (35ft); Wingspan: 15.24m (50ft); Height: 4.6m (15ft 1in)
WEIGHT:	Maximum take-off 4624kg (10,194lb)
POWERPLANT:	One 670kW (900hp) Pratt & Whitney R-1830-64 Twin Wasp 14-cylinder air-cooled radial piston engine
MAX SPEED:	332km/h (206mph)
RANGE:	1152km (715 miles)
CEILING:	5900m (19,500ft)
CREW:	3
ARMAMENT:	One 7.62mm (0.30in) Browning M1919 MG fixed forward-firing in cowling, one 7.62mm (0.30in) Browning M1919 MG flexibly mounted in rear cockpit; up to 453kg (1000lb) bombload or one 907kg (2000lb) Mk XIII torpedo

Painfully illustrating the pace of aviation development in the late 1930s, the Douglas TBD Devastator was an advanced carrier aircraft when it entered service in 1937, but once committed to action in early 1942 it proved to be outdated and vulnerable.

Right: A U.S. Navy Douglas TBD-1 Devastator of Torpedo Squadron Six (VT-6) pictured in flight. Note the squadron insignia, a Great White Albatross, on the fuselage beneath the cockpit. Established as VT-8S in 1937, the squadron was redesignated VT-6 that same year.

BOMBSIGHT
The aircraft was fitted with a Norden bombsight for level bombing, the bombardier utilizing this from a prone position under the pilot and aiming through a window just behind the engine cowling.

UNDERCARRIAGE
The undercarriage was semi-retractable – when raised the wheels protruded about 25cm (10in) below the lower surface of the wing – and was intended to minimize damage in the event of a wheels-up landing.

MARKINGS
The same aircraft as shown in the photograph above in typically colourful prewar USN markings, this TBD-1 served with VT-6 and wears the squadron's white albatross insignia under the windscreen.

COMBAT LOSSES
During the initial months of the Pacific War, the TBD proved combat effective, damaging the carrier *Shokaku* and sinking the carrier *Shoho* in concert with SBDs at the Battle of the Coral Sea. However, the June 1942 Battle of Midway exposed the shocking vulnerability of the TBD and the appalling unreliability of the Mk XIII torpedo. Of the 41 aircraft committed to action, only six came back.

TORPEDO RUN
A TBD flies low and flat to deliver a torpedo run. The TBD was reliable, easy to fly, and thanks to its generous wing area and large flaps possessed a landing speed of merely 95km/h (59mph), rendering it simple to land on the deck of a carrier.

Seversky P-35 (1937)

TYPE • Fighter COUNTRY • United States

The P-35 – the US Army's first cantilever monoplane with retractable undercarriage – was quickly surpassed as a fighter design. Those that saw combat proved too slow and unwieldy to survive against newer Japanese fighters.

SPECIFICATIONS (P-35A)

Dimensions:	Length: 8.18m (26ft 10in); Wingspan: 10.97m (36ft); Height: 2.97m (9ft 9in)
Weight:	Maximum take-off 3050kg (6724lb)
Powerplant:	One 780kW (1050hp) Pratt & Whitney R-1830-45 Twin Wasp air-cooled radial piston engine
Max speed:	470km/h (292mph)
Range:	1530km (950 miles)
Ceiling:	9600m (31,500ft)
Crew:	1
Armament:	Two 7.62mm (0.3in) M1919 Browning MGs in upper cowling and two 12.7mm (0.50in) Browning M2/AN MGs in wings; up to 160kg (350lb) of bombs

Above: Designed by Alexander Seversky, the P-35 was the first truly modern American fighter, although it was underarmed and lacked armour protection. This example from the 27th Pursuit Squadron wore temporary camouflage for 1940 war games.

Powerplant
The standard powerplant on a P-35A was a single 780kW (1050hp) Pratt & Whitney R-1830-45 Twin Wasp air-cooled radial piston engine.

Vulnerabilities
In combat, the P-35 not only had performance limitations but it was also highly vulnerable to enemy fire, as it was not fitted with armour or self-sealing fuel tanks.

Seversky P-35A
This P-35A was flown by Lt Boyd Wagner, the commanding officer of the 17th Pursuit Squadron, stationed at Nichols Field, Luzon, Philippines in 1941.

COMBAT FAILURE
Thrown into action in December 1941 to oppose the Japanese invasion, the P-35A's most spectacular success came when 1st Lt Samuel H. Marrett managed to destroy the minesweeper *W-10* with repeated strafing attacks. Some sources claim the P-35A managed to down one Japanese aircraft, while others say it did not down any aircraft. Either way, the P-35A had proved a dismal failure in air combat, being too slow and lacking the manoeuvrability necessary to engage Japanese fighters.

US AIR FORCE P-35
A P-35 on a training flight, clearly displaying its twin nose machine guns. The sole surviving Philippines-based P-35A made the type's last combat mission on 3 May 1942, strafing Japanese positions. The P-35As retained in the USA were subsequently transferred to Ecuador.

Dornier Do 24 (1937)

TYPE • *Seaplane* COUNTRY • *Germany*

SPECIFICATIONS (Do 24T-1)

DIMENSIONS:	Length: 22.05m (72ft 4in); Wingspan: 27.00m (88ft 7in); Height: 5.75m (18ft 10in)
WEIGHT:	18,400kg (40,565lb) maximum take-off
POWERPLANT:	3 x 746kW (1000hp) BMW-Bramo 323R-2 air-cooled radial engines
MAX SPEED:	331km/h (206mph)
RANGE:	4700km (2920 miles)
CEILING:	7500m (24,606ft)
CREW:	6
ARMAMENT:	1 x 7.92mm (0.31in) MG 15 machine gun in bow and stern turrets; 1 x 20mm (0.79in) Hispano-Suiza 404 cannon in the dorsal turret

The Dornier 24 originated with a 1935 Dutch specification for an aircraft to patrol the East Indies. Production continued in the Netherlands and France after the German occupation. In Luftwaffe service, its main function was as an air-sea rescue craft for recovering downed aircrew.

Above: This Dutch-built Do 24K-2 was completed for the Luftwaffe for air-sea rescue duties.

NOSE GUN
The nose and tail turrets were each equipped with a 7.92mm (0.31in) MG 15 machine gun. For mooring operations, a crew member was positioned on the deck between the nose turret and the cockpit.

ENGINES
Production Do 24s were initially equipped with Wright Cyclone engines until supplies of these dried up in the autumn of 1941 and the BMW-Bramo 323R replaced them.

DORSAL TURRET
Aft of the wing trailing edge the dorsal turret carried a 20mm (0.79in) Hispano-Suiza cannon for greater hitting power.

DUTCH RAID
The Do 24 would ironically have its biggest impact as a maritime patrol aircraft while operated by the Allies. On 17 December 1941, in response to Japanese landings on Borneo, three Dornier Do 24K-2s of the Dutch MLD mounted an attack on the invasion fleet. The destroyer *Shinonome* was hit by two bombs while a third exploded in the sea abeam the rear 12.5cm (5in) turret. This triggered a massive magazine explosion. Within minutes, the Shinonome had disappeared beneath the waves.

DO 24 SEA TAKE-OFF
With the *Seenotsdienst*, the Do 24 would carry out rescue missions from the Arctic to the Mediterranean, recovering downed airmen from the waters off the Scilly Isles at the extreme south-west of Britain, and on another occasion, 560km (348 miles) out into the Atlantic Ocean. Able to operate in waters up to a sea state 6, they were ideally suited to the role.

Henschel Hs 126 (1937)

TYPE • Reconnaissance **COUNTRY** • Germany

SPECIFICATIONS (HS 126B-1)

DIMENSIONS: Length: 10.85m (35ft 7in); Wingspan: 14.5m (47ft 7in); Height: 3.75m (12ft 4in)
WEIGHT: Maximum take-off 3270kg (7209lb)
POWERPLANT: 1 x 634kW (850hp) BMW-Bramo Fafnir 323A-1 air-cooled radial engine
MAX SPEED: 310km/h (193mph)
RANGE: 580km (360 miles)
CEILING: 8000m (26,247ft)
CREW: 2
ARMAMENT: 1 x 7.92mm (0.31in) MG 17 in upper forward fuselage; 1 x 7.92mm (0.31in) MG 15 in the rear cockpit; bombload of 150kg (330lb)

An evolution of the Hs 122, the Hs 126 was designed to fulfil the Army Co-operation role and despite its early success fading as the Luftwaffe faced serious aerial opposition, it would continue to find employment until the end of the war.

Left: One of the ten Hs 126A-0 pre-production aircraft, D-ODBT is seen here pre-war; the spats on the main gear were frequently removed in service.

ENGINE
The Hs 126 replaced the Siemens Sh 22B radial used on the Hs 122 with the BMW-Bramo Fafnir 323A-1, improving top speed and giving sufficient power for the Hs 126 to be used as a glider tug.

FUSELAGE
Despite its relatively dated appearance, the Hs 126's fuselage was of all-metal monocoque construction with the outer skin carrying structural loads that allowed the stringers and frames to be lighter.

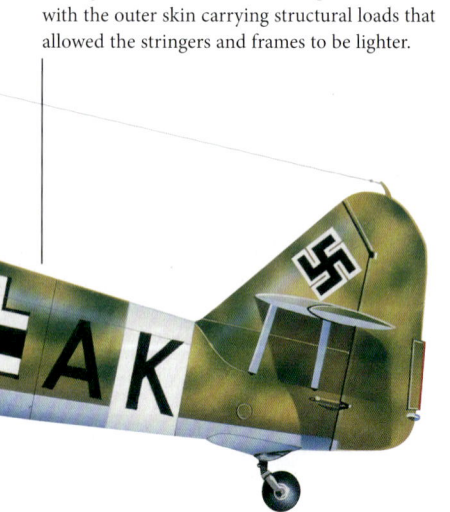

GLIDER TUGS
Hs 126s would see service as glider tugs. not least during Operation Eiche, the mission to free Italian dictator Benito Mussolini, who had been deposed in July 1943. He was held captive in a hotel 2000m (6562ft) above sea level in the Gran Sasso mountains east of Rome. On 12 September, 10 Hs 126s launched from Pratica di Mare air base each towed a DFS 230 carrying nine soldiers in addition to the pilot. The Henschels deployed the rescue force and Mussolini was liberated, at least temporarily.

ANGLED MARKINGS
The angled markings on the fuselage were to assist the observer in judging the correct moment to release bombs in level flight.

HS 126 COLUMN OVERFLIGHT
Around 810 Hs 126s were produced between 1938 and 1941. As well as with Spain, Greece and Germany, they also saw service with the Estonian Air Force, five of whose aircraft were captured and used by the Soviet Union before being destroyed in German bombing raids.

Blohm & Voss BV 138 (1938)

TYPE • *Maritime patrol* **COUNTRY** • *Germany*

A twin-boom, three-engined flying boat, the BV 138 did not look conventional but it would eventually prove to be a rugged patrol aircraft that would play an important role shadowing Allied convoys.

SPECIFICATIONS

DIMENSIONS:	Length: 19.9m (65ft 4in); Wingspan: 27.0m (88ft 7in); Height: 5.9m (19ft 4in)
WEIGHT:	Maximum take-off 17,650kg (38,912lb)
POWERPLANT:	3 x 656kW (880hp) Jumo 205D 6-cylinder opposed piston two-stroke diesel
MAX SPEED:	285km/h (177mph)
RANGE:	5000km (3107 miles)
CEILING:	5000m (16,404ft)
CREW:	6
ARMAMENT:	2 x 20mm (0.79in) MG 151 cannon; 1 x 13mm (0.51in) MG 131; 1 x 7.92mm (0.31in) MG 15; plus a bombload of 6 x 50kg (110lb) bombs or 4 x 150kg (330lb) depth charges

This page: A BV 138 MS. Although the majority of BV 138s served on conventional coastal patrol duties, a small number were converted for mine-hunting tasks. This aircraft served with 6. Staffel/Minensuchgruppe 1 at Grossenbrode in 1945.

GUN ARMAMENT
The BV 138A was fitted with an LB 204 bow turret housing an MG 204 cannon, but both proved troublesome and were replaced in the BV 138B by a redesigned turret with a single MG 151. The BV 138 MS aircraft had all gun armament deleted.

OFFENSIVE ARMAMENT
Offensive stores were carried under the wing roots. The BV 138C could carry up to six 50kg (110lb) bombs or four 150kg (330lb) depth charges.

DEGAUSSING LOOP
The BV 138 MS was fitted with a circular degaussing loop made of dural. An auxiliary motor generated power for producing a strong magnetic field in the loop, sufficient to explode mines as the aircraft passed overhead.

'FLYING CLOG'
Designed to fulfil a requirement for a medium-range flying boat capable of operating in the open ocean, Blohm & Voss's first flying boat featured distinctive looks that earned it the nickname of Der Fliegende Holzschuh or the 'flying clog'. Work for what would become the BV 138 started in 1934, the initial design being a two-engine gull-wing monoplane trials, a long way from the final three-engine BV 138A.

BLOHM & VOSS BV 138B-1
A BV 138B-1, unusually seen over the countryside rather than the open sea. From this angle, the open gun position in the rear of the central engine nacelle is clearly visible.

Vought SB2U Vindicator (1938)

TYPE • Dive-bomber **COUNTRY** • United States

SPECIFICATIONS

DIMENSIONS:	Length: 10.36m (34ft); Wingspan: 12.77m (41ft 11in); Height: 4.34m (14ft 3in)
WEIGHT:	Maximum take-off 4273kg (9421lb)
POWERPLANT:	One 615kW (825hp) Pratt & Whitney R-1535-02 Twin Wasp Junior 14-cylinder air-cooled radial piston engine
MAX SPEED:	391km/h (243mph)
RANGE:	1800km (1120 miles)
CEILING:	7200m (23,600ft)
CREW:	2
ARMAMENT:	One 12.7mm (0.5in) M2 Browning machine gun in starboard wing, one 12.7mm (0.5in) M2 Browning machine gun in rear cockpit; up to 453kg (1000lb) bombload under fuselage

The first monoplane dive-bomber in US Navy service, the SB2U Vindicator was still in use with Marine Corps units in 1943 and saw combat at the Battle of Midway as well as in French hands during the Battle for France.

Above: Serving aboard USS Ranger *with VS-41 in August 1941, immediately before the aircraft's withdrawal from frontline operations, this SB2U-1 was engaged in scouting missions over the Atlantic.*

ENGINE
The Vindicator was powered by a single 615kW (825hp) Pratt & Whitney R-1535-02 Twin Wasp Junior 14-cylinder air-cooled radial piston engine.

CANOPY
The long greenhouse-style canopy was one of the key distinguishing features of the Vindicator. The canopy housed the aircraft's two-man crew, which consisted of a pilot and a gunner seated in tandem.

FRENCH AIRCRAFT
This aircraft was assigned to Escadrille 10 of AB3 at Cuers in July 1940. Attrition of the French V-156s was high, the aircraft proving vulnerable to enemy fighters and ground fire. At the armistice, only eight aircraft survived of 40 originally ordered by the Aeronavale.

FRENCH SERVICE
The Vindicator also attracted the attention of the French *Aeronavale*, who would be the first to take the aircraft into action. The French operated the Vought V-156-F, an export model equivalent to the US SB2U-2 but featuring several items of French equipment and instrumentation. Despite training on the carrier *Bearn*, the V-156s would operate solely from land bases when the war broke out.

CARRIER LAUNCH
A Vought SB2U-1 Vindicator takes off from USS *Saratoga* (CV-3), 8 February 1938. Featuring an increased fuel capacity for greater range (compared to previous variants), the SB2U-3 would be the only variant in US service to see combat during the war.

Short Sunderland Mk III (1938)

TYPE • *Seaplane* COUNTRY • *United Kingdom*

Just as the Short S.23 C-class 'Empire' flying boat marked a startling advance on all previous civil transport aircraft in Imperial Airways service, so its military derivative, the Sunderland, marked an equally great advance on marine aircraft in the RAF.

SPECIFICATIONS (Sunderland Mk III)

DIMENSIONS:	Length: 26.01m (84ft 4in); Wingspan: 34.38m (112ft 9.5in); Height: 10.02m (32ft 10.5in)
WEIGHT:	Empty 15,649kg (34,500lb)
POWERPLANT:	4 × Bristol Pegasus XVIII 9-cylinder air-cooled radial piston engines, 794kW (1065hp) each
MAX SPEED:	340km/h (210mph) at 2000m (6500ft)
RANGE:	2860km (1800 miles)
CEILING:	5200m (17,200ft)
CREW:	9–11
ARMAMENT:	Up to 12 × 7.7mm (0.303in) Browning machine guns; up to 2000 lb (910kg) of bombs, mines and depth charges internally

Above: Sometimes nicknamed 'The Pig' by its crews, on account of its ungainly appearance, the Sunderland was also dubbed 'The Flying Porcupine' by Luftwaffe pilots who tried to attack it.

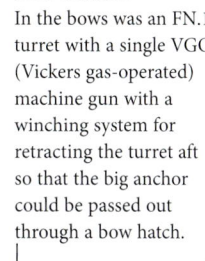
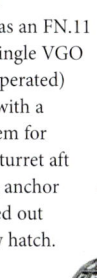

BOW TURRET
In the bows was an FN.11 turret with a single VGO (Vickers gas-operated) machine gun with a winching system for retracting the turret aft so that the big anchor could be passed out through a bow hatch.

FUEL LOAD
Fuel was housed in six vertical drum tanks between the spars with a capacity of 9206 litres (2025 Imp gals), later increased to 11,602 litres (2552 Imp gals) by four further cells aft of the rear spar.

HULL
Despite its great bulk, the hull was well shaped. Hydro-dynamically, a new feature was the bringing of the planing bottom to a vertical knife-edge at the rear (second) step, thereafter sweeping the bottom line smoothly up and back to the tail.

SPACIOUS CABIN
In the original Sunderland Mk I the normal crew was seven, accommodated basically on two decks with comprehensive provision for prolonged habitation, with six bunks, galley, workshops and stowage for a considerable quantity of equipment. At the upper level it was possible to walk aft from the two-pilot flight deck past the cubicles of the radio operator and navigator and through the deep front spar into the domain of the flight engineer.

PRODUCTION AIRCRAFT
The second production Sunderland Mk I, L2160 is seen here during RAF trials in June 1938. Production aircraft differed from the prototype in having the slightly swept wing and outwardly canted engines, both measures intended to restore the aircraft's centre-of-gravity after a powered rear turret was specified.

Handley Page Hampden (1938)

TYPE • Bomber COUNTRY • United Kingdom

SPECIFICATIONS	
DIMENSIONS:	Length: 16.33m (53ft 7in); Wingspan: 21.08m (69ft 2in); Height: 4.55m (14ft 11in)
WEIGHT:	Maximum take-off 10,206kg (22,500lb)
POWERPLANT:	Two 750kW (1000hp) Bristol Pegasus XVIII 9-cylinder air-cooled radial piston engines
MAX SPEED:	398km/h (247mph)
RANGE:	2770km (1720 miles)
CEILING:	5800m (19,000ft)
CREW:	4
ARMAMENT:	One fixed forward-firing 7.7mm (0.303in) Browning machine gun in nose; up to five 7.7mm (0.303in) Vickers K machine guns in flexible mounts; up to 1800kg (4000lb) bombload or mines or one 457mm (18in) torpedo

Designed along with the Vickers Wellington to Air Ministry specification B.9/32, Handley Page Hampden was the result of a unique but flawed approach to pre-war bomber design, its decent performance compromised by terrible ergonomics for the crew.

Above: Formerly operated by 455 Squadron RAAF (Royal Australian Air Force) and wearing standard RAF night-bomber camouflage with painted-out roundels and squadron codes, Hampden TB.1 'White 30' was based at Vaenga near Murmansk in 1942.

CREW COMPARTMENT
Crew fatigue was common due to the cramped accommodation and the aircraft's defensive firepower was totally inadequate, leading to the decision in October 1939 to operate the Hampden under the cover of darkness.

LAYOUT
The Hampden was highly original in its approach to the Ministry specification for a twin-engined bomber, with tapered wings and, most distinctively, the rear half of the fuselage tapering into a slim boom carrying the twin-tail empennage.

SLIM TAIL
The innovative design of the Hampden cut down significantly on the weight of the aircraft, and enabled the crew to be brought together in a compact group in the deeper front fuselage.

MARITIME PATROLS
After the mauling of Convoy PQ-17 in July 1942, urgent means were sought to protect future convoys venturing through the Arctic Sea. To this end Operation *Orator* saw RAF Coastal Command fly their Hampden TB.1 torpedo bombers to Vaenga, near Murmansk, to supply anti-shipping cover to the next convoy PQ-18, on its approach to the USSR, deterring German surface raiders.

HANDLING
Although the Hampden offered excellent handling qualities and almost fighter-like manoevrability, its shortcomings were exposed early on, when five out of 11 Hampdens from No. 144 Sqn were lost to German fighters on a reconnaissance mission over the Heligoland Bight area on 29 September 1939.

Brewster F2A Buffalo (1938)

TYPE • *Fighter* COUNTRY • *United States*

SPECIFICATIONS

DIMENSIONS:	Length: 7.92m (26ft); Wingspan: 10.67m (35ft); Height: 3.56m (11ft 8in)
WEIGHT:	Maximum take;off 3247kg (7159lb)
POWERPLANT:	One 700kW (940hp) Wright R-1820-34 Cyclone nine-cylinder air-cooled radial engine
MAX SPEED:	501km/h (311mph)
RANGE:	2486km (1545 miles)
CEILING:	10,100m (33,200ft)
CREW:	1
ARMAMENT:	Two 12.7mm (0.5in) M2 Browning machine guns fixed forward-firing in wings, one 12.7mm (0.5in) M2 Browning machine gun and one 7.7mm (0.3in) .30 AN/M2 Browning fixed forward-firing in cowling

The US Navy's first monoplane fighter delivered a significant upgrade in performance over the biplanes that preceded it. Unfortunately, in operational contexts the F2A failed to live up to its initial promise.

Above: VMF-221 took the F2A into action with American forces for the first and only time on 4 June 1942 at Midway. Losses were catastrophic, only three of the 20 F2As that took off to intercept Japanese bombers were serviceable by 6 June.

ARMAMENT
The F2A-2 had heavy armament of four 12.7mm (0.5in) Browning M2 machine guns, mounted in the wings. This was strong firepower, but the substantial guns contributed to the aircraft's weight issues.

DESIGN
The Buffalo possessed a fully flush-riveted, stressed-skin construction, split flaps and a hydraulically powered retractable undercarriage.

EXCESS WEIGHT
The Buffalo suffered from excessive weight, which often proved too much for the Buffalo's landing gear. Undercarriage failures were commonplace, resulting in the withdrawal of the F2A from US carriers. F2As were still aboard USS *Saratoga* at the time of Pearl Harbor. However, all had been transferred to shore-based roles within a month. The Buffalo also saw British and Belgian service (it was the British who bestowed the 'Buffalo' nickname.

CARRIER COMBAT
The F2A's only major engagement in US service occurred during the Battle of Midway, when poor tactics saw Marine Corps unit VMF-221 lose 13 out of 20 F2As committed. Subsequently, all F2As were relegated to advanced fighter training, a role for which they were well suited.

Vickers Wellington (1938)

TYPE • *Bomber* **COUNTRY** • *United Kingdom*

SPECIFICATIONS

DIMENSIONS:	Length: 18.54m (60ft 10in); Wingspan 26.26m (86ft 2in); Height: 5.33m (17ft 6in)
WEIGHT:	12,927kg (28,500lb) loaded
POWERPLANT:	2 x 783kW (1050hp) Bristol Pegasus XVIII 9 cylinder radial engines
MAX SPEED:	378km/h (235mph)
RANGE:	2905km (805 miles)
CEILING:	5486m (18,000ft)
CREW:	6
ARMAMENT:	Two 7.7mm (.303in) Browning MGs in both nose and tail; two 7.7mm (0.303in) beam guns; 2041kg (4500lb) bombload

An immensely strong warplane capable of surviving crippling battle damage, the Vickers Wellington was the Royal Air Force's most advanced bomber at the outbreak of World War II and was at the forefront of the British bomber effort for the first half of the war.

Above: The Vickers Wellington Mk I was built to Air Ministry Specification B.9/32, which called for an aircraft capable of delivering a bombload of 54kg (1000lb) and with a range of 1159km (720 miles).

CREW POSITIONS
The Wellington Mk IC bomber had a crew of six: a pilot, radio operator, navigator/bomb aimer, observer/nose gunner, tail gunner and waist gunner.

VICKERS WELLINGTON MK IC
This aircraft flew with No. 99 Squadron, No. 3 Group, RAF Bomber Command, Newmarket, Cambridgeshire, in 1940. Features of the Wellington Mk IC were two beam guns in place of one ventral gun, and larger main wheels that extended below the engine nacelle when retracted.

BOMB BAY
The Mk IC had a strengthened bomb bay beam, which allowed the aircraft to carry a 1814kg (4000lb) High Capacity bomb, known as a 'Blockbuster' or 'Cookie'.

MARITIME VARIANT
Although as early as late 1941 Coastal Command had employed modified Wellington Mk ICs for torpedo and mine-laying work, it was during the spring of 1942 that the first true general reconnaissance Wellingtons entered service with RAF Coastal Command. The first land-based maritime patrol and anti-submarine aircraft were GR.Mk VIIIs, powered by Pegasus VIII radial engines and equipped with ASV Mk II radar.

TRANSPORTER
A number of early bomber Wellingtons, including this Mk IA (N2887), underwent conversion as Mk XV interim transports for service with RAF Transport Command. Note that the defensive guns have been removed.

Mitsubishi A6M Reisen 'Zeke' (Zero) (1938)

TYPE • *Fighter* COUNTRY • *Japan*

Popularly known as the 'Zero', the Mitsubishi A6M was the world's most capable carrier-based fighter at the time of its appearance, out-performing all land-based contemporaries. Latterly outclassed, it remained in service until the end of the war.

SPECIFICATIONS

DIMENSIONS:	Length: 9.12m (29ft 11in); Wingspan: 11m (36ft 1in); Height: 3.51m (11ft 6in)
WEIGHT:	2733kg (6025lb) maximum take-off
POWERPLANT:	Nakajima NK2F Sakae 21 radial piston engine, 820kW (1100hp)
MAX SPEED:	565km/h (351mph)
RANGE:	1143km (710 miles)
CEILING:	11,740m (38,517ft)
CREW:	1
ARMAMENT:	2 × 20mm (0.79in) cannon and 2 × 13.2mm (0.52in) machine guns

Above: The Zero aircraft was notably agile, but was slower in the roll, and its acceleration in a dive was inferior to Allied counterparts.

ARMAMENT
In addition to the two 20mm (0.79in) cannon that were carried in the wings, a pair of Type 97 machine guns were mounted in the decking of the upper fuselage, forward of the cockpit.

COCKPIT
The pilot of the A6M2 was not provided with armour plating, a bulletproof windshield or even a jettisonable hood. The multi-panel canopy made use of simple flat plates of glass.

AIRFRAME
In order to achieve the eight-hour endurance for Pacific theatre operations, the Zero employed a very light construction. The result was an airframe that was vulnerable to hits by even rifle-calibre weapons. On occasions, structural failure could result from a high-speed dive.

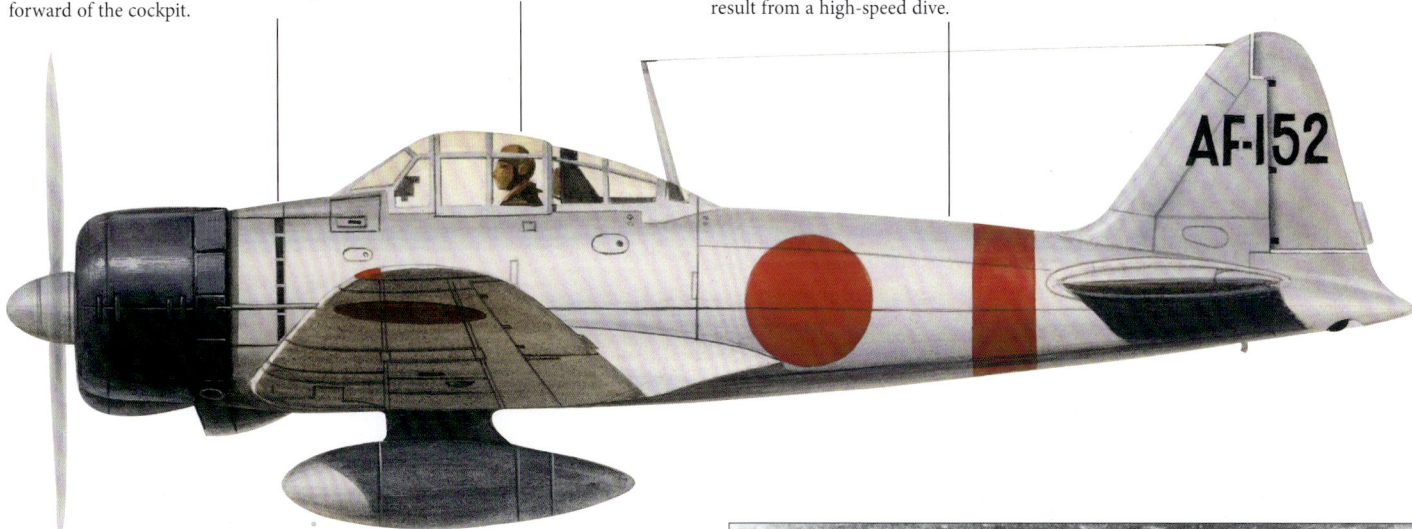

CHINA TESTING
The Mitsubishi A6M *Reisen* (Zero) fighter first flew on 1 April 1939; after 15 aircraft had been evaluated in combat in China, the type was accepted for service with the Japanese Naval Air Force in July 1940, entering full production in November as the A6M2 Model 11. Sixty-four Model 11s were completed, and were followed by the Model 21 with folding wingtips.

CARRIER ATTACK
A6M2s turn their engines on the *Hiryu* prior to the attack on Pearl Harbor. The A6M2 soon showed itself to be clearly superior to any fighter the Allies could put into the air in the early stages of the Pacific war. In later years, however, the aircraft and their pilots were totally outclassed. In 1942, the Americans had allocated the code name 'Zeke' to the A6M, but as time went by the name 'Zero' came into general use.

Curtiss SBC Helldiver (1938)

TYPE • *Dive-bomber* **COUNTRY** • *United States*

SPECIFICATIONS

DIMENSIONS:	Length: 8.58m (28ft 2in); Wingspan: 10.36m (34ft); Height: 3.18m (10ft 5in)
WEIGHT:	Maximum take-off 33462kg (7632lb)
POWERPLANT:	One 630kW (850hp) Wright R-1820-34 Cyclone 9-cylinder air-cooled radial piston engine
MAX SPEED:	377km/h (234mph)
RANGE:	652km (405 miles)
CEILING:	7300m (24,000ft)
CREW:	2
ARMAMENT:	2 x 7.62mm (0.3in) M1919 Browning MG, one in nose and one in rear cockpit; up to 454kg (1000lb) bombload

The last frontline combat biplane in US service, the SBC Helldiver was the last in a distinguished line of Curtiss carrier biplanes. Surprisingly, the design originally flew as a monoplane fighter, which then reverted to a biplane.

Above: This SBC-3 wears the post-May 1942 white star inside a blue roundel, in reserve squadron service. All upper surfaces are blue-grey and under surfaces were light grey.

PAINT SCHEME
This SBC-4 (BuAer No. 1287) served as a command liaison and transport aircraft with the 1st Marine Aviation Wing, San Diego, 1941. In line with the order of 30 December 1940 applying to all shipboard aircraft, the plane is painted in a light grey.

ARMAMENT
Combat variants of the Helldiver featured a single 7.62mm (0.3in) M1919 Browning machine gun flexibly mounted in the rear cockpit, as well as one forward-firing machine gun in the nose.

WINGS
Adding a second wing to the monoplane design increased the aircraft's strength, although at the expense of the wing-folding ability.

CURTISS MODEL 73
The Helldiver started life as the Curtiss Model 73, a two-seat parasol monoplane carrier fighter with a folding wing. Rejected for the fighter role, the aircraft was proposed as a scout bomber, but the parasol wing failed during dive-bombing tests, although the pilot was able to land the damaged machine. In response Curtiss proposed reworking the aircraft as a biplane.

SBC-4 FLIGHT TESTING
A Curtiss SBC-4 Helldiver assigned to US Marine Corps observation squadron VMO-151 undergoes flight testing. The USN ordered 83 production examples of the Helldiver, which began to appear in 1937. A further 124 of the improved SBC-4 were ordered in January 1938.

Vought OS2U Kingfisher (1938)

TYPE • Seaplane COUNTRY • United States

The US Navy's standard shipboard floatplane for the majority of the war, the unassuming OS2U Kingfisher proved reliable and versatile and served for considerably longer than had been planned at the outset.

SPECIFICATIONS (OS2U-3)

DIMENSIONS:	Length: 10.24m (33ft 7in); Wingspan: 10.94m (35ft 11in); Height: 4.47m (14ft 8in)
WEIGHT:	Maximum take-off 2722kg (6000lb)
POWERPLANT:	One 340kW (450hp) Pratt & Whitney R-985-AN2 Wasp Junior 9-cylinder, air-cooled radial piston engine
MAX SPEED:	275km/h (171mph)
RANGE:	1461km (908 miles)
CEILING:	5500m (18,200ft)
CREW:	2
ARMAMENT:	2 x 7.62mm (0.3in) Browning M1919 MG, one in forward fuselage and one in rear cockpit; up to 295kg (650lb) bombload

Left: A Vought OS2U Kingfisher floatplane, of Observation Squadron One (VO-1), being towed on a mat by USS Arizona (BB-39) during recovery after a flight in the Hawaiian Islands on 6 September 1941.

AUSTRALIAN KINGFISHERS
Australia received several Kingfishers originally intended for use by the Netherlands East Indies, putting them into RAAF service during late 1942.

POWERPLANT
The standard engine was the Pratt & Whitney R-985-AN2 Wasp Junior 9-cylinder, air-cooled radial.

FLOATS
The Kingfisher featured an interchangeable float or wheeled undercarriage and operated from both catapult-equipped ships and shore bases throughout its career.

US NAVY AIRCRAFT

The Kingfisher, as it became known in 1941, quickly became the standard aircraft operating in the gunnery-spotting observation and air-sea rescue role from the US Navy's capital ships. Battleships normally carried three aircraft, although the powerful new *Iowa* class battleships featured four OS2Us apiece when they were commissioned. The OS2U also exclusively equipped the Inshore Patrol Squadrons engaged in anti-submarine warfare in coastal waters.

OS2U OFF HAWAII
In the same aircraft as the photograph above, Radioman 2nd Class E.L. Higley prepares to go out on the plane's wing to hook up the aircraft to the battleship's crane for recovery.

Blackburn Skua (1938)

TYPE • *Torpedo bomber* COUNTRY • *United Kingdom*

The first monoplane to serve with the Fleet Air Arm, the Skua was intended to fulfil the roles both of dive-bomber and fighter. Although withdrawn from frontline use during 1941, the Skua was the first aircraft in history to sink an enemy capital ship during wartime.

SPECIFICATIONS

DIMENSIONS:	Length: 10.85m (35ft 7in); Wingspan: 14.07m (46ft 2in); Height: 3.81m (12ft 6in)
WEIGHT:	Maximum take-off 3732kg (8228lb)
POWERPLANT:	One 660kW (890hp) Bristol Perseus XII 9-cylinder, air-cooled, radial piston engine
MAX SPEED:	362km/h (225mph)
RANGE:	1220km (760 miles)
CEILING:	6200m (20,200ft)
CREW:	2
ARMAMENT:	Four 7.7mm (0.303in) Browning machine guns in wings, one 7.7mm (0.303in) Vickers K or Lewis machine gun flexibly mounted in rear cockpit; up to 227kg (500lb) bombload

Right: Despite the Skua's generally underwhelming performance as a fighter, the first Royal Navy 'ace' of the war was Skua pilot William Lucy who scored all five of his confirmed victories in the aircraft.

LENGTHENING
Stability problems in the prototypes resulted in the aircraft's nose being lengthened by 73cm (2ft 5in) and the horizontal tail surfaces were also extended.

BLACKBURN SKUA MK II
The Skua Mk II was the production variant of the Mk I prototype. Although designed as a carrierborne dive-bomber, the Skua doubled as a naval fighter, but was ineffective in both roles and withdrawn from front-line service in 1941.

WINGTIPS
Blackburn fitted the Skua with distinctive upturned wingtips to improve stability. When the aircraft was subsequently developed into the Roc turret fighter, these were discarded.

DOUBLE FIRST

The Skua's finest moment came in April 1940, when it became simultaneously the first aircraft to sink a major enemy warship in wartime and the first aircraft to sink a ship by dive-bombing. A total of 16 Skuas flew from RAF Hatston in the Orkney Islands and were central to dive-bombing and sinking the cruiser *Königsberg* at anchor in Bergen harbour. Further dive-bombing attacks were made as part of the Norwegian campaign and heavy losses were incurred during a raid on the battleship *Scharnhorst* in Trondheim.

IN FORMATION
Blackburn Skua Mk IIs of 803 Squadron in formation over the southern coast of England, 1939. The Skua had been designed with potential second-line duties in mind and subsequently served as a target tug, as did the Roc turret fighter, with both the Navy and the RAF. It served in far more units in second-line roles than it ever had operationally.

Curtiss P-36 & Hawk Model 75 (1938)

TYPE • Fighter COUNTRY • United States

SPECIFICATIONS (P-36A)	
Dimensions:	Length: 8.69m (28ft 6in); Wingspan: 11.38m (37ft 4in); Height: 2.57m (8ft 5in)
Weight:	Maximum take-off 2726kg (6010lb)
Powerplant:	One 780kW (1050hp) Pratt & Whitney R-1830-17 Twin Wasp 14-cylinder air-cooled radial piston engine
Max Speed:	504km/h (313mph)
Range:	1006km (625 miles)
Ceiling:	10,000m (32,800ft)
Crew:	1
Armament:	One 7.62mm (0.3in) M1919 MG and one 12.7mm (0.5in) M2 Browning MG in upper front fuselage decking; optional hardpoint under each wing capable of carrying up to 69kg (152lb) each

Although it scored the first USAAF air-to-air victories of World War II, the P-36 saw very little frontline use in US service. However, the aircraft proved hugely successful in French and Finnish hands, fighting both for and against the Allies.

Above: This P-36A was part of the 79th Pursuit Squadron, 20th Pursuit Group, based at Moffett Airfield, California, November 1939. The tail designator indicates it is the 21st aircraft.

Specification
The aircraft was designed to compete for an Army Air Corps specification issued in 1934 for an all-metal low-wing monoplane design with a top speed of at least 482km/h (300mph).

Performance
In flight, the P-36 had a creditable performance, with a tight turning circle and a good rate of climb from the powerful engine married with a low weight.

Curtiss P-36A Hawk
This example (38-92) is from the 47th Pursuit Squadron, 15th Pursuit Group, based at Haleiwa Field, Hawaii, in December 1941, from where Lt Harry Winston Brown flew against the Pearl Harbor attackers. Wearing 'pajama tops, tuxedo trousers, house-shoes, flight helmet and goggles', he claimed two victories, for which he gained the Silver Star. He ended the war with seven kills.

INTRODUCTION
The USAAC placed a contract for 210 P-36As in July 1937, the biggest single order the Air Corps had placed since World War I. Deliveries of the P-36A began in April 1938, with the 20th Pursuit Group being the first to receive the aircraft. Most P-36As were completed with the higher compression R-1830-17 engine, resulting in a maximum speed increase to 504km/h (313mph).

HAWK 75
French Air Force Curtiss Hawk 75A fighters patrol French airspace, during 1939–40. The Hawk 75 was a simplified export version with fixed landing gear and a Wright Cyclone engine.

World War II

World War II changed both the reality and the perception of air warfare for ever. At the beginning of the conflict, air power was something of an adjunct to the more senior services, the Army and the Navy. But the new monoplane aircraft types, with their enclosed cockpits, retractable undercarriage, purpose-designed armament and powerful engines, offered performance and combat capability that transformed the very conduct of warfare itself. Massed fleets of such aircraft were now battle-winning, indeed potentially war-winning, force multipliers, whether they were achieving air superiority over deep battlefields or raising enemy cities through devastating strategic bombing. The war also saw the first jet aircraft enter service, heralding a new era.

PAIR OF SPITFIRES
The Supermarine Spitire is one of those legendary types that have come to define both the aerial and the cultural struggle of World War II, an aircraft that embodied British national resistance as much as it did technical performance.

Ilyushin Il-2 Shturmovik (1939)

TYPE • Bomber COUNTRY • Soviet Union

SPECIFICATIONS

DIMENSIONS:	Length: 11.6m (38ft 1in); Wingspan: 14.6m (47ft 11in); Height: 4.2m (13ft 9in)
WEIGHT:	6380kg (14,065lb) maximum take-off
POWERPLANT:	Mikulin AM-38F V-12 liquid-cooled piston engine, 1285kW (1720hp)
MAX SPEED:	414km/h (257mph)
RANGE:	720km (450 miles)
CEILING:	5500m (18,045ft)
CREW:	2
ARMAMENT:	2 × 23mm (0.9in) forward-firing cannon, 2 × 7.62mm (0.3in) forward-firing machine guns, 1 × 7.62mm machine gun in rear cockpit, plus up to 600kg (1320lb) of disposable stores underwing

Built in greater numbers than any other military aircraft in history, the Il-2 was a war-winner for the Soviets, with more than 36,000 built between 1941 and 1955. The ground-attacker was capable of defeating the best-protected German tanks.

Above: 'Red 1' served with Soviet Occupation Forces, probably 2 GvShAP, in Germany in 1950. The overall aluminium finish was unusual, even in peacetime.

COCKPIT ARMOUR
The pilot was seated in an armoured tub with a thickness of 5–12mm (0.2–0.5in); this extended to protect the engine. The armour was capable of defeating all small-arms fire.

MARKINGS
During 1945 the Il-2s of 56 GvShAP participated in making a film called *Victory*. For the film, the inscription 'Stalingrad–Berlin' was painted over the existing fuselage markings.

ENGINE
The Il-2M3 was powered by a Mikulin AM-38F liquid-cooled V-12, developing 1285kW (1720hp). Introduced in July 1942, it was capable of running on motor fuel.

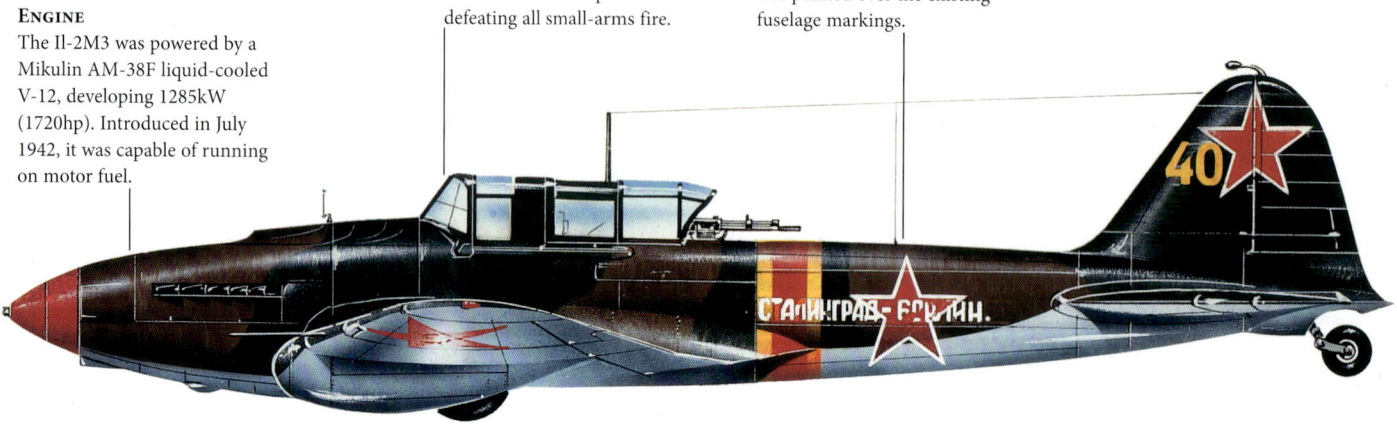

FOREIGN SERVICE
The Il-2 was quite possibly the most important aircraft to serve the VVS – over 36,000 were built and more than 10,000 were lost in combat. In addition to the USSE, IL-2s served with Czech, Polish and Yugoslavian forces during the war, with Mongolia, Hungary and Bulgaria flying them after 1945. Post-war, the aircraft received the NATO reporting name 'Bark'.

ROCKET AIRCRAFT
From mid-1941, Il-2s were equipped for the use of the RS-132 rocket on the production lines, although RS-82 weapons are being carried by aircraft '8' in this winter 1941/42 photograph.

North American B-25 Mitchell (1939)

TYPE • Bomber COUNTRY • United States

SPECIFICATIONS

DIMENSIONS:	Length: 16.12m (52ft 11in); Wingspan: 20.6m (67ft 7in); Height: 4.82m (15ft 10in)
WEIGHT:	15,880kg (35,000lb) maximum take-off
POWERPLANT:	2 × Wright R-2600-13 radial piston engines, 1268kW (1700hp)
MAX SPEED:	457km/h (284mph)
RANGE:	2414km (1500 miles)
CEILING:	6462m (21,200ft)
CREW:	5
ARMAMENT:	6 × 12.7mm (0.5in) machine guns plus a maximum bombload of 1361kg (3000lb)

The USAAF's definitive light/medium bomber of World War II, the B-25 may have served in more campaigns than any other type in that conflict. The aircraft also gave good service in US Navy and Allied hands, and flew the 'Doolittle raid' against Japan in 1942.

Above: Based at Sfax, Tunisia, this B-25C aircraft was one of those equipping the 487th Bombardment Sqn, 340th Bombardment Group (Medium), Ninth Air Force, in August 1943.

COLOUR SCHEME
The first B-25s to arrive in North Africa wore standard USAAF olive drab over neutral grey colours. These soon faded in the harsh desert conditions and were replaced by the 'desert pink'/ azure blue scheme depicted on this aircraft.

UPRATED ENGINES
Although the B-25C was fitted with uprated R-2600-17 engines, rated at 1270kW (1700hp), the addition of new combat equipment, especially the dorsal turret, made the B-25C 61km/h (38mph) slower than the first B-25s.

IMPROVEMENTS
The B-25C incorporated changes to the Mitchell found necessary following combat experience with the earlier models. Extra armour, defensive armament and self-sealing fuel tanks were among the features added.

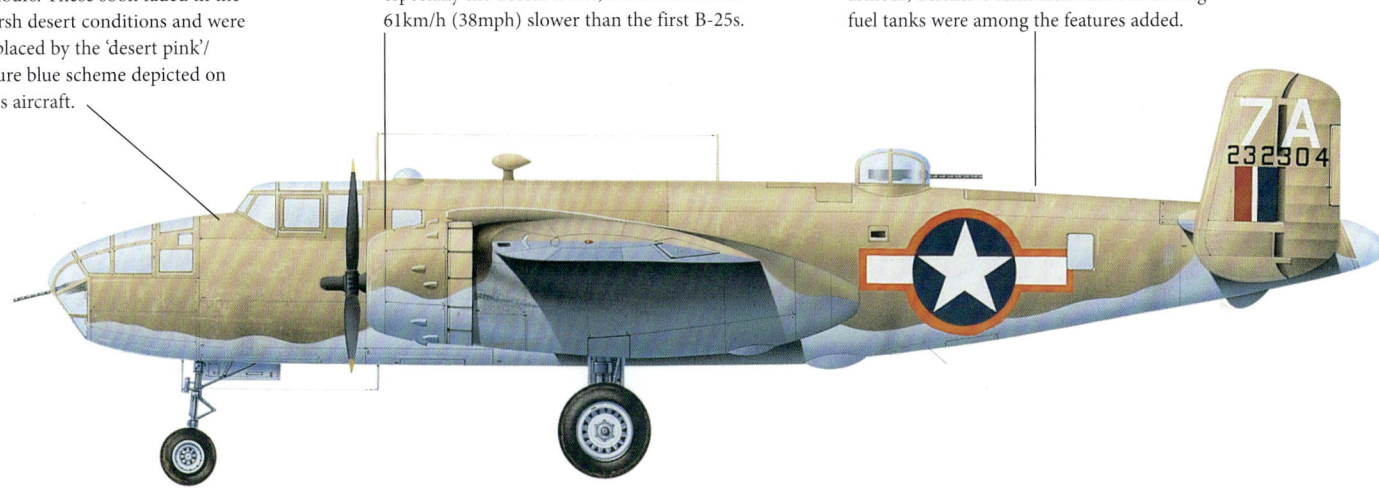

FIREPOWER
The B-25 was the principal American medium bomber of the war. With the addition of extra guns, armour and bomb capacity – but only modest improvements in engine power – later models were slower than their predecessors. Gun armament, however, grew to be the heaviest of any US bomber, and the Mitchell excelled in the anti-shipping and low-level attack roles.

STRAFER VARIANT
The B-25J was the major production Mitchell variant and was built with either a so-called 'glass nose' or an eight-gun, solid 'strafer nose' (seen here). Its other main distinctive feature was the relocation of the top turret to a position behind the cockpit. For strafing attacks, the solid gun nose was supplemented by 'package' guns on the fuselage sides.

Dewoitine D. 520 (1939)

TYPE • Fighter COUNTRY • France

SPECIFICATIONS

DIMENSIONS:	Length: 8.76m (28ft 8.75in); Wingspan: 10.20m (33ft 5.5in); Height: 2.57m (8ft 5.25in)
WEIGHT:	2783kg (6134lb) loaded
POWERPLANT:	686kW (920hp) Hispano-Suiza 12Y-45 liquid-cooled V-12
MAX SPEED:	535km/h (332mph)
RANGE:	900km (553 miles)
CEILING:	11,000m (36,090ft)
CREW:	1
ARMAMENT:	One 20mm (0.78in) Hispano-Suiza HS-404 cannon and four 7.5mm (0.295in) MAC1934 MGs

Its early development marred by official indifference, the D.520 was scarcely a match for the Bf 109, yet its pilots fought with great bravery and skill to bring the type respectability during the Battle of France in 1940.

This page: This aircraft wears the black panther badge of the 4e Escadrille, Groupe de Chasse II/7, while the unit was based at Gabes, Tunisia in 1942, shortly before the Torch landings.

MARKINGS
Vichy Air Force aircraft were adorned with bold yellow/red stripe markings as a means of identification.

ARMAMENT
By comparison with other French fighters, the D.520 was well armed with one Hispano-Suiza HS404 20mm (0.79in) cannon with 60 rounds, and four wing-mounted 7.5mm (0.295in) MAC 1934 M39 machine guns, each with 675 rounds.

COCKPIT
The cockpit was set well aft, giving excellent downward vision in flight. However, taxiing could be tricky, requiring much weaving to retain a semblance of forward visibility.

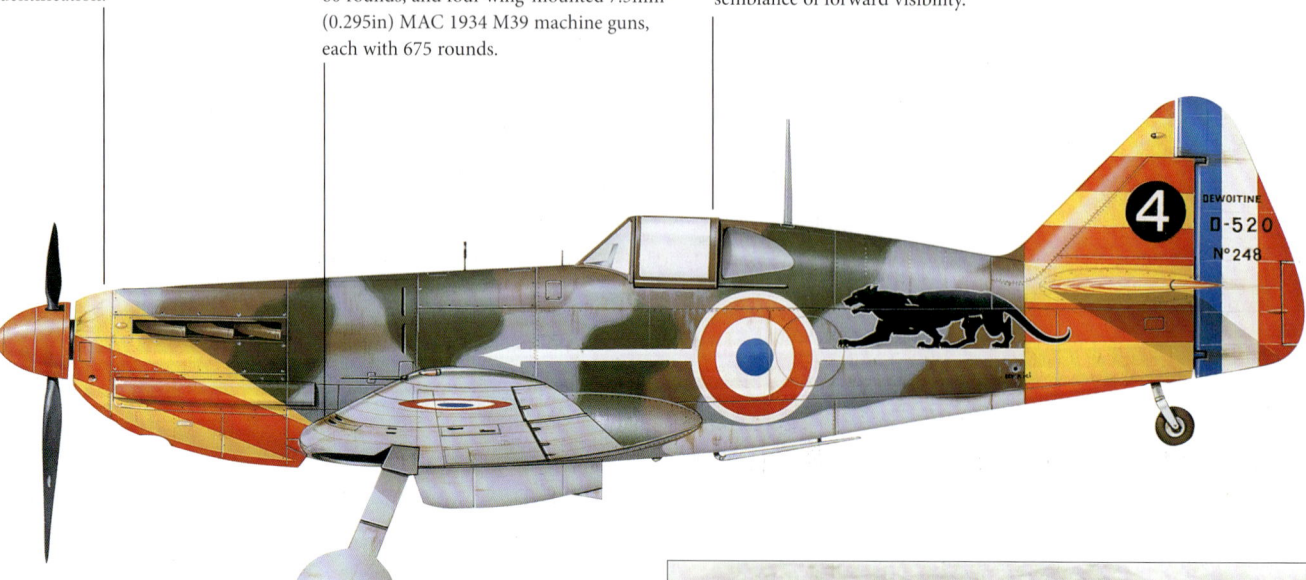

AXIS ALLEGIANCE
After the fall of France, and the establishment of the collobarationist Vichy regime, large numbers of D.520s were sent to North Africa, some being dispatched to participate in the 1941 Syrian campaign against British and Free French forces. Flying alongside Luftwaffe fighters, the D.520s were involved in the fight against the Allies during the Operation *Torch* landings of November 1942.

NORTH AFRICA
A D.520 on anti-Allied operations over North Africa. After the Allied victory, however, many D.520 pilots elected to rejoin the Allied cause, including the Armée de l'Air's top World War II ace, Pierre Le Gloan, who had scored kills against the Germans and Italians in France, and against the British in Syria. He was killed in a crash in a P-39 on 13 August 1943.

Arado Ar 196 (1939)

TYPE • *Seaplane* COUNTRY • *Germany*

SPECIFICATIONS

DIMENSIONS:	Length: 11m (36ft); Wingspan: 12.40m (50ft 9.5in); Height: 4.45m (14ft 7in)
WEIGHT:	Empty 2572kg (5670lb); maximum take-off weight 3730kg (8223lb)
POWERPLANT:	One 970hp (723kW) BMW 132K nine-cylinder single-row radial engine
MAX SPEED:	320km/h (199mph)
RANGE:	1070km (665 miles)
CEILING:	7000m (22,960ft)
CREW:	2
ARMAMENT:	Two 20mm fixed forward-firing cannon in wing, one 7.92mm fixed forward-firing machine gun in starboard side of forward fuselage, and one 7.92mm trainable rearward-firing MG in rear of cockpit, plus external bombload of 100kg (220lb)

Although it exerted only a minor influence on World War II, the Arado Ar 196 was nevertheless an important type. Evolving through a number of float configurations, the aircraft soon entered service, replacing the venerable He 60.

Left: The Arado Ar 196 had a broad-chord wing with an unswept leading edge and a slight taper on the trailing edge.

CONTROL SURFACES
The aircraft had wide-span ailerons outboard, with relatively small flaps inboard. All control surfaces were fabric-covered, the rest of the wing being a metal-skinned two-spar structure.

DEFENSIVE GUNS
The MG 81Z installation, mounted in the rear cockpit of the Ar 196A-5, paired two MG 81 7.9mm (0.31in) machine guns on a single mount, with a maximum combined load of 2000 rounds.

FLOAT ARRANGEMENT
Four Ar 196 prototypes were carefully evaluated in 1937–38, but there was indecision over the preferred float arrangement. The central float was considered preferable in operations from choppy water, but the stabilizing floats could easily dip into the sea during take-off, resulting in pronounced asymmetric drag. It was decided to standardize on the twin-float arrangement.

CATAPULT LAUNCH
The first and second prototypes of the Ar 196 had conventional twin floats, while the third, fourth and fifth prototypes (for the B-series) had a single main float on the centreline, with stabilizing floats under the wings. The unarmed second prototype is shown here undergoing a catapult launch.

Dornier Do 17 (1939)

TYPE • Bomber COUNTRY • Germany

SPECIFICATIONS (Do 17P)

DIMENSIONS:	Length: 116.25m (53ft 3in); Wingspan: 18m (59ft); Height: 4.32m (14ft 2in)
WEIGHT:	7040kg (15,520lb) maximum take-off
POWERPLANT:	Two 652kW (875hp) BMW 132N radial engines
MAX SPEED:	425km/h (263mph)
RANGE:	1160km (721 miles)
CEILING:	8150m (26,740ft)
CREW:	4
ARMAMENT:	Two 7.92mm (0.31in) MG 15 machine guns; provision for 1000kg (2205lb) bombload

With poor defensive armament, the Do 17/215 proved vulnerable to enemy fighters in the bombing role. The aircraft did, however, achieve more success in the high-altitude reconnaissance and glider-towing roles.

This page: The Dornier Do 17P was essentially the Do 17M but adapted for aerial reconnaissance, powered by two 652kW (875hp) BMW 132N engines and with a variety of different camera options.

CAMERAS
The Do 17P had either Rb 20/30 and Rb 50/30 or Rb 20/18 and Rb 50/18 cameras, mounted in the heavily glazed nose of the aircraft.

FUSELAGE
The Do 17 was instantly recognizable from its twin-stabilizer tailplane and its very slender fuselage, which earned it the nickname 'flying pencil'.

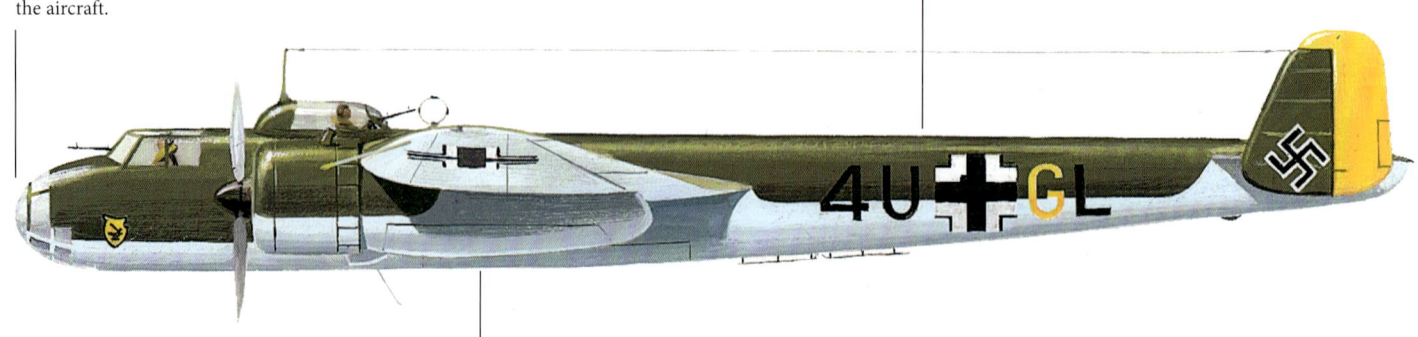

MAILPLANE DESIGN
The Dornier Do 17 emerged from an original Deutsche Lufthansa requirement for a high-speed six-passenger mailplane drafted in 1933, but the company saw the military potential of the aircraft. As introduced to series production, the Do 17E-1 version was intended for use as a medium bomber with provision for a 500kg (1102lb) bombload (increased to 750kg/1653lb for short-range missions), while the Do 17E-2 was a reconnaissance aircraft. Both types made their combat debut in the Spanish Civil War.

BOMBLOAD
The Do 17P still had offensive capability, despite its photo-reconnaissance purpose, with provision for a 1000kg (2205lb) bombload.

VARIANTS
The Do 17Z-2 first appeared in 1939 and these two examples were among the last in frontline Luftwaffe service, fighting on the Eastern Front in 1942 with 15.(Kroat)/KG 53. The Do 17 went through numerous variants, including trainers, long-range reconnaissance, glider tugs and night-fighters.

Aichi D3A (1939)

TYPE • *Dive-bomber* COUNTRY • *Japan*

SPECIFICATIONS

DIMENSIONS:	Length: 10.2m (33ft 5in); Wingspan: 14.37m (47ft 2in); Height 3.8m (12ft 8in)
WEIGHT:	Maximum 4122kg (9100lb)
POWERPLANT:	One 969kW (1300hp) Mitsubishi Kinsei 54 radial piston engine
MAX SPEED:	430km/h (267mph)
RANGE:	1352km (840 miles)
CEILING:	10,500m (34,450ft)
CREW:	2
ARMAMENT:	Two 7.7mm (.303in) Type 97 machine guns and one 7.7mm (.303in) Type 92 machine gun; up to 250kg (550lb) of bombs

One of the outstanding Japanese warplanes of World War II, the D3A 'Val' played a crucial role in the attack on Pearl Harbor and actually sank more tonnage of Allied shipping than any other Axis type of aircraft.

This page: Pictured here is an Aichi D3A1 of the Soryu Air Group, 1941. The Soryu *formed part of the 1st Air Fleet and was one of the carriers in the task force that attacked Pearl Harbor.*

POWERPLANT
The original Nakajima Hikari 1 radial engine was replaced in production aircraft by the 745kW (1000hp) Kinsei 43 or 798kW (1070hp) Kinsei 44.

DORSAL FIN
Directional instability problems were eradicated by the addition of a large dorsal fin. This made the aircraft highly manoeuvrable.

INDIAN OCEAN
Although it looked ungainly with its fixed spatted undercarriage, the D3A was, in its heyday, a highly effective aircraft capable of carrying a single 250kg (551lb) bomb under the fuselage and two 60kg (132lb) bombs beneath the wings. Between 4 and 9 April 1942, D3A1s in the Indian Ocean sank not only the British cruisers HMS *Cornwall, Dorsetshire* and the aircraft-carrier *Hermes*, but also two destroyers, a corvette, an auxiliary vessel, two oilers and 11 merchant vessels.

FIXED UNDERCARRIAGE
Despite the inherent drag of a fixed undercarriage, this was retained as retractable landing gear would have added to the weight.

PEARL HARBOR ATTACK
A D3A1 with the 33rd Kokutai. The fixed-gear Type 99 dive-bomber carried a fairly small load not very fast, but in the opening campaigns of the Pacific War, particularly at Pearl Harbor, it was used to devastating effect. Of the 135 D3A1s (known as 'Vals' to the Allies) used on 7 December, 15 were shot down, but they caused great destruction at Wheeler Field air base as well as to the US battleships.

Nakajima Ki.43 Hayabusa (1939)

TYPE • *Fighter* COUNTRY • *Japan*

SPECIFICATIONS

DIMENSIONS:	Length: 8.92m (29ft 3.25in); Height: 3.27m (10ft 8.33in)
WEIGHT:	Empty 1910kg (4211lb); maximum take-off 2925kg (6450lb)
POWERPLANT:	858kW (1150hp) Nakajima Ha.115 14-cylinder radial
MAX SPEED:	530km/h (329mph) at 5000m (16,405ft)
RANGE:	3200km (1990 miles)
CEILING:	11,200m (36,750ft)
CREW:	1
ARMAMENT:	Two 12.7mm (0.50in) machine guns plus external bombload of 500kg (1102lb)

Like its naval counterpart, the Mitsubishi Zero, the Nakajima Ki.43 *Hayabusa* (Allied code name Oscar) was in action from the first day of Japan's war until the last, by which time it was woefully outclassed by the latest Allied fighters.

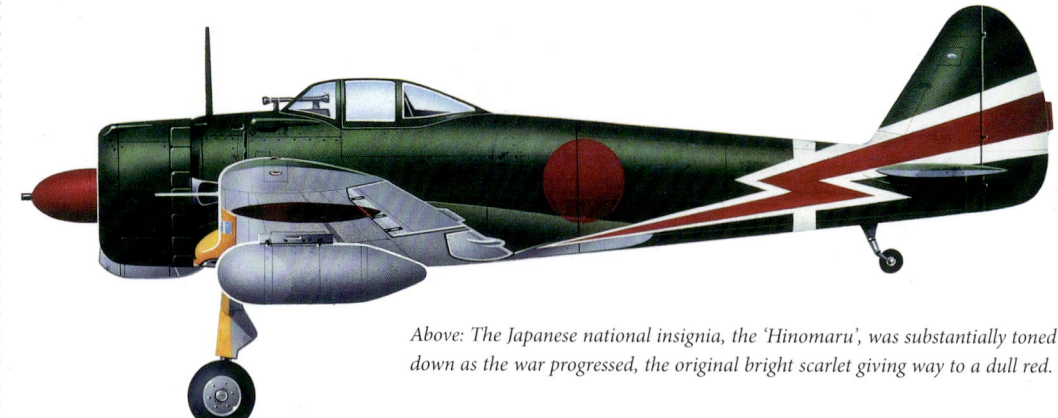

Above: The Japanese national insignia, the 'Hinomaru', was substantially toned down as the war progressed, the original bright scarlet giving way to a dull red.

PROPELLER
The initial production variant, designated Army Type 1 Fighter Model 1A, was fitted with a fixed-pitch, two-blade wooden propeller, but this was soon replaced by a two-pitch, two-blade metal unit.

BODY
An excellent and versatile fighter, the *Hayabusa*'s main drawback was its lack of adequate armament and any form of armour protection, which made it extremely vulnerable.

TAIL MARKINGS
This aircraft was flown by the 64th Sentai during initial Japanese attempts to cut off China from Allied forces in India and Burma. The blue tail marking indicates an aircraft of the unit's headquarters (Chutai).

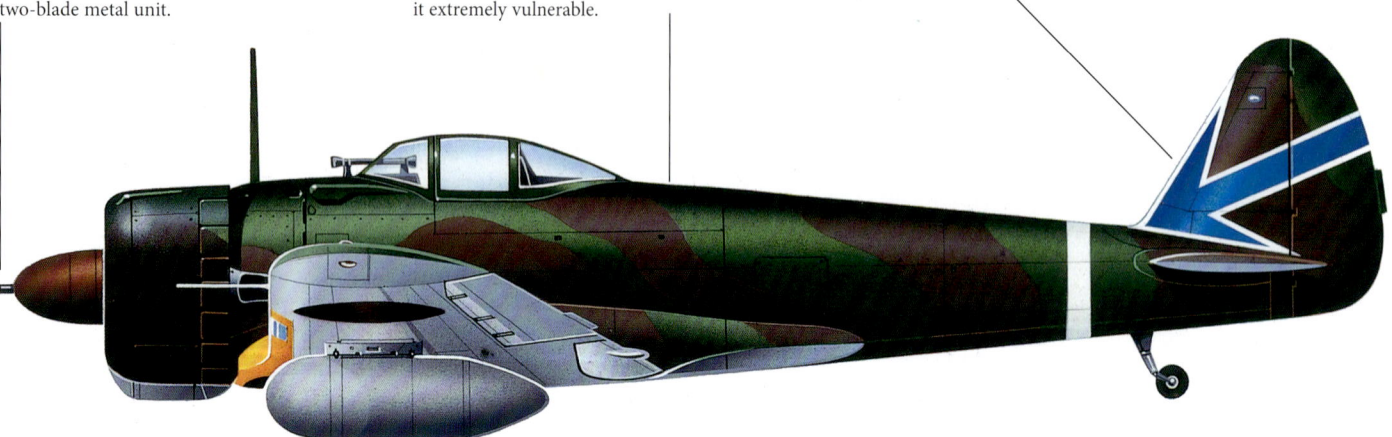

VARIANTS
The prototype Ki.43 flew in January 1939 and 716 early production models were produced: the Ki.43-I, K.43-Ia, Ki.43-Ib and Ki.43-Ic, the last two having a better armament. They were followed in 1942 by a much improved model, the Ki.43-II; this appeared in three subvariants, the Ki.43-IIa and -IIb, and the Ki.43-Kai, which adopted all the refinements incorporated in the earlier models. The final model was the Ki.43-III, the only variant to include cannon in its armament.

CHINA THEATRE
Shining under the Chinese sun, this Ki-43-II-Otsu of the 2nd Chutai (red diagonal tail stripe), 25th Sentai, proves that the application of green mottle over natural metal was less than effective.

Polikarpov I-153 (1939)

TYPE • Fighter COUNTRY • Soviet Union

SPECIFICATIONS

DIMENSIONS: Length: 6.17m (20ft 3in); Wingspan 10.00m (32ft 9.5in); Height 2.80m (9ft 2.25in)

WEIGHT: Empty 1348kg (2972lb); maximum take-off 2110kg (4652lb)

POWERPLANT: 746kW (1000hp) Shvetsov M-62 9-cylinder radial

MAX SPEED: 444km/h (276mph) at 3000m (9845ft)

RANGE: 880km (547 miles)

CEILING: 10,700m (35,105ft)

CREW: 1

ARMAMENT: 4 x 7.62mm (0.30in) machine guns, plus a light bombload or six air-to-ground rockets

In the 1930s the Soviet designer Nikolai N. Polikarpov was at the forefront of Soviet fighter design, and at the outbreak of the war in the east it was aircraft of his design that were the mainstay of the Soviet Air Force's fighter squadrons.

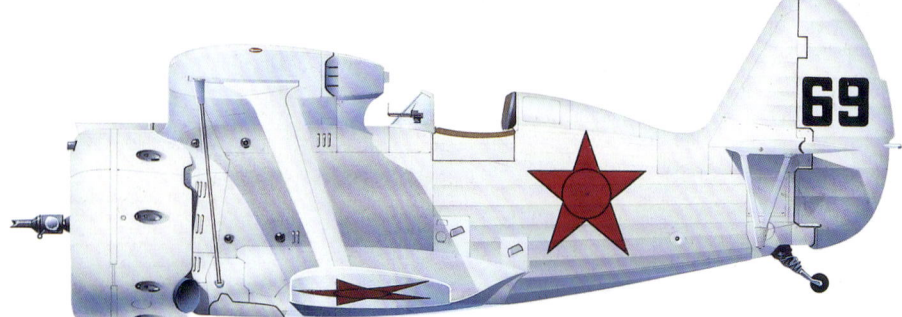

Above: 'Black 69' wears the standard silver and light grey livery in which most pre-war I-153s left the factory. This aircraft was flying with an unknown unit when it was captured following a forced landing during the summer of 1941.

WINGS
The I-153's progenitor, the I-15, had featured a gull-type upper wing, whereas the next variant, the I-15bis (I-152), was fitted with a straight wing. The I-153 reverted to the gull-wing arrangement; its manoeuvrability surpassed that of all other contemporary biplanes.

ARMAMENT
The I-153 was armed with four synchronized machine guns, firing along canals lying between the engine cylinders. Small numbers were later fitted with two 20mm (0.79in) cannon.

THE SEAGULL
The I-153, dubbed *Chaika* (Seagull) because of its distinctive wing shape, was a first-rate combat aircraft and was subsequently to prove its worth in air fighting, being able to out-turn almost every aircraft that opposed it in action. The I-153 saw its first action in the 1939 Sino–Soviet incident, and was heavily involved in the 'Winter War' between Russia and Finland in 1939–40. The type was quickly withdrawn.

UNDERCARRIAGE
The I-153's retractable undercarriage was a novel feature in a biplane. Both legs retracted rearward into the underside of the fuselage, the wheels being turned 90 degrees during the process until they lay flat. As with many other Soviet aircraft, there was provision to fit skis instead of wheels.

RESTORATION
The Alpine Fighter Collection's I-153 restoration project has been aided by the availability of Shvetsov ASh-62IR radial engines (Polish-built as the ASz-62), the engine fitted to the Antonov An-2 utility transport. The ASh-62 is itself a modernized derivative of the I-153's original M-62.

Curtiss P-40 Kittyhawk (1939)

TYPE • Fighter COUNTRY • United States

The Curtiss P-40 Kittyhawk performed excellent service with RAF, RAAF and South African Air Force squadrons of the Desert Air Force in North Africa, and contributed greatly to the Allies' eventual victory in that theatre.

SPECIFICATIONS (P-40D Kittyhawk)

DIMENSIONS:	Length: 10.16m (33ft 4in); Wingspan: 11.38m (37ft 4in); Height: 3.76m (12ft 4in)
WEIGHT:	Empty 2722kg (6000lb); maximum take-off 5171kg (11,400lb)
POWERPLANT:	1014kW (1360hp) Allison V-1710-81 V-12 engine
MAX SPEED:	609km/h (378mph) at 3200m (10,500ft)
RANGE:	386km (240 miles)
CEILING:	11,580m (38,000ft)
CREW:	1
ARMAMENT:	6 x 12.7mm (0.50in) machine guns, plus a bombload of up to three 227kg (500lb) bombs

Above: This P-40E Warhawk from the 11th Squadron, 343rd Fighter Group, was stationed in the Aleutian Islands in 1942.

ARMAMENT
The P-40's armament of six 12.7mm (0.50in) machine guns was standard on most American-built fighters of World War II. Each gun was provided with 235 rounds.

COCKPIT
Visibility from the cockpit was poor, but adequate for the aircraft's primary role of ground attack. In aerial combat, the P-40 was no match for aircraft like the Messerschmitt Bf 109F.

AIR INTAKES
The inlet duct above the Allison engine allowed air into the carburettor, which was situated at the rear of the engine.

RAF TOMAHAWK
The Curtiss XP-40 prototype flew for the first time in October 1938. The US Army Air Corps ordered 524 aircraft, later reduced to 200, and in 1940 the RAF received 140, originally destined for France. Although considered unsuitable for operational use by Fighter Command, the P-40s were fitted with four wing-mounted 7.7mm (0.303in) machine guns and were allocated to Army Co-operation Command as the Tomahawk I, for the tactical reconnaissance role.

RAAF KITTYHAWKS
Considerable numbers of Kittyhawks supplied to the RAF under Lend-Lease were transferred to the RNZAF and RAAF (seen here) for service in the Pacific theatre. The Kittyhawk was a member of the tactical fighter family known to the US Army Air Forces as the P-40. The type operated almost exclusively in the low-level fighter-bomber role, with weapons such as a 227kg (500lb) bomb under the fuselage.

Lockheed Hudson Mk.1 (1939)

TYPE • *Reconnaissance* **COUNTRY** • *United States*

SPECIFICATIONS

DIMENSIONS:	Length: 13.50m (44ft 3.33in); Wingspan: 19.96m (65ft 6in); Height: 3.32m (10ft 10.5in)
WEIGHT:	Empty 5817kg (12,825lb); maximum take-off 8845kg (19,500lb)
POWERPLANT:	Two 820kW (1100hp) Wright GR-1820-G102A Cyclone radial engines
MAX SPEED:	357km/h (222mph) at sea level
RANGE:	3154km (1960 miles)
CEILING:	6400m (21,000ft)
CREW:	6
ARMAMENT:	7 x 7.7mm (0.303in) machine guns plus internal bombload of 612kg (1350lb)

Developed in response to a British requirement for a maritime patrol aircraft, the Lockheed Hudson reached the squadrons of RAF Coastal Command in time to combat the developing U-boat threat against the UK's merchant fleet.

This page: This Hudson Mk 1 served with No. 206 Squadron, RAF Coastal Command, at RAF St Eval, Cornwall, from 1940 to 1942.

GLAZED NOSE
The glazed nose was occupied by the navigator, who had a seat and a table for his charts. Under his seat was a flat pane window for bomb-aiming.

ENGINES
Hudsons were powered by either Wright Cyclone or Pratt & Whitney Twin Wasp radial engines. The scoop above the engine is for the carburettor, while the intake under the engine is to cool the oil.

DORSAL TURRET
The main defence against enemy fighters was provided by the dorsal turret, a Boulton Paul 'C' type mounting a pair of Browning 7.62mm (0.30in) machine guns.

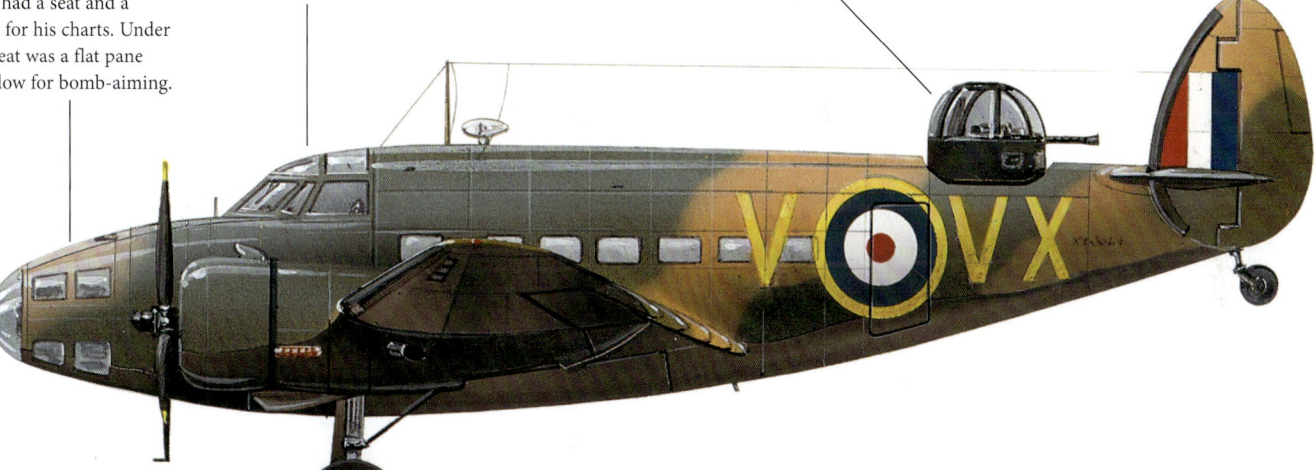

CLANDESTINE OPS
The Lockheed Hudson was primarily a maritime patrol aircraft, developed from the Lockheed Model 14 twin-engined commercial airliner. Hudsons were also used, however, for clandestine operations, ferrying agents to and from France. No. 161 (Special Duties) Squadron RAF used several Hudsons in this capacity until the end of the war, latterly dropping supplies to agents in Germany itself.

COASTAL COMMAND
RAF No. 269 Sqn, Coastal Command, was an early recipient of the Hudson, replacing its Ansons with Hudson Mk Is from March 1940. The squadron was equipped with the Hudson until July 1945.

Ilyushin Il-4 (1939)

TYPE • Bomber COUNTRY • Soviet Union

Starting life as the DB-3F, the Il-4 represented a complete redesign and after initial engine problems it served effectively for the duration of the Great Patriotic War, Berlin becoming a regular destination for the bombers in the later war years.

SPECIFICATIONS

DIMENSIONS:	Length: 14.76m (48ft 5in); Wingspan: 21.44m (70ft 4in); Height: 4.10m (13ft 6in)
WEIGHT:	Maximum take-off 12,120kg (26,720lb)
POWERPLANT:	Two Tumanksy M-88B 14-cylinder air-cooled radial engines
MAX SPEED:	410km/h (250mph)
RANGE:	3800km (2400 miles)
CEILING:	8700m (28,500ft)
CREW:	4
ARMAMENT:	One 7.62mm (0.3in) ShKAS machine gun in nose, one 12.7mm (0.5in) Berezin UBT machine gun in dorsal turret, one 7.62mm (0.3in) ShKAS machine gun in ventral hatch; up to 2700kg (6000lb) of bombs or mines

Right: Ultimately the time taken to build an IL-4 would drop to 12,500 man hours, fewer than that taken to construct the smaller Pe-2.

CAMOUFLAGE
This Il-4 from an unknown unit, possibly 815 BAP, appeared in a Soviet newsreel of 1944. The basic black-and-green camouflage has been field modified with light brown.

DEFENSIVE ARMAMENT
The aircraft's primary defence against a rear attack was one 12.7mm (0.5in) Berezin UBT machine gun set in a dorsal turret.

NOSE
Compared to the DB-3, the IL-4 had a significantly elongated nose, which provided a more comfortable accommodation for the navigator/bombardier while also reducing drag.

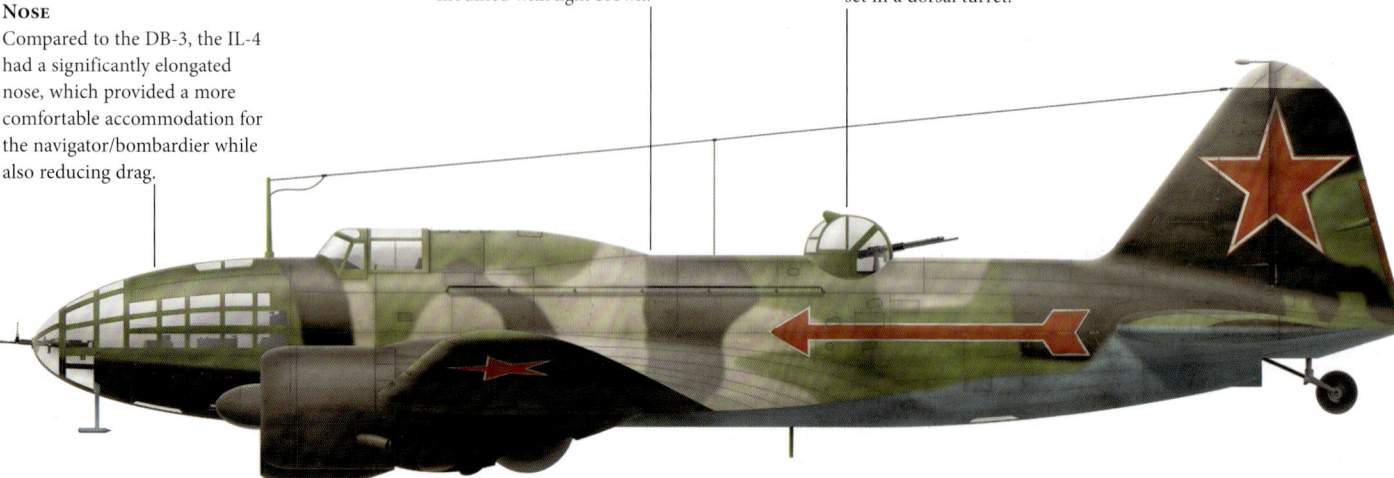

REDESIGN
The DB-3 had given the VVS its first all-metal long-range bomber, but it was difficult to build and had proved unreliable in service. Ilyushin sought to eradicate these problems while improving performance with the DB-3F. It was a totally new design featuring a completely rebuilt structure, greatly influenced by the latest US construction methods. The DB-3 was redesignated IL-4 in spring 1942.

CHANGING TACTICS
Operating in the tactical role for which the IL-4 had not been designed, and required to fly by day – which rendered them vulnerable to fighters – aircraft losses in 1941 were severe. By late 1941 missions were invariably flown at night.

Focke-Wulf Fw 189 Uhu (1939)

TYPE • Bomber COUNTRY • Germany

SPECIFICATIONS

DIMENSIONS: Length: 11.9m (39ft 1in); Wingspan: 18.4m (60ft 4in); Height: 3.1m (10ft 2in)

WEIGHT: Maximum take-off 3950kg (8708lb)

POWERPLANT: Two 342kW (459hp) Argus As 410A-1V-12 inverted-V air-cooled piston engines

MAX SPEED: 344km/h (214mph)

RANGE: 940 km (580 miles)

CEILING: 7000m (23,000ft)

CREW: 3

ARMAMENT: Two flexible 7.92mm (0.31in) MG 15 machine guns (dorsal and rear positions), two fixed 7.92mm (0.31in) MG 17 machine guns (wing roots), plus one Schräge Musik upward-firing 20mm (0.8in) MG FF autocannon

Unconventional in appearance for the time, the Focke-Wulf Fw 189 Uhu (Eagle Owl) would prove to be a tough, manoeuvrable, multi-purpose aircraft that would see the majority of its action on the Eastern Front.

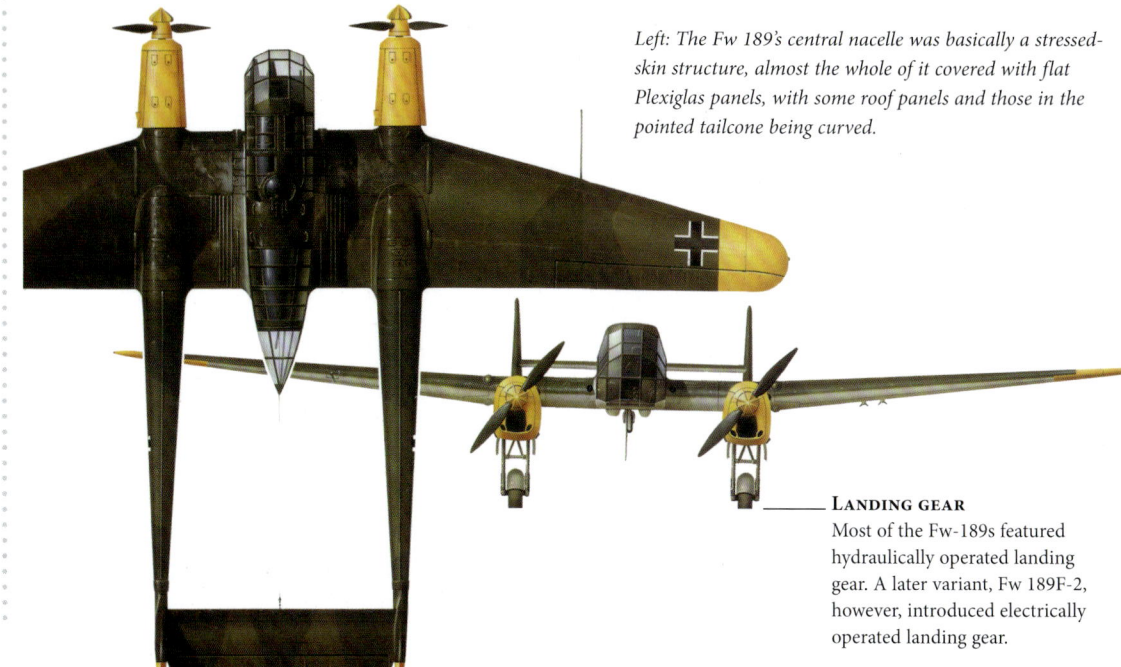

Left: The Fw 189's central nacelle was basically a stressed-skin structure, almost the whole of it covered with flat Plexiglas panels, with some roof panels and those in the pointed tailcone being curved.

LANDING GEAR
Most of the Fw-189s featured hydraulically operated landing gear. A later variant, Fw 189F-2, however, introduced electrically operated landing gear.

FUSELAGE
Everything about the Fw 189 was slender, especially the wings and tail booms. Despite this, it was an immensely strong aircraft and able to absorb large amounts of battle damage.

LAYOUT

Apart from the twin-boom configuration, the Fw 189 was actually quite conventional. The all-metal stressed-skin structure had a smooth flush-riveted exterior. The chosen engine was the Argus As 410A-1, an inverted-V with 12 air-cooled cylinders, which was very smooth and easy to start even in a Russian winter. A single fuel tank of 110-litre (24-Imp gal) capacity was in each tail boom. The aircraft was also produced as a night-fighter variant, with upward-firing 20mm (0.79in) cannon.

TEST FLIGHT

Fw 189 V1 D-OPVN was the first example to fly, taking to the air in July 1938. Kurt Tank – the famous German aeronautical engineer and test pilot – himself took the controls for the first flight. Due to the soundness of the design, the production example differed little from the V-1.

Boulton Paul Defiant (1939)

TYPE • Fighter **COUNTRY** • United Kingdom

SPECIFICATIONS

DIMENSIONS:	Length: 10.77m (35ft 4in); Wingspan: 11.99m (39ft 4in); Height 4.39m (14ft 5in)
WEIGHT:	Empty 2849kg (6282lb); maximum 3900kg (8600lb)
POWERPLANT:	One 954kW (1280hp) Rolls-Royce Merlin XX V-12 piston engine
MAX SPEED:	504km/h (313mph)
RANGE:	748km (465 miles)
CEILING:	10,242m (33,600ft)
CREW:	2
ARMAMENT:	Four 7.7mm (0.303in) Browning machine guns in rear turret

The Defiant was designed as a bomber destroyer with a four-gun powered turret, but no forward-facing armament. It was heavier than a Spitfire or Hurricane but had much the same engine. Defiants were withdrawn as day fighters after suffering heavy losses in August 1940.

Above: This Defiant Mk I was flown by Squadron Leader P.A. Hunter. The Defiant scored some initial successes when the nature of its turreted armament was not grasped, but once the Germans knew that the aircraft lacked both fixed forward-firing guns and agility, they were able to shoot the type down with ease.

MODIFICATIONS
The Defiant II had a larger fin and a more powerful Merlin engine, but most importantly, it possessed AI Mk IV air intercept radar for electronically guided vectoring.

TURRET FIGHTER
The Defiant's gun turret was electrically powered and equipped with four 7.7mm (0.303in) Browning machine guns, primarily for upward firing.

BOULTON PAUL DEFIANT MK II
Though never a truly effective fighter, the Defiant was better operating at night than as a conventional day fighter. This Mk II served with No. 151 Squadron at Wittering, Cambridgeshire.

VULNERABILITIES
The Defiant was developed in the mid-1930s, but didn't actually enter service with the RAF until the end of 1939, just months after the British declaration of war against Germany. As a day fighter against far more manoeuvrable Me 109 opponents, it was extremely vulnerable. During one action in May 1940 five out of six Defiants were destroyed in a head-on pass by the Messerschmitts.

NIGHT FIGHTER
When operating as night fighters – a role to which they were best suited – Defiants tended to position themselves below and ahead of the enemy bomber, raking the bomber with its upward-firing turret guns.

Douglas A-20 Havoc (1939)

TYPE • Medium bomber COUNTRY • United States

SPECIFICATIONS

DIMENSIONS:	Length: 14.63m (48ft); Wingspan: 18.69m (61ft 4in); Height: 5.51m (18ft 1in)
WEIGHT:	Maximum take-off 10,795kg (23,800lb)
POWERPLANT:	Two 1194kW (1600hp) Wright R-2600-11 Twin Cyclone 14-cylinder air-cooled radial piston engines
MAX SPEED:	563km/h (350mph)
RANGE:	1328km (825 miles)
CEILING:	8717m (28,600ft)
CREW:	3
ARMAMENT:	6 x 12.7mm (0.5in) Browning M2 MGs, two 7.62mm (0.3in) Browning M1919 MG; 1089kg (2400lb) bombload

Despite initial indifference from the US Army Air Corps, the fast and agile A-20 Havoc proved itself to be an exceptional attack bomber, which saw widespread service in the air arms of many nations.

Below: An A-20B (41-3134) from 84th Bomb Squadron, 47th Bomb Group, based at Mediouna, Morocco, December 1942.

'Joker'
This A-20G-35-DO 'Joker' wears the markings of the 647th Bombardment Squadron, 410th Bomb Group, USAAF, around the time of the Allied invasion of Europe in June 1944. Entering combat in May 1944, the group's four squadrons of A-20s attacked targets in France in preparation for the Normandy landings, before focusing on lines of communication after D-Day.

Nose armament
Most A-20s were armed with six forward-firing 12.7mm (0.5in) Browning machine guns.

BOMBLOAD
An A-20G was able to carry 1814kg (4000lb) in an internal bay – twice the load of the A-20C.

A-20B
The first variant to be ordered in large numbers by the USAAC (a total of 999) was the A-20B, which, although slightly faster, was in most regards a backwards step when compared to the A-20A. Based on the earlier DB-7A airframe, the A-20B lacked self-sealing fuel tanks and carried less armour protection. Following experience in the Pacific with other, better-protected A-20 variants, most of the A-20Bs were supplied through Lend-Lease channels to the USSR.

NINTH AIR FORCE A-20A
The A-20A was powered by two Wright R-2600-3 engines. The type entered service in the spring of 1941. The USAAF used the British name Havoc for the A-20A, while the RAF referred to them as Bostons.

Fairey Albacore (1940)

TYPE • Torpedo bomber **COUNTRY** • United Kingdom

SPECIFICATIONS	
Dimensions:	Length: 12.13m (39ft 10in); Wingspan: 15.24m (50ft); Height: 4.65m (15ft 3in)
Weight:	Maximum take-off 5820kg (12,830lb)
Powerplant:	One 794kW (1065hp) Bristol Taurus II or one 840kW (1130hp) Bristol Taurus XII 14-cylinder, air-cooled, radial piston engine
Max Speed:	272km/h (169mph)
Range:	1497km (930 miles)
Ceiling:	66,309m (20,700ft)
Crew:	2 or 3
Armament:	One 7.7mm (0.303in) Browning MG in starboard wing, one or two 7.7mm (0.303in) Vickers K MGs in rear cockpit; up to 907kg (2000lb) bombload or one 730kg (1610lb) torpedo

Intended to replace the Swordfish on British carriers, the Albacore never entirely supplanted its famous forebear. Nonetheless it saw much intense action and proved effective in a number of demanding roles and theatres.

Left: A Fleet Air Arm Fairey Albacore has its wings folded for stowage aboard a British carrier during World War II. The Albacores operated from both HMS Victorious *and HMS* Formidable.

Taurus engine
Early problems with the Taurus engine were largely solved, but an unfortunate reputation for poor reliability would dog the Taurus for the rest of its service.

Floatplane
During testing the Albacore showed itself prone to severe porpoising, even in calm conditions, with water being thrown up onto the engine and propeller. The floatplane requirement was duly dropped.

Flight characteristics
The Albacore proved generally pleasant to fly with a superb view for the pilot and good deck landing characteristics, but was less agile than the Swordfish.

Defence of Malta
From late 1941 until the summer of 1943 Albacores played a significant role in the defence of Malta, seeing relentless action attacking Axis shipping, mostly at night to avoid enemy fighters. It was kept in the air only through the Herculean efforts of ground crew in a theatre in which spares and equipment were extremely limited. By the end of 1943 the aircraft was generally being replaced by the monoplane Barracuda in most roles, although it persisted in limited frontline use into 1945.

Torpedo practice
The 'Applecore', as it was rather predictably nicknamed, was never as popular as its predecessor the Swordfish, but performed well. This example was photographed at the moment of releasing a practice torpedo.

Martin PBM Mariner (1940)

TYPE • *Seaplane* COUNTRY • *United States*

SPECIFICATIONS (PBM-3)

Dimensions: Length: 24.38m (80ft), Wingspan: 35.97m (118ft), Height: 8.23m (27ft)

Weight: Maximum take-off 26,253kg (57,878lb)

Powerplant: Two 1194kW (1600hp) Wright R-2600-6 14-cylinder radial engines

Max Speed: 338km/h (210mph)

Range: 3597km (2235 miles)

Ceiling: 6095m (20,000ft)

Crew: 7–8

Armament: Eight 12.7mm (0.5in) MGs in nose, dorsal turrets, waist and tail; plus up to 746kg (1646lb) of bombs, torpedoes and depth charges

In the early 1930s, the Martin Aircraft Company turned away from the maritime patrol market in favour of flying-boat airliners. The PBM Mariner was built as a patrol and bomber flying boat, with variants for the Coast Guard and surplus aircraft sold to several nations.

Above: A US Navy Martin PBM-3 Mariner of Patrol Squadron 74 (74-P-3), 1942. The same aircraft is shown in the photograph below.

Search radar
This aircraft is painted in a Pacific War colour scheme. The large search radar is apparent above and behind the main cockpit, enclosed in a streamlined fairing.

Wing floats
Like other flying boats, the PBM needed to keep its engines high above saltwater spray as it would damage them. To this end, it used a high gull wing and retractable balance floats that were replaced with fixed floats on later models.

Defensive weapons
Defensive armament initially consisted of a single 12.7mm (0.5in) machine gun in the nose and dorsal turrets plus others in waist and tail positions. Turret guns were increased to dual-mount weapons in later versions.

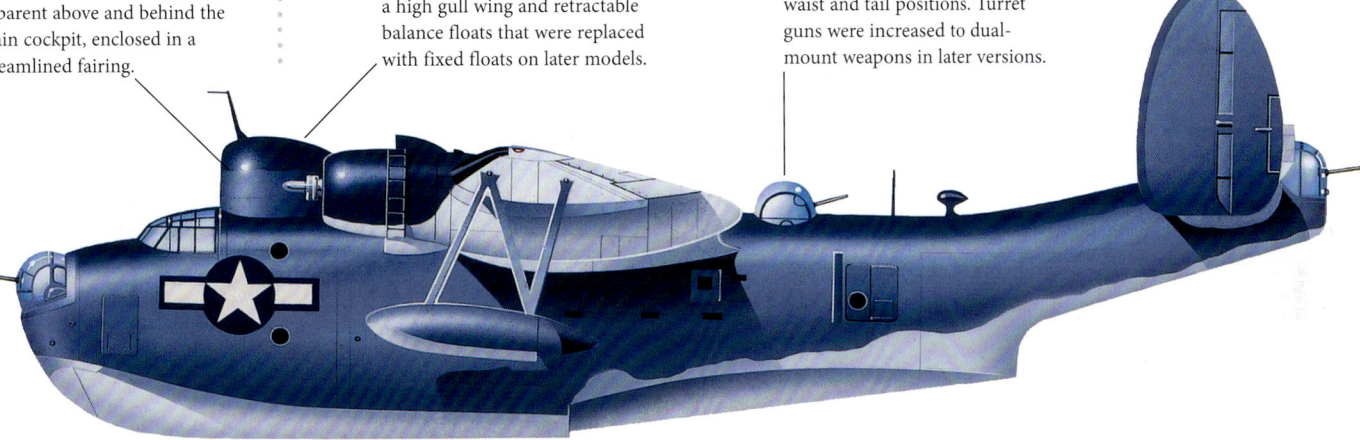

U-BOAT BATTLE
A U-boat commander sighting an approaching aircraft had a critical choice to make: either crash-dive and hope to slip away or remain on the surface and fight. A misjudged dive would leave the boat with its stern out of the water as the aircraft made its attack run. Several Mariners were lost to U-boats whose captains decided or were forced to stay on the surface. Nevertheless, Mariners sank 10 U-boats during the course of the war.

PBM-3 IN FLIGHT
Various specialist versions were created, notably the PBM-3S long-range variant. Greater endurance was achieved by removing as much weight as possible, while the PBM-3R had its armour and defensive armament removed to create a transport aircraft.

North American P-51 Mustang (1940)

TYPE • Fighter **COUNTRY** • United States

SPECIFICATIONS	
DIMENSIONS:	Length: 9.85m (32ft 4in); Wingspan: 11.28m (37ft); Height: 3.71m (12ft 2in)
WEIGHT:	5262kg (11,600lb) maximum take-off
POWERPLANT:	Packard Rolls-Royce Merlin V-1650-7 V-12 liquid-cooled piston engine, 1112kW (1490hp)
MAX SPEED:	704km/h (437mph)
RANGE:	3347km (2080 miles)
CEILING:	12,770m (41,900ft)
CREW:	1
ARMAMENT:	6 × 12.7mm (0.5in) machine guns plus provision for up to 2 × 454kg (1000lb) bombs or 6 × 127mm (5in) rockets

With a strong claim to be the finest piston-engined fighter of World War II, the superlative P-51 emerged from potential obscurity when re-engined with the British-designed Merlin powerplant. It became a war-winning long-range escort after its service entry in late 1943.

Above: The A-36A was an attack aircraft variant of the Mustang fighter, but was actually in service before the P-51. The A-36A shared the Mustang's airframe but never had the upgrade to the Rolls-Royce Merlin engine.

COCKPIT
Up to and including the P-51C, visibility from the Mustang's cockpit canopy left a lot to be desired; much improved with the one-piece transparent canopy introduced with the P-51D.

COLOUR SCHEME
By 1944, Mustangs were being delivered unpainted, so that the fighter groups could set to work on their own particular brand of decoration. The 332nd FG employed red tails and red spinners, with red wingtips and code outlines.

OPERATIONS

P-51Bs of the 354th Fighter Group flew their first operational escort mission from England in December 1943. The first P-51Ds arrived in England in the late spring of 1944 and quickly became standard equipment for the USAAF Eighth Fighter Command. In the Pacific, Mustangs operating from the captured Japanese islands of Iwo Jima and Okinawa had the task of escorting B-29s to their targets and neutralizing the Japanese air force on the ground.

DROP TANKS
The Mustang's range, already impressive, was greatly extended by the use of drop tanks. Standard metal drop tanks were of 284-litre (62.5 Imp gal) capacity, but most aircraft in the European theatre carried the 409-litre (90 Imp gal) tank.

P-51D MUSTANG
The most numerous Mustang variant was the P-51D, with a one-piece sliding cockpit canopy. This canopy has been compared to that of a high-class sports car, and for general all-round comfort, visibility and ease of operation the P-51 was rated as being the finest American fighter of World War II.

Yakovlev Yak-1 (1940)

TYPE • Fighter COUNTRY • Soviet Union

The Russians were late in developing really effective monoplane fighters that were in the same class as Britain's Hurricane and Spitfire and Germany's Bf 109, but Aleksandr Yakovlev's attractive designs soon redressed the situation.

SPECIFICATIONS
(Yak-1M late production)

DIMENSIONS:	Length: 8.48m (27ft 10in); Wingspan: 10m (32ft 10in); Height: 2.64m (8ft 8in)
WEIGHT:	Maximum take-off 2884kg (6358lb)
POWERPLANT:	One 940kW (1260hp) Klimov M-105PF V-12 liquid-cooled piston engine
MAX SPEED:	600km/h (373mph) at 3500m (11,482ft)
RANGE:	700km (435 miles)
CEILING:	10,000m (32,810ft)
CREW:	1
ARMAMENT:	One 20mm (0.8in) ShVAK cannon and one 12.7mm (0.5in) Berezin UBS machine gun

Above: High-scoring Yak-1 pilot, and twice Hero of the Soviet Union, Sergey Lugansky scored 32 of a total 37 victories, as recorded in the laurel wreath.

ARMAMENT
The standard armament of the Yak-1 was a pair of 12.7mm (0.50in) machine guns in the upper front fuselage, and a 20mm (0.79in) ShVAK cannon in the nose, firing through the propeller boss, with 120 rounds of ammunition.

IMPROVEMENTS
From October 1942, Yak-1s were built to an improved standard with bubble canopy, lowered rear fuselage and many internal changes. This example, presented by the 'Collective Workers of Shatovskovo Village Soviet' was flown by Hero of the Soviet Union, 35-kill ace, Aleksey Reshetov.

AIRFRAME
The airframe of the Yak-1 was lightweight, giving rise to a generation of fast and manoeuvrable interceptors; on the other hand, the more robust Yak-7A was developed into a succession of heavier tactical fighters like the Yak-9.

WINNING DESIGN
The Yak-1 *Krasavyets* (Beauty) made its first public appearance at an air display on 7 November 1940. It was Aleksandr Yakovlev's first fighter design, and it earned him the Order of Lenin and a prize of 100,000 roubles. The fighter was powered by a 746kW (1000hp) M-105PA engine and carried an armament of one 20mm (0.79in) ShVAK cannon, two 7.62mm (0.30in) ShKAS machine guns and sometimes six RS-82 rockets.

PRODUCTION AIRCRAFT
The Yak-1 was of mixed construction, fabric and plywood covered; it was simple to build and service, and a delight to fly. Production was slow because of the relocation of factories after the German invasion, and so it was decided to convert a trainer variant of the Yak-1, the Yak-7V, into a single-seat fighter by covering the second cockpit with metal sheeting. In this new guise the aircraft was designated Yak-7A.

SBD-3 Dauntless (1940)

TYPE • *Dive-bomber* **COUNTRY** • *United States*

SPECIFICATIONS

DIMENSIONS:	Length: 10.06m (33ft); Wingspan: 12.65m (41ft 6in); Height: 3.94m (12ft 11in)
WEIGHT:	4924kg (10,855lb) maximum take-off
POWERPLANT:	(SBD-5) Wright R-1820-60 radial piston engine, 895kW (1200hp)
MAX SPEED:	394km/h (245mph)
RANGE:	1795km (1115 miles)
CEILING:	7407m (24,300ft)
CREW:	2
ARMAMENT:	2 × 12.7mm (0.5in) machine guns, 2 × 7.62mm (0.3in) machine guns plus 1020kg (2250lb) of bombs

By the end of the war in the Pacific, the venerable Dauntless dive-bomber was showing its age. Nonetheless, its contribution to victory in a succession of key naval battles cannot be overstated and its tally of Japanese shipping is unmatched.

Above: The French made extensive use of the SBD. This SBD-5 was serving with Aeronvale unit Flotille 4FB at Cognac in January of 1945 for operations against pockets of German resistance in southern France.

PILOT POSITION
The pilot sat high in the cockpit with an armoured backplate, but no bulletproof windscreen. A telescopic sight was used for aiming both bombs and guns.

MARKINGS
The yellow outline to the national markings on this SBD denotes its participation in Operation Torch, the Allied invasion of northern Africa in November 1942. This VS-41 Dauntless was operating from USS *Ranger*.

POWERPLANT
The Dauntless was powered by a Wright R-1820-32 Cyclone engine. The oil was cooled via an under-nose intake, while the large intake on top of the engine cowling fed air into the carburettor.

MARINE DAUNTLESS
Delivery of the SBD-1 to the US Marine Corps began in mid-1940 and this version was followed by the SBD-2 and SBD-3, with extra fuel tankage, protective armour and autopilot. The attrition rate of the Dauntless squadrons was the lowest of any US carrier aircraft in the Pacific, thanks to the SBD's ability to absorb an astounding amount of battle damage.

ATTACK MISSION
USMC SBDs, each carrying centreline and underwing bombs, head for a Japanese target on the island of Rabaul in 1944. Dauntless production ended on 22 July 1944, the type being supplanted by the Curtiss SB2C Helldiver.

Focke-Wulf Fw 200 Condor (1940)

TYPE • Bomber COUNTRY • Germany

Designed as a four-engined long-range airliner, the Fw 200 was adapted to fill a Luftwaffe requirement for a maritime patrol aircraft. Only built in small numbers, the Fw 200's effect on Allied shipping earned it the sobriquet 'Scourge of the Atlantic'.

SPECIFICATIONS

DIMENSIONS: Length: 23.85m (78ft 3in); Wingspan: 32.84m (107ft 8in); Height: 6.30m (20ft 8in)

WEIGHT: Empty 17,005kg (37,496lb); maximum take-off 22,700kg (50,044lb)

POWERPLANT: Four 895kW (1200hp) BMWBramo 323R-2 Fafnir nine-cylinder radial engines (FW 200C-3/U4)

MAX SPEED: 360km/h (224mph) at 4700m (15,420ft)

RANGE: 4440km (2759 miles)

CEILING: 6000m (19,685ft)

CREW: 5

ARMAMENT: 4 x 7.92mm (0.31in) MG 15 machine guns, one 20mm (0.8in) MG 151 cannon; 1000kg (2200lb) bombs internally or up to 5400kg (11,900lb) externally

Above: The forward gondola gun position mounted an Oerlikon 20mm (0.79in) FF cannon. The protruding object is a Lofte 7D bombsight.

FORWARD CUPOLA
The forward-firing 7.92mm(0.31in) MG 15 machine gun in the fully enclosed forward cupola was manned by the co-pilot when the aircraft was threatened by frontal attack.

MISSILE CARRIER
The final operational variant of the Condor was the Fw 200C-6, developed from the C-3 to carry a Henschel Hs 293B air-to-surface missile under each outer engine nacelle, the underwing bomb racks being removed. The combination of Hs 293 and Fw 200 was first used operationally on 28 December 1943. The total number of Condors produced during the war years was 252 aircraft.

GONDOLA
The Condor's weapons bay was situated in the ventral gondola; the rear gun position immediately behind it was manned by the flight engineer. On armed reconnaissance missions, four 250kg (551lb) bombs were normally carried, two on the outboard engine nacelles and two on wing racks.

STORAGE
The Condor's rear fuselage was used as a storage area for small stores like flares, light buoys or direction-finding buoys. These were dropped through a hatch in the fuselage underside.

SHIP HUNTER
The Fw 200C-3/U2 was readily identified by the bulge in the gondola for the Lofte 7D bombsight. Fitting this accurate device necessitated a reorganization of the ventral armament. In the early years of World War II, the long-range Focke-Wulf Condor was a far greater threat to Allied shipping than the German U-boats.

Mitsubishi Ki.46 Dinah (1940)

TYPE • *Reconnaissance* COUNTRY • *Japan*

SPECIFICATIONS (Ki.46 III)	
DIMENSIONS:	Length: 11.00m (36ft 1in); Wingspan: 14.70m (48ft 23in); Height: 3.88m (12ft 8.75in)
WEIGHT:	Empty 3263kg (7194lb); maximum take-off 5800kg (12,789lb)
POWERPLANT:	Two 787kW (1055hp) Mitsubishi Ha.102 14-cylinder radials
MAX SPEED:	604km/h (375mph) at 8000m (26,245ft)
RANGE:	2474km (1537 miles)
CEILING:	10,720m (35,170ft)
CREW:	2
ARMAMENT:	One 7.7mm (0.303in) machine gun

Known to the Allies by the code name 'Dinah', the Mitsubishi Ki.46 was one of the best reconnaissance aircraft of World War II, and aerodynamically one of the most perfect aircraft produced by any of the belligerents.

Above: Production of all versions totalled 1783 aircraft, including three Ki.46-IVa machines with turbocharged engines.

CREW POSITIONS
The Ki.46's pilot and gunner were seated in two cockpits separated by a large fuel tank. To meet performance requirements, the aircraft's designers adopted a fuselage of small diameter.

HEAVY CANNON
The 37mm (1.46in) fixed upward-firing cannon was a devastating weapon, if it could be brought to bear. Aircraft of this type were known as the Type 100 in Japanese Army Service.

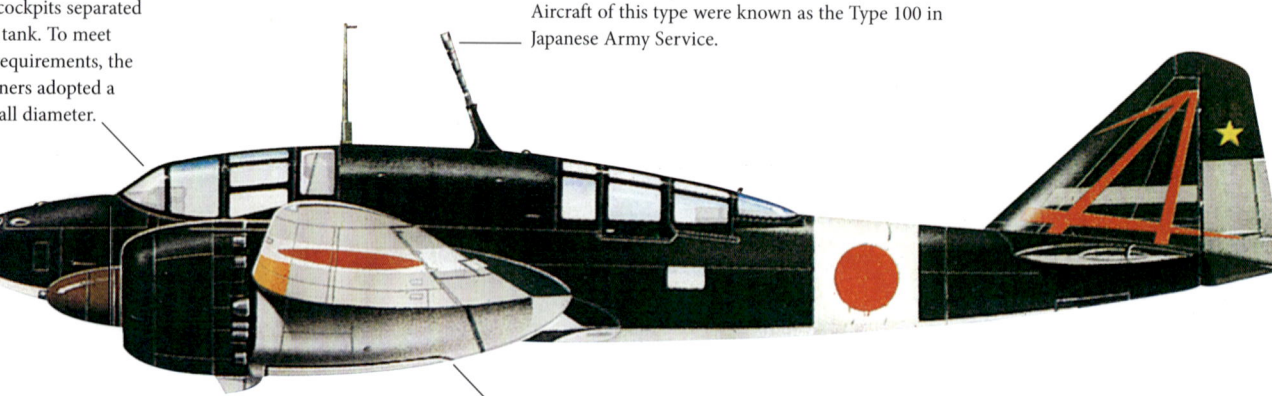

ENGINE LAYOUT
Special close-fitting cowlings were developed for the Ki.46's engines. This resulted in a substantial improvement in the pilot's sideways vision and also brought about a reduction in drag, as well as contributing to the design of the fully retractable landing gear.

HEAVY FIGHTER
The Ki.46 was designed specifically as a high-altitude reconnaissance aeroplane to meet a 1937 requirement, and the prototype made its maiden flight in November 1939. With a strong powerplant, it was initially virtually impossible to intercept. Versions of the Ki.46 developed as heavy fighters for the air defence of Japan featured a 37mm (1.46in) Ho-203 cannon mounted obliquely in the upper fuselage, designed to fire forward and upward into the underside of a B-29 bomber.

STRATEGIC RECONNAISSANCE
Anticipating that Allied fighters capable of intercepting the Ki.46-II would appear in the later part of 1942, the Imperial Japanese Army Air Force had ordered an improved version. The Ki.46-III was without doubt the best strategic reconnaissance aeroplane used in significant numbers in World War II, offering long range as well as high speed and a good service ceiling.

Petlyakov Pe-8 (1940)

TYPE • Bomber COUNTRY • Soviet Union

The only modern four-engined bomber developed by the USSR, the Pe-8 was an impressive performer. But like the Luftwaffe and the Italian Air Force, the Soviet Air Force in WWII was geared to tactical support, and paid little attention to strategic bombers.

SPECIFICATIONS

DIMENSIONS:	Length: 22.47m (73ft 8.5in); Wingspan: 39.94m (131ft 0.25in); Height: 6.10m (20ft)
WEIGHT:	Empty 18,420kg (40,616lb); maximum take-off 33,325kg (73,481lb)
POWERPLANT:	Four 1007kW (1350hp) Mikulin AM-35A V-type
MAX SPEED:	438km/h (272mph) at 7600m (24,935ft)
RANGE:	5445km (3383 miles)
CEILING:	9750m (31,988ft)
CREW:	11
ARMAMENT:	One 20mm (0.79in) cannon; one 12.7mm (0.50in) machine gun and two 7.62mm (0.30in) machine guns; bombload of up to 4000kg (8820lb)

Above: The Pe-8 was the only Soviet strategic heavy bomber to see service in WWII. First flown on 27 December 1936, the Pe-8 entered service in 1940.

COLOURING
This Pe-8 of the Soviet Long-range Aviation is camouflaged in the standard pattern of dark green upper surfaces and pale blue under surfaces. Apart from the application of temporary winter camouflage, this was retained throughout the war.

GUN POSITIONS
An unusual feature of the Pe-8 was that it had machine gun positions built into the rear of the two inboard engine nacelles. Gunners found these positions cramped and smelly, but they had the advantage of being warm.

BOMB BAY
The Pe-8's bomb bay could accommodate a 5000kg (11023lb) FAB-5000NG blast bomb. On the eve of the Battle of Kursk in July 1943, Pe-8s were used to drop these weapons on concentrations of German tanks and other armoured vehicles.

LONG DISTANCE

The Pe-8 was dogged by engine difficulties throughout its career and various powerplants were tried, including M-30B diesel engines. From 1943, production Pe-8s were fitted with Mikulin M-82FN fuel injection engines, but the problems persisted and production ended in 1944 after 79 examples had been built. Despite its troubles, the Pe-8 made some notable long-distance flights, including a round trip of more than 17,700km (11,000 miles) from Moscow to Washington.

COMBAT LOSSES
Early in its service life the Pe-8 had a very good performance and was difficult to intercept, the loss rate in 1942 being only one per 106 missions. However, as the war progressed, German defences improved and losses more than doubled by 1944.

Avro Manchester (1940)

TYPE • Bomber COUNTRY • United Kingdom

The Avro Manchester first flew in July 1939, becoming operational in November 1940. The Manchester heavy bomber was not successful in itself, largely because of the chronic unreliability of its two underdeveloped Rolls-Royce Vulture engines, but it paved the way for the Lancaster.

SPECIFICATIONS (Manchester Mk I)

DIMENSIONS:	Length: 21.14m (69ft 4.25in); Wingspan: 27.46m (90ft 1in); Height: 5.94m (19ft 6in)
WEIGHT:	Empty 13,350kg (29,432lb); maximum take-off 22,680kg (50,000lb)
POWERPLANT:	Two 1312kW (1760hp) Rolls-Royce Vulture 24-cylinder X-type engines
MAX SPEED:	426km/h (265mph) at 5180m (17,000ft)
RANGE:	2623km (1630 miles)
CEILING:	5850m (19,200ft)
CREW:	7
ARMAMENT:	8 x 7.7mm (.303in) MGs (two each in nose and dorsal turrets, four in tail turret); up to 4695kg (10,350lb) of bombs

Above: The Manchester's three-bladed propellers were driven by Rolls-Royce Vulture engines, which consistently failed to deliver full power at crucial moments.

FRONT TURRET
Early trials revealed that when the nose turret was rotated, the airflow along the fuselage sides was disturbed. The problem was solved by moving the turret's axis of rotation slightly forward.

WING STRUCTURE
The Manchester had a very strong wing structure, which avoided problems when the aircraft was altered to a four-engine configuration to become the Lancaster.

TOP TURRET
The Manchester's Frazer-Nash FN.7 mid-upper gun turret was extremely uncomfortable for its occupant, especially on long trips.

OPERATIONAL
The prototype Manchester flew on 25 July 1939, followed by a second aircraft on 26 May 1940. The Manchester Mk I, which featured a central tail fin as well as twin fins and rudders, went operational with No. 207 Squadron in November 1940. The first 20 aircraft were followed by 200 Manchester IAs with the central fin removed. The Manchester pictured here served with No. 83 Squadron at RAF Scampton, Lincolnshire, in March 1942; it failed to return from its 15th operational sortie to Hamburg.

TAIL FIN
The Manchester Mk I originally had a central tailfin; this was removed from the main batch of production aircraft.

ENGINE TROUBLES
As many Manchesters were lost to mechanical failure as to enemy action. As well as failing to deliver their promised power, Vulture engines were notoriously unreliable. Rolls-Royce had no spare capacity available for the rectification of the serious faults affecting the engine.

Bristol Beaufighter TF.Mk X (1940)

TYPE • Torpedo bomber **COUNTRY** • United Kingdom

SPECIFICATIONS

DIMENSIONS:	Length: 12.70m (41ft 8in); Wingspan: 17.63m (57ft 10in); Height: 4.82m (15ft 10in)
WEIGHT:	Empty 7076kg (15,600lb); maximum take-off 11,431kg (25,200lb)
POWERPLANT:	Two 1320kW (1770hp) Hercules XVII 14-cylinder radials
MAX SPEED:	512km/h (318mph)
RANGE:	2366km (1470 miles)
CEILING:	4572m (15,000ft)
CREW:	2
ARMAMENT:	Four 20mm (0.79in) cannon and one 7.7mm (0.303in) machine gun; one 748kg (1650lb) or 965kg (2127lb) torpedo, two 227kg (500lb) bombs, eight 76.2mm (3in) rocket projectiles

The Beaufighter TF.Mk X pictured here served with No. 489 Squadron, RNZAF, which was part of the Coastal Command Strike Wing based at Dallachy, Scotland. Apart from night-fighting, the role with which the Beaufighter became most associated was that of maritime strike.

Above: The 45.7cm (18in) torpedo carried by the Beaufighter was fitted with a Mono Air Tail (MAT) that stabilized the torpedo after it was dropped. The MAT was released upon impact with the water.

ASV RADAR
Coastal Command Beaufighters often carried ASV (air to surface vessel) radar to locate shipping targets. This aircraft is fitted with ASV Mk III, which was of limited use due to the blanking effect of the engine nacelles.

INVASION STRIPES
This No. 489 Squadron aircraft is depicted with 'invasion stripes' over the standard Coastal Command scheme of dark sea grey over grey. Most Torbeau operations were directed against German convoys off Norway.

TORPEDO CONFIGURATION
The Torpedo Beaufighter, known as the Torbeau, was fitted with Fairey-Youngman wing dive brakes installed between the ailerons and the wing root fairings.

ANTI-SHIP AIRCRAFT
The TF. Mk X torpedo-bomber and the Mk XIC, which was not equipped to carry torpedoes, were fitted with 1320kW (1770hp) Hercules XVII engines and had a dorsal cupola containing a rearward-firing 7.7mm (0.303in) machine gun. Production of the TF. Mk X, which was the most important British anti-shipping aircraft from 1944 to the end of the war, totalled 2205 aircraft, while 163 aircraft were completed to Mk XIC standard.

MK I BEAUFIGHTER
The first strike fighter variant of the Beaufighter was the Mk IC, 300 of which were produced. No. 252 Sqn, RAF was the first Coastal Command unit equipped with Beaufighter Mk Is, from December 1940. On this example, the early straight tailplane can be seen.

Handley Page Halifax (1940)

TYPE • Bomber **COUNTRY** • United Kingdom

Second of the four-engined heavy bombers to enter service with the RAF, in November 1940, the Handley Page Halifax was one of the famous triad, comprising the Halifax, Avro Lancaster and Short Stirling, which mounted Bomber Command's night-bombing offensive against Germany.

SPECIFICATIONS

DIMENSIONS:	Length: 21.82m (71ft 7in); Wingspan 30.07m (98ft 8in); Height 6.32m (20ft 9in)
WEIGHT:	Empty 17690kg (39,000lb); maximum take-off 30,845kg (68,000lb)
POWERPLANT:	Four 1204kW (1615hp) Bristol Hercules VI or XVI 14-cylinder two-row radial engines
MAX SPEED:	454km/h (282mph) at 4115m (13,500ft);
RANGE:	3194km (1985 miles)
CEILING:	7315m (24,000ft)
CREW:	7
ARMAMENT:	Five 7.62mm (0.30in) machine guns, plus an internal bombload of 6577kg (14,500lb)

This page: Pictured here is Halifax B.VII PN230 'Vicky the Vicious Virgin' of No. 408 Squadron RCAF, RAF Linton-on-Ouse, Yorkshire, 1945.

DORSAL TURRET
The dorsal turret was a Boulton Paul A Mk III mid-upper turret, armed with four 7.62mm (0.30in) guns with 1160 rounds each. The teardrop fairing on top of the fuselage between cockpit and dorsal turret housed the direction finder aerial.

ENGINES
The Halifax Mk III was fitted with Bristol Hercules engines, but the earlier marks, with Rolls-Royce Merlins, had a longer range, and these were retained by the special duties squadrons for infiltrating agents into enemy territory.

RADOME
The later marks of Halifax carried a large radome under the fuselage housing H2S ground mapping radar. Both Halifax and Lancaster were originally intended to have a ventral gun turret in this position.

ENGINE UPGRADE
In 1943 the Halifax's Merlin engines were replaced by four 1204kW (1615hp) Bristol Hercules XVI radial engines in the Halifax Mk III, which remained in the front line up to the end of the war. The next operational variants were the Mks VI and VII, the former powered by the 1249kW (1675hp) Hercules 100 and the latter using the MK III's Hercules XVI. These were the ultimate bomber versions, and were produced in relatively small numbers.

BOMB DAMAGE
Flight Sergeant D. Cameron and his crew pose in their damaged No. 158 Squadron Halifax. The damage was caused by a 'friendly' bomb that hit the aircraft during a mission from RAF Lisset in 1943. After being repaired, the aircraft was returned to service.

Short Stirling (1940)

TYPE • Bomber **COUNTRY** • United Kingdom

SPECIFICATIONS

Dimensions:	Length: 26.59m (87ft 3in); Wingspan: 30.20m (99ft 1in); Height: 6.93m (22 ft 9in)
Weight:	Empty 21,274kg (46,900lb); maximum take-off 31,752kg (70,000lb)
Powerplant:	Four 1230kW (1650hp) Bristol Hercules XVI 14-cylinder radials
Max Speed:	434km/h (270mph) at 4420m (14,500ft)
Range:	3235km (2010 miles)
Ceiling:	5180m (17,000ft)
Crew:	6
Armament:	6 x 7.7mm (0.303in) machine guns plus internal bombload of 6350kg (14,000lb)

Throughout its operational life, the Short Stirling suffered from an Air Ministry instruction which dictated that its wing span should be reduced so that the aircraft would fit inside existing hangars. Its altitude performance suffered accordingly.

Above: No. 196 Sqn Stirling Mk V PJ887, coded 'ZO-H', is shown as it appeared in early 1946, shortly after it entered service. Though it had operated in the Far East during World War II, No. 196 Sqn was flying mail runs on the Continent by 1946.

Wings
Though stable in flight and surprisingly manoeuvrable, thanks to its high wing loading, the Stirling had a poor operational ceiling when loaded, often being hard pushed to climb above 3660m (12,000ft).

Code letters
This Stirling carries the code letters of No. 7 Squadron, the first to equip with the type in August 1940. The squadron was then at RAF Leeming, in Yorkshire, but in October 1940 it moved to Oakington, near Cambridge, and remained there for the rest of the war.

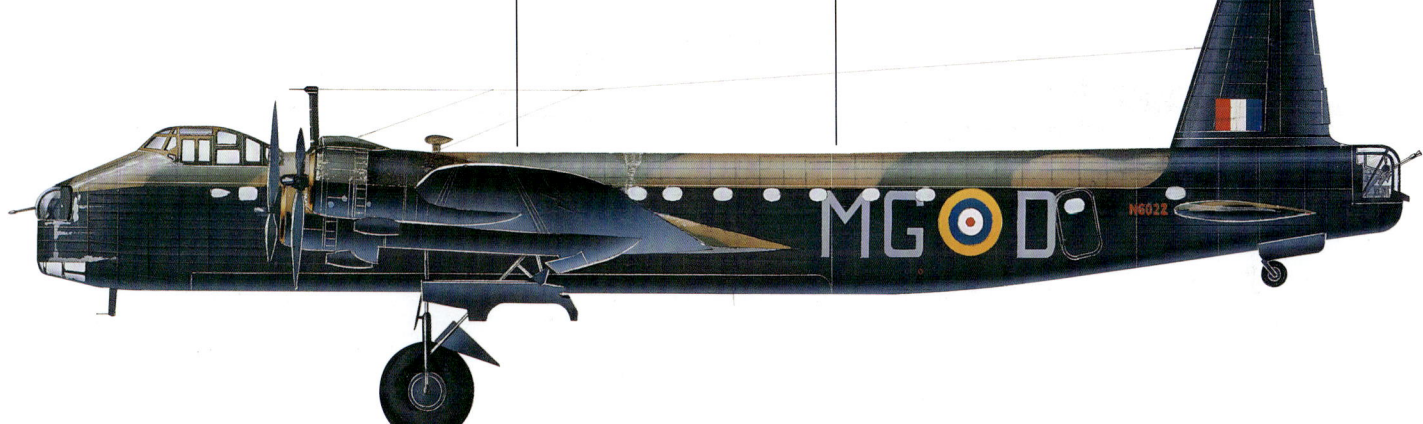

FIRST STIRLINGS
The Stirling eventually equipped seven squadrons in No. 3 Group, RAF Bomber Command. No. 7 Sqn was the first of these to receive Stirlings in August 1940, the unit taking the new bomber on its first operation to attack an oil storage depot at Rotterdam on the night of 10/11 February 1941. The main aircraft on this page – N3641, coded 'MG-D' – was the machine to join No. 7 Sqn.

Undercarriage
To shorten take-offs and landings the Stirling was fitted with a very tall undercarriage. This improved lift during the take-off run, but made ground handling tricky.

MISSION BRIEFING
A No. 149 Sqn crew confers with a meteorological officer prior to another sortie. Dorsal turret-equipped Stirling bombers generally carried a crew of six. The rear turret seen here carried four 7.7mm (.303in) Browning machine guns for rear defence.

Douglas DB-7/A-20 Havoc/Boston (1940)

TYPE • Bomber COUNTRY • United States

Blooded in battle in every theatre of war, sighted on every continent, the Douglas Havoc/Boston was a fast and agile aircraft that proved to be an exceptional attack bomber and saw widespread service in the air arms of many nations.

SPECIFICATIONS (A-20G Havoc)

DIMENSIONS:	Length: 14.63m (48ft); Wingspan: 18.69m (61ft 4in); Height: 5.51m (18ft 1in)
WEIGHT:	Maximum take-off 13,608kg (30,000lb)
POWERPLANT:	Two 1194kW (1600hp) Wright R-2600-23 Twin Cyclone 14-cylinder air-cooled radial piston engines
MAX SPEED:	523km/h (325mph)
RANGE:	1521km (945 miles)
CEILING:	7200m (23,700ft)
CREW:	3
ARMAMENT:	Six 12.7mm (0.50in) Browning M2 fixed forward-firing MGs, and three similar weapons in power-operated dorsal turret, and one rearward-firing through ventral tunnel; up to 1814kg (4000lb) of bombs

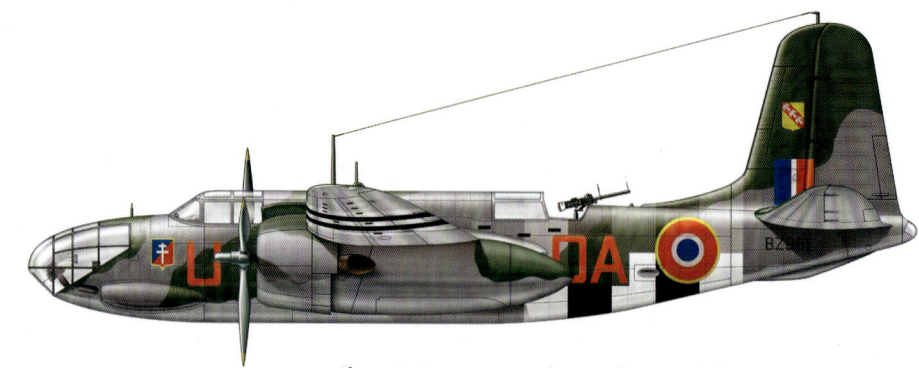

Above: By January 1940, the French Armée de l'Air began operating the DB-7 aircraft type with its 19th and 32nd Air Groups, initially in Morocco and later in metropolitan France.

VENTRAL TURRET
The power-driven ventral turret held two Browning 12.7mm (0.50in) machine guns, while beneath, the bomb bay held up to 1814kg (4000lb) of bombs.

FORWARD ARMAMENT
The A-20G was armed with six forward-firing 12.7mm (0.50in) Browning machine guns, which were ideal for heavy strafing runs against ground targets.

NINTH AIR FORCE
This Douglas A-20G-35-DO Havoc flew with the US 646th Bomb Squadron, 410th Bomb Group, Ninth Air Force.

NAMING

The Boston/Havoc was in USAAC, RAF, French and Soviet service. Throughout its career, there was confusion and overlap between the company team DB-7 and the USAAF designation A-20, and between the popular names Boston and Havoc. Many RAF machines officially had both names, the UK's involvement beginning with a 20 February 1940 contract for 150 aircraft.

VETERAN AIRCRAFT
This A-20G-20 served with a unit of the Ninth Air Force, which took its place alongside the Eighth Air Force in operations in the ETO, and flew more than 100,000 combat missions in the run up to D-Day.

Grumman F4F Wildcat (1940)

TYPE • Fighter COUNTRY • United States

The F4F was the US Navy's most important fighter at the time of the United States's entry into World War II in December 1941 after the Japanese attack on Pearl Harbor, and it remained in production right through the war.

SPECIFICATIONS (F4F-3)

DIMENSIONS:	Length: 8.76m (28ft 9in); Wingspan: 11.58m (38ft); Height: 3.61m (11ft 10in)
WEIGHT:	Empty 2471kg (5448lb); maximum take-off 3607kg (7953lb)
POWERPLANT:	895kW (1200hp) Pratt & Whitney R-1830-66 radial engine
MAX SPEED:	512km/h (318mph) at 5913m (19,400ft)
RANGE:	1239km (770 miles)
CEILING:	10,638m (34,900ft)
CREW:	1
ARMAMENT:	6 x 12.7mm (0.50in) machine guns plus external bombload up to 91kg (200lb)

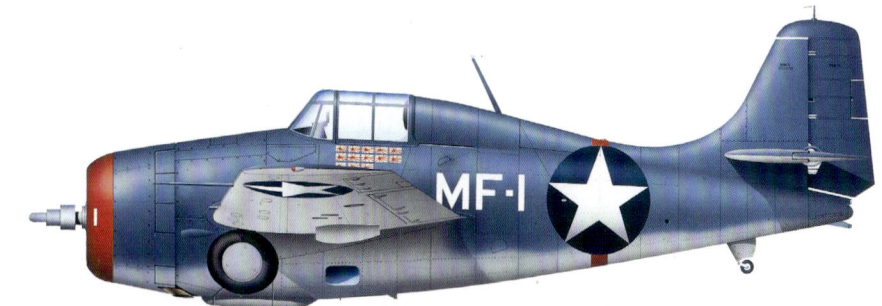

Above: This Grumman F4F-3 MF-1, evidently displaying many kills, was part of VMF-224, based at Guadalcanal in the summer of 1942.

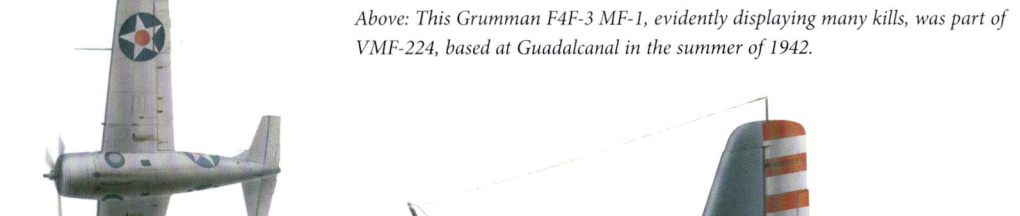

COLOUR SCHEME
The Wildcat is painted in the standard shipboard colour scheme of the early Pacific War years – an unspectacular blue/grey with a light grey underside sea-blue finish.

FIREPOWER
Whereas the F4F-3 had a rather inadequate armament of four Browning 12.7mm (0.50in) machine guns, the F4F-4 carried six. It could also carry two 113kg (250lb) bombs on underwing racks.

US NAVY
The first US Navy Wildcats entered service in December 1940. In the USA, delivery of the Wildcat was slow, and at the time of the Japanese attack on Pearl Harbor only 183 F4F-3s and 65 F4F-3As were in service with the US Navy and Marine Corps. The principal production Wildcat, the F4F-4, which featured folding wings, entered service at the end of 1941.

CONTOURS
The portly Wildcat should have been no match for the better performing and more manoeuvrable Zero, yet its ruggedness, better guns and the innovative tactics of its pilots carried the day. The F4F-4 variant featured two extra 12.7mm (0.5in) guns and introduced folding wings for the first time, utilizing Grumman's patented 'sto-wing' system, which allowed the wings to swing through 90 degrees to be stored pointing backwards alongside the fuselage.

Lavochkin-Gorbunov-Gudkov LaGG-3 (1940)

TYPE • *Fighter* **COUNTRY** • *Soviet Union*

SPECIFICATIONS

DIMENSIONS:	Length: 8.81m (28ft 11in); Wingspan: 9.8m (32ft 2in); Height: 2.54m (8ft 4in)
WEIGHT:	Maximum take-off 3190kg (7033lb)
POWERPLANT:	One 924kW (1239hp) Klimov M-105PF liquid-cooled V-12 engine
MAX SPEED:	589km/h (366mph)
RANGE:	1000km (620 miles)
CEILING:	9700m (31,800 ft)
CREW:	1
ARMAMENT:	One 12.7mm (0.50in) Berezin BS machine gun, one 20mm (0.78in) ShVAK cannon; two 50kg (110lb) bombs or six RS-82 or RS-132 rockets

The LaGG-3 in the form it first rolled off the production line in 1940 was, at best, an immature combat aircraft. The overweight and underpowered aircraft was unloved by those who flew it and derided by its enemies.

Above: In service the LaGG was unpopular. Despite significant improvements to its handling and a concentrated effort to lighten the airframe, handling remained less than sparkling.

PERFORMANCE
Compared to its great rival, the Yak-1, the LaGG-3 was marginally inferior in speed and climb rate but enjoyed a slightly better range.

AIRFRAME
The LaGG-3's only real advantage remained the immensely strong, fire-resistant airframe that could absorb remarkable amounts of punishment and remain airworthy.

FLAWED FIGHTER
Design flaws meant than the LaGG-3 had a collection of serious flight challenges. It was liable to develop a violent spin without warning in a steep banking turn, it had a tendency to nose-up on approach to land and would stall at the slightest provocation. The undercarriage was prone to collapse, the view from the cockpit was poor (a situation exacerbated by the opacity of Russian acrylic at the time), the hydraulics were unreliable, brakes seized, engines leaked oil, the gun-firing mechanism often failed and initial build quality was bad.

HANDLING
The wing had leading-edge slats to improve behaviour near the stall and handling was supposedly improved by the addition of rather crude drop-shaped mass balance weights at the top and bottom of the rudder

FINNISH SERVICE
This LaGG-3 of 524 IAP was shot down by Finnish anti-aircraft fire on 6 March 1942. Repaired by the Finns, this aircraft, coded LG-1, scored the only confirmed kill by a LaGG-3 in Finnish service when it shot down a Soviet LaGG-3 on 16 February 1944.

Fairey Fulmar (1940)

TYPE • Fighter COUNTRY • United Kingdom

Despite the performance limitations imposed by its two-seat design, and the visual awkwardness of its design, the Fulmar two-seat shipborne and land-based fighter proved remarkably successful in British service and shot down more enemy aircraft than any other Fleet Air Arm fighter.

SPECIFICATIONS (Fulmar Mk II)

Dimensions: Length: 12.24m (40ft 2in); Wingspan: 14.14m (46ft 4.5in); Height: 3.25m (10ft 8in)

Weight: Maximum takeoff 4627kg (10,200lb)

Powerplant: One 970kW (1300hp) Rolls-Royce Merlin 30 V-12 liquid-cooled piston engine

Max Speed: 440km/h (272mph)

Range: 1255km (780 miles)

Ceiling: 8300m (27,200ft)

Crew: 2

Armament: Eight 7.7mm (0.303in) or four 12.7mm (0.5in) Browning machine guns fixed forward-firing in wings; up to 226kg (500lb) bombload

Left: The two-seat layout of the Fulmar allowed for the aircraft to readily take on reconnaissance and spotting duties.

Stressed airframe
The fact that the aircraft was stressed for dive-bombing meant that it possessed sufficient structural strength to withstand the rigours of arrested carrier landings and catapult launches.

Mk I aircraft
The aircraft here is Mk I Fulmar N1860, shown as it appeared when it was serving with 808 Squadron at Dhekalia, Egypt in March or April 1941.

COMBAT DEBUT
The Fulmar first flew on 4 January 1940 and the first example was delivered to 778 Squadron for deck landing trials on 10 May of the same year. The first frontline units began receiving the aircraft in June. The Fulmar made its combat debut on 2 September 1940, when four Italian SM.79 torpedo bombers and a Cant Z.501 were all shot down by Fulmars operating off HMS *Illustrious* in the Mediterranean.

Mk II variant
From January 1941 production switched to the Mk II variant, which featured a more powerful Merlin 30 engine that raised the maximum speed slightly and improved load carrying.

HMS *ARK ROYAL*
In accordance with its multi-purpose nature, much use was made of the Fulmar as both a convoy escort and a (relatively) high-speed reconnaissance machine; in the former role external fuel tanks conferred the excellent patrol endurance of five-and-a-half hours.

Vought F4U Corsair (1941)

TYPE • Fighter **COUNTRY** • United States

The F4U had a troubled introduction to service in World War II, but by the end of the conflict it was challenging for a place among the best single-seat fighters of the war. It remained a viable ground-attack aircraft and night-fighter during the subsequent fighting in Korea.

SPECIFICATIONS (F4U-5N)

DIMENSIONS:	Length: 110.5m (34ft 7in); Wingspan: 12.49m (41ft); Height: 4.49m (14ft 9in)
WEIGHT:	Maximum takeoff 6398kg (14,106lb)
POWERPLANT:	One 1790kW (2400hp) Pratt & Whitney R-2800-32W Double Wasp 18-cylinder air-cooled radial piston engine
MAX SPEED:	756km/h (470mph)
RANGE:	1790km (1120 miles)
CEILING:	11,247m (36,900ft)
CREW:	1
ARMAMENT:	Four 20mm (0.79in) AN/M3 fixed forward firing in wings; up to 1452kg (3200lb) bombload or eight 127mm (5in) rockets

Above: Through modifications to the undercarriage and adoption of new landing techniques, the British Fleet Air Arm approved the Corsair for service at sea. This early 'birdcage' canopy Corsair Mk I served with No. 1835 Squadron in late 1943.

PRODUCTION RUN
The Corsair remained in production from 1942 to 1953, the longest production run of any piston-engine fighter. The F4U-5N here was one of the later versions and saw extensive service over Korea in 1950–53.

ARMAMENT
The Corsair had been fitted with four 20mm (0.79in) M-3 cannon starting with the F4U1C in mid 1943, but this armament was not initially popular and suffered from teething issues. The F4U-5N could also carry two 454kg (1000lb) bombs or napalm tanks, eight 127mm (5in) HVAR rockets or eight 300mm (11.75in) Tiny Tim rockets.

CORSAIR CONSTRUCTION
Apart from the highly cranked wing, which could be folded for storage below the carrier deck, the fighter utilized a broadly conventional airframe of all-metal construction. The FG-1 version differed in having fixed rather than folding wings.

DEPLOYMENTS
The first F4U-1 was delivered to the USN on 31 July 1944. Carrier trials began in September 1942 and the first Corsair unit, Marine Fighting Squadron VMF-214, was declared combat-ready in December, deploying to Guadalcanal in February 1943. After trials with VF-12, the Corsair became operational with Navy Fighting Squadron VF-17 in April 1943, deploying to a land base in New Georgia in September.

LETTER CODES
Vought F4U-5 Corsair fighters From USS *Tarawa* (CV-40) fly in formation over the Mediterranean, 15 December 1952. The US Navy introduced tail letter codes for its aircraft in July 1945, replacing the geometrical symbols used since January of the same year.

Lockheed P-38 Lightning (1941)

TYPE • *Heavy fighter* **COUNTRY** • *United States*

SPECIFICATIONS (P-38J)

DIMENSIONS:	Length: 11.53m (37ft 10in); Wingspan: 15.85m (52ft); Height: 2.99m (9ft 10in)
WEIGHT:	5806kg (12,800lb) (empty)
POWERPLANT:	Two 1063kW (1425hp) Allison V-1710-91 12-cylinder V-type
MAX SPEED:	667km/h (414mph) at 7620m (25,000ft)
RANGE:	3600km (2260 miles)
CEILING:	13,400m (44,000ft)
CREW:	1
ARMAMENT:	1 × 20mm (0.79in) cannon, 4 × 12.7mm (0.5in) machine guns in the nose, plus a bombload of 2 × 726kg (1600lb) bombs or 10 × 70mm (2.75in) rockets

The mighty P-38 was something of an anomaly among the US Army Air Force in World War II: a genuinely successful heavy fighter that was equally capable in the long-range escort role or as a hard-hitting ground-attack aircraft in both European and Pacific theatres.

This page: The P-38J resulted in a new lease of life for the Lightning, particularly during the 1943 daylight raids by USAAF B-17 and B-24 bombers over Europe.

RAISED COCKPIT
From the raised cockpit of the P-38 the pilot had an excellent view forward, unobstructed by a propeller. The canopy hinged backwards and had downward-winding side windows.

SUPERCHARGERS
The P-38 Lightning had supercharged engines. In the early part of the war the RAF wanted to buy substantial numbers of P-38s, but orders were cancelled when the Americans refused to fit the superchargers, which were deemed to be secret.

TWIN BOOMS
The Lightning's twin tail booms were the type's distinctive recognition feature, and led the Germans to nickname it the *Gabelschwanzteufel* or 'Fork-Tailed Devil'.

INTERCEPTOR
The P-38 was designed to meet a 1937 USAAC specification calling for a high-altitude interceptor capable of 580km/h (360mph) at 6100m (20,000ft) and 467km/h (290mph) at sea level. While the P-38J here is equipped with the standard 'fighter' nose, a number of J-models were adapted for use in the light bomber role, for which they were fitted with an alternative glazed nose to the centre nacelle for use by a bomb aimer.

P-38H FLIGHT
P-38Hs were the first to have the bar added to the national insignia. Uprated 1063kW (1425hp) V-1710-89/-91 engines powered the 375 P-38Hs, which also introduced automatic oil radiator flaps to solve a major engine overheating problem. This factory-fresh example is seen on a test flight from Lockheed's Burbank facility in California, prior to delivery to the USAAF.

Focke-Wulf Fw 190 (1941)

TYPE • Fighter   COUNTRY • Germany

SPECIFICATIONS (Fw 190A)

DIMENSIONS:	Length: 8.84m (29ft); Wingspan: 10.50m (34ft 5.5in); Height: 3.96m (13ft)
WEIGHT:	Empty 3170kg (7000lb); maximum take-off 4900kg (10,805lb)
POWERPLANT:	1566kW (2100hp) BMW 801D-2 radial engine
MAX SPEED:	654km/h (406mph) at 6000m (19,685ft)
RANGE:	1470km (915 miles)
CEILING:	11,400m (37,401ft)
CREW:	1
ARMAMENT:	Two 7.92mm (0.31in) machine guns and up to four 20mm (0.79in) cannon in wings, plus provision for under-fuselage and underwing bombs and rockets

At the time of its combat appearance, the Fw 190 was the most capable fighter in service, offering a winning combination of performance and manoeuvrability. Further development ensured it retained its prowess until the end of the war.

Above: The wide-track undercarriage of the Fw 190 made it much more suitable than the Bf 109 for operations from rough and semi-prepared airstrips, and gave it more forgiving handling characteristics in the hands of inexperienced pilots.

ARMAMENT
As built, early production Fw 190s were armed with four Rheinmetall Borsig MG.17 machine guns mounted in the upper fuselage and wing roots.

FIGHTER ACE
This aircraft is a Focke-Wulf Fw 190A-1 of 6/JG26 Schlageter, which was based in the Pas de Calais in November 1941. The machine was flown by Staffelkapitan Walter Schneider, an ace with 20 kills (note the tally on the tail). He was killed in December 1941 when his fighter hit high ground in fog.

TAILFIN
The robust fin was comprised of two spars; one vertical along the rear, and one angled along the leading edge. The rudder ran the full length of the fin.

NIGHT-FIGHTER
Night-fighting was a specialist role for which the Fw 190A-5/U2 was developed. This was used for *Wilde Sau* (Wild Boar) tactics in which day fighters were used at night, especially in order to counter the 'window'-jamming employed by RAF bombers. The prime exponent of *Wilde Sau* was JG 300, which operated both Fw 190s and Bf 109Gs.

FOCKE-WULF FW 190A-4
The Fw 190 spawned many variants between its entry into service in August 1941 and the end of the war, such as the A-4s seen here. The A-1 was followed into production by the A-2, with a longer span and heavier armament, and the A-3 fighter bomber. The Fw 190A-4 had a methanol-waterpower boost system. The final variant was the Fw 190A-8.

Republic P-47 Thunderbolt (1941)

TYPE • *Fighter* COUNTRY • *United States*

The pugnacious P-47 was one of the stand-out Allied fighters of World War II, equally adept as a long-range bomber escort over occupied Europe or as a potent ground-attack aircraft in theatres that ranged from the Mediterranean to Burma.

SPECIFICATIONS (P-47D)

DIMENSIONS:	Length: 10.99m (36ft 1in); Wingspan: 12.43m (40ft 9in); Height: 4.44m (14ft 7in)
WEIGHT:	Maximum take-off 7938kg (17,500lb)
POWERPLANT:	One 1500kW (2000hp) Pratt & Whitney R-2800-59 18-cylinder air-cooled radial piston engine
MAX SPEED:	686km/h (426mph)
RANGE:	1660km (1030 miles) with external tanks
CEILING:	13,000m (42,000ft)
CREW:	1
ARMAMENT:	Eight 12.7mm (0.5in) M2 Browning machine guns; up to 1100kg (2500lb) of bombs or six zero-length rockets under wings with drop tanks or 10 rockets without drop tanks

Above: This aircraft (43-25429) was part of the 19th Fighter Squadron, 218th Fighter Group, flying over Saipan in July 1944.

ENGINE EVALUATION
The aircraft had powerful new Pratt & Whitney R-2800 engines and a large 3.9m (13ft) diameter Curtiss Electric propeller.

CANOPY
The Block 25 P-47D introduced a bubble cockpit canopy as standard. As a result, the pilot's all-round vision was much improved compared to that found on earlier 'razorback' models.

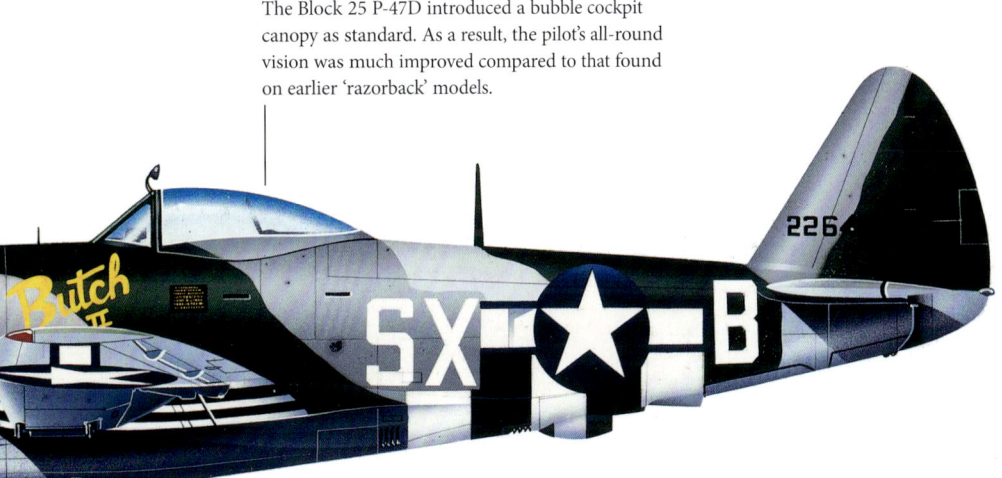

GROUND ATTACK
In all, P-47 production amounted to 15,675 aircraft, making it the most numerous American-made fighter in history. Of this total, a significant number were adapted for ground-attack duties, the P-47D featuring provision for underwing racks that could carry a pair of 454kg (1000lb) bombs, in addition to the 568-litre (125-Imp gal) drop tank under the fuselage. P-47Ds from later production batches had increased external stores capacity, including up to 10 127mm (5in) rockets.

WING GUNS
The characteristic staggered wing guns of the P-47D comprised four 12.7mm (0.5in) weapons in each wing, which were useful for strafing softer ground targets.

LADY RUTH
Evansville-built P-47D-15-RA 42-23289 *Lady Ruth* of the 19th FS, 318th FG is prepared for gun harmonization after repairs at its base at Ie Shima in the Marianas, July 1945. The P-47D-15 introduced important improvements, such as wing pylons for the carriage of fuel or a 454-kg (1000-lb) bomb, plus extra internal fuel tankage.

De Havilland Mosquito (1941)

TYPE • *Fighter-bomber* **COUNTRY** • *United Kingdom*

SPECIFICATIONS (NF.Mk 30)

DIMENSIONS:	Length: 12.73m (41ft 9in); Wingspan: 16.51m (54ft 2in); Height: 4.65m (15ft 3in)
WEIGHT:	9798kg (21,600lb) maximum take-off
POWERPLANT:	2 x 1275kW (1700hp) Rolls-Royce Merlin 76 V-12 liquid-cooled piston engines
MAX SPEED:	655km/h (407mph)
RANGE:	1143km (710 miles)
CEILING:	11,885m (39,000ft)
CREW:	2
ARMAMENT:	4 x 20mm (0.79in) cannon

Immortalized as the 'Wooden Wonder' on account of its construction, the Mosquito was the RAF's most flexible warplane of World War II, excelling in a wide variety of tactical roles and seeing service well into the 1950s.

Above: The Eighth Air Force, USAAF operated the second-largest fleet of Mosquitoes during World War II. RF992/'R' is an aircraft of the 654th Bomb Sqn, 25th Bomb Group (Reconnaissance), 325th Photo Wing based at RAF Watton in March 1945.

RADAR AND ARMAMENT
Mosquito NF.Mk IIs carried AI.Mk IV radar, with its so-called 'bow-and-arrow' nose antenna and two pairs of dipoles on each outer wing. Armament consisted of four Hispano 20-mm (0.79-in) cannon in a pack under the forward fuselage, and four Browning 7.7mm (0.303in) machine guns in the nose.

WING PLANFORM
One of the Mosquito's major recognition features was its distinctive wing planform, with engine nacelles extending forward of the fuselage nose.

NIGHT-FIGHTER
SW4079 was one of the original batch of 50 Mosquitoes, 21 of which were completed to the NF.Mk II night-fighter standard. It was issued to No. 157 Sqn, the first Mosquito night-fighter unit, in June 1942.

MULTIROLE AIRCRAFT
Built as a private venture, and allowed to go ahead only because it used mainly 'non-strategic' materials (spruce and balsa wood), the Mosquito was one of the most versatile aircraft of the war. The bomber and reconnaissance variants used their speed rather than armament for self-protection, while the night-fighter, fighter-bomber and anti-shipping models packed a heavy punch with machine guns, cannon and rockets. The Mosquito was licence-built in Canada and Australia.

SHIP STRIKE
On 12 August 1944, Nos. 235 and 248 Sqns of the RAF's Portreath Strike Wing visited the Gironde Estuary near Bordeaux in their Mosquito FB.Mk VIs. Using bombs and rockets they sank two enemy minesweepers.

Avro Lancaster (1941)

TYPE • Bomber   COUNTRY • United Kingdom

The most celebrated British heavy bomber of World War II, the Lancaster found fame for daring missions such as the 'Dambusters Raid' and the attack on the *Tirpitz*, but achieved greatest impact through its harrowing night-time strategic bombing campaign against Germany.

SPECIFICATIONS
(Lancaster B.Mk I)

DIMENSIONS:	Length: 21.18m (69ft 6in); Wingspan: 31.09m (102ft); Height: 6.10m (20ft)
WEIGHT:	31,751kg (70,000lb) maximum take-off
POWERPLANT:	4 x 1223kW (1640hp) Rolls-Royce Merlin XXIV V-12 piston engines
MAX SPEED:	462km/h (287mph)
RANGE:	4070km (2530 miles) with bombload
CEILING:	7470m (24,500ft)
CREW:	7
ARMAMENT:	8 x 7.7mm (0.303in) machine guns, plus bombload comprising one bomb of up to 9979kg (22,000lb) or smaller bombs up to a total weight of 6350kg (14,000lb)

Above: The special Lancaster Mk Is were powered by Rolls-Royce Merlin 28 engines, developing more power than the Merlin 20s that powered the unmodified Mk I aircraft.

REAR GUNS
As a weight-saving measure in this variant, the normal tail armament of four machine guns was reduced to two in the hydraulically operated Frazer-Nash gun turret.

NOSE TURRET
In the Lancaster B.I (Special) the nose turret, normally situated above the bomb-aimer's prone position, was removed in order to save weight.

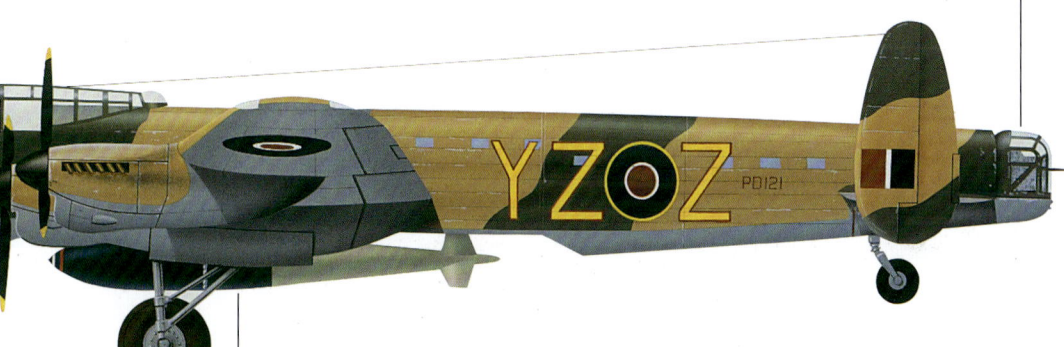

DAMBUSTERS
In early 1943, Wing Commander Guy Gibson, No. 5 Group RAF, was selected to recruit the best Bomber Command pilots to form a new, elite squadron for a special mission. The result was No. 617 Squadron, which was tasked with perfecting low-altitude flying before being assigned to target three dams in the heart of Germany's industrial Ruhr region, using specialized 'bouncing bombs'. The dams raid of May 1943 was a propaganda success, and one of the most daring bombing missions in history, but its strategic results were limited.

GRAND SLAM BOMB
In August 1943 No. 617 (Dambusters) Squadron began to receive Lancaster Mk Is specially modified to carry the Tallboy penetration bomb. Later, further modifications enabled the aircraft to carry the 10-tonne (9.8 ton) Grand Slam bomb, seen here.

DAY STRIKE
Although Lancasters are most commonly identified with the night-time strategic bombing campaign against Germany, they were also used in daytime raids, especially against targets over occupied Europe in the months immediately before the D-Day landings.

Consolidated B-24 Liberator (1941)

TYPE • Bomber   COUNTRY • United States

SPECIFICATIONS (B-24D)

DIMENSIONS:	Length: 20.22m (66ft 4in); Wingspan: 33.52m (110ft); Height: 5.46m (17ft 11in)
WEIGHT:	Maximum take-off: 29,029kg (64,000lb)
POWERPLANT:	Four 895kW (1200hp) Pratt & Whitney R-1830-43 Twin Wasp 14-cylinder air-cooled radial piston engines
MAX SPEED:	488km/h (303mph)
RANGE:	3700km (2300 miles)
CEILING:	9876m (32,400ft)
CREW:	10
ARMAMENT:	One 12.7mm (0.5in) Browning M2 MG in nose, ventral tunnel, and one in each of the left and right waist positions; two 12.7mm (0.5in) Browning M2 MGs in both top turret and tail turret; up to 5806kg (12,800lb) of bombs

Although never as famous as the Flying Fortress, the B-24 was an altogether more versatile machine, and in some respects more capable. More B-24s were built than any other US combat aircraft in history.

Left: The very long wingspan of the USAAF B-24s gave the aircraft excellent long range and good performance at high altitude.

REAR GUNNER
The rear gunner sat in a Consolidated or Motor Products turret manning two 12.7mm (0.5in) machine guns. These were fed from a magazine situated amidships.

DORSAL TURRET
The dorsal turret behind the flight deck was manned by the radio operator, to defend the aircraft against attack from above.

PLOESTI

The first major production version, the B-24D with R-1830-43 engines, appeared late in 1941. A policy decision to concentrate B-24s primarily in the Pacific theatre resulted in most of the 2738 B-24Ds being deployed against Japan. However, the Eighth and Ninth Air Forces in Europe and North Africa also received the aircraft, one of their outstanding raids being the attack on the Ploesti oil refineries in Romania on 1 August 1943.

FUSELAGE
The deep fuselage of the B-24 was less streamlined than that of the B-17 Flying Fortress, and the Liberator tended to burn more easily when seriously hit.

LONG-RANGE CAPABILITY
The B-24 Liberator was notable for its long range, which made it ideal for operations in the Pacific and for attacks on targets in the Balkans from bases in North Africa.

Mikoyan-Gurevich MiG-3 (1941)

TYPE • *Fighter* **COUNTRY** • *Soviet Union*

SPECIFICATIONS

DIMENSIONS:	Length: 8.25m (27ft 1in); Wingspan: 10.2m (33ft 6in); Height: 3.3m (10ft 10in)
WEIGHT:	Maximum take-off 3355kg (7397lb)
POWERPLANT:	One 1007kW (1350hp) Mikulin AM-35A V-12 liquid-cooled piston engine
MAX SPEED:	640km/h (400mph)
RANGE:	820km (510 miles)
CEILING:	12,000m (39,000ft)
CREW:	1
ARMAMENT:	One 12.7mm (0.5in) Berezin UBS machine gun, two 7.62mm (0.3in) ShKAS machine guns; two 100kg (220lb) bombs or up to eight RS-82 or rockets

Too specialized in concept for the combat conditions prevalent on the Eastern Front, the MiG-3 generally fought at a disadvantage. Despite this, several VVS aces scored their first kills with the aircraft type.

Above: Still in frontline service in 1944, this MiG-3 was operating with 7 IAP of the Black Sea Fleet, in concert with P-40s in the Kuban bridgehead area.

WING COLOUR
'Red 02' was a MiG-3 serving 12 GvIAP (Guards Fighter Aviaation Regiment). Red panels on the wing surfaces were intended to assist search aircraft in the event the aircraft was forced down.

BUILD QUALITY
Performance suffered on early production machines due to poor build quality, a situation dealt with by the authorities in a brutally simple fashion by removing a number of senior managers on trumped-up charges of sabotage.

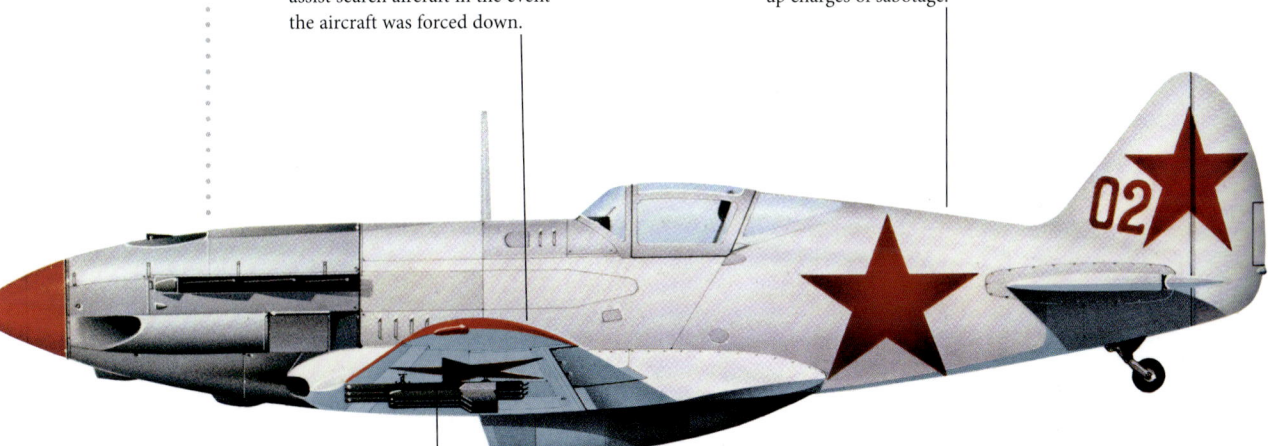

MIG-1 TO MIG-3

The MiG-3 was an evolution of the earlier MiG-1, which had been found severely wanting as a combat aircraft. Changes were introduced in an attempt to remedy the worst of its flaws. Improvements included increased fuel capacity, better cockpit visibility and armour for the pilot. First flown in October 1940, the MiG-3 had gained 250kg (551lb) in loaded weight over the MiG-1 and though the MiG-3 was faster at all altitudes, its climb rate had decreased and it now required a longer field length to get airborne.

ROCKETS
MiG-3s carried RS-82 rockets under their wings. Up to eight could be carried, as an alternative to a pair of 100kg (220lb) FAB-100 bombs in the fighter-bomber role.

READY FOR SERVICE
By the time Operation Barbarossa was launched in June 1941, the VVS had received 1289 MiG-3s, though these aircraft formed only about 10 per cent of its frontline fighter force.

Petlyakov Pe-2 (1941)

TYPE • *Bomber* COUNTRY • *Soviet Union*

SPECIFICATIONS (Pe-2FT)

DIMENSIONS: Length: 12.78m (41ft 11in); Wingspan: 17.11m (56ft 1.33in); Height: 3.42m (11ft 2.5in)

WEIGHT: Empty 5950kg (13,119lb); maximum take-off 8520kg (18,783lb) loaded

POWERPLANT: Two 940kW (1260hp) Klimov VK-105PF 12-cylinder V-type

MAX SPEED: 580km/h (360mph) at 4000m (13,125ft)

RANGE: 1315km (817 miles)

CEILING: 8800m (28,870ft)

CREW: 3

ARMAMENT: 6 x 7.62mm (0.3in) or 12.7mm (0.5in) machine guns; 1600kg (3527lb) of bombs

Soviet combat aircraft design made massive strides in the early 1940s, and nowhere was this better revealed than in the elegant Petlyakov Pe-2, which performed a multitude of tasks on the Eastern Front.

Above: With its winter camouflage nearly worn off, this early production Pe-2 'White 9' of 46 BAP was operational in the Moscow area during the dark days of the winter of 1941/42.

AIR BRAKES
The Pe-2 was equipped with slatted dive-brakes in its wings to slow it to an acceptable speed during a dive-bombing attack.

REAR GUNNER
A gunner manned a 12.7mm (0.5in) Berezin UB machine gun in the rear of the cockpit for defensive purposes.

PETLYAKOV PE-2FT
This Pe-2FT was flown by Lieutenant Colonel Vasili I. Rakov, commander of the 12th Guards Bomber Air Regiment in the Gulf of Finland in 1944.

HIGH PERFORMANCE

In its initial production variant, the Pe-2 was an impressive performer. Its level speed was identical to the Bf 109E, its excellent manoeuvrability belied its fighter origins and it was capable of carrying an internal bombload of 600kg (1323lb), plus a further 400kg (882lb) externally. In a dive-bombing attack the nose glazing allowed the pilot to visually acquire the target before initiating the dive, recovery being automatic.

PE-2 FLIGHT
The clean lines of the Pe-2 are clearly evident in this view of Soviet Air Force Pe-2FTs. Fighter escorts were seldom provided for Pe-2 bombers after 1942, the aircraft relying instead upon their defensive armament.

Supermarine Seafire (1941)

TYPE • *Naval fighter* **COUNTRY** • *United Kingdom*

SPECIFICATIONS (Mk III)

Dimensions:	Length: 9.21m (30ft 2.5in); Wingspan: 11.23m (36ft 10in); Height: 3.42m (11ft 2.5in)
Weight:	Empty 2814kg (6204lb); maximum take-off 3465kg (7640lb)
Powerplant:	1193kW (1600hp) Rolls-Royce Merlin 55m 12-cylinder V-type
Max Speed:	560km/h (348mph) at 3050m (10,000ft)
Range:	890km (553 miles)
Ceiling:	7315m (24,000ft)
Crew:	1
Armament:	Two 20mm cannon and four 7.7mm (0.303in) machine guns; external bomb or rocket load of 227kg (500lb)

With its narrow-track undercarriage and long nose, the Supermarine Seafire was far from ideal for aircraft carrier operations, but despite many accidents it served the Royal Navy well in all theatres of war.

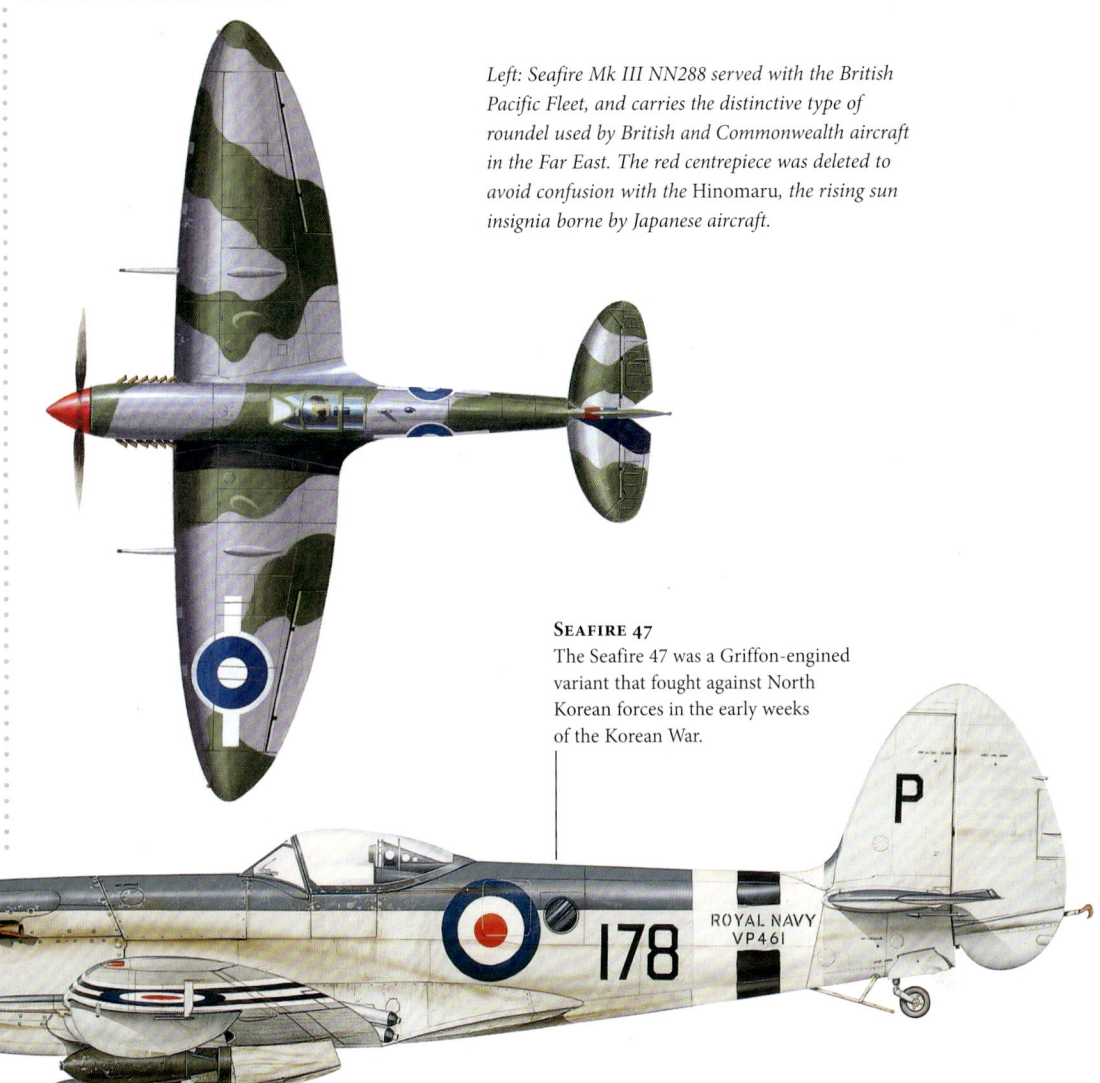

Left: Seafire Mk III NN288 served with the British Pacific Fleet, and carries the distinctive type of roundel used by British and Commonwealth aircraft in the Far East. The red centrepiece was deleted to avoid confusion with the Hinomaru, *the rising sun insignia borne by Japanese aircraft.*

Seafire 47
The Seafire 47 was a Griffon-engined variant that fought against North Korean forces in the early weeks of the Korean War.

FLEET AIR ARM
In Fleet Air Arm service, the Spitfire became the Seafire. Following trials aboard HMS *Illustrious* in late 1941, the first production Seafire appeared as the Mk IB, based on the land-based Mk V. In May 1942 it was followed by the Seafire Mk IIC, replacing the 'B'-type wing with the 'C'-type cannon wing. The Seafire L.Mk IIC was intended for low-altitude operations, while the F.Mk III and low-level LF.Mk III introduced a manually folding wing.

Underwing bombs
Seafires could carry two 113kg (250lb) bombs under the wings or a single 227kg (500lb) bomb under the fuselage. Provision was also made for the fitting of a 136-litre (30-Imp gal) drop tank in place of the single bomb.

ELLIPTICAL WINGS
The Seafire had the same elegant lines of the Spitfire. Some Seafire IIIs were delivered with non-folding wings. In other aircraft, the wing folded just inboard of the cannon mounting, the hinges being formed at the top boom of the front spar and at the rear spar.

Martin B-26 Marauder (1941)

TYPE • Bomber COUNTRY • United States

Dubbed the 'Widow Maker' after a series of early crashes, the B-26 went on to become one of the USAAF's most important medium bombers. By 1944, Ninth Air Force examples had the lowest loss rate in the European Theatre of Operations.

SPECIFICATIONS (B-26C)

DIMENSIONS:	Length: 17.75m (58ft 3in); Wingspan: 21.64m (71ft); Height: 6.05m (19ft 10in)
WEIGHT:	Maximum take-off: 16,783kg (37,000lb)
POWERPLANT:	Two 1491kW (2000hp) Pratt & Whitney R-2800-43 Double Wasp air-cooled radial piston engines
MAX SPEED:	454km/h (282mph)
RANGE:	1850km (1150 miles)
CEILING:	6614m (21,700ft)
CREW:	7
ARMAMENT:	11 x 12.7mm (0.50in) machine guns plus bombload of up to 1814kg (4000lb)

Above: The Marauder's box-type wing structure, formed by two heavy main spars with heavy-gauge skin, was reinforced by spanwise members to provide torsional stiffness.

WING REDESIGN
Later-production B-26Bs and the B-26C incorporated a redesign of the wing, increasing the span and reducing wing loading. Despite decreasing the maximum speed slightly, the new wing led to a considerably shortened take-off distance.

CREW
The Marauder's seven-man crew included two pilots, navigator, radio operator, front gunner/bombardier, turret and tail gunners.

B-26C MARAUDER
This B-26C-45-MO Marauder (42-107812/KS-J), nicknamed 'Baby Bumps II', was part of the 557th Bombardment Squadron, 387th Bomber Group, circa 1944.

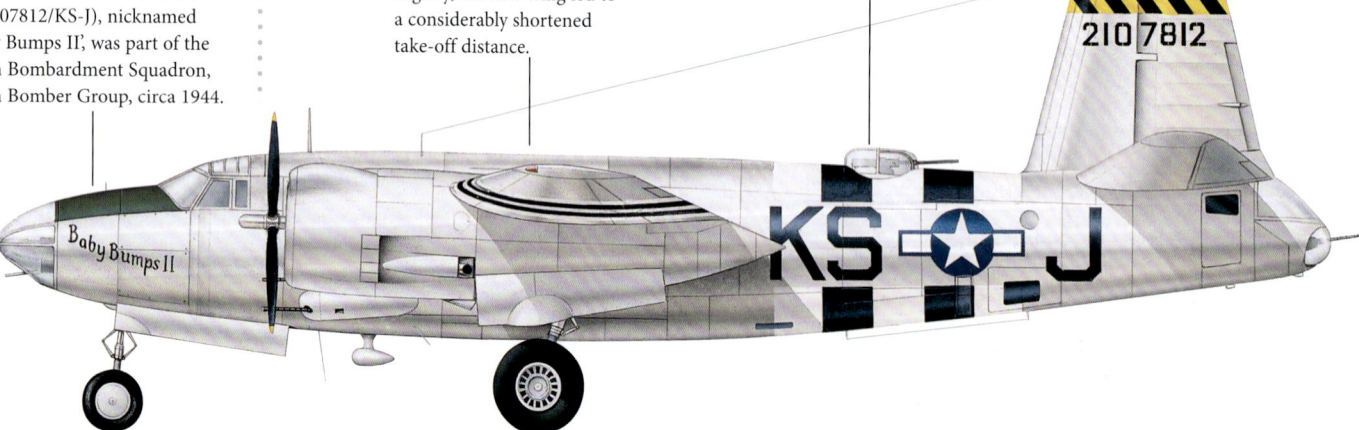

DESIGN CHALLENGE
The Marauder was ordered in response to the March 1939 USAAC Circular Proposal 39-640, calling for a twin-engined medium bomber with a max speed of 560km/h (350mph), a range of 4800km (3000 miles) and a bombload of 910kg (2000lb). This was a tough specification, and the Martin design team responded with an aerodynamically advanced design that featured a tricycle undercarriage and a notably small wing.

ETO MARAUDERS
'PN' codes identify these B-26Bs as aircraft of the 449th Bomb Sqn, 322nd Bomb Group. The 322nd was the first B-26 Group assigned to the European theatre.

Piper L-4 Grasshopper (1941)

TYPE • *Reconnaissance* **COUNTRY** • *United States*

The military version of the civilian Piper Cub light aircraft, the versatile L-4 Grasshopper saw widespread service in the US armed forces during World War II and after, appreciated for its role flexibility and useful flight characteristics.

SPECIFICATIONS

DIMENSIONS:	Length: 6.70m (22ft); Weight: 2.03m (6ft 8in)
WEIGHT:	Empty 290kg (640lb); maximum take-off 553kg (1220lb)
POWERPLANT:	48.5kW (65hp) Continental O-170-3 4-cylinder air-cooled engine
MAX SPEED:	137km/h (85mph)
RANGE:	402km (250 miles)
CEILING:	2835m (9300ft)
CREW:	2
ARMAMENT:	None

Above: The Grasshopper's undercarriage was a fixed tailwheel type, with two side Vs and two half axles hinged to a cabane below the fuselage. The wheeled undercarriage could be replaced by skis or floats.

COCKPIT
The Grasshopper had a fully enclosed cockpit, seating two crew members in tandem behind dual controls. There was a large door on the right-hand side, and sliding windows on the left.

BRACED WING
The L-4 was a braced high-wing monoplane, with steel-tube V-bracing struts each side. The wing was an aluminium structure covered with fabric, and was fitted with plain aluminium ailerons and flaps with fabric covering. There were no trim tabs.

TAIL UNIT
The tail unit was a wire-braced structure of welded steel tubes and channels, covered with fabric, reflecting the similar construction of the fuselage.

WORKHORSE
The military variant of the Piper Cub, the L-4 Grasshopper was selected for service in the USAAC in 1941, largely because of its impressive ability to operate from virtually any terrain and in very confined spaces. Over 5500 were built, serving in all theatres of war. The Grasshopper operated in numerous capacities, including casualty evacuation, resupply of troops in forward areas, liaison and frontline reconnaissance.

JUNGLE OPS
The Piper L-4 had a unique ability to operate from rough jungle airstrips. It remained in service for many years after the war, until the helicopter gradually took over its roles.

Messerschmitt Me 210 (1941)

TYPE • *Fighter* COUNTRY • *Germany*

SPECIFICATIONS (Me 210A-0)

DIMENSIONS: Length: 12.2m (40 ft); Wingspan: 16.3m (53ft 6in); Height: 4.2 m (13ft 9in)

WEIGHT: Maximum take-off 9705kg (21,396lb)

POWERPLANT: Two 783kW (100hp) Daimler-Benz DB 601A engines

MAX SPEED: 463km/h (288mph)

RANGE: 1818km (1130 miles)

CEILING: 8900m (29,200ft)

CREW: 2

ARMAMENT: 2 x 13mm (0.51in) MG 131 machine guns and 2 x 20mm (0.79in) MG 151 cannon firing forward; 2 x 13mm (0.51in) MG 131 firing to the rear

As early as 1938, the Reichsluftfahrtministerium (German Ministry of Aviation, RLM) were planning for a replacement for the Bf 110 heavy fighter. In the summer of 1938, Messerschmitt was awarded a contract for its Me 210 design.

Above: A pre-series Me 210A-0 of 3./ Schnellkampfgeschwader 210, circa 1942. This aircraft belly-landed at Tours, France, on 9 January 1942, after engine failure.

COCKPIT
The crew of two sat in tandem with the observer facing aft. To assist with aiming the rear-firing MG 131 gunsoptically flat panels at the rear of the canopy gave a clear view down and aft.

TAIL
Initially fitted with a Bf 110-style twin tail unit, this was swapped for a large single vertical stabilizer after the first flight of the prototype Me 210 in an attempt to improve the handling.

ME-210A1
Messerschmitt Me 210A-1, 2H+AA was operated by Erprobungsstaffel 210, the dedicated test and development unit based at Soesterberg. This included bombing missions over the United Kingdom.

ADVANCED DESIGN
The Me 210 was a considerable advance on the design of its predecessor with the cockpit right at the front of the fuselage sitting above the gun armament and a small bomb bay. The Daimler-Benz DB 601F engines were mounted on the front of the low-set mainplane, the propellers sitting forwards of the fuselage. On either side of the rear fuselage was a MG 131 13mm (0.51in) machine gun, controlled remotely by the gunner.

HANDLING
With its neatly cowled engines and purposeful nose contours, the Me 210 looked the part, but it was plagued with vicious and unpredictable handling qualities. This aircraft is one of those fitted with a longer rear fuselage, which largely cured the design's major handling faults.

Bell P-39 Airacobra (1941)

TYPE • Fighter COUNTRY • United States

The Airacobra was an unusual design that proved disappointing in US and British service. By chance, it turned out to be particularly well-suited to conditions on the Eastern Front and was by far the most successful Lend-Lease fighter to serve the USSR.

SPECIFICATIONS
(P-39Q Airacobra)

DIMENSIONS:	Length: 9.19m (30ft 2in); Wingspan: 10.36m (34ft); Height: 3.78m (12ft 5in)
WEIGHT:	Maximum take-off 3810kg (8400lb)
POWERPLANT:	One 890kW (1200hp) Allison V-1710-85 V-12 liquid-cooled piston engine
MAX SPEED:	626km/h (389mph)
RANGE:	845km (525 miles)
CEILING:	11,000m (35,000ft)
CREW:	1
ARMAMENT:	One 37mm (1.45in) M4 cannon, four 12.7mm (0.5in) Browning M2 machine guns; up to 230kg (500lb) bombload

Right: The engine was mounted immediately behind the pilot's seat, with a long shaft running under the seat and through the cockpit connecting the gear box to the propeller shaft.

COCKPIT DOOR
One of the more unusual features of the Bell Cobras was their car-type entry doors, a feature only found elsewhere among wartime fighter aircraft on early models of the Hawker Typhoon. The jettisonable door unit was much the same as that fitted to an automobile, with the normal opening handle and a winding handle for opening the window.

MARKINGS
The Guards badge on the pilot's door of this P-39 narrows it down to one of the Guards IAPs (Istrebitel'nyy Avia Polk –Fighter Regiments).

UNDERCARRIAGE
In the light of combat experience with the P-39, the nose undercarriage leg fork was strengthened, beginning with the P-39L model. This was a non-steerable, self-castoring unit.

LAYOUT
The unconventional Airacobra was designed around its main gun, a powerful 37mm (1.45in) Colt M4 cannon. The P-39 reversed normal practice by placing the engine behind the pilot and driving the propeller via a 3-m (9ft 11-in) long shaft that ran between the pilot's legs. This freed up the entire nose of the aircraft for weapons.

P-39Q
The P-39Q represented the final attempt to mould the Airacobra into a world-class fighter. This is a P-39Q-20, most of which were built for the USSR and delivered without guns, although this example was destined for the USAAF and retained its armament.

Yermolaev Yer-2 (1941)

TYPE • Bomber COUNTRY • Soviet Union

SPECIFICATIONS

DIMENSIONS:	Length: 16.42m (53ft 10in); Wingspan: 23m (75ft 6in) Height: 4.82m (15ft 10in)
WEIGHT:	Maximum take-off 18,580kg (40,962lb)
POWERPLANT:	Two 1100kW (1500hp) Charomskiy ACh-30B V-12 liquid-cooled diesel piston engines
MAX SPEED:	507km/h (315mph)
RANGE:	5500km (3417 miles)
CEILING:	7200m (23,600ft)
CREW:	4
ARMAMENT:	One 12.7mm (0.5in) Berezin UBT flexibly mounted in nose, one 12.7mm (0.5in) Berezin UBT on flexible ventral mount, one 20mm (0.78in) ShVAK cannon in dorsal turret; up to 5000kg (11,023lb) bombload

The Yermolaev Yer-2 was derived from a highly advanced prototype trainer. Despite being built in relatively small numbers, the Yermolaev Yer-2 was an important design that saw intensive combat on the Eastern Front.

Above: The original airliner was redesigned as the DB-240, standing for Dalny Bombardirovshcik *(long-range bomber), but became the Ye-2 on production.*

DESIGN
The Yer-2 was developed from the Stal-7 – designed by Italian-born engineer Roberto Ludvigovich Bartini, development of which had passed to his deputy, Vladimir Yermolaev.

ARMAMENT
The DB-240 offered an excellent range with a 1000kg (2205lb) bombload and the defensive armament was considered good, with multiple heavy machine guns and a 20mm (0.79in) cannon in the dorsal turret.

ENGINES
Performance was not quite as good as expected due to the aircraft being forced to use Klimov M-105s rather than the higher-powered M-106s it was designed for, as the M-106 was suffering from insuperable cooling problems. The aircraft was later fitted with efficient Charomskiy ACh-30 diesel engines.

FIRST MISSION
The Yer-2's initial production life was short, with only 128 constructed when in August 1941 Zavod 18 switched to building the more urgently required IL-2. In the same month the Yer-2 entered combat for the first time, when on the 10th three examples bombed Berlin, one subsequently being shot down in error by I-16s on its return to Soviet airspace.

LATE PRODUCTION
After a pause in production in August 1941, Yer-2 production was reinitiated towards the end of 1943. Production difficulties delayed service entry and the first bombing mission of the new Yer-2 occurred on 7 April 1945.

Northrop P-61 Black Widow (1941)

TYPE • *Heavy fighter* COUNTRY • *United States*

SPECIFICATIONS
(P-61B Black Widow)

DIMENSIONS:	Length: 15.11m (49ft 7in); Wingspan: 20.12m (66ft); Height: 4.46m (14ft 8in)
WEIGHT:	13,472kg (29,700lb) maximum take-off
POWERPLANT:	2 x 1491kW (2000hp) Pratt & Whitney R-2800-65 radial piston engines
MAX SPEED:	589km/h (366mph)
RANGE:	4506km (2800 miles)
CEILING:	10,090m (33,100ft)
CREW:	3
ARMAMENT:	Four 20mm (0.79in) cannon, plus 4 x 12.7mm (0.5in) machine guns in later aircraft, and provision to carry up to 4 x 726kg (1600lb) bombs

Without doubt one of the finest night-fighters of the war, the P-61 was one of the few Allied aircraft that was specifically designed for the task. The Black Widow served with distinction in the Pacific, often as a night intruder on offensive missions against land and sea targets.

Above: The streamlined four-gun dorsal turret caused aerodynamic buffeting when fitted to the P-61A, but a slight lengthening of the nose in the P-61B cured the problem.

RADAR
The SCR-720 radar was an advanced piece of equipment, having anti-jamming features that would seek out an enemy aircraft even if the latter were using countermeasures.

NIGHT CAMO
Wartime research at MIT discovered that the best camouflage for rendering a night fighter 'invisible' when caught in searchlight beams was found to be a very glossy black.

INVASION STRIPES
P-61A-5 42-5564, *Jukin Judy*, was flown by the 422nd NFS during its work-up period at RAF Scorton, Yorkshire. In addition to its 'shark's mouth', the aircraft wore partial invasion stripes.

WIDOW INTRUDERS
The first production P-61A Black Widow aircraft appeared in early 1941. Although always classed as a night-fighter, the P-61B version that followed saw increasing use as a night intruder, and was capable of carrying weapons loads including four 726kg (1600lb) bombs, or four 1136-litre (300-Imp gal) drop tanks under the wings. Serving in the Far East, some aircraft were further adapted in field in order to carry 127mm (0.5in) rocket projectiles.

AERIAL AGILITY
Northrop installed flaps over almost the whole span of the wing. The conventional ailerons were very small, but roll control was backed up when needed by four sections of differential spoiler on each wing. This enabled the P-61 to be amazingly agile considering its size and weight.

Yakovlev Yak-9 (1942)

TYPE • Fighter   COUNTRY • Soviet Union

SPECIFICATIONS (Yak-9)	
DIMENSIONS:	Length: 8.48m (27ft 10in); Wingspan: 10m (32ft 10in); Height: 2.64m (8ft 8in)
WEIGHT:	Maximum take-off 2844kg (6269lb)
POWERPLANT:	One 1014kW (1360hp) Klimov M-105PF-3 V-12 liquid-cooled piston engine
MAX SPEED:	597km/h (371mph)
RANGE:	1400km (870 miles)
CEILING:	10,400m (34,100ft)
CREW:	1
ARMAMENT:	One 37mm (1.45in) Nudelman-Suranov NS-37 cannon, one 12.7mm (0.5in) Berezin UBS machine gun

Two distinct 'heavyweight' and 'lightweight' lines of the basic Yak fighter emerged and the various Yak-9 models represented the culmination of the development of the more versatile 'heavy' version derived from the Yak-7.

Above: The tricolour spinner and French roundel beneath the windscreen identify this Yak-9D as operating with the French-manned Normandie-Niemen Regiment.9

COCKPIT
The Yak had a four-piece cockpit canopy, which gave the pilot poor visibility. Cockpit equipment was rudimentary; the gunsight was primitive, there were no blind flying instruments and no fuel gauges.

FUEL CAPACITY
the Yak-9 differed from the Yak-7 primarily in its increased fuel capacity in the wings that were now primarily constructed of light alloy.

ARMAMENT
The standard armament of the Yak-9 was a 37mm (1.45in) NS-37 cannon in the nose, firing through the propeller boss.

AIR SUPERIORITY
The Russians were late in developing really effective monoplane fighters that were in the same class as Britain's Hurricane and Spitfire and Germany's Bf 109, but Aleksandr Yakovlev's attractive designs soon redressed the situation. The Yak-1, which made its first public appearance on 7 November 1940, evolved into the improved Yak-7 and Yak-9; the latter could take on the best of late-war Luftwaffe fighter aircraft.

YAK-9 FORMATION
The lead aircraft in this Yak-9 formation carries insignia denoting both the Order of the Red Banner (forward) and that the aircraft belongs to a Guard's Unit (aft). This machine was flown by Hero of the Soviet Union M.V. Avdyeyev over the Crimea in May 1944.

Grumman F6F Hellcat (1942)

TYPE • *Naval fighter* COUNTRY • *United States*

Credited with more combat victories than any other carrier fighter in history, the large and rugged F6F Hellcat fighter effectively established the US Navy's aerial ascendancy over their Japanese foes.

SPECIFICATIONS
(F6F-5 Hellcat)

DIMENSIONS: Length: 10.24m (33ft 7in); Wingspan: 13.05m (42ft 9.66in); Height: 3.99m (13ft 1in)

WEIGHT: Maximum take-off 7025kg (15,487lb) loaded

POWERPLANT: 1491kW (2000hp) Pratt & Whitney R-2800-10W radial engine

MAX SPEED: 612km/h (380mph) at 7132m (23,400ft)

RANGE: 1521km (945 miles)

CEILING: 11,369m (37,300ft)

CREW: 1

ARMAMENT: Six 12.7mm (0.50in) machine guns, or two 20mm (0.79in) cannon and four 12.7mm (0.50in) machine guns, plus provision for two 453kg (1000lb) bombs or six 127mm (5in) rocket

Above: Three auxiliary air intakes under the engine cowling fed cooling air for the engine oil (centre) and supercharger (two side intakes). Cooling gills on the engine cowling could be opened to increase the airflow over the cylinders.

MIDNIGHT BLUE
The Hellcat shown here is painted in overall 'midnight blue' camouflage, which replaced the standard Pacific colour scheme of sea blue and pale grey under-surfaces later in the war.

CANOPY
The pilot sat under a sliding canopy and was well-protected by armour, particularly to the rear. However, no rear-view mirror was provided and rearward visibility was lacking.

ARRESTOR HOOK
The Hellcat's arrestor hook was a 'Sting' unit projecting from the extreme rear of the fuselage.

PACIFIC FIGHTER
The Hellcat made its first flight in June 1942. In the Pacific, the Hellcat played a key role in all US naval operations. A more powerful variant, the F6F-5, was fitted with a Pratt & Whitney R-2800-10W engine, capable of developing an emergency power of 1640kW (2200hp) by using water injection. The F6F-5 began to reach the Pacific task forces in the summer of 1944, and 6436 examples of this variant had been built when production ended in November 1945.

POST-WAR HELLCAT
A radar-equipped F6F-5N night-fighter seen in post-war US Navy service. This aircraft was typical of Hellcats that served with 13 reserve units around the United States after VJ-Day. Aboard carriers, the last US Navy Hellcats bowed out in 1948, although the type made a brief return to sea in 1953.

Hawker Typhoon (1942)

TYPE • *Fighter-bomber* **COUNTRY** • *United Kingdom*

SPECIFICATIONS (Mk IB)

DIMENSIONS:	Length: 9.73m (31ft 11in); Wingspan: 12.67m (41ft 7in); Height: 4.66m (15ft 4in)
WEIGHT:	4010kg (8840lb)
POWERPLANT:	Napier Sabre IIC liquid-cooled H-24 piston engine, 1685kW (2260hp)
MAX SPEED:	663km/h (412mph)
RANGE:	821km (510 miles)
CEILING:	10,729mm (35,200ft)
CREW:	1
ARMAMENT:	4 × 20mm (0.79in) Hispano Mk II cannon; 8 × RP-3 air-to-ground rockets; 2 × 227kg (500lb) or 2 × 454kg (1000lb) bombs

An interceptor that failed, the Hawker Typhoon was nearly cancelled before it blossomed into the finest close-support aircraft of World War II. With its pugnacious snub nose, four long-barrelled cannon and whining Sabre engine, the big fighter-bomber wreaked havoc on its foes.

Above: The Typhoon's wing was built around an immensely strong two-spar structure. Mounted in the port wing just outboard of the landing light was a gun camera to record rocket and cannon strikes.

CODE LETTERS
This Typhoon IB carries the code letters of No. 181 Squadron. In the early part of 1944 the squadron's Typhoons carried out many attacks on V-1 flying bomb sites before moving to France after D-Day in support of the Allied armies.

AIR INTAKE
In the centre of the large air intake was a ram air inlet for the supercharger, surrounded by an annular radiator for oil cooling. The engine radiator was outside this.

PROBLEMS
The Hawker Typhoon was designed to combat heavily armed and armoured escort fighters like the Messerschmitt Bf 110. The first of two prototypes flew for the first time on 24 February 1940. However, the aircraft suffered from constant teething troubles, and the first Typhoon squadron did not become operational until May 1942.

HARDPOINTS
The Typhoon's underwing hardpoints could accommodate eight 27kg (60lb) rockets, as illustrated, or two bombs of up to 453kg (1000lb) in weight. The four 20mm (0.79in) Hispano cannon were armed with 120 rounds per gun.

GROUND-ATTACK
The Typhoon MK 1A, which was armed with 12 7.7mm (0.303in) machine guns, soon gave way to the Mk 1B (seen here), whose four 20mm (0.79in) cannon proved effective in the ground-attack role and which was powered by the more reliable 1625kW (2180hp) Sabre IIA engine. By the end of 1943 the Typhoon was heading for its place in history as the most potent Allied fighter-bomber.

Kawasaki Ki.45 Toryu (1942)

TYPE • Bomber **COUNTRY** • Japan

After a protracted development programme, the Kawasaki Ki.45 *Toryu* (Dragon Slayer) became one of the most effective Japanese combat aircraft of World War II, but despite its performance its full potential was never realized.

SPECIFICATIONS (Ki.45 Kai-c)

DIMENSIONS:	Length: 11.00m (36ft 1in); Wingspan: 15.02m (49ft 3in); Height: 3.70m (12ft 1.33in)
WEIGHT:	Empty 4000kg (8820lb); maximum take-off 5500kg (12,125lb) loaded
POWERPLANT:	Two 805kW (1080hp) Mitsubishi Ha-102 14-cylinder radials
MAX SPEED:	540km/h (336mph) at 5000m (16,405ft)
RANGE:	2000km (1243 miles)
CEILING:	10,000m (32,810ft)
CREW:	2
ARMAMENT:	One 37mm (1.46in) cannon, two 20mm (0.79in) cannon and one 7.92mm (0.31in) machine gun

Above: Early in 1937, Kawasaki initiated the design and development of a twin-engined heavy fighter that would be suitable for long-range operations over the Pacific. The result was the Kawasaki Ki.45.

ENGINES
The Ki.45 Kai-c was powered by two reliable Mitsubishi HA-102 radial engines. The nacelles housing these were longer than those of earlier variants.

TOP CANNON
The night-fighter version of the *Toryu* was equipped with a pair of 20mm (0.79in) cannon, obliquely mounted in the upper fuselage between the two cockpits, enabling the pilot to attack an enemy bomber from below.

CAMOUFLAGE
This *Toryu* is camouflaged light grey, with olive green blotches on the upper surface. The rising sun national insignia (*Hinomaru*) is superimposed in this case on a white band, rather than having a white outline. Tail markings identifying the unit were always colourful, as in this instance.

HISTORY
The Kawasaki Ki.45 *Toryu* prototype first flew in January 1939. The development programme continued slowly and the aircraft did not enter service until the autumn of 1942 as the Ki.45 Kai-a fighter and the Ki.45 Kai-b ground-attack and anti-shipping strike aircraft, the 'Kai' suffix denoting 'improved'. The Ki.45 Kai-c was a night-fighter version, while the Kai-d was an improved ground-attack/anti-shipping variant.

IN COMBAT
Although the Ki.45 was a capable aircraft, many of this type were expended in *kamikaze* attacks during the final months of the war. Total production of the *Toryu* was 1675 aircraft, of which 477 were night-fighters. The aircraft received the Allied code name 'Nick'.

Kawasaki Ki.61 Hien (1942)

TYPE • *Fighter* **COUNTRY** • *Japan*

SPECIFICATIONS (Ki-61-I-KAIc)

DIMENSIONS:	Length: 8.75m (28ft 8.5in); Wingspan: 12m (39ft 4.25in); Height: 3.7m (12ft 1.75in)
WEIGHT:	Empty 2210kg (4872lb); maximum take-off weight 3250kg (7165lb)
POWERPLANT:	One 1175hp (876kW) Kawasaki Ha-40 (Army Type 2) 12-cylinder inverted-V engine
MAX SPEED:	592km/h (368mph)
RANGE:	1100km (684 miles)
CEILING:	11,600m (37,730ft)
CREW:	1
ARMAMENT:	Two 20mm cannon and two 12.7mm (.5in) machine guns; up to 500kg (1102 lb) of bombs

The only frontline Japanese combat aircraft to deviate from the radial engine, the Kawasaki Ki-61 *Hien* (Swallow) was dubbed the 'Tony' by the Allies. It was often mistaken for the Messerschmitt Me 109, although all it had in common was the (Kawasaki-built) DB 601 engine.

Above: The Ki.61 was unique among Japanese fighter aircraft in that it marked the first attempt by the Japanese Army Air Force to incorporate armour protection and self-sealing fuel tanks into the design from the outset.

ENGINE
The Ki.61 Daimler-Benz DB.601A was chosen because reports from the air fighting in Europe seemed to indicate that the liquid-cooled engine was superior to the air-cooled variety.

WING RATIO
Another feature that made the Ki.61 stand out from other Japanese fighters was its high aspect ratio wing; this showed the influence of Dr Vogt, a German designer who had been closely involved with Kawasaki and who became chief designer of Blohm & Voss.

VARIANTS
Designed to replace the Nakajima Ki.43 *Hayabusa* (Oscar) in Japanese Army service, the Kawasaki Ki.61 began to reach frontline air units in August 1942. The principal versions were the Ki.61-I (1380 aircraft built in two subvariants, differentiated by their armament); the Ki.61 Kai, with a lengthened fuselage and different armament fits (1274 built); and the Ki.61-II, optimized for high altitude operation with a Kawasaki Ha.140 engine (374 built).

ARMAMENT
The standard *Hien* was armed with four 12.7mm (0.50in) machine guns, two of which were mounted in the upper forward fuselage. Some Ki.61-I fighters were armed with German 20mm (0.79in) Mauser MG 151/20 cannon, 400 of which were imported.

LAST STAND
This Ki-61-I of the 37th Sentai was among those that fought in the last stages of the defence of the Philippines, before being forced to redeploy to Formosa and Okinawa in the last year of the war.

Curtiss SB2C Helldiver (1942)

TYPE • *Naval bomber* COUNTRY • *United States*

The Curtiss SB2C Helldiver weathered a protracted development, problematic service introduction, and an appalling reputation to become one of the most successful naval aircraft of the entire war, making a signal contribution in the Pacific Theatre.

SPECIFICATIONS
(Curtiss SB2C-1C)

DIMENSIONS:	Length: 11.18m (36ft 8in); Wingspan: 15.16m (49ft 9in); Height: 4.01m (13ft 2in)
WEIGHT:	Maximum takeoff 7388kg (16287lb)
POWERPLANT:	One 1400kW (1900hp) Wright R-2600-20 Twin Cyclone 14-cylinder air-cooled radial piston engine
MAX SPEED:	462km/h (287mph)
RANGE:	1786km (1110 miles)
CEILING:	7370m (24179ft)
CREW:	2
ARMAMENT:	Two 20mm (0.79in) AN/M2 cannon in wings, two 7.62mm (0.3in) M1919 Browning machine guns flexibly mounted in rear cockpit; up to 910kg (2000lb) bombload or one 910kg (2000lb) Mark XIII torpedo

Above: A Curtiss SB2C operating as part of Task Force 58. The Helldivers assigned to USS Hancock flew strikes against Japanese ground targets in support of the landings on Iwo Jima during February 1945.

COLOURING
The Helldiver is finished in standard US Navy 'midnight blue' upper surfaces and sea-grey under surfaces; paint schemes varied from theatre to theatre.

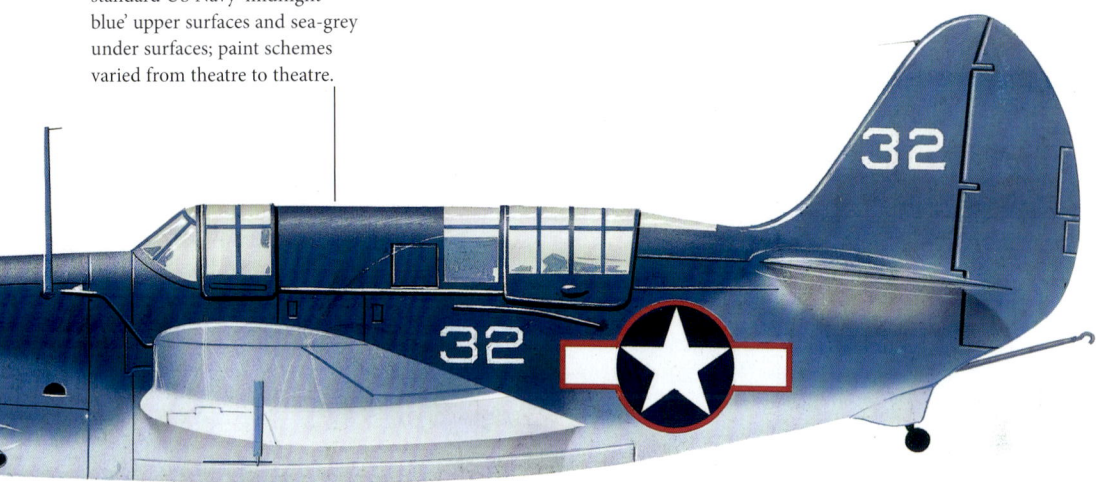

POWERPLANT
The Helldiver was powered by the Wright Cyclone radial engine, a very reliable powerplant on which crews could depend during their long over-water missions.

BOMB BAY
The Helldiver could accommodate one torpedo or a bombload of 454kg (1000lb) in its fuselage weapons bay; 454kg (1000lb) of bombs could also be carried on underwing racks.

RABAUL ATTACK
The SB2C Helldiver was designed as a replacement for the SBD Dauntless, which had achieved fame in the Battle of Midway in June 1942. Because of delays caused by the crash of the prototype, the first production Helldiver did not fly until June 1942 and although the aircraft entered service with the US Navy in December 1942, it did not make its operational debut until 11 November 1943, in an attack on the Japanese-held island of Rabaul.

US NAVY HELLDIVER
The Helldiver proved initially disappointing, with little improvement in speed compared to the Dauntless, and better range and load-carrying abilities spoiled by poor handling and low serviceability. Yet, despite a rather unpromising start, the Helldiver was of such great value in the Pacific theatre that the US Navy absorbed almost the entire production of over 7000 aircraft.

Grumman TBF Avenger (1942)

TYPE • *Dive-bomber* **COUNTRY** • *United States*

SPECIFICATIONS (TBF-1C)

Dimensions:	Length: 12.19m (40ft); Wingspan: 16.51m (54ft 2in); Height: 5m (16ft 5in)
Weight:	Empty 4788kg (10,555lb); maximum take-off 7876kg (17,364lb)
Powerplant:	1268kW (1700hp) Wright R-2600-8 Cyclone 14-cylinder radial engine
Max Speed:	414km/h (257mph) at 3660m (12,000ft)
Range:	1780km (1105 miles)
Ceiling:	6525m (21,400ft)
Crew:	3
Armament:	Three 12.7mm (0.50in) machine guns and one 7.62mm (0.30in) machine gun; torpedo, bomb and rocket load up to 1134kg (2500lb)

Although it had a disastrous start to its operational career at the Battle of Midway in June 1942, when five out of six aircraft were shot down in an attack on the Japanese task force, the Avenger went on to become one of the best shipborne torpedo-bombers of World War II.

Above: The TBF's cantilever undercarriage oleo legs were hinged at the extremities of the centre section and retracted outwards into the undersides of the outer wing.

Fuselage
The Avenger's fuselage was of oval section and semi-monocoque construction, built up from a series of angle frames and stamped bulkheads, all covered by a smooth metal skin.

Crew
The TBF normally carried a crew of three: pilot, navigator/bomb aimer and radio operator. NZ2518 was flown by Flight Lieutenant Fred Ladd of No. 30 Sqn, RNZAF at Piva, Bougainville, in 1944.

Insignia
The insignia adopted by the RNZAF in the Pacific theatre were a hybrid of British and US styles.

SUBVARIANTS

Subvariants of the Avenger included the TBF-1C, some of which were fitted with two wing-mounted 20mm (0.79in) cannon, the TBF-1B, which was supplied to the Royal Navy under Lend-Lease and was initially known as the Tarpon, the TBF-1D and TBF-1E with ASV radar, and the TBF-1L with a searchlight in the bomb bay. Final wartime versions were the camera-equipped TBM-3P and the TBM-3H, with search radar.

SINGLE-ENGINE DESIGN
As the largest single-engined aircraft to serve in World War II, the Avenger represented a massive step up in capability and complexity compared to the TBD Devastator. Production of the TBF-1 and TBM-1, including subvariants, amounted to 2290 and 2882 aircraft respectively.

Dornier Do 217 (1942)

TYPE • *Fighter-bomber* COUNTRY • *Germany*

Drawing heavily on the Do 17 design, Dornier produced a bigger, heavier bomber in the shape of the Do 217. Although it was overshadowed in its primary role by the He 111 and Ju 88, it nevertheless proved adept at anti-shipping strikes.

SPECIFICATIONS (Do 217N-1)

DIMENSIONS:	Length: 17.1m (56ft 3in); Wingspan: 19m (62ft 4in); Height: 4.8 m (15ft 8in)
WEIGHT:	Maximum take-off 15,000kg (33,000lb)
POWERPLANT:	Two 1380kW (1850hp) DB603A piston V12 aero engines
MAX SPEED:	525km/h (326mph)
RANGE:	1755km (1090 miles)
CEILING:	8400m (27,600ft)
CREW:	4
ARMAMENT:	4 x 20mm (0.79 in) MG 151 cannon and 4 x 7.92mm (0.31 in) MG 17 machine guns (nose); one 13mm (0.51 in) MG 131 machine gun (dorsal turret) and one 7.92mm (0.31 in) MG 13 (ventral position)

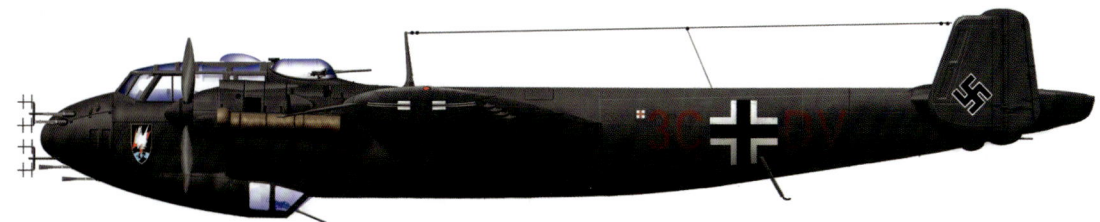

Above: Do 217N-1 3C+DV was operated by II./ Nachtjagdgeschwader 4 from a base in Germany in late 1943.

NOSE RADAR
The radar was usually the FuG 202 or 212 Lichtenstein C-1 with its characteristic *Matratzen* (mattress) antenna array.

SCHRÄGE MUSIK ARMAMENT
The Do 217N dispensed with the rear-facing armament of earlier night-fighter versions, with a consequent weight-saving put to good use by the fitment of two pairs of MG 151/20 20mm (0.79in) cannon in the rear fuselage, firing upwards at an angle of 70 degrees.

MARKINGS
This aircraft wears typical mottled grey night-fighter colours. The four-letter codes were a pre-delivery factory radio call sign.

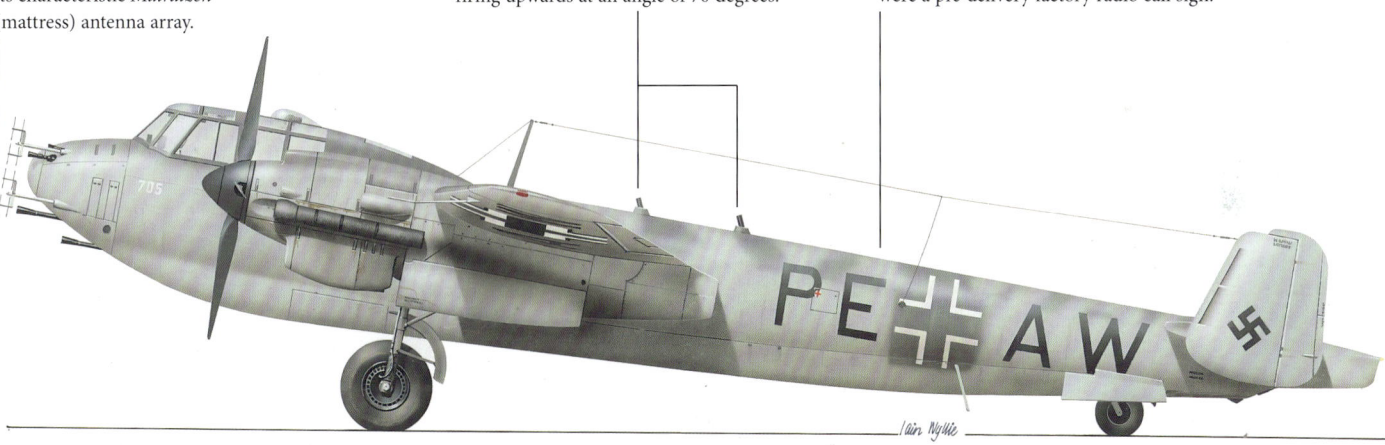

DIVERSIFICATION
The Do 217 was Dornier's response to a 1937 requirement for a long-range warplane optimized for the heavy level and dive-bombing roles. Some 800 Do 217Es were built, before being succeeded by 950 Do 217Ks. Variants included bombers, anti-shipping bombers, high-altitude reconnaissance aircraft and a missile-launching aircraft. The aircraft also proved able to carry some of the more exotic weapons fielded by the Luftwaffe.

NIGHT-FIGHTER
Following some success with night-fighting Do 17s and 215s, Dornier fitted a new nose to its Do 217 bomber to produce an interim night-fighter variant. Few wartime aircraft packed a harder punch – this N-2 shows the four-cannon, four-gun nose armament, as well as the antennas for the FuG 212 Lichtenstein C-1 radar.

Heinkel He 177 (1942)

TYPE • Bomber COUNTRY • Germany

SPECIFICATIONS (He 177A-5)

DIMENSIONS:	Length: 22m (72ft 2in); Wingspan: 31.44m (10ft 2in); Height: 6.4m (21ft)
WEIGHT:	Maximum take-off 31,000kg (68,300lb)
POWERPLANT:	Two 2133kW (2860hp) Daimler-Benz DB 610 24-cylinder liquid-cooled piston engines
MAX SPEED:	565km/h (351mph)
RANGE:	1540km (960 miles)
CEILING:	8000m (26,000ft)
CREW:	6
ARMAMENT:	One 7.92mm (0.31) MG 81 MGs, two 20mm (0.79in) MG 151 cannon, four 13mm (0.51in) MG 131 MGs; 2000kg (4410lb) bombload, LT 50 torpedos or Hs 293 missiles

The He 177 heavy bomber traces its origins back to the P.1041 project for a long-range bomber that emerged out of mid-1930s plans to field a strategic bombing force for the Luftwaffe.

Right: A captured Heinkel He 177, with French markings. German crews nicknamed the aircraft Luftwaffenfeuerzeug ('flying lighter') because of its tendency to catch fire.

POWERPLANT
The main advantage of the A-5 was that it introduced the more powerful DB 610 engine, and as the weights were only fractionally heavier than those of the first versions the performance was improved, especially in ceiling.

'BOMBER A' REQUIREMENT
This challenging requirement that led to the He 177, demanded a maximum speed of 540km/h (335mph) and the ability to carry a 2000kg (4410lb) bombload over a radius of 1600km (995 miles) at a cruising speed of 500km/h (310mph).

FIRE RISK
In addition to the fire risks associated with the He 177's oil and fuel systems, the engines of the A-1 were installed so tightly up to the main spar that there was no room for a firewall. The piping, electric cables and other services were also jammed in so tightly that, especially when soaked in leaking fuel and oil, the fire risk was terrifying.

HE 177 WITH HS-293 GUIDED MISSILE
The heavy-lifting power of the He 177 meant that it was able to lift all manner of experimental and advanced Luftwaffe weaponry/weapons carried by the He 177A-3 including the Hs 293 radio-controlled missile and, in the A-3/R7 and all A-5 versions, a range of anti-ship torpedoes, including the LT 50 glider torpedo.

Lavochkin La-5 (1942)

TYPE • Fighter **COUNTRY** • Soviet Union

When the La-5 appeared in service it was the first Soviet fighter with superior performance to the Bf 109G. The most widely used series of radial-engined fighters in the Soviet air force inventory, 9920 were built during the Great Patriotic War.

SPECIFICATION (La-5FN)

Dimensions:	Length: 8.67m (28ft 5in); Wingspan: 9.8m (32ft 2in); Height: 2.54m (8ft 4in)
Weight:	Maximum take-off 3402kg (7500lb)
Powerplant:	One 1380kW (1850hp) Shvetsov M-82FN 14-cylinder air-cooled radial engine
Max Speed:	648km/h (403mph)
Range:	765km (475 miles)
Ceiling:	11,000m (36,000ft)
Crew:	1
Armament:	Two 20mm (0.79in) ShVAK cannon; two bombs up to 100kg (220lb) each

Above: Vitaly Popkov of 2 GvIAP scored the majority of his 40 victories with the La-5FN. 'Yellow 01' was his personal aircraft in the summer and autumn of 1944 while operating over the Ukrainian Front.

COOLING
The radial engine of the fighter was extremely closely cowled with sufficient engine cooling being achieved through the use of fans mounted in front and behind the engine.

CANOPY
The canopy of the La-5 offered excellent all-round vision, but was difficult to open at high speeds. Consequently, some pilots flew with their canopies open.

LAVOCHKIN LA-5FN
This Lavochkin La-5FN was flown by Kapitan Petr Yakovlevich Likholetov of the 159th Fighter Aviation Regiment. Likholetov's final score was 30 enemy aircraft destroyed. Seriously injured in a car crash at the end of 1944, he died of his injuries on 13 July 1945.

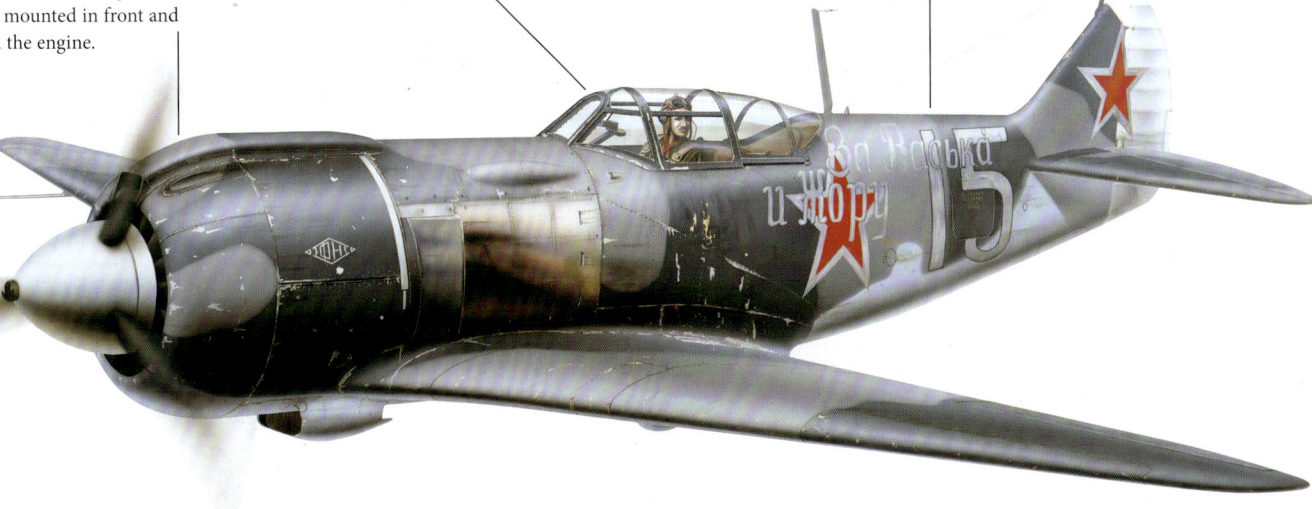

IMPROVEMENTS
An extensive conversion of the LaGG-3, the early La-5 aircraft suffered serious performance shortfalls due to shoddy workmanship. However, once these difficulties had been overcome, the aircraft began to take on Fw 190s and Bf 109Gs on equal terms. Many improvements were gradually introduced, the most important being to cut down the rear fuselage and eventually to introduce the more powerful ASh-82F in the La-5F.

CZECH FLYERS
These Lavochkin La-5FNs, operated by the 1st Czech Fighter Regiment, were photographed on 11 September 1944 immediately prior to a flight from Proskurov in the Ukraine to Stubno.

Fairey Barracuda (1942)

TYPE • Naval bomber **COUNTRY** • United Kingdom

SPECIFICATIONS (Mk II)	
DIMENSIONS:	Length: 12.12m (39ft 9in); Wingspan: 14.99m (49ft 2in); Height: 4.62m (15ft 2in)
WEIGHT:	Maximum take-off 6396kg (14100lb)
POWERPLANT:	One 1223kW (1640hp) Rolls-Royce Merlin 32 V-12 liquid-cooled piston engine
MAX SPEED:	390km/h (240mph)
RANGE:	1850km (1150 miles)
CEILING:	4900m (16,000ft)
CREW:	3
ARMAMENT:	Two 7.7mm (0.303in) Vickers K machine guns flexibly mounted in rear cockpit; one 735kg (1620lb) torpedo or four 205kg (450lb) depth charges or up to 660kg (1500lb) bombload

Through no fault of its own, the Barracuda was seriously delayed and underpowered. Despite this, it proved tough and effective in various roles and played a major part in Fleet Air Arm operations for the last two years of the war.

Above: Differing from the Mk I in being powered by a more robust 1223kW (1640hp) Merlin 32, driving a four-bladed propeller, the Barracuda Mk II (pictured) was the first major production variant.

ENGINE UPGRADE
To deal with excessive weight issues, after the first 30 production aircraft had been built subsequent aircraft, known as Barracuda Mk IIs, had a more powerful 1223kW (1640hp) Merlin 32 engine installed.

WING FLAPS
Large flaps were provided to give additional wing area when set in the neutral position, and for take-off, these were lowered by 20 degrees to increase lift.

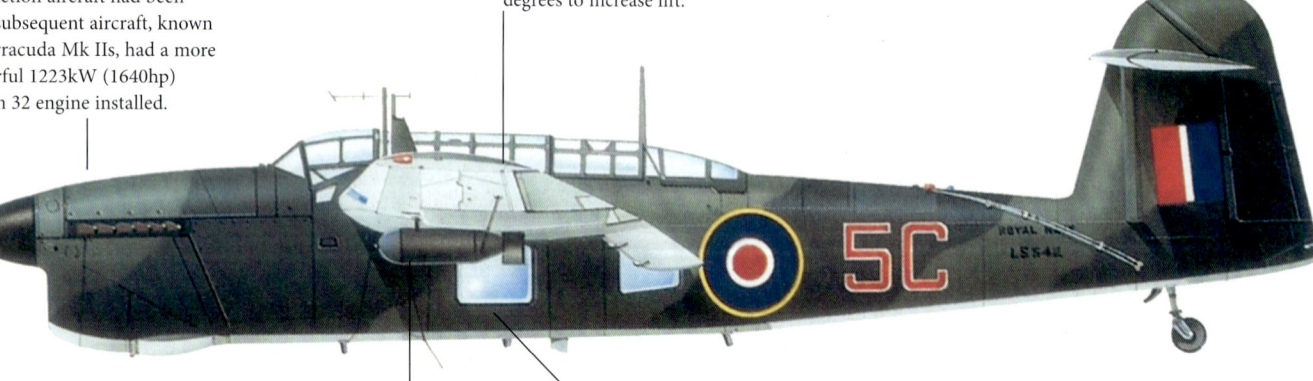

WEAPONRY
Although the Barracuda could carry underwing bombs, the type's primary weapon was a 735kg (1620lb) torpedo slung under the belly.

MK II FEATURES
In this view of a Mk II, some of the unique features of the Barracuda are evident, including the navigator's window below the shoulder-mounted mainplane, and the high-set tailplane.

OPERATION TUNGSTEN
The Barracuda's first combat operations were flown during early 1944, mixed flights of torpedo- and bomb-armed Barracudas conducting a series of anti-shipping strikes along the Norwegian coast with impressive results; by the end of May Barracudas had accounted for 14 ships sunk and 18 damaged. Operation Tungsten was a significant early mission when four Barracuda squadrons attacked the German battleship *Tirpitz*, inflicting heavy damage.

DIVE-BRAKES
The Barracuda used its Youngman flaps as dive-brakes. These can be seen in this dramatic press photograph behind the trailing edge of the wing, angled slightly up to control speed in the dive. When angled down the flaps increased lift at low speed making the Barracuda a docile deck-landing aircraft.

Messerschmitt Me 323 Gigant (1943)

TYPE • *Transport* COUNTRY • *Germany*

With the sudden realization that existing assets would be insufficient to the task of invading Great Britain, the Luftwaffe ordered a giant aircraft to transport men, supplies and vehicles. The immense Gigant family dwarfed all other contemporary aircraft, but suffered innumerable problems, leading to its eventual demise.

SPECIFICATIONS (323D-1 Gigant)

DIMENSIONS:	Length: 28.2m (92ft 6in); Wingspan: 55.2m (181ft 1in); Height: 10.15m (33ft 4in)
WEIGHT:	Maximum take-off 43,000kg (94,799lb)
POWERPLANT:	Six 868kW (1164hp) Gnome-Rhône 14N-48 14-cylinder radial piston engines
MAX SPEED:	285km/h (177mph)
RANGE:	1000km (620 miles)
CEILING:	4000m (13,000ft)
CREW:	5, plus 120 troops or 9750kg (21,495lb) load
ARMAMENT:	Two 7.92mm (0.31in) MG 15, MG 81 or 13mm (0.51in) MG 131 machine guns in cockpit, two 7.92mm (0.31in) MG 15s in waist (optional)

This page: The ultimate standard form of Gigant was the Me 323E-2; this is an E-2 of I./TG 5, which was desperately overworked on the collapsing Eastern Front from late 1943.

CARGO DOORS
The front of the aircraft was hinged in the middle, allowing it to open for loading and unloading of cargo, vehicles or troops.

ENGINES
A number of Me 323E-2s were given six 1007-kW (1350-hp) Junkers Jumo 21 1R engines, with the resulting conversion redesignated Me 323F-1.

ARMAMENT
The E-2's defensive armament, comprised two MG 131s in the front doors, another MG 131 firing aft from the radio compartment, two 20-mm (0.79in) MG 151s in low-drag DL 151 turrets behind the outboard engines, and four single MG 131s firing from front and rear beam positions.

GLIDER CONVERSION
The Me 323 Gigant actually began life as the audacious Me 321, an enormous invasion transport glider with a wingspan of 55m (180ft) and a capacity for loads of up to 20,000kg (44,092lb). Continued problems on take-off led to the decision (it was made airborne by a bomber aircraft tug and rocket assistance) to develop a powered version, the Me 232. The Me 323D began to be produced in series from August 1942.

HEAVY TRANSPORTER
The Me 232 made its operational debut in support of Axis troops in North Africa in November 1942. It could carry a load of 9750kg (21,495lb), or 120 fully equipped troops, or 60 stretcher patients plus attendants, over a distance of more than 1000km (621 miles).

Fiat G.55 Centauro (1943)

TYPE • *Fighter* COUNTRY • *Italy*

SPECIFICATIONS	
DIMENSIONS:	Length: 9.37m (30ft 9in); Wingspan: 11.85m (38ft 10.5in); Height: 3.77m (12ft 4in)
WEIGHT:	Empty 2630kg (5799lb); maximum take-off 3718kg (8197lb)
POWERPLANT:	1100kW (1475hp) Daimler-Benz DB 605A 12-cylinder V-type
MAX SPEED:	620km/h (385mph) at 7400m (24,300ft)
RANGE:	1650km (1025 miles)
CEILING:	12,700m (41,700ft)
CREW:	1
ARMAMENT:	Two 12.7mm (0.50in) machine guns and three 20mm (0.79in) cannon

The Fiat G.55 *Centauro* (Centaur) fighter aircraft was developed by Fiat for the Italian Air Force in World War II. It was generally able to fight well, but all of its variants had defects that made them vulnerable to Allied fighters.

This page: The G.55 Centauro was Fiat's successor to its G.50 fighter, but with a higher performance and equipped with an inline engine.

ARMAMENT
The G.55 carried three 20mm (0.79in) cannon and two 12.7mm (0.50in) machine guns, the latter mounted in the upper forward fuselage. One of the three cannon was centrally mounted to fire through the propeller boss.

COCKPIT
Unlike its predecessor, the G.50, the Fiat G.55 had a fully enclosed cockpit. Downward visibility was poor, and the long nose seriously restricted the pilot's view ahead while on the ground.

LATECOMER
Fitted with a DB.605A engine and featuring an enclosed cockpit, the G.55 was undoubtedly the best fighter produced in Italy during World War II, but it did not enter production until 1943, with the result that only a few had been delivered before the Armistice. Production continued after this, however, and most of the 130 or so aircraft that were completed served with the pro-German Italian Socialist Republic forces.

AIR INTAKE
The G.55 was fitted with an air intake mounted below the fuselage, in much the same configuration as that of the P-51 Mustang (for which it was sometimes mistaken).

GERMAN CENTAUR
A G.55 *Centauro* painted in German markings. The G.55s gave a good account of themselves against Allied fighters like the Spitfire and Mustang, but problems ranging from armament layout to problems with balance impacted on otherwise excellent performance characteristics.

Bell P-63 Kingcobra (1943)

TYPE • Fighter **COUNTRY** • United States

Although bearing a close resemblance to the P-39 and sharing the same unconventional layout, the Kingcobra was in fact a completely new design. Over two-thirds of all P-63s produced were supplied to the Soviet Union.

SPECIFICATIONS (P-63A)

Dimensions:	Length: 9.96m (32ft 8in); Wingspan: 11.68m (38ft 4in); Height: 3.84m (12ft 7in)
Weight:	Empty 2892kg (6375lb); maximum take-off 4763kg (10,500lb)
Powerplant:	988kW (1325hp) Allison V-1710-95 V-type
Max Speed:	657km/h (408mph) at 7452m (24,450ft)
Range:	724km (450 miles)
Ceiling:	13,105m (43,000ft)
Crew:	1
Armament:	One 37mm (1.46in) gun and four 12.7mm (0.50in) guns; provision for three 227kg (500lb) bombs

This page: The Kingcobra went on to have international post-war service. In 1949 this P-63C-5 was among those equipping GC II/6 'Normandie-Niemen' at Tan Son Nhut AB, Indochina.

Armament
The P-63 carried five guns: a single 37mm (1.5in) M10 cannon in the propeller hub, a pair of 12.7mm (0.5in) machine guns above the nose and a second pair of podded '50 calibres' under the wings.

Ventral fin
To correct 'unacceptable' directional stability in the P-63A, the P-63C was equipped with a ventral fin at the rear of the aircraft.

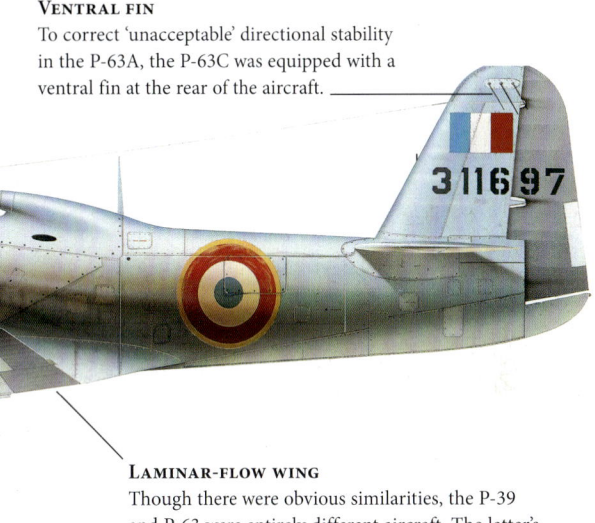

Laminar-flow wing
Though there were obvious similarities, the P-39 and P-63 were entirely different aircraft. The latter's enlarged vertical fin and four-bladed propeller were obvious visual differences, but perhaps more significant was the more efficient laminar flow wing.

LEND-LEASE FIGHTERS
The aircraft was intended to succeed the P-39 in the fighter and fighter-bomber roles, but only 332 were in fact delivered to the USAAF and were used as ground-based gunnery targets. No Airacobra ever saw combat with the USAAF. Of the 1725 P-63A and 1227 P-63C Kingcobras built, 2421 were supplied to the USSR under Lend-Lease, proving to be excellent ground-attack aircraft, and 300 were allocated to the Free French Air Force.

P-63 VARIANTS
The Bell P-63 Kingcobra was a wholly new fighter design using the essential layout of the P-39 Airacobra and introducing the laminar-flow wing and taller tail tested on the XP-39E. In this view of a P-63A (foreground) and a P-39Q the main external differences are clear. The Kingcobra was about 12 per cent larger than the P-39.

Messerschmitt Me 410 Hornisse (1943)

TYPE • *Fighter* COUNTRY • *Germany*

The Me 410 *Hornisse* (Hornet) was a thoroughly improved development of the Me 210, featuring a lengthened rear fuselage of new design and a wing fitted with automatic leading-edge slats. It entered service in early 1943.

SPECIFICATIONS
(Me 410A-3)

DIMENSIONS:	Length: 12.4m (40 ft 11.5in); Wingspan: 16.3m (53ft 7.7in); Height: 4.3m (14ft)
WEIGHT:	Maximum take-off 9651kg (21,276lb)
POWERPLANT:	Two 1300kW (1750hp) Daimler-Benz DB 603A V-12 inverted-V piston engines
MAX SPEED:	624km/h (388 mph)
RANGE:	1200km (746 miles)
CEILING:	10,000m (33,000 ft)
CREW:	2
ARMAMENT:	2 x 20mm (0.8 in) MG 151/20 cannon, 2 x 7.92mm (0.31 in) MG 17 machine guns, two remote controlled rear-firing 13mm (0.51 in) MG 131 machine guns

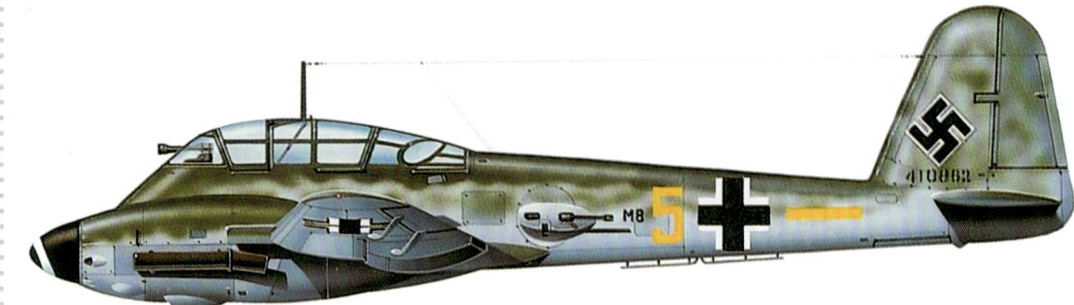

Above: The Me 410 Hornisse *was a thoroughly improved development of the Me 210, featuring a lengthened rear fuselage of new design and a wing fitted with automatic leading-edge slats.*

COCKPIT
The crew of two sat in tandem with the observer facing aft. To assist with aiming the rear-firing MG 131 guns, optically flat panels at the rear of the canopy gave a clear view down and aft.

ME 410A-2
This Me 410A-2 of 9./ ZG 1 was based at Gerbini, Sicily, before the German withdrawal. Note the 13mm (0.51in) machine gun mounted in the barbette on the side of the fuselage.

TAIL
The large single vertical stabilizer was implemented on the Me 210 to improve the handling, and carried through to the Me 410.

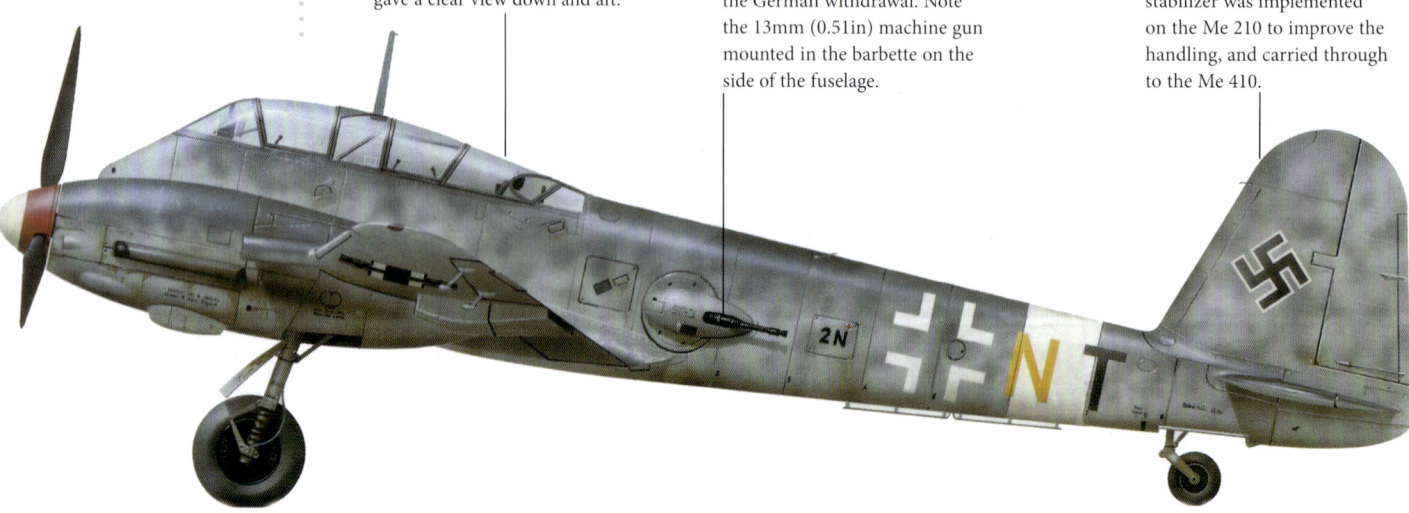

ARMAMENT OPTIONS
A range of armament options were available for the Me 410. Aircraft with a /U2 suffix carried an additional two MG 151s in the bomb bay while the /U2/R2 instead carried a pair of 30mm (1.18in) MK 103s or 108s. Meanwhile, the /U2/R5 carried a total of six MG 151 20mm (0.79in) cannon with the additional four again in the bomb bay. The /U4 variant carried a single 50mm (1.97in) cannon, the BK 5, in the bomb bay.

CANNON INSPECTION
Surrounded by Soviet troops after its capture, this Me 410A-2/U4 is armed with a BK 5 50mm (1.97in) cannon. The formidable BK 5 was intended to be used to destroy Allied bombers, with the Me 410 being able to carry 21 rounds for the weapon.

Fairey Firefly (1943)

TYPE • Naval fighter COUNTRY • United Kingdom

Superficially resembling the Fulmar, the Fairey Firefly was a considerably more powerful and formidable aircraft that saw much action during the last two years of war before enjoying a long post-war career.

SPECIFICATIONS (Mk I)

DIMENSIONS: Length: 11.45m (37ft 7in); Wingspan: 13.55m (44ft 6in); Height: 4.15m (13ft 7in)
WEIGHT: Maximum take-off 6375kg (14,054lb)
POWERPLANT: One 970kW (1735hp) Rolls-Royce Griffon IIB V-12 liquid-cooled piston engine
MAX SPEED: 509km/h (316mph)
RANGE: 2100km (1305 miles)
CEILING: 8535m (28,000ft)
CREW: 2
ARMAMENT: Four 20mm (0.79in) Hispano cannon fixed forward-firing in wings; up to 908kg (2000lb) bombload or eight 27kg (60lb) rockets

Right: A Firefly Mk I of No. 1770 Sqn, the first FAA Firefly unit to take the type into action, is seen landing after a sortie in the Pacific theatre. Note the Youngman area-increasing flaps, fully extended for landing, and underwing rails for rocket projectiles.

POWERPLANT
The Rolls-Royce Griffon was selected for the engine. Initially, it gave 1295kW (1735hp) at sea level, but later versions were capable of over 1565kW (2100hp).

CANOPY
The first production Fireflies such as Z2035 featured a low profile canopy but the restricted headroom this provided led to the introduction of a taller canopy that was fitted to the majority of Fireflies.

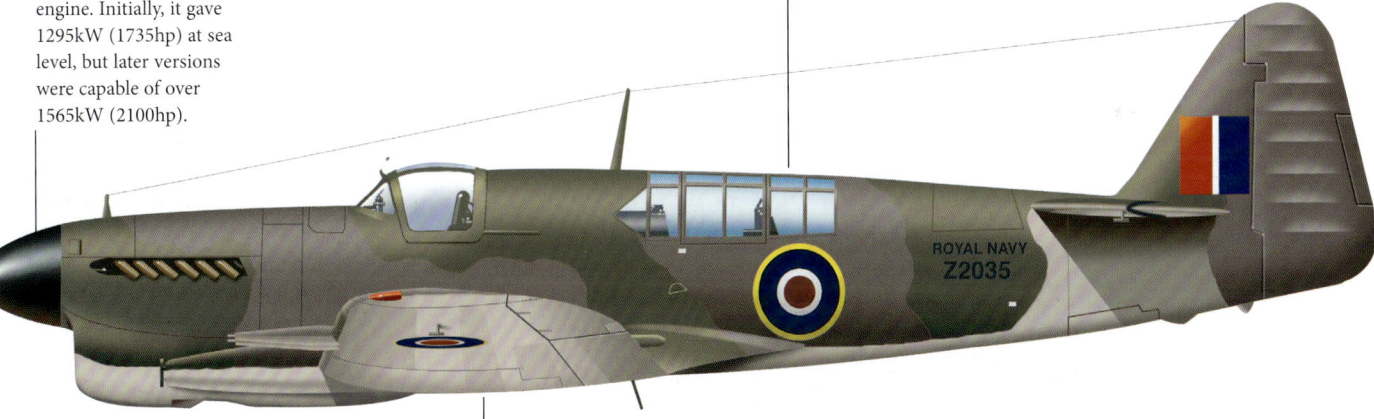

AGILITY
The Firefly would always require a significant physical effort to perform aerobatics, especially to a pilot used to the beautifully light controls of the Seafire. Nevertheless, the aircraft was considered agile, the patented Fairey-Youngman flaps allowed for excellent low-speed handling and during tests with the US Navy during 1944 the Firefly I was shown to be capable of out-turning the F6F Hellcat.

CONTROL SURFACES
The most significant change made to the airframe before it was cleared for service was a switch from fabric to metal skinning on the ailerons and elevators to alleviate control heaviness and handling issues.

POST-WAR AIRCRAFT
This view of a post-war Firefly landing demonstrates well the cleaned up nose sported by the Mk 4, 5 and 6 Fireflies. Twin radiators in the leading edge of the inner wings replaced the beard radiator sported by earlier variants.

Messerschmitt Me 262 (1944)

TYPE • *Fighter* **COUNTRY** • *Germany*

Assured its place in history as the first jet fighter to enter service, the Luftwaffe's Me 262 was the most advanced fighter to reach operational status during World War II, and ushered military aviation into the jet age.

SPECIFICATIONS

DIMENSIONS:	Length: 10.61m (34ft 9in); Wingspan: 12.50m (41ft); Height: 3.83m (12ft 7in)
WEIGHT:	6775kg (14,936lb) maximum take-off
POWERPLANT:	2 × Junkers Jumo 004B-1 turbojets, 8.8kN (1890lb)
MAX SPEED:	870km/h (541mph)
RANGE:	845km (525 miles)
CEILING:	11,000m (36,090ft)
CREW:	1
ARMAMENT:	4 × 30mm (1.18in) cannon

Left: This Me 262A-1a carries the insignia of Jagdgeschwader (JG) 7, the Luftwaffe's first and only fully operational jet fighter Geschwader. Its pilots included some of the world's first jet fighter aces.

RADAR HOMING
The 262B-1a/U1 was a modified B-1a conversion trainer, fitted with FuG 218 Neptun V radar with a *Hirschgeweih* (Antler) antenna and FuG 350 ZC Naxos, for homing in to emissions from British H2S radar equipment.

CONTROL SURFACES
Control surfaces included fabric-covered elevators, replaced with stronger metal skins on later production aircraft. The powerful rudder was required to maintain directional stability.

NIGHT-FIGHTER ROLE
The Me 262 saw some limited success in the night-fighter role. A dedicated night-fighter model appeared before the end of the war, in the form of the radar-equipped Me 262B-1a/U1. As a daytime bomber-destroyer, the Me 262 could also be armed with 24 underwing R4M unguided rockets. Ultimately, however, the Me 262 was a case of 'too little, too late'.

JET ENGINES
Power was provided by a pair of Junkers Jumo 004B-1 axial-flow turbojets. These suffered from poor reliability and limited service life.

ME 262B-1A/U1
The Me 262B-1a/U1 night-fighter was created on the basis of the Me 262B-1a dual-control trainer. First trials were undertaken in October 1944, using Lichtenstein SN-2 radar. The production version featured a radar operator in the rear seat and a FuG 218 Neptun V radar with nose-mounted antenna array.

Boeing B-29 Superfortress (1944)

TYPE • Bomber **COUNTRY** • United States

SPECIFICATIONS (B-29B)

DIMENSIONS:	Length: 30.18m (99ft); Wingspan: 43.36m (142ft 3in); Height: 9.01m (29ft 7in)
WEIGHT:	64,003kg (141,100lb) maximum take-off
POWERPLANT:	4 × Wright R-3350- 57 radial piston engines, 1641kW (2200hp)
MAX SPEED:	576km/h (358mph)
RANGE:	6598km (4100 miles)
CEILING:	9695m (31,800ft)
CREW:	10-11
ARMAMENT:	12 × 12.7mm (0.5in) machine guns in remote-controlled turrets, 1 × 20mm (0.79in) cannon and 2 × 12.7mm (0.5in) machine guns in tail, plus bombload of up to 9072kg (20,000lb)

Even without the atomic missions that the type flew against Japan in August 1945, the B-29 – the most advanced heavy bomber of the war – can be considered a war-winner on the devastating raids it conducted against the Japanese mainland in the preceding months.

Above: The Superfortress's long, cylindrical fuselage stretched the bomber's length to 30.18m (99ft), and was topped by the huge tailfin that was the B-29's main recognition feature.

PRESSURIZED
The B-29 was the first aircraft in quantity production with a pressurized fuselage, enhancing crew comfort during long flights.

GUN TURRETS
All gun turrets were remotely controlled, reducing their size and therefore cutting down drag. Everything possible was done to ensure a smooth airflow.

DEVASTATING RAIDS
During their initial nine months of service, the B-29s were employed mainly for high-level daylight raids. However, tactics switched in March 1945, when they began low-level night attacks from the Marianas Islands. These were the most destructive raids of the war in terms of casualties, with the first night-time incendiary raid on Tokyo killing around 80,000 people.

BOMBING RADAR
The B-29 was equipped with the very accurate AN/APQ-13 bombing radar located between the two bomb bays. This was often deleted from wartime photos for security reasons.

COMBAT HEAVYWEIGHT
Renton-built B-29A-5-BN serial 42-93869 pictured soon after leaving the production line. The huge Wright Duplex Cyclone engines drove 5.05m (16ft 7in)-diameter four-bladed propellers. Developed as a result of prolonged design studies to be a 'super bomber' capable of attacking targets at extreme ranges, the Superfortress was the heaviest combat aircraft of WWII.

Arado Ar 234 (1944)

TYPE • Bomber   **COUNTRY** • Germany

SPECIFICATIONS (Ar 234B-2)

DIMENSIONS:	Length: 12.64m (41ft 5in); Wingspan: 14.41m (46ft 3in); Height: 4.29m (14ft 1in)
WEIGHT:	Empty 5200kg (11,466lb); 9850kg (21,715lb) loaded
POWERPLANT:	Two 800kg (1764lb) thrust BMW 003A-1 turbojets
MAX SPEED:	742km/h (461mph) at 6000m (19,685ft)
RANGE:	1630km (1013 miles)
CEILING:	10,000m (32,810ft)
CREW:	1
ARMAMENT:	2 x 20mm (0.79in) MG 151 in belly pod; 2 x rearward-firing 20mm (0.79in) cannon

The Arado Ar 234 was the world's first jet bomber. Entering service in the reconnaissance role in July 1944, the Blitz made flights over Britain and France almost undetected. As a bomber it was also quite successful, but operations were hampered by attacks on airfields and fuel supplies.

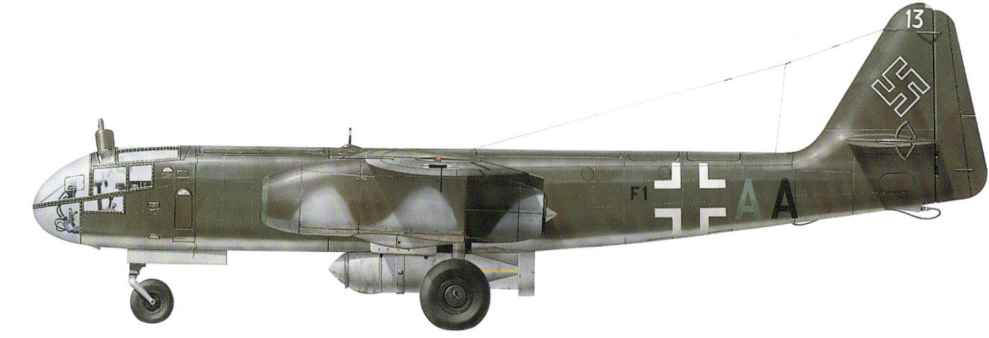

Above: The undercarriage was mounted in the fuselage, leaving the wing clean. This narrow-track arrangement caused some instability when taxiing, but was otherwise satisfactory.

BOMBLOAD
A normal bombload comprised two SC 250 bombs, one slung under each engine nacelle. Rocket-assisted take-off pods could also be fitted under the outer wings.

COCKPIT
The pilot's cockpit was equipped with an autopilot with overriding control, enabling the pilot to swing the control column clear so that he could use his tachometric bombsight. A rear-view periscope was fitted.

ENGINE LIFE
The mass-produced engines proved vulnerable to cracks in the impeller, turbine wheels and vane rings, and engine life was only 25 hours.

BRIDGE RAIDS

The Arado 234B-2 was the world's first jet bomber, and followed the unarmed Ar 234B-1 reconnaissance version into service. The Ar 234B-2 flew its first combat missions with KG 76 during the German offensive in the Ardennes in December 1944. The jet bombers were extremely active in the early weeks of 1945, one of their most notable missions being a ten-day series of attacks on the Ludendorff Bridge at Remagen.

ROCKET ASSISTED
The Ar 234B-2 was far more versatile than its predecessor, the Ar 234B-1, being capable of bombing, pathfinding or reconnaissance missions. This model is equipped with *Rauchgeräte* take-off assistance rockets outboard of the engine nacelles.

Messerschmitt Me 163 Komet (1944)

TYPE • *Fighter* **COUNTRY** • *Germany*

SPECIFICATIONS (Me 163B-1a)

DIMENSIONS:	Length: 5.85m (19ft 2in); Wingspan: 9.4m (30ft 7in); Height: 2.76m (9ft)
WEIGHT:	Empty 1900kg (4190lb); maximum take-off 4310kg (9502lb)
POWERPLANT:	1700kg (3749lb) thrust Walter 109-509A-2 rocket motor
MAX SPEED:	955km/h (593mph)
RANGE:	35.5km (22 miles)
CEILING:	12,000m (39,370ft)
CREW:	1
ARMAMENT:	Two 30mm (1.18in) cannon

The remarkable and revolutionary Me 163 rocket-powered interceptor was yet another example of German ingenuity, but despite its impressive (albeit short-range) performance, it came too late to alter the course of the air war.

Above: The first production interceptor designated Me 163B was the red-painted V41, first flown on 14 May 1944 by Major Wolfgang Spät

COCKPIT
The pilot sat in a primitive cockpit with a reflector gunsight for aiming the cannon. Rear visibility was limited, but this was of little importance when the rocket motor was running.

ENGINE
The aircraft was powered by a single Walter HWK 509A-2 rocket motor, which would run for about six minutes on full throttle.

ARMAMENT
The Komet was armed with a pair of MK 108 30mm (1.18in) cannon in the wingroots, which gave it a mighty punch. The high closing speed of the aircraft gave the pilot approximately three seconds of firing time.

ROCKET INTERCEPTOR
The Me 163 was perhaps the most radical and futuristic of World War II aircraft. The concept of the short-endurance, high-speed interceptor powered by a rocket engine was certainly valid. Bereft of a horizontal tail and with an extremely short fuselage, the Me 163 was propelled by two extremely volatile liquids. By May 1944, these tiny aircraft were devastating US bomber formations.

LANDING GEAR
As the main wheels were detached from the aircraft shortly after take-off, the Me 163 relied on an extending skid for landing.

PROTOTYPE
Two of the Me 163B prototypes, V6 and V18, were modified with prototypes of the HWK 509C-1 motor with main and cruising thrust chambers, to give much better flight endurance. Here the V6 blows steam through its propellant lines in the summer of 1944. Note the repositioned retractable tailwheel.

Mitsubishi Ki.67 Hiryu (1944)

TYPE • *Bomber* COUNTRY • *Japan*

Although classed as a heavy bomber, the Mitsubishi Ki.67 was roughly in the same class as the Martin B-26 Marauder. An excellent design, it appeared too late to have a decisive effect on the Pacific air war.

SPECIFICATIONS (Ki.67-I)

DIMENSIONS: Length: 18.70m (61ft 4.25in); Wingspan: 22.5m (73ft 10in) Height: 7.70m (25ft 3in)

WEIGHT: Empty 8650kg (19,073lb); maximum take-off 13,765kg (30,347lb)

POWERPLANT: Two 1417kW (1900hp) Mitsubishi Ha.104 18-cylinder radials

MAX SPEED: 537km/h (334mph) at 6000m (19,685ft)

RANGE: 3800km (2361 miles)

CEILING: 9470m (31,069ft)

CREW: 6–8

ARMAMENT: One 20mm (0.79in) cannon and five 12.7mm (0.5in) machine guns; bomb or torpedo load of 1070kg (2359lb)

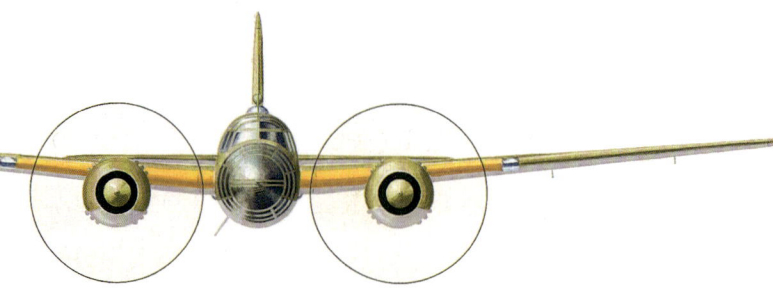

Above: The Ki.67 Hiryu *(Flying Dragon) was unquestionably the best bomber to see service with the Imperial Japanese Army, combining excellent performance with good defensive firepower.*

CONTROL SURFACES
The Ki.67's control surfaces were very responsive; in fact, under certain flight conditions control was excessively sensitive. Without bombs, the Ki.67 could easily make loops and vertical turns.

REAR ARMAMENT
Late-production Ki.67-Is had the single 12.7mm (0.5in) tail machine gun replaced by a twin mounting, starting with the 451st Mitsubishi-built machine.

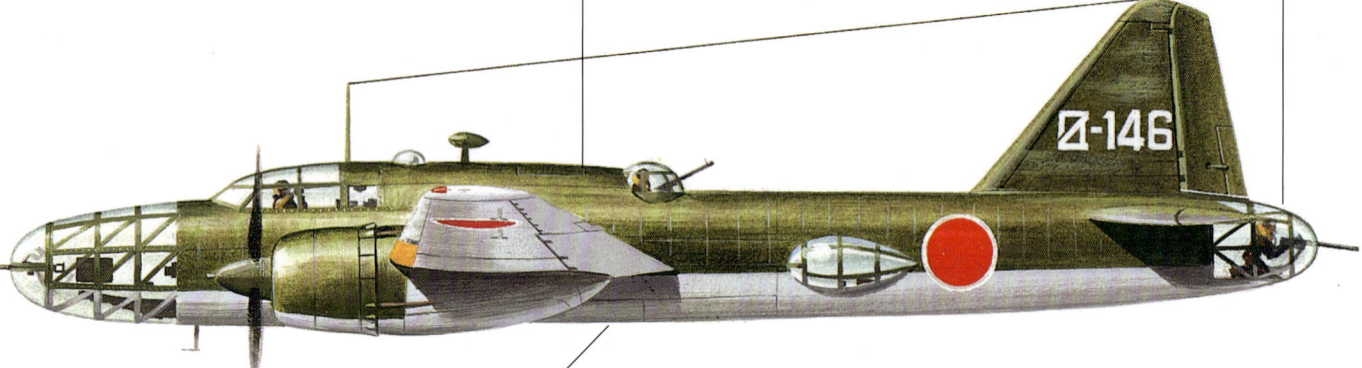

BOMBLOAD
The *Hiryu* had a respectable bombload and a capacious weapons bay. Ki.67s operated in China, and in 1945, using Iwo Jima as a refuelling point, they made repeated attacks on B-29 bases in the Marianas.

MANY VARIANTS

Production of the Ki.67 totalled 698 aircraft, serving as bombers, torpedo-bombers, reconnaissance aircraft, interceptors, ground-attack aircraft and suicide bombers. The interceptor was designated Ki.109 and was not a success, with only 22 being completed. As a torpedo-bomber, the Ki.67 was particularly active during the air-sea battle off Formosa in October 1944 and during the American landings on Okinawa in 1945.

BOMBER LAYOUT
The wing and tail surfaces of the *Hiryu* bore a strong resemblance to those of the G4M (Betty) bomber; apart from that, the Ki.67 was a completely separate design. Its designer, Chief Engineer Ozawa, insisted on all fuel and oil tanks being of the self-sealing type with armour protection.

Yokosuka MXY7 Ohka (1944)

TYPE • *Fighter* **COUNTRY** • *Japan*

SPECIFICATIONS

DIMENSIONS:	Length: 6.07m (19ft 10.33in); Wingspan: 5.12m (16ft 9.5in); Height: 1.16m (3ft 9.5in)
WEIGHT:	Empty 440kg (970lb); loaded 2140kg (4718lb)
POWERPLANT:	Three solid-fuel Type 4 Mk 1 Model 20 rockets with total thrust of 800kg (1764lb)
MAX SPEED:	927km/h (576mph) in terminal dive
RANGE:	37km (23 miles)
CREW:	1
ARMAMENT:	One 1200kg (2646lb) warhead

The notion of a pilot deliberately sacrificing his life as an act of war was foreign to Western minds in World War II, but the Japanese brought the concept to grim reality in their Special Attack Corps, whose ultimate development was the Ohka (Cherry Blossom) suicide bomb.

Above: The wingspan of the Ohka Model 22, an improved version, was less than that of the Model 11, and its warhead limited to 600kg (1323lb). This version was to have been powered by a turbojet engine.

WARHEAD
The initial version of the Ohka was fitted with a 1200kg (2646lb) warhead and was designed to be transported to within a few miles of its target in the bomb bay of a specially modified Navy Type 1 Attack Bomber Model 24J (G4M2e).

CONSTRUCTION
The tiny aircraft was built of wood and non-critical metal alloys. Great care was taken in planning to enable it to be mass-produced by unskilled labour.

COCKPIT
As the aircraft was to be flown on its one-way mission by pilots with only limited flying experience, instruments were kept to a minimum and good manoeuvrability was required to achieve accuracy. Once the cockpit hood was bolted into place, there was no escape for the pilot.

ROCKET MOTORS
After release from its parent aircraft, the Ohka accelerated towards its target under the power of three solid-propellant rockets mounted in the tail, which could be fired either singly or together.

COMBAT SUCCESS
First success for the Ohka came on 1 April 1945, when Ohkas damaged the US battleship *West Virginia* and three transport vessels. The first ship to be destroyed by an Ohka was the destroyer *Mannert L. Abele*, lost off Okinawa on 12 April. Several further versions of the Ohka were proposed, including the turbojet-powered Model 33, but none materialized in operational form before the end of the war.

DEFEATED JETS
US personnel inspect captured Ohka jets at the end of the war. In objective military terms the concept of the Ohka was sound enough, but the lack of suitable carrier aircraft meant the weapon consistently failed to break through the dense Allied fighter screen.

Lavochkin La-7 (1944)

TYPE • *Fighter* **COUNTRY** • *Soviet Union*

SPECIFICATIONS

DIMENSIONS:	Length: 8.6m (28ft 3in); Wingspan: 9.8m (32ft 2in); Height: 2.54m (8ft 4in)
WEIGHT:	Maximum take-off 3400kg (7496lb)
POWERPLANT:	1380kW (1850hp) Shvetsov M-82FN 14-cylinder air-cooled radial
MAX SPEED:	661km/h (411mph)
RANGE:	635km (395 miles)
CEILING:	10,450m (34,280ft)
CREW:	1
ARMAMENT:	Two cowl-mounted 20mm (0.79in) ShVAK cannons or three 20mm (0.79in) Berezin B-20 cannons; up to 200kg (440lb) bombload

The ultimate in Soviet WWII piston-engined fighters, the Lavochkin La-7 was produced in large numbers during the last year of the war, and was popular with its pilots. The La-7 maintained the hard-won Soviet air superiority over the Eastern Front.

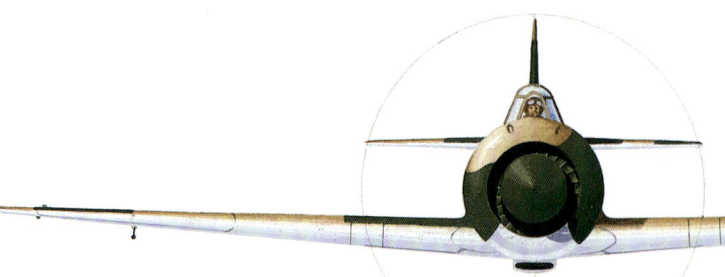

Above: The La-7 was not a new design but it represented a relatively modest aerodynamic refinement of the proven La-5 airframe.

SYNCRONIZED GUNS
The La-7 carried two guns in its upper fuselage decking, synchronized to fire through the propeller disc. The guns were 20mm (0.79in) ShVAK cannon and had 200 rounds of ammunition each.

COMBAT ROLES
The primary role of the La-5/La-7 series was as a low- and medium-level fighter, but it was occasionally assigned ground-attack missions. For these the aircraft could carry a variety of rocket and bombloads on underwing pylons.

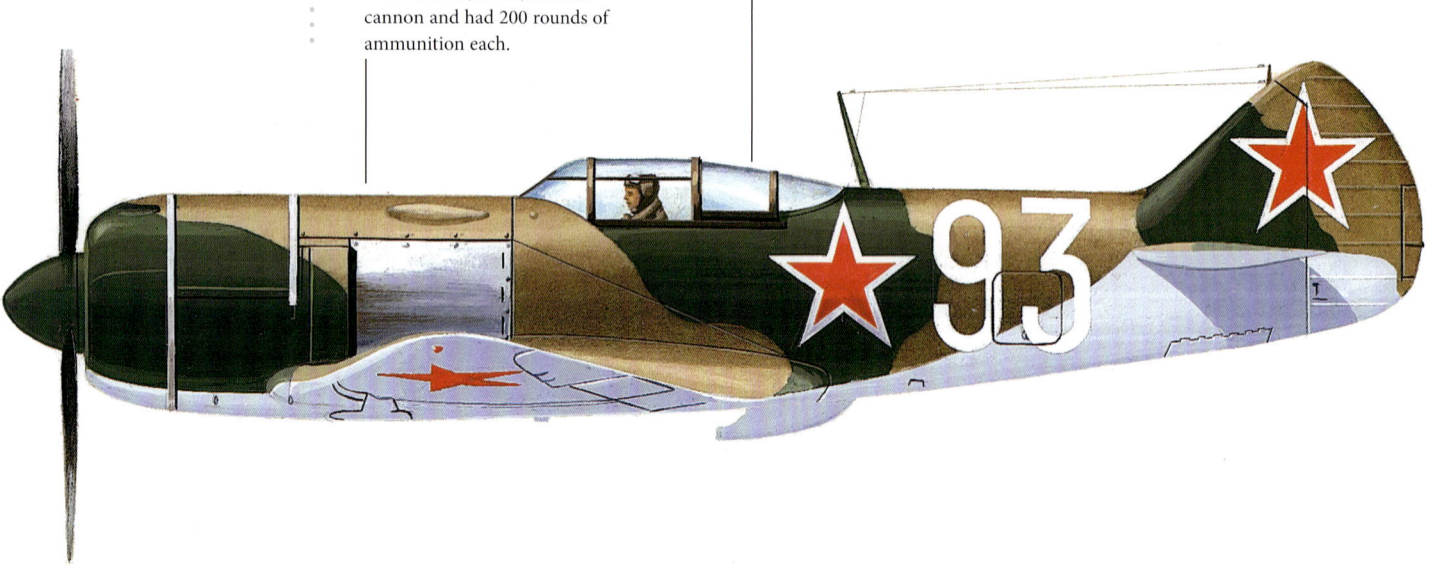

AIR SUPERIORITY

Although able to carry a 100kg (220lb) bomb under each wing, the La-7s exemplary air-to-air abilities meant that virtually all of the 5753 built saw service exclusively in the air superiority role. A further 582 two-seaters were built for training as the La-7UTI. As well as Soviet service, the La-7 saw brief service with Czech forces shortly before the end of the war.

LA-7 PILOTS
A squadron of Lavochkin La-7 pilots scramble to their aircraft. Compared to the La-5FN, the La-7 offered usefully improved performance and handling, with its dog-fighting capability at low altitude being regarded as unbeatable.

Tupolev Tu-2 (1944)

TYPE • Bomber COUNTRY • Soviet Union

The Tu-2 was tailored to meet a requirement for a high-speed bomber or dive-bomber, with a large internal bombload, and speed similar to that of a single-seat fighter. Designed to challenge the Ju 88, the Tu-2 proved equally versatile, and was produced in torpedo, interceptor and reconnaissance versions.

SPECIFICATIONS (Tu-2S)

DIMENSIONS: Length: 13.8m (45ft 3in); Wingspan: 18.86m (61ft 11in); Height: 4.13m (13ft 7in)

WEIGHT: Maximum take-off 11,768kg (25,944lb)

POWERPLANT: Two 1380kW (1850hp) Shvetsov ASh-82 14-cylinder air-cooled radial piston engines

MAX SPEED: 528km/h (328mph)

RANGE: 2020km (1260 miles)

CEILING: 9000m (30,000ft)

CREW: 4

ARMAMENT: Two fixed forward-firing 20mm (0.79in) ShVAK cannon in the wings, three 12.7mm (0.5in) flexibly mounted rear-firing Berezin UBT machine guns in rear cockpit canopy, dorsal and ventral hatches; up to 3000kg (6600lb) bombload

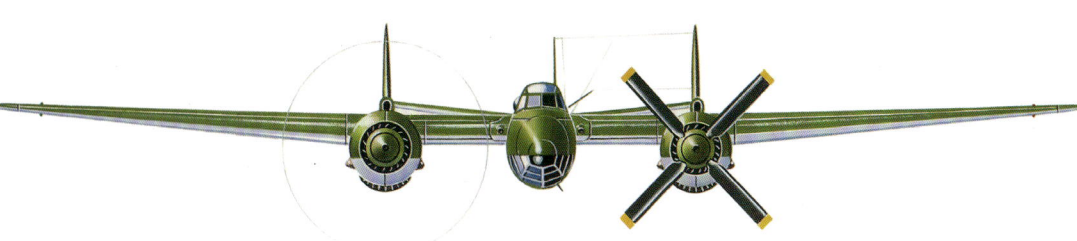

Above: The prototype Tu-2 (ANT-58) flew for the first time on 29 January 1941 and subsequent flight testing showed that the aircraft had an outstanding performance.

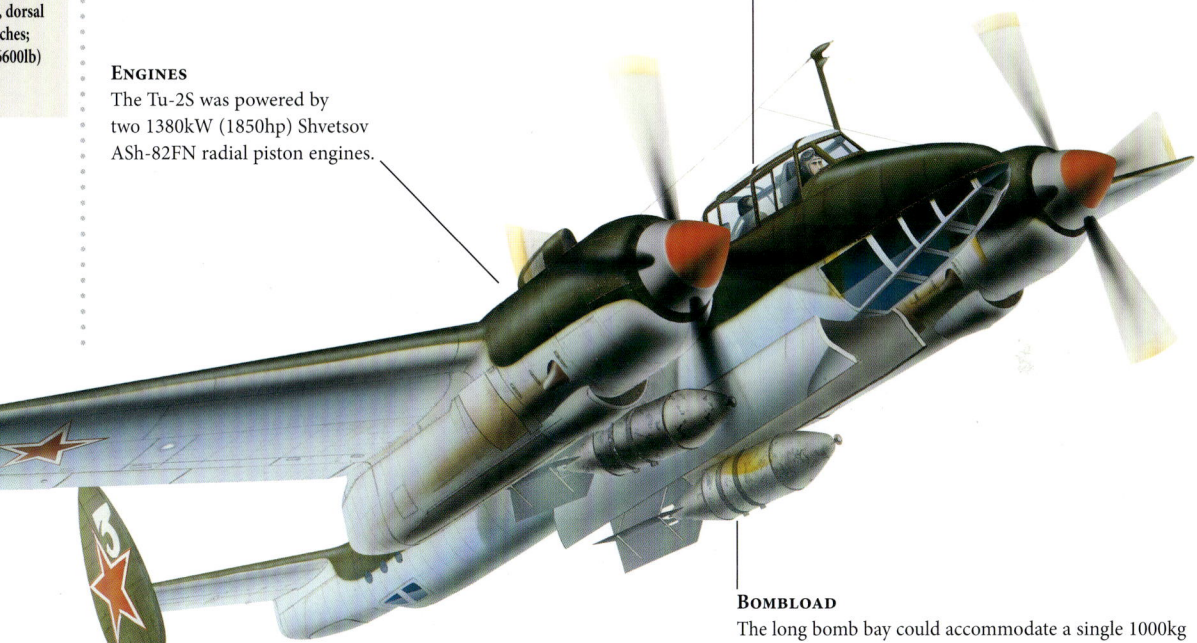

ENGINES
The Tu-2S was powered by two 1380kW (1850hp) Shvetsov ASh-82FN radial piston engines.

COCKPIT
The pilot and navigator sat back-to-back on the flight deck. The prominent mast attached to the canopy contained the pitot tube, as well as acting as an attachment for the radio aerial.

BOMBLOAD
The long bomb bay could accommodate a single 1000kg (2205lb) bomb or several smaller weapons. Additional large bombs were carried under the wing roots on racks, while smaller bombs could be carried on five racks on each side, outboard of the engine.

POST-WAR SERVICE
After World War II, the Tu-2 proved to be an ideal test vehicle for various powerplants, including the first generation of Soviet jet engines. Production continued after 1945, some 3000 aircraft eventually being delivered to various Soviet Bloc air forces. The last Tu-2 model was the ANT-68, a high-altitude version that saw limited service as the Tu-10.

PROTOTYPE 2
Samolyet 103U was the second Tu-2 prototype, with a raised canopy and lengthened fuselage with accommodation for a fourth crew member, firing a ShKAS from a lower rear fuselage hatch. The 103U had provision for 10 RS rockets underwing.

Gloster Meteor III (1944)

TYPE • *Fighter* COUNTRY • *United Kingdom*

Britain's first and only operational jet combat aircraft of World War II was Gloster's extraordinary Meteor, which would go on to serve with frontline RAF squadrons for a further 17 years.

SPECIFICATIONS (F. Mk III)

DIMENSIONS:	Length: 12.50m (41ft); Wingspan: 13.10m (43ft); Height 3.96m (13ft)
WEIGHT:	Empty 4771kg (10,520lb); maximum take-off 6314kg (13,920lb
POWERPLANT:	Two 906kg (2000lb) thrust Rolls-Royce Derwent 1 turbojets
MAX SPEED:	675km/h (415mph) at 3050m (10,000ft)
RANGE:	1580km (982 miles)
CEILING:	13,106m (43,000ft);
CREW:	1
ARMAMENT:	Four 20mm (0.79in) cannon

Above: The Meteor Mks I to IV featured a distinctive tail unit, as did the T.7 trainer version and the photo-reconnaissance PR.10. The last day-fighter version, the F.8, had a much modified and more angular tail, as did the night-fighter variants of the 1950s.

NOSE CANNON
The Meteor packed a powerful punch in the shape of four 20mm (0.79in) nose-mounted cannon. In the closing weeks of WWII Meteors carried out ground-attack operations in northwest Europe, but never met the Luftwaffe in action.

COCKPIT
The Meteor F.I had a hinged sideways-opening cockpit canopy, but the Mk III was fitted with a one-piece sliding canopy. Pilots coming to the Meteor from piston-engined fighters were astounded by the relative absence of noise in the cockpit.

CODE LETTERS
This Meteor Mk III bears the code letters 'YQ' of No. 616 Squadron, the first to equip with the Meteor. The squadron flew Meteors until August 1945, when it disbanded. Reformed in 1947, it operated Mosquitoes and then Meteor Mks 3, 4 and 8 before disbanding for the last time in 1957.

TURBOJET FIGHTER
The only Allied turbojet-powered aircraft to see action during World War II, the Gloster Meteor was designed by George Carter, whose preliminary study was given Air Ministry approval in November 1940 under Specification F.9/40. Its twin-engined layout was determined by the low thrust produced by the turbojet engines then available. The first operational jet fighter squadron was No. 616, which received its first two Meteor F.Mk Is on 12 July 1944.

616 SQUADRON METEORS
No. 616 Squadron Meteor F.Mk I/IIIs are seen here at RAF Colerne prior to the unit moving to Europe in January 1945. The first operational sorties were flown on 27 July 1944, and on 4 August, near Tonbridge, Flying Officer Dean destroyed the first V-1 flying-bomb to be claimed by a jet fighter.

Hawker Tempest (1944)

TYPE • Fighter **COUNTRY** • United Kingdom

A product from the drawing board of Hawker's prestigious designer, Sir Sydney Camm, the Tempest was built to remedy the problems of the earlier Typhoon. The resulting design proved to be the RAF's best low-/medium-altitude fighter.

SPECIFICATIONS

DIMENSIONS:	Length: 10.26m (33ft 8in); Wingspan: 12.50m (41ft); Height: 4.52m (14ft 10in)
WEIGHT:	Empty 4854kg (10,700lb); maximum take-off 6187kg (13,640lb)
POWERPLANT:	1685kW (2260hp) Napier Sabre IIA, IIB or IIC 24-cylinder H-type engine
MAX SPEED:	700km/h (435mph)
RANGE:	2092km (1300 miles)
CEILING:	10,975m (36,000ft)
CREW:	1
ARMAMENT:	Four 20mm (0.79in) cannon; bombs or rockets up to 907kg (2000lb)

Above: The Tempest's wing was so thin that special cannon (Hispano Mk V) had to be designed for it. Special ultra-thin tyres were also devised by Dunlop, and the spread of the undercarriage was increased to give extra stability on the ground.

COCKPIT
In redesigning the Typhoon to produce the Tempest, the cockpit was moved further aft to improve pilot visibility, which was one of the Typhoon's shortcomings. The pilot sat under a one-piece sliding canopy, its size reduced to the absolute minimum in order to reduce drag.

TAIL UNIT
The tail unit was completely redesigned, the area of the tailfin being doubled to ensure maximum stability at high speeds.

V-1 KILLER
In April 1944 the Tempest was the fastest and most powerful fighter in the world, and in the summer of 1944 it was thrown into battle against the V-1 flying bombs that were then being launched against London. Between them, the Tempest squadrons claimed the destruction of nearly 600 V-1s. The Tempest squadrons subsequently moved to the Continent with 2nd TAF and became a potent addition to the Allies' striking power.

WING SHAPE
The Tempest was fitted with an elliptical wing, giving excellent manoeuvrability. Flaps were fitted along practically the whole of the wing under-surface between the ailerons and wing root.

TEMPEST MK V
No. 501 Squadron received the Tempest Mk V in August 1944. During this month it absorbed the Fighter Interception Unit (FIU) and became the only Tempest unit specializing in night operations against the V-1 flying bombs.

Supermarine Spitfire FR. Mk XIVE (1944)

TYPE • Fighter COUNTRY • United Kingdom

SPECIFICATIONS	
DIMENSIONS:	Length: 9.11m (29ft 11in); Wingspan: 11.23m (36ft 10in); Height: 3.48m (11ft 5in)
WEIGHT:	Empty 2994kg (6600lb); maximum take-off 3856kg (8500lb)
POWERPLANT:	1529kW (2050hp) Rolls-Royce Griffon 65 12-cylinder V-type engine
MAX SPEED:	721km/h (448mph) at 7468m (24,500ft)
RANGE:	756km (470 miles)
CEILING:	13,565m (44,500ft)
CREW:	1
ARMAMENT:	Two 20mm (0.79in) cannon and two 12.7mm (0.50in) machine guns

The famous Supermarine Spitfire, which began its operational career as an interceptor fighter, went on to perform many varied tasks, one of which was low-level tactical reconnaissance – a role to which it was well suited.

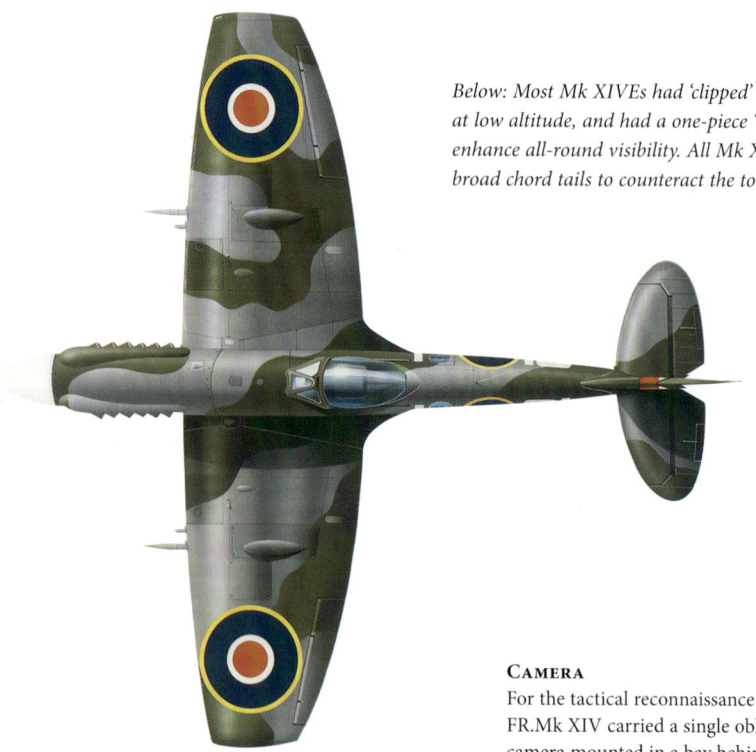

Below: Most Mk XIVEs had 'clipped' wings to improve agility at low altitude, and had a one-piece 'teardrop' cockpit canopy to enhance all-round visibility. All Mk XIVs were equipped with broad chord tails to counteract the torque of the Griffon engine.

PROPELLER
To absorb the extra power of the Griffon 65 engine, the Spitfire Mk XIV was fitted with a five-bladed Rotol propeller in place of the four-bladed one fitted to the Mk XII.

CAMERA
For the tactical reconnaissance role the FR.Mk XIV carried a single oblique camera mounted in a bay behind the cockpit and arranged to point to port or starboard, as required.

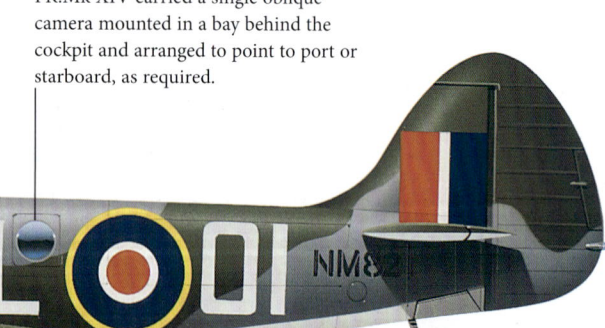

GRIFFON ENGINE
The Spitfire FR.Mk XIVE was one of the later Griffon-engined variants of Reginald Mitchell's famous design, the first of which, the Mk XII, was developed specifically to counter the low-level attacks by Focke-Wulf 190s. Only 100 MK XII Spitfires were built, but they were followed by the more numerous Mk XIV, the first Griffon-engined Spitfire variant to go into large-scale production.

FAR EAST SPITFIRE
Based at Tan Son Nhut, French Indochina, No. 273 Squadron operated FR.Mk XIVs briefly from November 1945. Mk XIV production reached 957, post-war operators including India, Belgium and Thailand.

Dornier Do 335 Pfeil (1944)

TYPE • Fighter COUNTRY • Germany

SPECIFICATIONS (Do 335A-0)

DIMENSIONS: Length: 13.85m (45ft 5in); Wingspan: 13.8m (45ft 3in); Height: 5m (16ft 5in)

WEIGHT: Maximum take-off 9600kg (21,164lb)

POWERPLANT: Two 1417kW (1900hp) Daimler-Benz DB 603E-1 V-12 liquid-cooled piston engines

MAX SPEED: 763km/h (474mph)

RANGE: 1395km (867 miles)

CEILING: 11,400m (37,400 ft)

CREW: 1

ARMAMENT: One engine-mounted 30mm (1.18in) MK 103 cannon plus two 20mm (0.79in) MG 151/20 cannon; up to 1000kg (2200lb) bombload in internal weapons bay

A tandem-engine layout had been patented by Claudius Dornier in 1937 and led to feasibility trials with the Goppingen Go 9 research aircraft designed by Ulrich Hütter and built by Schempp-Hirth in 1939.

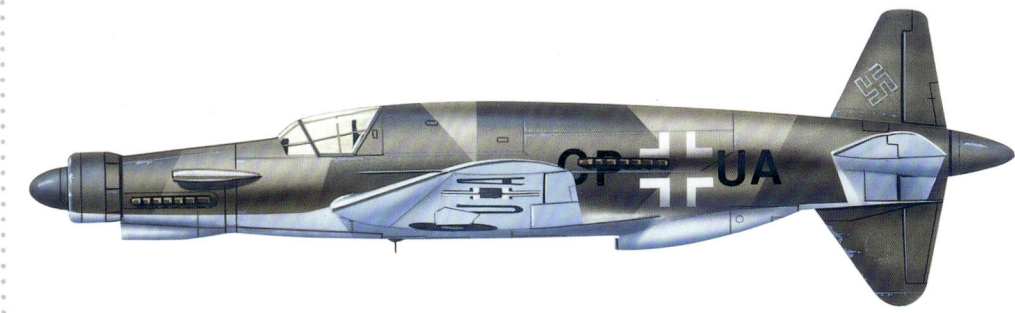

Above: The Do 335 V1 first prototype, bearing the Stammkennzeichen (factory radio code) of CP+UA, was flown on 26 October 1943 by test pilot Flugkapitän Hans Dieterle.

PRE-PRODUCTION
Below is the seventh of 10 pre-production aircraft, most of which went to Erprobungskommando (EK) 335 for evaluation.

LAYOUT
By placing the engines at the front and rear rather than on the wings, the power was delivered inline with the fuselage, increasing the aircraft's manoeuvrability.

POWERPLANT
Power was provided by two 1342kW (1800hp) Daimler-Benz DB 603 engines, one in the rear fuselage driving a three-bladed pusher propeller and the other in the nose with a three-bladed tractor propeller.

VARIANTS
The full-production Do 335A-1 first appeared in late autumn 1944, with the definitive DB 603E-1 engine. A single example of the unarmed Do 335A-4 reconnaissance version was also completed. The Do 335A-6 was next to appear, as the first of a night-fighter version. The last of the Do 335 variants were the Do 335A-10 and A-12 trainers, both of which featured the second cockpit that was introduced on the Do 335A-6 night-fighter.

WARTIME LATECOMER
A Do 335 Pfeil in flight in the late months of the war. The closest the Pfeil came to combat service was when it was issued to the operational test unit, Erprobungskommando 335, in spring 1945.

Lockheed F-80 Shooting Star (1944)

TYPE • Fighter COUNTRY • United States

The P-80 (later F-80) holds the distinction of being the first USAF jet aircraft to enter combat. Although it was subsequently outclassed by swept-wing fighters, it provided good service, especially in the fighter-bomber role in the Korean War.

SPECIFICATIONS (F-80C)

DIMENSIONS:	Length: 10.52m (34ft 6in); Wingspan: 12.15m (39ft 10.5in); Height: 3.45m (11ft 4in)
WEIGHT:	7646kg (16,856lb) maximum take-off
POWERPLANT:	1 x 24.01kN (5400lbf) Allison J33-A-35 turbojet
MAX SPEED:	956km/h (594mph)
RANGE:	1706km (1060 miles)
CEILING:	13,030m (42,750ft)
CREW:	1
ARMAMENT:	6 x 12.7mm (0.5in) machine guns and up to 907kg (200lb) of external ordnance or 10 x 127mm (0.5in) rockets

Below: P-80A-1-LO (44-85226) of the 412th Fighter Group, the first unit to equip with P-80 Shooting Stars in 1946.

P-80A SHOOTING STAR
Here we see a P-80A variant of the Shooting Star, this one operating in a Flight Test Division at Wright-Patterson Air Force Base, Ohio.

COCKPIT
The F-80 pilot sat on a simple ejection seat within a lightly pressurized cockpit. Without suitable air conditioning or ventilation, the cockpit soon became overheated in hot and humid climatic conditions.

POWERPLANT
The late-production F-80C was powered by an Allison J33 centrifugal flow turbojet, equipped with water injection. Access to the engine was achieved by the simple removal of the entire tube-like rear fuselage assembly.

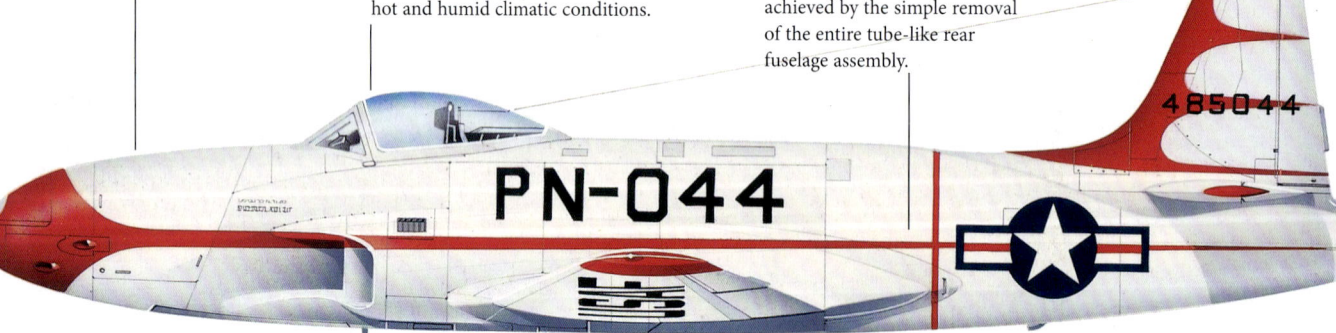

KOREAN WAR
The F-80C saw service in the Korean War as it was the most modern fighter type available to the Far East Air Force when the conflict broke out. In November 1950, an F-80C flown by Russell J. Brown of the 51st Fighter Interceptor Wing reportedly downed a MiG-15 in what the USA claimed was the first conclusive aerial jet combat between two jet fighters. Soviet records indicate that the MiG survived the encounter with Brown's aircraft. Against swept-wing opposition like the MiG-15, however, the Shooting Star was obsolescent.

P-80C FORMATION
The USAF's first operational jet fighter first flew as the XP-80 in January 1944. Crashes of early test and production aircraft killed several experienced pilots. Several P-80s were rushed to Europe in 1945 and two examples saw limited service in Italy before the war's end.

Gloster Meteor F.4 (1944)

TYPE • Fighter COUNTRY • United Kingdom

The Gloster Meteor was born in the late years of World War II, although its service to British forces extended into the 1950s. The aircraft also found widespread use in Commonwealth and export markets globally.

SPECIFICATIONS (F.Mk 8)

DIMENSIONS:	Length: 13.58m (44ft 7in); Wingspan: 11.32m (37ft 2in); Height 3.96m (13ft)
WEIGHT:	8664kg (19,100lb) loaded
POWERPLANT:	2 x 16.0kN (3600lbf) Rolls-Royce Derwent 8 turbojets
MAX SPEED:	962km/h (598mph)
RANGE:	1580km (980 miles)
CEILING:	13,106m (43,000ft)
CREW:	1
ARMAMENT:	4 x 20mm (0.79in) Hispano-Suiza cannon; up to 16 RP-3 rockets or eight 127mm (5in) HVAR rockets under outer wings, or two 1000lb (450kg) bombs

Below: The Meteor T7 was a two-seat trainer variant. It had a longer fuselage than the standard F.8 Meteor but the same tail configuration.

ENHANCEMENTS
The Meteor F.8 was a much-improved post-war variant. Enhancements included an ejector seat, a lengthened fuselage and increased fuel capacity.

METEOR F.8
This Meteor F.8 served with the Royal Australian Air Force (RAAF), which operated 104 Meteors between 1947 and 1949.

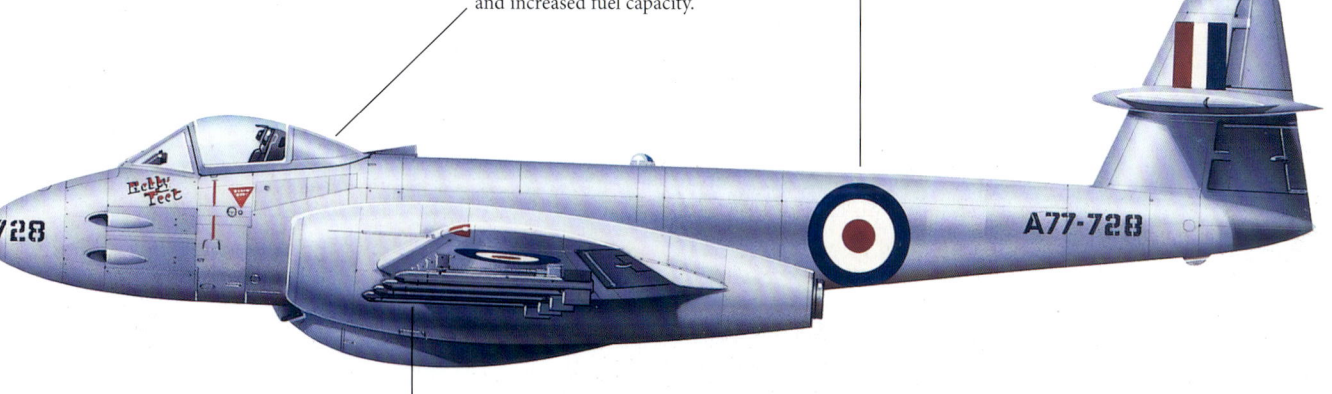

ROCKETS
This aircraft is equipped with eight 127mm (5in) High Velocity Aircraft Rockets (HVAR) under the wings.

EXPORT METEORS
The Meteor had a broad export distribution during the post-war years. Operators included Argentina, Australia, Belgium, Biafra, Brazil, Canada, Denmark, Ecuador, Egypt, Israel, the Netherlands, New Zealand, South Africa, Syria and even the United States (which bought a single test aircraft). The last major combat outing for the Meteor was during the Suez Crisis in 1956.

GLOSTER METEOR F.4
The Meteor F.4 was one of the major production versions, with more than 500 built for British service and 170 exported to international markets. It was powered by the Derwent 4 turbojet.

Douglas A-26 Invader (1944)

TYPE • Bomber COUNTRY • United States

SPECIFICATIONS (A-26A)	
Dimensions:	Length: 15.62m (51ft 3in); Wingspan: 21.34m (70ft); Height: 5.56m (18ft 3in)
Weight:	15,876kg (35,000lb) loaded
Powerplant:	Two 1491kW (2000hp) Pratt & Whitney radial piston engines
Max Speed:	600km/h (373mph)
Range:	2253km (1400 miles)
Ceiling:	6735m (22,100ft)
Crew:	3
Armament:	6 x 12.7mm (0.5in) MGs: two each in nose, dorsal and ventral positions; up to 1814kg (4000lb) of bombs

Representing a great technological leap in this class of aircraft, the A-26 entered service very late in World War II, but proved highly effective. It remained in frontline USAF service until 1972.

Below: The A-26 Invader reached the European theatre in late 1944. This particular aircraft, 'Stinky', was based in eastern France in April 1945.

Glazed nose
During World War II some 1355 A-26Bs were followed by 1091 A-26C machines with a bombardier's glazed nose.

B-26C Invader
The A-26 Invader was conceived by Douglas as a replacement for the DB-7 (A-20) series. It was similar in configuration to its predecessor, but more streamlined, powerful and better armed.

Four-man crew
The A-26C/B-26C had a four-man crew, consisting of pilot, navigator/radioman, co-pilot/bombardier and top-turret gunner.

POST-WAR SERVICE
The A-26 was redesignated B-26 in 1948, following the dropping of the 'A' for 'attack' prefix. The B-26 saw a great deal of active service during the Korean War and subsequently in Vietnam where, upgraded to B-26K standard, it was used as a highly effective counter-insurgency aircraft. B-26s were also used by many post-war air forces as well as the CIA, before being widely employed as a firefighting aircraft.

A-26C FORMATION
Pilots were delighted with the Invader's manoeuvrability and ease of handling, but the A-26 began life with a needlessly complex and fatiguing instrument array, a weak nose gear that collapsed easily and an early cockpit canopy that was difficult to hold in the 'open' position for bailout.

Ryan FR-1 Fireball (1945)

TYPE • *Naval fighter* **COUNTRY** • *United States*

SPECIFICATIONS

DIMENSIONS:	Length: 9.86m (32ft 4in); Wingspan: 12.2m (40ft); Height: 3.97m (13ft)
WEIGHT:	Maximum take-off 4810kg (10,595lb)
POWERPLANT:	One 7.1kN (1600lbf) thrust General Electric J31-GE-3 turbojet engine and one 1060kW (1350hp) Wright R-1820-72W Cyclone radial piston engine
MAX SPEED:	685km/h (426mph)
RANGE:	1657km (1030 miles)
CEILING:	13,145m (43100ft)
CREW:	1
ARMAMENT:	Four 12.7mm (0.5in) Browning MG 53-2 machine guns; up to eight 12.7mm (0.5in) rockets underwing and one 454kg (1000lb) bomb

The particular requirements of carrier aircraft resulted in the development of the mixed-power Ryan Fireball, the only fighter to enter service equipped with both a piston and a jet engine.

Below: Piloted by Lt Commander John Gray, this Ryan Fireball FR-1 was part of Fighter Squadron 66 (VF-66), based at NAS North Island, California, May 1944. The same aircraft is shown in the photograph below.

PISTON ENGINE
At the front of the aircraft was a 1060kW (1350hp) Wright R-1820-72W Cyclone radial piston engine.

PROTOTYPE
Ryan later produced a prototype of a developed version with a General Electric XT31-GE-2 turboprop in the nose and a J31 jet engine in the rear fuselage.

TURBOJET ENGINE
At the rear of the aircraft was a 7.1kN (1600lb) thrust General Electric J31-GE-3 turbojet engine.

HYBRID AIRCRAFT

The potential of the jet engine came at the cost of long take-off runs, high landing speeds and heavy fuel consumption, all elements that were less than ideal for a carrier aircraft. Therefore, the US Navy proposed utilizing a turbojet in the tail of a conventional piston-engined aircraft to act as a supplementary power unit. A requirement for such an aircraft was issued in December 1942, and Ryan responded with the FR-1.

KAMIKAZE CATCHER
The US Navy took a more cautious approach for its first jet-propelled aircraft, commissioning the Fireball in 1942. The jet engine was regarded as an adjunct to the radial to give the dash speed for catching kamikazes, but the Fireball was too late for operational war service.

Grumman F7F Tigercat (1945)

TYPE • *Naval fighter* **COUNTRY** • *United States*

One of very few twin-piston engine planes designed from the outset as a carrier aircraft, the potent Grumman F7F Tigercat demonstrated exceptional performance and saw the briefest of wartime service.

SPECIFICATIONS (F7F-1)

Dimensions:	Length: 13.85m (45ft 5in); Wingspan: 15.7m (51ft 6in); Height: 5.06m (16ft 7in)
Weight:	Maximum take-off 10730kg (23636lb)
Powerplant:	Two 1565kW (2100hp) Pratt & Whitney R-2800-22W Double Wasp 18-cylinder air-cooled radial engines
Max Speed:	687km/h (427mph)
Range:	1882km (1170 miles), 2880km (1790 miles) with 1136-litre (300-Imp gal) external tank
Ceiling:	11,040m (36,200ft)
Crew:	1
Armament:	4 x AN/M3 20mm (0.79in) cannon in wing roots, 4 x 12.7mm (0.5in) M2 Browning machine guns in fuselage nose; up to two 454kg (1000lb) bombs or 8 x 12.7mm (0.5in) rockets under wings, one 1136-litre (300-Imp gal) fuel tank or one 568-litre (150-Imp gal) napalm tank under fuselage

Left: Despite having missed any meaningful service in World War II, the Tigercat was utilized extensively in Korea.

FUSELAGE
The F7F was a shoulder-wing aircraft with large underslung nacelles for its two R-2800 engines. The fuselage – intended to present as small a frontal area as possible to minimize drag – was therefore notably slim.

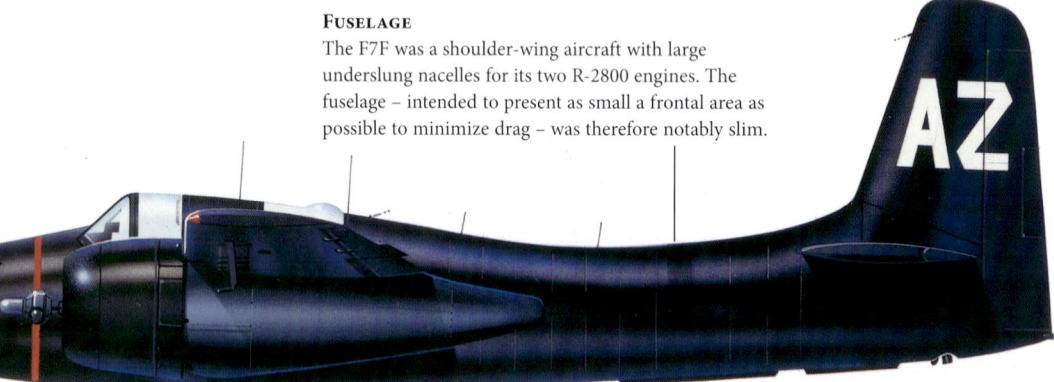

NOSE WHEEL UNDERCARRIAGE
First flown in December 1943, the F7F was the first aircraft with a nose wheel undercarriage ordered by the US Navy.

ARMAMENT
Armament was four machine guns in the nose and four 20mm (0.79in) cannons in the wing roots; the aircraft could also carry one 1136-litre (300-Imp gal) fuel tank or one 568-litre (150-Imp gal) napalm tank under the fuselage.

MARINE CORPS
Only 35 of the initial F7F-1 single-seat model were built before production switched to the F7F-2N, a two-seat night-fighter featuring a cockpit for a radar operator behind the pilot and an AN/APS-6 radar unit replacing the nose guns. The first Marine Corps unit to convert to the F7F-2N arrived at Okinawa on 14 August 1945, the day before the Japanese surrender.

POST-WAR TIGERCATS
Production of the Tigercat continued post-war with 250 of the improved single-seat F7F-3 variant, of which 60 were converted to F7F-3N night-fighters. Both of these variants would see combat in Korea before all Tigercats were withdrawn in 1952.

Bachem Ba 349 Natter (1945)

TYPE • *Fighter* COUNTRY • *Germany*

SPECIFICATIONS

DIMENSIONS:	Length: 6m (19ft 8in); Wingspan: 4m (13ft 1in); Height: 2.25m (7ft 5in)
WEIGHT:	Maximum take-off 2232kg (4921lb)
POWERPLANT:	One Walter HWK 109-509A-2 sustainer rocket; four Schmidding 109-553 solid-fuel boosters
MAX SPEED:	1000km/h (620mph)
RANGE:	55km (34 miles) after climbing at 6000m (19,685ft)
CEILING:	12,000m (39,000ft)
CREW:	1
ARMAMENT:	24 × 73mm (2.87in) Henschel Hs 297 Föhn rocket shells

The remarkable Bachem Ba 349 *Natter* (Grass Snake) emerged from an urgent requirement for a point-defence interceptor to protect Germany from the massed Allied bombing raids that were commonplace by early 1944.

SIMPLIFIED CONSTRUCTION
The *Natter*'s design allowed for easy construction by unskilled woodworkers, eliminating the need for complex jigs or tooling, ensuring swift production even in less industrialized settings.

RAPID LAUNCH
Equipped with a powerful Walter 109-509A-2 sustainer rocket and four Schmidding 109-553 solid-fuel boosters, the Natter was designed for vertical launch, ensuring rapid response against enemy aircraft formations.

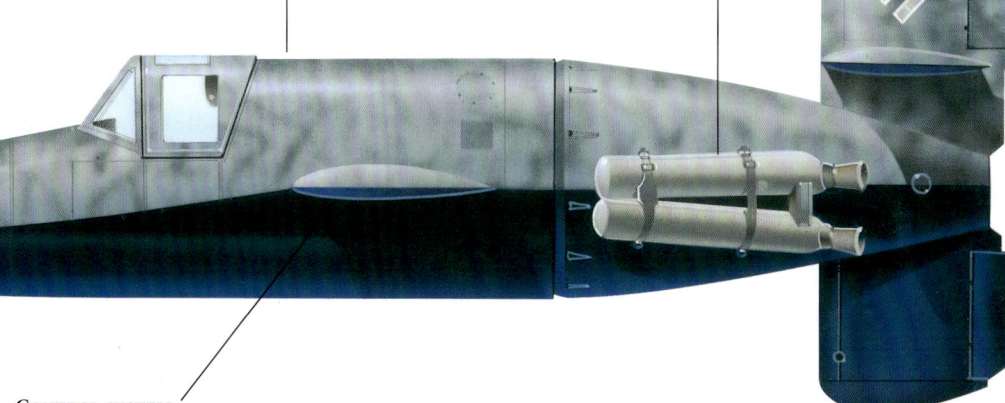

CONTROL SYSTEM
With lateral control provided by differential use of the elevators instead of traditional ailerons, the Natter showcased an innovative control system, maximizing manoeuverability while simplifying design and construction.

TACTICS

Operational tactics called for a vertical launch on autopilot, the pilot taking control of the aircraft when positioned above the enemy bomber stream. After assuming a shallow dive onto the target, the Ba 349 would have been armed by jettisoning the nose cone to expose a battery of 24 73mm (2.87in) Henschel Hs 217 Föhn or 33 55mm (2.17in) R4M unguided rockets. After these had been fired, the aircraft would break away from the battle and the pilot would prepare to bail out. This was achieved by jettisoning the nose section, and separating it from the rest of the fuselage. The pilot would be extracted by the deceleration of the rear section as it deployed a braking parachute, leaving him to descend under his own parachute.

VERTICAL LAUNCH
The Ba 349 was one of a number of ingenious projects intended to stem the Allied bombing campaign. It was intended to be vertically launched, the pilot and aircraft being recovered afterwards by parachute.

Heinkel He 162 (1945)

TYPE • *Fighter* COUNTRY • *Germany*

SPECIFICATIONS (He 162A-2)

DIMENSIONS:	Length: 9.05m (29ft 8in); Wingspan: 7.2m (23ft 7in); Height: 2.6m (8ft 6in)
WEIGHT:	Maximum take-off 2803kg (6180lb)
POWERPLANT:	One BMW-109-003E-1 Sturm axial-flow jet engine, 800kg (1764lb) static thrust
MAX SPEED:	889km/h (553mph) at sea level (with emergency boosted thrust)
RANGE:	975km (606 miles)
CEILING:	12,000m (39,000 ft)
CREW:	1
ARMAMENT:	Two 20mm (0.79in) MG 151/20 cannon

The Heinkel He 162 was developed and flown in an extraordinarily short period of time, the factory requiring only 38 days between receiving detailed drawings and the maiden flight of a prototype on 6 December 1944.

This page: This aircraft was captured by British forces at Leck. It had previously served with 3. Staffel, Einsatz-Gruppe I/JG 1 and was the personal aircraft of the Staffelkapitän.

STRUCTURAL DAMAGE
Early prototypes of the He 162A suffered from structural damage due to the use of acids in construction, which prompted design adjustments in subsequent models.

POWERPLANT
The powerplant was a single BMW 003 turbojet, mounted directly above the high-mounted wing, giving the aircraft a top-heavy profile.

AERODYNAMICS
The He 162A prototype faced aerodynamic shortcomings, notably lateral and directional instability, impacting manoeuvrability during tight turns, with left-hand turns being a notable problem.

HIGH PERFORMANCE
On its first flight, this turbojet-engined interceptor demonstrated a top speed of 840km/h (520mph). The aircraft had been devised by Nazi officials as a *Volksjäger* ('people's fighter'). From the outset, it was intended to be small, agile enough to outmanoeuvre numerically superior Allied fighter opposition, and require only limited skilled labour and minimal scarce strategic materials for its manufacture.

HE 162A-2
This He 162A-2 was allocated to 3. Staffel/JG 1 at its Leck base in May 1945. By this time, the 50 aircraft had been reorganized into one single Gruppe, Einsatz-Gruppe I./JG 1; many pilots from other fragmented units at Leck were absorbed by this new Gruppe.

Focke-Wulf Ta 152 (1945)

TYPE • *Fighter* **COUNTRY** • *Germany*

SPECIFICATIONS (Ta 152H-1)

DIMENSIONS:	Length: 10.82m (35ft 6in); Wingspan: 14.44m (47ft 5in); Height: 3.36m (11ft)
WEIGHT:	Maximum take-off 4727kg (10421 lb)
POWERPLANT:	One 1530kW (2050hp) Junkers Jumo 213E-1 liquid-cooled piston engine
MAX SPEED:	760km/h (472mph)
RANGE:	2000km (1200 miles)
CEILING:	15,100m (49500ft) with GM-1 boost
CREW:	1
ARMAMENT:	One 30mm (1.18in) MK 108 cannon, two 20mm (0.79in) MG 151/20 cannon

Reflecting Kurt Tank's overall design responsibility for the Fw 190 family, and a change in fighter aircraft designations ordered by the Reichsluftfahrtministerium (German Air Ministry), the Ta 152 name was introduced for the ultimate 'long-nose' development of the series.

Above: The 'H' variant of the Focke-Wulf Ta 152 was a high-altitude fighter. Small numbers were operated by Jagdgeschwader 301, primarily to provide cover over Messerschmitt Me 262 bases while the jets were taking off and landing.

AIR INLET
The Ta 152C-1's DB 603 engine had its air inlet on the left, diverging from the H-1's Jumo engine, with its inlet situated on the right.

DERIVATIVES
Other derivatives of the basic Ta 152 included the Ta 152S-1, a two-seat conversion trainer, and the Ta 152E that was planned as a reconnaissance fighter version of the Ta 152C, again with the standard wing.

ARMAMENT
C-series Ta 152s boasted four MG 151/20 guns, enhancing firepower but compromising speed and agility. Meanwhile, H-series variants prioritized speed, featuring only two cannons and the hub-firing weapon for bomber interception.

AIRFRAME INNOVATION
The Ta 152 was developed from the Fw 190D, with the aim of further improving performance at high altitude through a series of airframe refinements, but with the basic structure inherited from the 'Dora', except for the flap and undercarriage systems which featured hydraulic, rather than electrical actuation. Production versions with standard wing comprised the Ta 153C-1, the C-2 with improved radio equipment and the C-3 with revised armament.

HIGH-ALTITUDE FIGHTER
Long-span wings gave the final variant of the Fw 190/Ta 152 family, the Ta 152H, a superb high-altitude performance. Fortunately for the Allies, only a few made it into action.

Junkers Ju 88 Mistel (1945)

TYPE • Bomber   COUNTRY • Germany

SPECIFICATIONS
(*Mistel* 1)

DIMENSIONS: Length: 14.36m (47ft 1in); Wingspan: 20.08m (65ft 9in); Height: 13.97m (45ft 9in)

WEIGHT: 33,780kg (74,472lb) loaded

POWERPLANT: One 993kW (1332hp) DB 601E radial piston engine, two 1051kW (1410hp) Jumo 211 J-1 radial piston engines

MAX SPEED: 380km/h (236mph)

RANGE: 650km (1025 miles)

CEILING: 10,655m (34,950ft)

CREW: 1

ARMAMENT: 3800kg (8380lb) hollow-charge device

The *Mistel* (Mistletoe) was a composite aircraft consisting of an explosives-packed surplus bomber flown to a launch point by a fighter mounted on top. *Mistels* were inaccurate and caused negligible damage on the few occasions they were used.

Left: Operated by an unknown unit, this Mistel 2 combination consisted of a Ju 88G-1 and an Fw 190F-8/U3. The Fw 190F-8/U3 also carried a 1400kg (3100lb) BT-1400 heavy torpedo.

MISTEL 1
This *Mistel* 1 combination comprised a Ju 88A-4 with a Bf 109F-2. It was operated by IV./Kampfgeschwader 101.

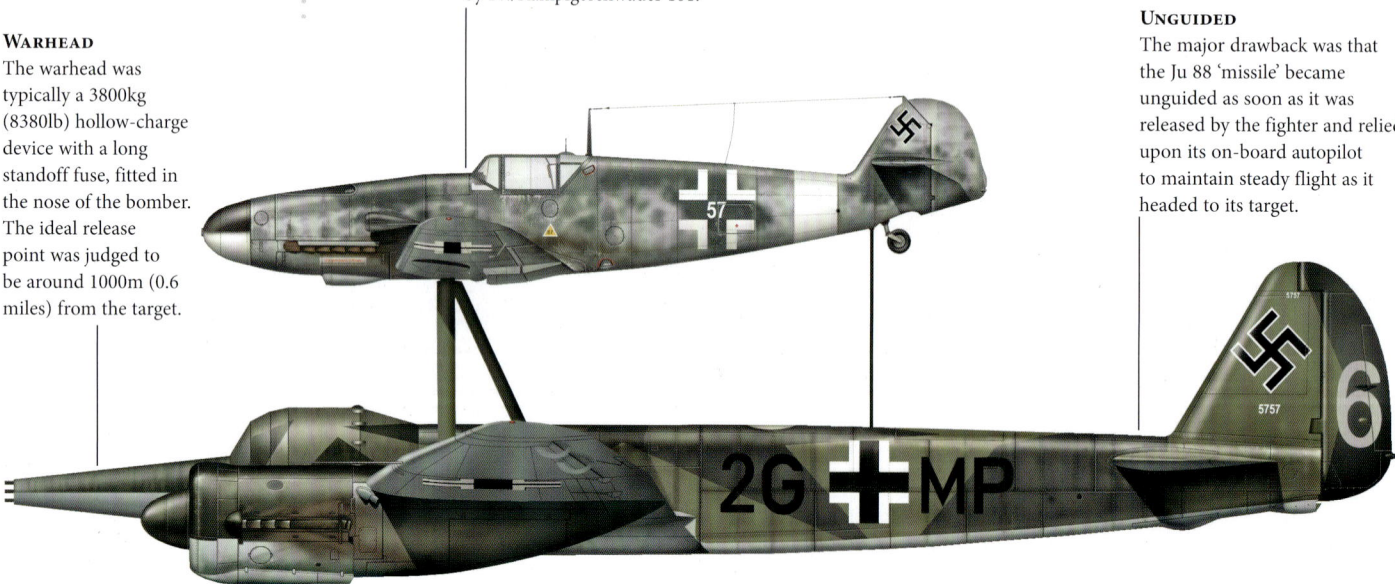

WARHEAD
The warhead was typically a 3800kg (8380lb) hollow-charge device with a long standoff fuse, fitted in the nose of the bomber. The ideal release point was judged to be around 1000m (0.6 miles) from the target.

UNGUIDED
The major drawback was that the Ju 88 'missile' became unguided as soon as it was released by the fighter and relied upon its on-board autopilot to maintain steady flight as it headed to its target.

COMPOSITE AIRCRAFT
Several different composites were completed, including the *Mistel* 1 that was used operationally, and which combined the Ju 88A-4 with a Bf 109F; it was also available in a training version as the *Mistel* S-1. The *Mistel* 2 (and S-2) combined the Ju 88G-1 with an Fw 190A-8 or Fw 190F-8/U3; while the *Mistel* 3a (and S-3a) were produced by combining the Ju 88A-6 with an Fw 190A-6.

TRAINING AIRCRAFT
American soldiers examine a captured *Mistel* composite aircraft at Bernberg, Germany, in May 1945. The Fw 190 has been mated with a Ju 88; the lack of warhead indicates this combination was indeed for training the pilots of the piggyback aircraft.

Grumman F8F Bearcat (1945)

TYPE • *Naval fighter* COUNTRY • *United States*

Lightweight and powerful, the F8F Bearcat was one of the fastest single-seat, piston-engined fighters ever built. Although in service by May 1945, it missed out on wartime combat. Highly regarded by pilots, the Bearcat had a short frontline career before its replacement by jets.

SPECIFICATIONS

DIMENSIONS:	Length: 8.62m (28ft 3in); Wingspan: 10.94m (35ft 10in), Height: 4.23m (13ft 10in)
WEIGHT:	Maximum take-off 5878kg (12,947lb)
POWERPLANT:	One 1565kW (2100hp) Pratt & Whitney R-2800-34W Double Wasp 18-cylinder air-cooled radial engine
SPEED:	677km/h (421mph)
RANGE:	1778km (1105 miles), 3162km (1965 miles) with external tanks
CEILING:	11,804m (38,700ft)
CREW:	1
ARMAMENT:	Four 12.7mm (0.5in) M2 Browning machine guns in wings; up to two 454kg (1000lb) bombs or four 12.7mm (0.5in) rockets under wings

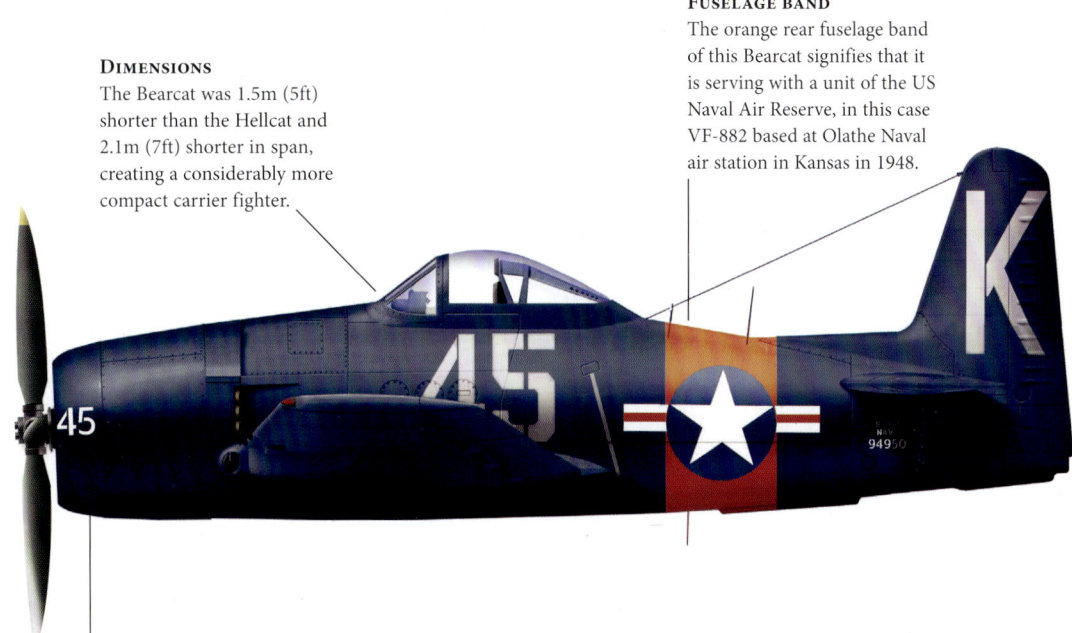

DIMENSIONS
The Bearcat was 1.5m (5ft) shorter than the Hellcat and 2.1m (7ft) shorter in span, creating a considerably more compact carrier fighter.

FUSELAGE BAND
The orange rear fuselage band of this Bearcat signifies that it is serving with a unit of the US Naval Air Reserve, in this case VF-882 based at Olathe Naval air station in Kansas in 1948.

POWERPLANT
The Bearcat drew its power from a 1565kW (2100hp) Pratt & Whitney R-2800-34W Double Wasp radial piston engine

HEIGHT RECORD

With no US engine available possessing greater power than the R-2800 of the Hellcat, the only way to achieve a significant improvement in climb performance in the Bearcat was to design the smallest and lightest possible airframe around the existing engine. The F8F was first flown in August 1944. It proved to be 80km/h (50mph) faster than the F6F and demonstrated a spectacular rate of climb – in 1946 a standard Bearcat would set a time-to-height record of 3028m (10,000ft) in 94 seconds, a record that would stand for 10 years.

PACIFIC BEARCAT
A member of the ground crew observes from the wing as the pilot turns over the engine of an F8F Bearcat. Orders for over 2000 of the new aircraft were placed and deliveries began in February 1945, with the first unit VF-19 receiving its aircraft in May.

Cold War

Aircraft development during the Cold War reshaped the imagined possibilities of military aviation. The first decade after World War II was characterized largely by jet equivalents to the wartime types, faster than their forebears but still reliant on direct-fire guns and dumb bombs. But the pace of technological change, fuelled by numerous conflicts, put innovation into a steep climb. Supersonic flight became commonplace, delivered by sleek airframes, new composite materials and immensely powerful afterburning powerplants. Computerization crept into cockpits and flight controls. Weaponry became as advanced as the aircraft, with precision-guided munitions and air-to-air missiles profoundly extending the range at which attacks could be made, and the accuracy with which they were delivered. And new avionics increased aircraft performance and pilot situational awareness equally.

SAAB 37 VIGGEN
The Saab 37 Viggen was one of the new generations of multirole aircraft that entered service in the 1960s and 1970s, extending the boundaries of speed, manoeuvrability and lethality.

De Havilland Vampire (1946)

TYPE • Fighter COUNTRY • United Kingdom

The De Havilland DH.100 Vampire was Britain's second jet fighter after the Gloster Meteor. The Meteor received priority for engines and did not enter service until after the war, despite having first flown in September 1943.

SPECIFICATIONS (FB.9)

DIMENSIONS:	Length: 11.58 m (30ft 9in); Wingspan: 11.58m (38ft); Height: 1.91m (6ft 3in)
WEIGHT:	5606kg (12,360lb) loaded
POWERPLANT:	One 19.57kN (4400lbf) thrust Rolls-Royce Goblin 2/2 turbojet
MAX SPEED:	853km/h (530mph)
RANGE:	1842km (1145 miles)
CEILING:	12,500m (41,000ft)
CREW:	1
ARMAMENT:	Four 20mm (0.79in) Hispano cannon, eight 27kg (60lb) rockets and two 227kg (500lb) bombs or two 454kg (1000lb) bombs

Below: Originally named Spider Crab, but fortunately changed to Vampire in April 1944, the De Havilland DH.100 was designed around the Halford H.1 centrifugal-flow turbojet and had a relatively tubby fuselage.

WING CONSTRUCTION
In contrast to the fuselage pod, the Vampire's wings, control surfaces, tailplane and tailbooms were metal.

CENTRIFUGAL FLOW
Although the Vampire's centrifugal flow design was almost universal initially, the German axial-flow engine design was to become dominant over time.

FUSELAGE POD
Sharing many components with the Mosquito, the Vampire's fuselage pod was of plywood and balsa construction with armour plate bulkheads.

VAMPIRE FB.5
The most important version was the FB.5, of which 888 were built from 1948. Strengthened for the ground-attack role, the FB.5 and related FB.50-series were exported to New Zealand, South Africa, France, Italy, India, Finland, Iraq and other countries. A number of countries licence-built Vampires. Australian production was undertaken by De Havilland Australia at Bankstown in Sydney.

RAF VAMPIRE
The Royal Auxiliary Air Force (RAuxAF) was integrated into the regular RAF during the war, but was re-formed in 1946. In May that year, No. 607 Squadron was re-established at Ouston in Yorkshire as a day fighter squadron with Mk 14 and Mk 22 Spitfires. In June 1951, Vampire FB.5s were received and these were supplemented by FB.9s from April 1956.

Douglas A-1 Skyraider (1946)

TYPE • Ground attack **COUNTRY** • United States

SPECIFICATIONS (A-1H)

DIMENSIONS:	Length: 11.84m (38ft 10in); Wingspan: 15.25m (50ft); Height: 4.78m (15ft 8in)
WEIGHT:	11,340kg (25,000lb) maximum
POWERPLANT:	1 x 2013kW (2700hp) Pratt & Whitney R-3350-26WA radial piston engine
MAX SPEED:	520km/h (320mph)
RANGE:	2115km (1315 miles)
CEILING:	8660m (28,500ft)
CREW:	1
ARMAMENT:	4 x 20mm (0.79in) cannon; up to 3600kg (8000lb) of bombs, rockets or other stores

Able to deliver an incredible array of ordnance, the Skyraider's origins began with the Douglas Dauntless of World War II fame. Incredibly strong with a good performance, the A-1 long outlived its anticipated service life.

Below: During the early years of the Vietnam War this A-1H served with VA-145 'Swordsmen' aboard USS Constellation.

WING STRENGTH
Massive strength was built into the wing. In combat, several Skyraiders made it home after taking hits from flak shells of up to 37mm (1.46in).

POWERPLANT
Much of the Skyraider's success was due to the superb Wright R-3350 two-row radial engine, which was both powerful and reliable.

A1-H SKYRAIDER
Built for the US Navy as an AD-6, this aircraft was redesignated as an A-1H in 1962.

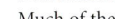

AD-1 SKYRAIDER
The mainstay of the US Navy's carrier-borne attack force in Korea, the AD-1 entered service in 1946 and was only finally retired in 1968 during the Vietnam War. During the Korean War, the Skyraiders proved that 'low and slow' still produced results in ground-attack work. The AD squadrons of the US Navy and Marine Corps gave outstanding air support to UN troops.

AD-4N SKYRAIDER, 1953
The AD-4 had many improvements, including an increase in gross weight to 10,886kg (24,000lb), a P-1 autopilot, a modified windscreen and an improved radar (APS-19A).

De Havilland Hornet (1946)

TYPE • Fighter **COUNTRY** • United Kingdom

The Hornet's more-than-passing resemblance to the Mosquito of World War II reflects the evolution of this aircraft from De Havilland. It was intended as a fast, tough fighter-bomber and operated in both the Royal Air Force and, as the Sea Hornet, the Royal Navy.

SPECIFICATIONS (F.MK III)

DIMENSIONS:	Length: 11.18m (36ft 8in); Wingspan: 13.72m (45ft); Height: 4.3m (14ft 2in)
WEIGHT:	9480kg (20,900lb) loaded
POWERPLANT:	2 x 1551kW (2080hp) Rolls-Royce Merlin 130/131 12-cylinder engine
MAX SPEED:	760km/h (472mph)
RANGE:	4828km (3000 miles)
CEILING:	10,668m (35,000ft)
CREW:	1
ARMAMENT:	4 x 20mm Hispano Mk. V cannons; 2 x 454kg (1000lb) bombs; 8 x rockets

Right: The underside of this Hornet clearly shows the mouth of the 20mm (0.79in) cannon. Each gun had 190 rounds.

NOSE ARMAMENT
Beneath the nose of the aircraft four 20mm (0.79in) Hispano Mk. V cannons were housed, used to deliver powerful strafing attacks.

HORNET F.MK 3
Normally stationed at Tengah, this De Havilland Hornet single-seat fighter was deployed to RAF Butterworth during the Malayan Emergency.

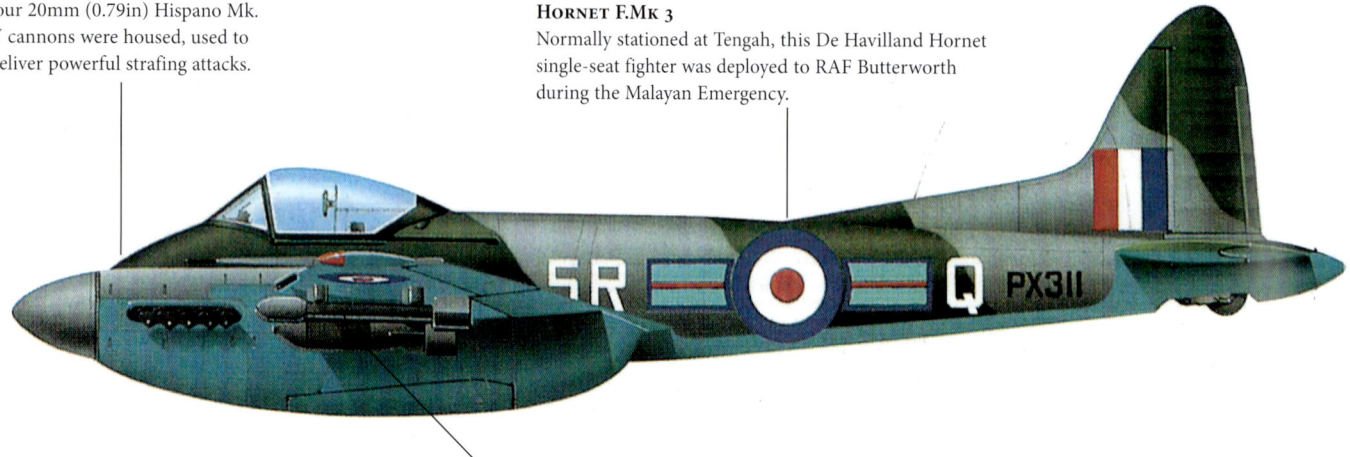

WEAPONRY
The aircraft is armed with underwing rockets, typical weapons used in the campaign waged against the guerrillas, who exploited the dense jungle of Malaya to their advantage.

MALAYAN EMERGENCY
By 1950 the RAF had 160 aircraft in the Malayan theatre, while the guerrillas strengthened their presence in the rural areas, operating out of village strongholds. Guerrilla movements off the coast were tracked by RAF Sunderland flying boats. Hornet fighter-bombers were in action from early 1951 as successors to the Tempest, delivering ground support to British troops and hitting remote insurgent bases and trails.

F. MK III IN FLIGHT
Development of the Hornet began in the 1940s. Although the aircraft appears very similar to the Mosquito, the aircraft was in fact a new design from De Havilland. This particular aircraft was destroyed in 1949 following in-flight mechanical failure.

Mikoyan-Gurevich MiG-9 (1946)

TYPE • *Fighter* COUNTRY • *Soviet Union*

SPECIFICATIONS

DIMENSIONS:	Length: 9.83m (32ft 3in); Wingspan: 10m (32ft 10in); Height: 3.22m (10ft 7in)
WEIGHT:	Empty 3420kg (7540lb); maximum 5500kg (12,125lb)
POWERPLANT:	Two 7.8kN (1533lbf) thrust Kolesov RD-20 afterburning turbojet engines
MAX SPEED:	909km/h (565mph)
RANGE:	800km (495 miles)
CEILING:	13,000m (42,650ft)
CREW:	1
ARMAMENT:	One 37mm (1.45in) NL-37 cannon, two NS-23 23mm (0.9in) cannon

The MiG-9 was the first Soviet jet to fly, powered by Russian copies of the German BMW 003 turbojet. It was built in relatively large numbers (610 examples) and received the reporting name 'Fargo', but was never regarded as a satisfactory fighter despite its heavy cannon armament.

Below: This side view of a MiG-9 shows to good effect the extremely heavy cannon armament fitted in the nose.

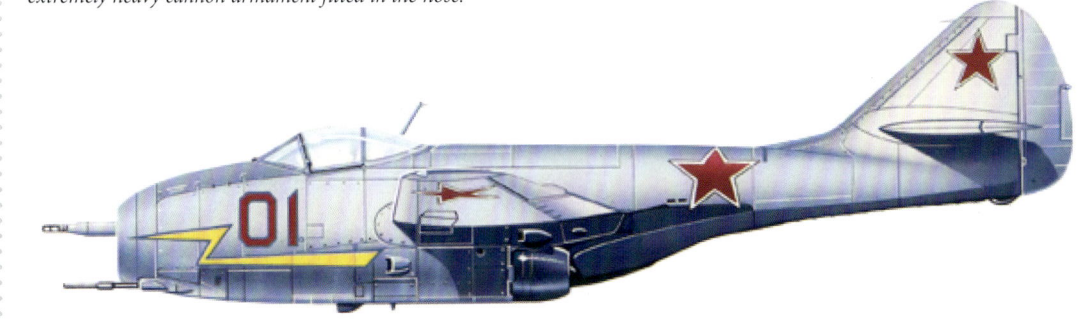

NOSE-MOUNTED INTAKE
As with all early MiG jet fighters, the air intake was mounted in the nose of the aircraft, which facilitated sleek layouts, albeit creating some problems with airflow.

LAYOUT
The unusual layout of the aircraft was known as 'pod-and-boom'. This layout was particularly easy for making landings, but the high angle of the rear fuselage underside was necessary to protect the airframe from jet blast.

SOVIET SERVICE
The MiG-9 was a relatively short-lived aircraft in the history of Mikoyan-Gurevich. It made its first flight in April 1946 and entered production that year, with a total of 610 aircraft manufactured between 1946 and 1948. It quickly fell out of Soviet service with the introduction of the far more capable MiG-15, although many MiG-9s subsequently went into communist Chinese hands.

RD-20 TURBOJETS
The RD-20 turbojets that powered the MiG-9 were copies of the BMW 003 engine, examples of which had been captured by the Soviets in World War II.

SOVIET MIG-9
A MiG-9 sits on the tarmac at a Soviet air base. Several variants of the MiG-9 were produced, including test beds for advanced anti-shipping missiles and two-seat trainers.

McDonnell FH-1 Phantom (1947)

TYPE • *Fighter* COUNTRY • *United States*

SPECIFICATIONS

DIMENSIONS:	Length: 11.35m (37ft 3in); Wingspan: 12.42m (40ft 9in); Height 4.32m (14ft 2in)
WEIGHT:	4552kg (10,035lb) loaded
POWERPLANT:	2 x 7.1kN (1600lbf) Westinghouse J30-WE-20 turbojets
MAX SPEED:	771km/h (479mph)
RANGE:	1120km (695 miles)
CEILING:	12,525m (41,100ft)
CREW:	1
ARMAMENT:	4 x 12.7mm (0.50in) MGs

The McDonnell FH-1 Phantom was the first US jet aircraft to take off and land on a US aircraft carrier, completing its carrier qualification on 21 July 1946. It was also the first all-jet aircraft to be ordered into production by the US Navy.

Below: This aircraft served with VF-171 (the former VF-17A), the only US Navy unit to take the aircraft aboard carriers, before the Phantom was demoted to Reserve and Marine Corps service.

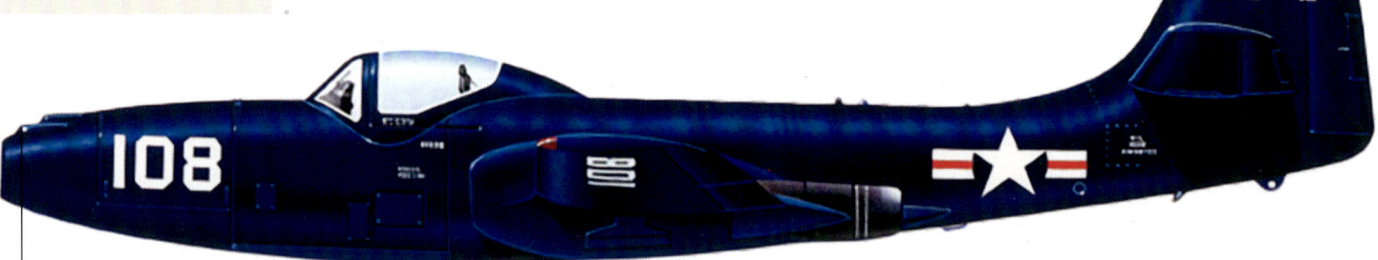

TAIL
The horizontal stabilizer on the tail was placed high to give it some distance above the back-thrust of the Westinghouse turbojet engines.

NOSE ARMAMENT
The FH-1 Phantom's armament was confined to its nose, in the form of four 12.7mm (0.50in) fixed forward-firing machine guns.

WINGS
The FH-1 had an unswept wing design. The wings were fully folding for stowage on board an aircraft carrier.

FH-1 LANDING
A McDonnell FH-1 Phantom lands on USS *Saipan* (CVL-48) in May 1948. The success of the FH-1 Phantom meant that it was almost inevitable that McDonnell would be asked to submit a design to succeed the Phantom in service. The new aircraft – the F2H-1 – was larger, incorporating folding wings and a lengthened fuselage to accommodate more fuel and more powerful engines.

FLEET DELIVERIES
Work on a jet-powered fighter for the US Navy actually began back in the early 1940s. The McDonnell Aircraft Corporation became involved in this project in August 1943, teaming up with Westinghouse Electric Corporation, who would develop the turbojet engines. The prototype aircraft that emerged impressed the Navy evaluators, and fleet squadron deliveries began in July 1947.

Lockheed P2V-1 Neptune (1947)

TYPE • *Maritime patrol* COUNTRY • *United States*

SPECIFICATIONS

DIMENSIONS:	Length: 22.9m (75ft 4in); Wingspan: 30.4m (100ft); Height 8.6m (28ft 6in)
WEIGHT:	26,3030kg (58,000lb) loaded
POWERPLANT:	2 x 1715kW (2300hp) Wright R-3350-8A engines
MAX SPEED:	487km/h (303mph)
RANGE:	6618km (4110 miles)
CEILING:	8230m (27,000ft)
CREW:	9–11
ARMAMENT:	Up to 16 x 127mm (5in) rockets and up to 3628kg (8000lb) of bombs, depth charges or torpedoes

The first post-war US Navy patrol aircraft to see widespread service had been the P2V (later P-2) Neptune, which served with many NATO allies for at least three decades, until the arrival of more capable equipment such as the P-3 Orion.

Below: With increasing Soviet submarine activity in the seas surrounding its islands from the mid-1970s, Japan's ASW capacity rested with both fixed-wing and helicopter assets. Built by Kawasaki, the P-2J was an advanced local development of the Neptune, with a lengthened fuselage and a mixed turboprop/turbojet powerplant.

ACCOMMODATION
The Neptune's ease of handling and manoeuvrability, with its unusually large rudder and its spacious accommodation made it popular with US Navy crews.

DEFENCES
The aircraft carried a weapons bay for two torpedoes or up to twelve depth charges, and six defensive machine guns.

P2V-1 VP-8, US NAVY / C. 1949
The West's dominant land-based maritime air patrol platform for the first three decades after World War II, the first of the Neptunes to enter service was the P2V-1, from March 1947.

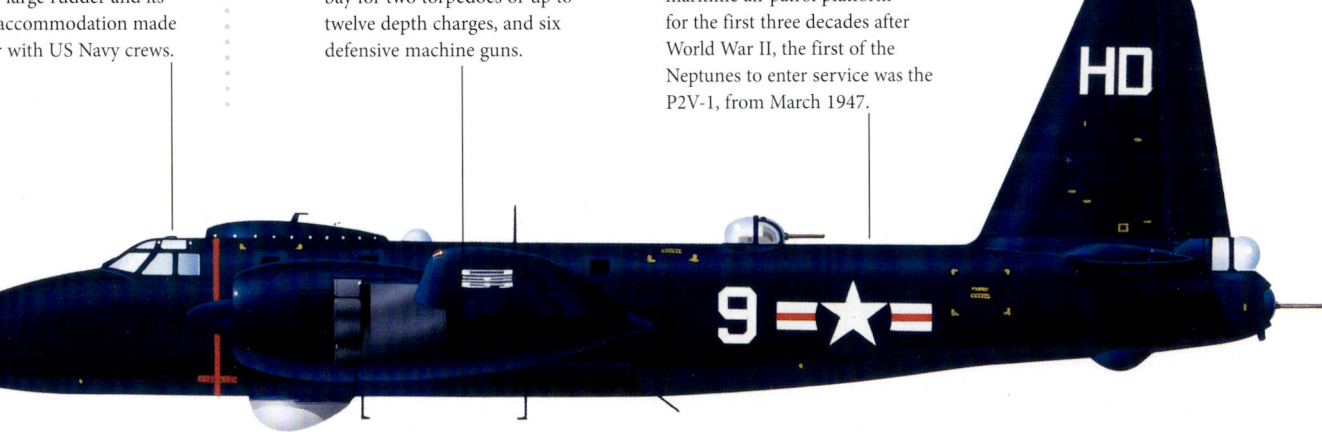

ENDURANCE
Perhaps the most famous achievement of this maritime patrol aircraft was a distance record that has stood the test of time. Piloted by Commander Thomas P. Davies in September 1946, a P2V-1 named *The Truculent Turtle* flew 18,227km (11,326 miles) non-stop from Perth, Australia, to Columbus, Ohio, in 55 hours and 17 minutes. This record-breaking flight demonstrated the Neptune's superb range and endurance.

LOCKHEED XP2V-1 PROTOTYPE
An XP2V-1 protoype in flight in the mid-1940s. One of the greats of naval aviation, the Lockheed P2V (P-2 from 1962) Neptune was the West's answer to the Soviet Union's awesome submarine threat during the first half of the Cold War. This superb land-based maritime patrol aircraft not only searched for submarines, but also fulfilled an anti-surface vessel role.

Hawker Sea Fury (1947)

TYPE • Naval fighter **COUNTRY** • United Kingdom

SPECIFICATIONS (FB.Mk 11)

DIMENSIONS:	Length: 10.6m (34ft 8in); Wingspan: 11.7m (38ft 4in); Height 4.9m (16ft 1in)
WEIGHT:	5670kg (12,500lb) loaded
POWERPLANT:	1 x 1850kW (2480hp) Bristol Centaurus XVIIC 18-cylinder twin-row radial engine
MAX SPEED:	740km/h (460mph)
RANGE:	1127km (700 miles)
CEILING:	10,900m (35,800ft)
CREW:	1
ARMAMENT:	4 x 20mm (.79 in) Hispano Mk V cannon; 12 x 76.2mm (3in) rockets or 907kg (2000lb) of bombs

Derived from the Tempest Mk II, the Sea Fury was one of the fastest and best post-war piston-engined fighters. Canada, Australia and the Netherlands also flew carrier-based Sea Furies, and land-based versions without folding wings or arrestor hooks were exported internationally.

Below: A Dutch Sea Fury FB.50. The Netherlands was actually the first customer for export versions of the Sea Fury.

POWERPLANT
This Cuban Sea Fury was powered by a 1850kW (2480hp) Bristol Centaurus XVIIC 18-cylinder twin-row radial engine, driving a five-bladed propeller.

CUBAN SEA FURY FB.MK 11
The small force of British-supplied Sea Furies inherited from the Batista regime saw some action in the hands of the Fuerza Aérea Revolucionaria (FAR) during the abortive Bay of Pigs invasion.

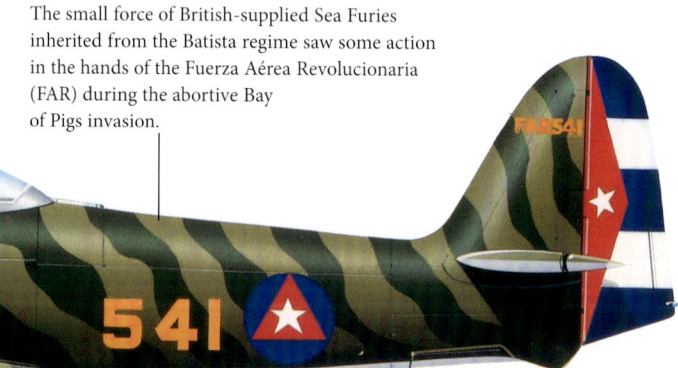

ROLL RATE
The Fury had an exceptional roll rate of 100 degrees per second. It could also climb up to (20,000ft) in under five minutes.

SPEED MACHINE
Suggested by its sleek aerodynamic lines, the Hawker Fury was one of the fastest military piston-engine aircraft of the 20th century. Powered by its Bristol Centaurus engine, it was able to achieve a maximum speed of 740km/h (460mph), which when combined with its exceptional manoeuvrability meant that it was a capable threat to many contemporary jets.

NAVAL SEA FURY
A Hawker Sea Fury FB.11, flown by the Royal Navy (WF619). Despite its pioneering work introducing jets to carriers, the Fleet Air Arm deployed only piston-engined aircraft in Korea, where the Sea Fury achieved kills of MiG fighters.

Mikoyan-Gurevich MiG-15 (1948)

TYPE • Fighter   COUNTRY • Soviet Union

The MiG-15 was very much the equivalent of the F-86 Sabre, with advantages such as heavy armament and weaknesses such as a poorer turn radius.

SPECIFICATIONS

DIMENSIONS:	Length: 10.86m (35ft 7in); Wingspan: 10.8m (33ft); Height: 3.7m (12ft 2in)
WEIGHT:	6045kg (13,327lb)
POWERPLANT:	Kilmov VK-1 turbojet, 26.5kN (5950lbf)
MAX SPEED:	1075km/h (668mph)
RANGE:	2520km (1565 miles)
CEILING:	15,500m (50,840ft)
CREW:	1
ARMAMENT:	1 × 37mm (1.45in) cannon, 2 × 23mm (0.9in) cannon

Below: The MiG-15 was designed using German research (captured by the USSR in 1945) into swept wings, and was one of the first operational jets of this type.

ENGINE
The MiG-15bis used an engine developed from a copy of the Rolls-Royce Nene engine, which powered several contemporary fighters.

WING FENCES
Note the prominent wing fences, which reduce the tendency of swept-wing aircraft to stall due to spanwise (rather than front-to-back) airflow over the wings.

TAILFIN
The huge wide-chord swept tailfin houses a gyro-compass in its base, and has room for a radar warning receiver.

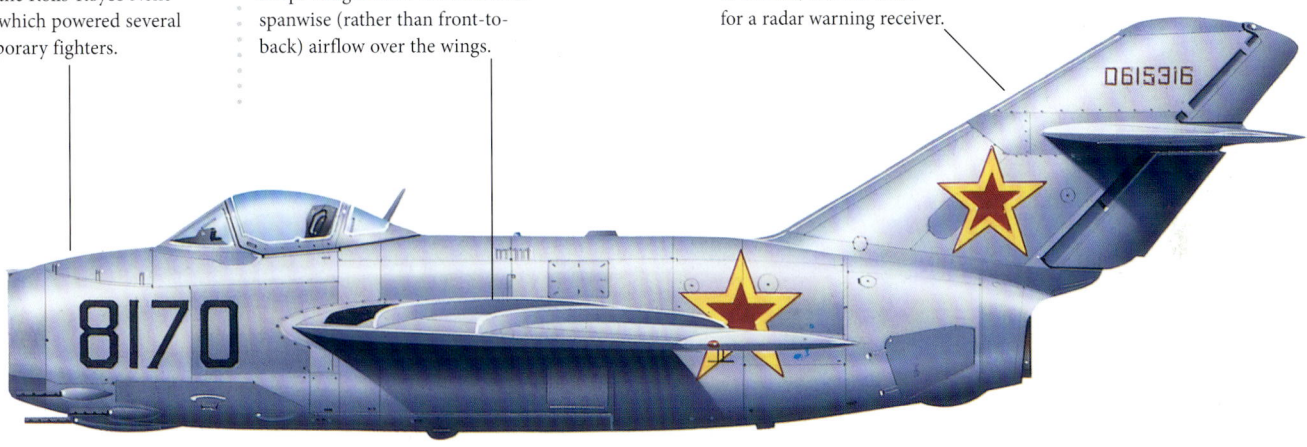

MiG vs SABRE
The MiG-15, which first flew weeks after the F-86 Sabre, proved that the Soviet Union could design, produce and use a jet fighter as modern as any in the world. The MiG-15 had better climb, ceiling, rate of roll and turn radius than early F-86s, and only with late versions of the F-86F, introduced in 1953, did the Soviet fighter inescapably meet its superior. The Sabre's success against the MiG was due less to the aircraft than to the skill and tactics of American pilots.

VINTAGE MiG
A surviving two-seat MiG-15 on display at a modern air show. Some air forces are known to have retained their MiG-15s almost up to the end of the 1990s. The MiG-15 was primarily built in the Soviet Union, but also in Czechoslovakia and Poland. It served with air forces of various Warsaw Pact nations until the late 1960s, with some remaining in service as trainers.

McDonnell F2H Banshee (1948)

TYPE • Naval fighter **COUNTRY** • United States

The success of the FH-1 Phantom meant that it was almost inevitable that McDonnell would be asked to design a successor. The new aircraft was larger, incorporating folding wings and a lengthened fuselage to accommodate more fuel and more powerful engines.

SPECIFICATIONS (F2H-2)

Dimensions: Length: 14.68m (48ft 2in); Wingspan: 12.73m (41ft 9in); Height 4.42m (14ft 6in)
Weight: Empty 5980kg (13,183lb); maximum take-off weight 11,437kg (25,214lb)
Powerplant: One 14.5kN (3250lbf) Westinghouse turbojet
Max Speed: 933km/h (580mph)
Range: 1883km (1170 miles)
Ceiling: 14,205m (46,600ft)
Crew: 1
Armament: Four 20mm (0.79in) cannon; underwing racks for two 227kg (500lb) or four 113kg (250lb) bombs

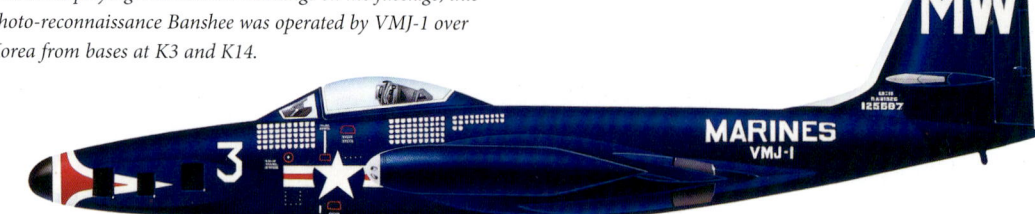

Below: Displaying 122 mission markings on the fuselage, this photo-reconnaissance Banshee was operated by VMJ-1 over Korea from bases at K3 and K14.

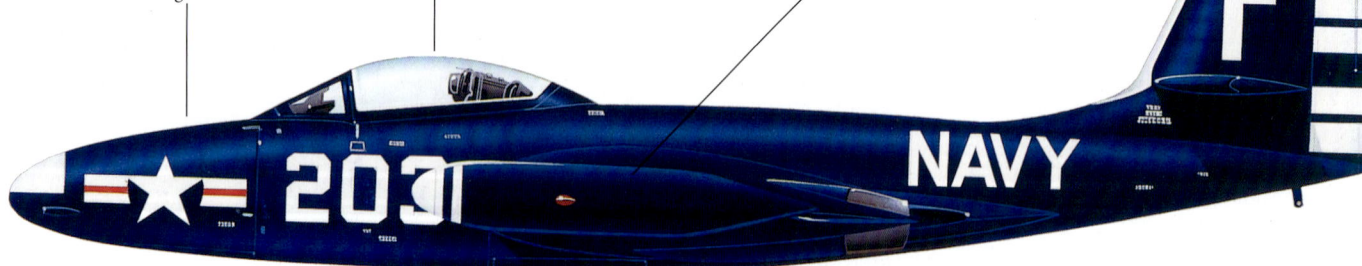

High altitude
The Banshee could achieve an altitude of more than 14,000m (46,000ft). This operational advantage made the aircraft extremely challenging to intercept when it was at its ceiling.

Cockpit
The cockpit was fully pressurized and air conditioned, maximizing the pilot's comfort during long operations. The front of the canopy also had bulletproof glass.

Wingtip fuel tanks
The wings of the Banshee were specially strengthened to hold 760-litre (200-Imp gal) wingtip fuel tanks, which significantly extended the aircraft's range.

Photo-recon
The Banshee's high-altitude performance made it ideal for adaptation to photo-reconnaissance duties as the F2H-2P. Six high-resolution cameras were fitted in a specially elongated nose, the angle of the cameras controlled by the pilot from the cockpit. The camera bay was also heated to prevent frosting in the sub-zero flight conditions.

F2H on carrier elevator
A Banshee is taken below deck on the carrier USS *Essex*, its wings in the folded position. The Banshee's straight-wing design significantly reduced the aircraft's top speed compared to the latest swept-wing, land-based fighters then emerging.

Convair B-36 Peacemaker (1948)

TYPE • Bomber   COUNTRY • United States

The extraordinary Convair B-36 had no fewer than 10 engines: four piston engines and six jets. Although designed to carry some of the extremely large atomic and hydrogen bombs of the time, it also had the capacity to drop up to 80 conventional bombs.

SPECIFICATIONS (B-36J)

Dimensions: Length: 9.40m (162ft 1in); Wingspan: 70.1m (230ft); Height: 14.25m (46ft 9in)

Weight: Empty 77,580kg (171,035lb); maximum 186,000kg (410,000lb)

Powerplant: 6 x 2500kW (3800hp) Pratt & Whitney R-4360-53 radial and 4 x 23kN (5200lbf) thrust General Electric J47-GE-19 turbojet engines

Max Speed: 685 km/h (420 mph)

Range: 10,945km (6800 miles)

Ceiling: 15,000m (48,000ft)

Crew: 15

Armament: 16 x 20mm (0.79in) cannon; up to 39,010kg (86,000lb) of bombs

Above: A force of almost 400 B-36s formed the backbone of SAC from 1948 to 1959. The type was the largest bomber to serve with USAF, the RB-36D illustrated here being a strategic reconnaissance version, with a forward bomb bay adapted to carry 14 cameras. The crew was increased from 15 to 22.

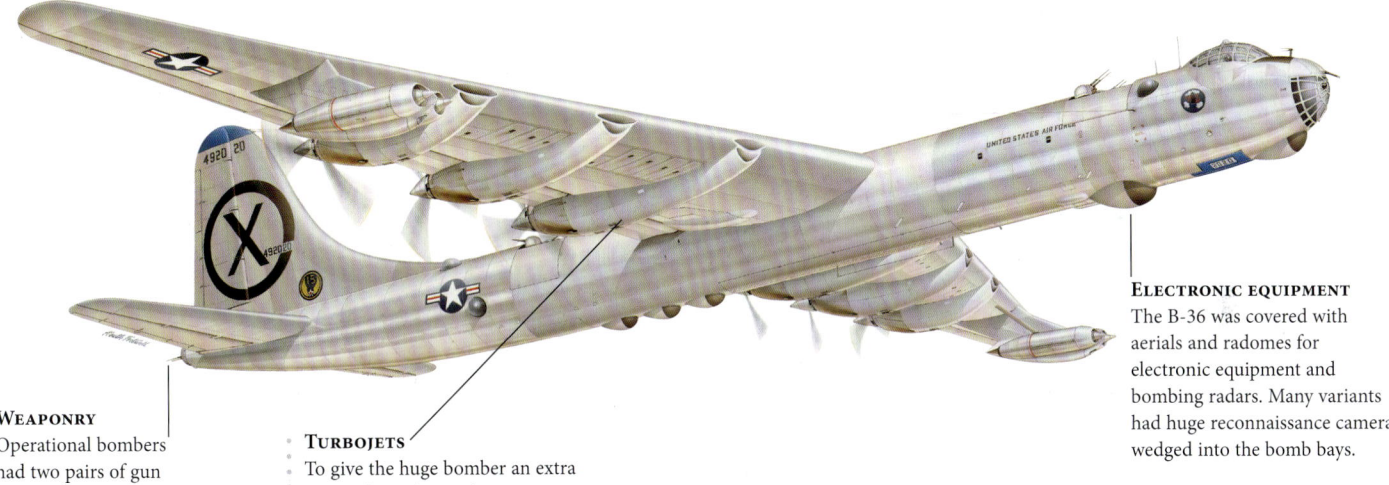

Weaponry
Operational bombers had two pairs of gun turrets in the rear fuselage, operated remotely from observation posts by

Turbojets
To give the huge bomber an extra burst of speed over the target area, the B-36 was fitted with four J47 turbojets to augment the six huge piston engines driving the propellers.

Electronic equipment
The B-36 was covered with aerials and radomes for electronic equipment and bombing radars. Many variants had huge reconnaissance cameras wedged into the bomb bays.

SAC BOMBER

The B-36 was America's most spectacular Cold War deterrent; 383 of these bombers formed the backbone of the mighty Strategic Air Command (SAC) from 1948 to 1959. The largest warplanes to fly in the West, they carried the biggest hydrogen bombs ever built and girdled the globe on nuclear alert or highly dangerous spying missions. At one stage they even carried their own fighter aircraft.

HIGH-ALTITUDE FLIGHT

At high altitude, the vast wings of the B-36 clawed so much air that the bomber was more manoeuvrable than jet fighters. Missions in this giant lasted as long as 40 hours. No other American bomber ever approached the B-36's size, weight and bomb-carrying capacity.

North American F-86 Sabre (1949)

TYPE • Fighter   COUNTRY • United States

Perhaps the finest jet fighter of its generation, the F-86 (initially P-86) won its spurs over Korea and went on to enjoy unprecedented success in a variety of combat roles with dozens of different air arms around the world, serving for close to 50 years.

SPECIFICATIONS (F-86H)

DIMENSIONS:	Length: 11.84m (38ft 10in); Wingspan: 11.93m (39ft 1.5in); Height: 4.57m (15ft)
WEIGHT:	9912kg (21,852lb) maximum take-off
POWERPLANT:	1 x 39.7kN (8920lbf) General Electric J33-GE-3D turbojet
MAX SPEED:	1114km/h (692mph)
RANGE:	835km (519 miles)
CEILING:	15,485m (50,800ft)
CREW:	1
ARMAMENT:	4 x 20mm (0.79in) cannon and 2 x 227kg (500lb), 340kg (750lb) or 454kg (1000lb) bombs, or 16 x 127mm (5in) rockets

Below: In addition to the USAF, Sabres were operated over Korea by the South African AF, which replaced its F-51Ds with F-86Fs that were based at K55. The South African jets were primarily used in a fighter-bomber role, 2 Sqn being operated under the command of the USAF's 18th Fighter Bomber Group.

RADAR
Adding radar in the D-model demanded a redesign of the entire nose, which now incorporated an engine intake in the chin position, with air being ducted below the cockpit.

COCKPIT
The pilot sat below an aft-sliding canopy on a North American ejector seat, in a pressurized cockpit with air conditioning. The display was dominated by a screen for the fire control system.

520TH FIS
This F-86D was flown by the 'Geiger Tigers' of the 520th Fighter Interceptor Squadron, Air Defense Command, based at Geiger Field, Washington, in 1955. The three stripes on the fuselage indicate this is the commander's aircraft.

ALL-WEATHER F-86D
While the day fighters are the best-known Sabre variants today, the most numerous member of the family was the F-86D all-weather fighter that was intended for service with Air Defense Command. Originally designated the F-96 on account of its considerable changes, the F-86D incorporated an interception radar and fire control system and carried an armament of 24 70mm (2.75in) Mighty Mouse rockets in a retractable ventral tray.

P-86A FLIGHT
The North American Aviation P-86A-1-NA Sabre 47-605 is historic as the first production Sabre aircraft. The first operational F-86As were delivered to the 1st Fighter Group early in 1949. During the Korean War, Sabres claimed the destruction of 810 enemy aircraft, 792 of them MiG-15s.

Ilyushin Il-28 Beagle (1949)

TYPE • Bomber COUNTRY • Soviet Union

Still in service half a century after it was designed, the Il-28 was a stunning aircraft when it first flew, the equal of the British Canberra. The 'Beagle' was exported to many nations and hundreds were based in Warsaw Pact countries.

SPECIFICATIONS

DIMENSIONS:	Length: 17.65m (57ft 10.75in); Wingspan: 21.4m (70ft 4in); Height: 6.7m (21ft 11.8in)
WEIGHT:	21,200kg (46,738lb) loaded
POWERPLANT:	2 x 26.3kN (5952lbf) Klimov VK-1 turbojets
MAX SPEED:	902km/h (560mph)
RANGE:	2180km (1355 miles)
CEILING:	12,300m (40,355ft)
CREW:	3
ARMAMENT:	4 x 23mm (0.9in) cannon; internal bomb capacity 1000kg (2205lb), maximum bomb capacity 3000kg (6614lb)

Below: The success of the 'Beagle' was widely recognized abroad, with sales to most Eastern bloc states, Egypt, Finland, Indonesia, Somalia and Yemen. China licence-built the Il-28 as the Hongzhaji-5 (H-5).

COCKPIT
The pilot sat in a fighter-style cockpit with the canopy hinging to the right. Both pilot and navigator/bombardier sat on ejection seats.

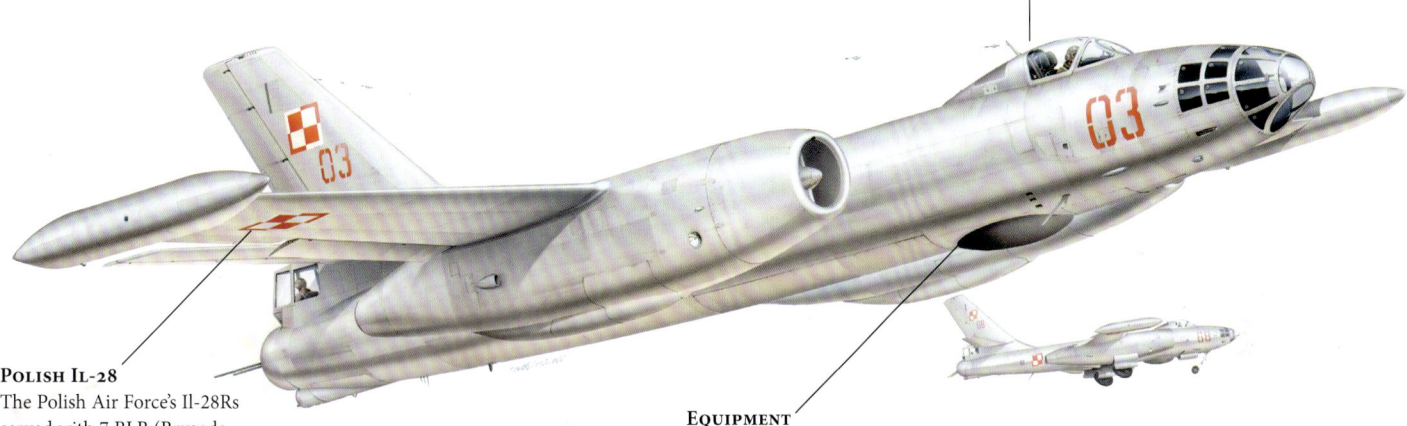

POLISH IL-28
The Polish Air Force's Il-28Rs served with 7 BLB (Brygada Lotnictwa Bombowego – Bomber Brigade) at Powidz Air Base from 1953.

EQUIPMENT
The equipment kit included radar-warning receiver, instrument landing system, gun camera, autopilot, VHF Omni-Range, distance-measuring equipment and radar altimeter.

TACTICAL AIRCRAFT
Ilyushin's Il-28 was the mainstay of the Warsaw Pact tactical bomber force in the early Cold War period. Design began in 1947, benefiting from advanced British engine technology (the Rolls-Royce Nene engine), which had just been sold to the Soviet Union. The aircraft first flew in 1948 and entered squadron service in 1950. More than 1500 had been built by 1955.

CZECH IL-28 BEAGLE
The Beagle had much to recommend it to export markets. It was fast and carried an effective warload, including the TN nuclear weapon. The Il-28 reconnaissance version also flew in 1950, followed by the long-range nuclear strike Il-28D and other versions for target towing, electronic warfare and systems development.

Mikoyan-Gurevich MiG-17 (1950)

TYPE • *Fighter* **COUNTRY** • *Soviet Union*

Scourge of the US air arms operating over Vietnam, the MiG-17 (NATO codename 'Fresco') was a refinement of the Korean War-era MiG-15. The result was an agile fighter and fighter-bomber that also gave good service in battles fought by Arab air forces against Israel into the 1970s.

SPECIFICATIONS (MiG-17F)

DIMENSIONS:	Length: 11.26m (36ft 11in); Wingspan: 9.63m (31ft 7in); Height: 3.8m (12ft 5in)
WEIGHT:	5350kg (11,770lb) maximum take-off
POWERPLANT:	Klimov VK-1F afterburning turbojet, 22.5kN (5046lbf)
MAX SPEED:	1145km/h (711mph)
RANGE:	2060km (1280 miles) with drop tanks
CEILING:	16,600m (54,450ft)
CREW:	1
ARMAMENT:	1 × 37mm (1.45in) cannon and 2 × 23mm (0.9in) cannon, plus up to 500kg (1102lb) of external stores on underwing pylons

Below: A Hungarian MiG-17PF. The MiG-17PF was a minimum-change conversion of the basic MiG-17, incorporating a radar and an afterburning engine. Addition of the radar altered the nose profile, with a scanning antenna in the upper part of the engine air intake, and a tracking/ranging antenna in a radome in the middle of the intake.

ENGINE INTAKE
The prominent intake divided the incoming airflow as it entered before it was ducted below the cockpit to feed the engine, roughly in line with the wing trailing edge.

EGYPTIAN MiG-17
Egypt had four operational squadrons and a training unit of MiG-17s in 1967. Syria and Iraq also committed theirs to the fighting and lost a number in air combat. In all, the Arab air forces lost 89 MiG-15s and -17s, 90 per cent of them on the ground.

VIETNAM IMPACT
Thought at first by Western observers to be just an improved MiG-15, the MiG-17 was in fact a new design, incorporating a number of aerodynamic refinements that included a new tail on a longer fuselage and a thinner wing with different section and planform, and with three boundary layer fences to improve handling at high speed. In the hands of the North Vietnamese during the Vietnam War, the MiG-17 made full use of its excellent agility, diminutive size and hard-hitting cannon.

UNDERWING TANKS
Endurance was extended through the use of 400-litre (105-Imp gal) drop tanks that could be carried underwing. In definitive form, four pylons were provided underwing, although drop tanks could only be carried inboard.

MiG-17F 'FRESCO'
The first major improvement to the MiG-17 line was manifest in the MiG-17F that added an afterburning engine, the VK-1F. First flown in September 1951, the MiG-17F began to replace the basic version in November 1952 and subsequently became the most important production version.

Sikorsky H-19 Chickasaw/S-55/HRS-1 (1950)

TYPE • Helicopter   COUNTRY • United States

SPECIFICATIONS (HRS-1)

DIMENSIONS:	Length: 62ft 7in (19.1m); Rotor diameter: 16.16m (53ft); Height 4.07m (13ft 4in)
WEIGHT:	3266kg (7200lb) loaded
POWERPLANT:	1 x 450kW (600hp) Pratt & Whitney R-1340-57 radial engine
MAX SPEED:	163km/h (101mph)
RANGE:	652km (405 miles)
CEILING:	3200m (10,500ft)
CREW:	2
ARMAMENT:	None

The Sikorsky HRS-1 was a US Marine Corps designation for the H-19 Chickasaw (company number S-55), a successful utility helicopter that entered service in April 1950. It did not retire from US service until the late 1960s.

Below: This US Marine Corps HRS-1 served in the Korean War.

HSS-1 SEABAT
The HSS-1 Seabat was an anti-submarine model of the H-34 'Choctaw'/S-58, which was in turn a lengthened, more powerful version of the S-55.

WEIGHT ADVANTAGE
A benefit of the forward-positioned engine was that it was easier to maintain aircraft balance under different loading conditions.

S-55 VERSATILITY
The S-55 was the first in a long line of successful large Sikorsky helicopters. Starting life as a piston-powered machine with a limited load-carrying capacity, the S-55 later received a turbine engine and considerably more capability. It could perform various roles, including airlifting troops, air-sea rescue work, air taxiing, cargo-hauling and anti-submarine patrol. The S-55 took part in the Korean War, but was still providing useful service to military and civil users in the 1980s.

ENGINE
The Pratt & Whitney engine was located forward and beneath the crew compartment. The large, bulky radial was faired under twin clamshell doors. Engineers loved this as it meant easy access to the engine.

HRS-1 IN KOREA
An assault transport with capacity for up to eight troops, 60 examples of the HRS-1 served with nine US Marine Corps transport helicopter squadrons before the end of the war in Korea. The HRS-1 was also used for rescue missions, and operations were based out of K18 from September 1951.

Douglas C-124C Globemaster II (1950)

TYPE • Transport   COUNTRY • United States

In times of tension, the sea lines of communication between the USA and Europe were reinforced by airlift. The C-124 was one of the aircraft involved in the deployment of troops, equipment and fighters to Europe during the Berlin Crisis of 1961.

SPECIFICATIONS (C-124C)

DIMENSIONS:	Length: 40m (130ft); Wingspan: 53.06m (174ft 1in); Height: 14.7m (48ft 4in)
WEIGHT:	98,000kg (216,000lb) loaded
POWERPLANT:	4 x 1834kW (3800hp) Pratt & Whitney R-4360-63A Wasp Major radial piston engines
MAX SPEED:	520km/h (320mph)
RANGE:	3500km (2175 miles)
CEILING:	10,000m (34,000ft)
CREW:	6 or 7
ARMAMENT:	None

Right: A C-124A Globemaster, powered by four R-4360-20WA engines. Many of these models were retrofitted with the APS-42 radar.

WEATHER RADAR
The bulbous projection from the nose of the C-124C housed the addition of the APS-42 weather radar, which provided radar mapping, responder beacon operations, terrain detection and weather mapping.

CREW
The C-124C Globemaster was manned by six or seven personnel: aircraft commander, pilot, navigator, flight engineer, radio operator, 1/2 loadmasters.

CARGO CAPACITY
The C-124C had the capacity to transport 200 fully equipped troops or 168 casualties (123 on litters) with 15 medical staff or 34,000kg (74,000lb) of cargo.

NUCLEAR TRANSPORT
The Douglas C-124 Globemasters provided US forces with a global heavy-lift capability for nearly 25 years. They served in the Korean War (1950–53) and Vietnam War (1963–73), but supplied deployed US troops wherever they found themselves, including on missions to Antarctica. C-124s also delivered Thor nuclear ballistic missiles to the United Kingdom between 1959 and 1961.

US AIR FORCE C-124A
The US Air Force was the last operator of the C-124 Globemaster, although the last unit to retire the aircraft was the Georgia Air National Guard in 1974.

English Electric Canberra (1951)

TYPE • Bomber **COUNTRY** • United Kingdom

First flown in 1949, the Canberra became one of the great post-war successes of the British aircraft industry, with 782 built for the RAF and RN, and another 120 for export customers. Canberras saw action in numerous wars in Africa, Southeast Asia and South America.

SPECIFICATIONS (B. Mk 2)

- **DIMENSIONS:** Length: 19.96m (65ft 6in); Wingspan: 29.49m (63ft 11in); Height: 4.78m (15ft 8in)
- **WEIGHT:** 24,925kg (54,950lb) loaded
- **POWERPLANT:** 2 x 28.9kN (6500lbf) Rolls-Royce Avon Mk 101 turbojets
- **MAX SPEED:** 917km/h (570mph)
- **RANGE:** 4274km (2656 miles)
- **CEILING:** 14,630m (48,000ft)
- **CREW:** 2
- **ARMAMENT:** Bomb bay with provision for up to 2727kg (6000lb) of bombs, plus 909kg (2000lb) of underwing pylons

Below: English Electric Canberra B(I)Mk 58 was a tropicalized Canberra supplied to the Indian Air Force.

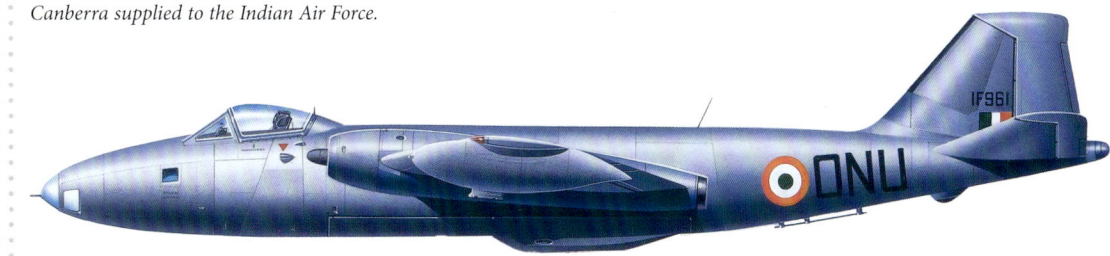

CANBERRA B.MK 2
The Canberra bombers of RAF 10 Sqn were normally based at Honington, as evidenced by the Honington Wing emblem on the fin. For the Suez operation, RAF Canberra B.Mk 2s and B.Mk 6s were based at Hal Far, Luqa and Nicosia and took part in the initial raids on Egyptian airfields, flown at high altitude.

WOODEN TAIL
The wooden tail of the Canberra belonged to an earlier era, but the aircraft was still flying as late as Operation Iraqi Freedom in 2003.

IN SERVICE
The first Canberras went into British service in 1951, with an improved version appearing in 1954. The first operational use was during the Malayan Emergency and the Suez Crisis, where the Canberra was sufficiently impressive that the US Air Force adopted it. The arrival of the Valiant, which was the first of the 'V-Bombers' – capable of delivering a larger payload over a longer range – allowed the Canberra to move from a high-altitude bombing role to low-level strike operations.

B-57 CANBERRA
Convinced of its merits, the US Air Force ordered more than 400 Canberras, which were licence-built by Glenn L. Martin Company as the B-57 Canberra. Here a B-57 shows its 2727kg (6000lb) internal bombload.

Boeing B-52 Stratofortress (1952)

TYPE • Bomber COUNTRY • United States

Still in service after 60 years, the mighty B-52 remains a defiant symbol of American military power that has operated as both a nuclear deterrent and as a frontline type in conflicts extending from Vietnam to the Global War on Terror. B-52s are set to serve at least until 2040.

SPECIFICATIONS (B-52G)

DIMENSIONS:	Length: 48m (157ft 7in); Wingspan: 56.4m (185ft); Height: 12.4m (40ft 8in)
WEIGHT:	221,500kg (448,000lb) loaded
POWERPLANT:	8 x 61.1kN (13,750lbf) Pratt & Whitney J57-P-43W turbojets
MAX SPEED:	1014km/h (630mph)
RANGE:	13,680km (8500 miles)
CEILING:	16,765m (55,000ft)
CREW:	5
ARMAMENT:	4 x 12.7mm (0.5in) MGs; normal internal bomb capacity 12,247kg (27,000lb); external pylons for 2 x Hound Dog missiles

Below: Operating out of Guam during the Linebacker offensive, this B-52G lacks the 'Big Belly' modification that was found on converted B-52Ds. As a result, its warload was decreased. The B-52G also lacked certain ECM equipment, so was assigned less heavily defended targets. Nonetheless, the variant offered increased range.

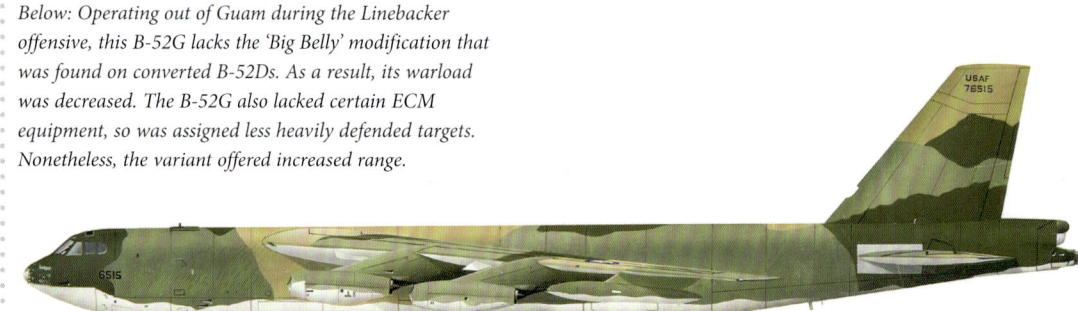

SENSOR TECHNOLOGY
For low-level penetration missions, the later B-52G was retrofitted with undernose blisters containing low light level TV and forward-looking infrared sensors.

B-52D
B-52D 55-0069 is illustrated here as it appeared at the height of the war in Vietnam. Operating out of Guam, the bomber received a coat of black paint to reduce its conspicuity for nocturnal missions over Southeast Asia.

FUEL
B-52s carry a huge amount of fuel – around 147,112 litres (38,863 Imp gal) – in large internal tanks in the fuselage and wings. The J57-powered models also carry tanks for the water injection system used by the engines during take-off to increase thrust.

VETERAN AIRCRAFT
A total of 744 B-52s were built with the last, a B-52H, dispatched in October 1962. The first of 102 B-52Hs was delivered to Strategic Air Command in May 1961. The H-model was developed after the cancellation of a planned successor, the North American B-70 Valkyrie. Today, survivors can carry up to 20 AGM-86 air-launched cruise missiles (ALCMs), and the original tail gun fitting has been deleted altogether.

CONVENTIONAL BOMBER
A solo USAF B-52 Stratofortress from the 40th Expeditionary Bomb Squadron (EBS) flies back to its home station after striking multiple targets deep in Iraqi territory, in support of Operation Iraqi Freedom.

Mil Mi-4 (1952)

TYPE • Helicopter **COUNTRY** • Soviet Union

SPECIFICATIONS (Mi-4A)

DIMENSIONS:	Length: 25m (82ft 1in); Rotor diameter: 21m (68ft 10in)
WEIGHT:	Empty 6626kg (14,608lb) maximum take-off 7534kg (16,610 lb)
POWERPLANT:	One 1268kW (1700hp) Shvetsov ASh-82V radial piston engine
MAX SPEED:	200km/h (124mph) at 1000m (3300ft)
RANGE:	250km (155 miles)
CEILING:	5486m (18,000ft)
CREW:	1–2
ARMAMENT:	One 12.7mm (0.50in) machine gun in ventral gondola

With its uncanny and possibly less than coincidental resemblance to the Sikorsky S-55, the Mi-4 appeared in 1952. Taking barely a year to reach flying status from initial design, the 'Hound', as NATO codenamed it, became widely used at home and overseas.

Left and below: An Mi-4A. The Mi-4T was the initial military variant, with Mi-4As following on as an armed assault variant. Yellow 36 is an early example of later Soviet markings.

ROTORS
It was not until 1954 that Mil was able to increase main rotor blade life to above 300 hours. From 1960, metal blades were fitted to production aircraft, replacing the early Bakelite and wood examples.

POWERPLANT
Shvetsov's reliable ASh-82V 14-cylinder radial engine powered the Mi-4, in a similar layout to the Sikorsky S-55. Developed from the ASh-62, the ASh-82V itself derived from the American Wright Cyclone.

MI-4A 'HOUND'
Depending on configuration, the Mi-4 was capable of carrying between 10 and 14 people, or eight stretcher cases (plus attendant) in the air ambulance role. The rear clamshell doors allowed easy loading of the latter and other cargo.

CONFIGURATION
The Mi-4's basic configuration mirrored that of the Sikorsky S-55 (although the Mi-4 was larger) using a scaled-up Mi-1 rotor with four Bakelite-and-plywood blades. The tried-and-tested Shvetsov ASh-82V 14-cylinder radial engine provided power. Improved versions followed, including anti-submarine warfare and assault variants (with appropriate armament), a supercharged high altitude model in 1965 and an electronic warfare platform.

MI-4 LANDING
First and foremost a military helicopter, the Mi-4 also had a civil role and was adapted for agricultural use and as a firefighter. Soviet production ended in 1964 by which time more than 3200 had been produced.

Avro Vulcan (1952)

TYPE • Bomber COUNTRY • United Kingdom

The most successful of Britain's three V-bombers, the Vulcan enjoyed an impressively long service career, starting out as a high-level strategic nuclear bomber and later going into combat as a conventional bomber during the Falklands campaign.

SPECIFICATIONS (B.Mk 2)

DIMENSIONS:	Length: 30.45m (99ft 11in); Wingspan: 33.83m (111ft); Height: 8.28m (27ft 2in)
WEIGHT:	113,398kg (250,000lb) loaded
POWERPLANT:	4 x 88.9kN (20,000lbf) Olympus Mk.301 turbojets
MAX SPEED:	1038km/h (645mph)
RANGE:	7403km/h (4600 miles)
CEILING:	19,810m (65,000ft)
CREW:	5
ARMAMENT:	Internal weapon bay for up to 21,454kg (47,198lb) bombs, or a Blue Steel nuclear missile

Below: The Vulcan B.Mk 2 XM605 entered service with the RAF in 1969. It is currently a static display at the Castle Air Museum, California.

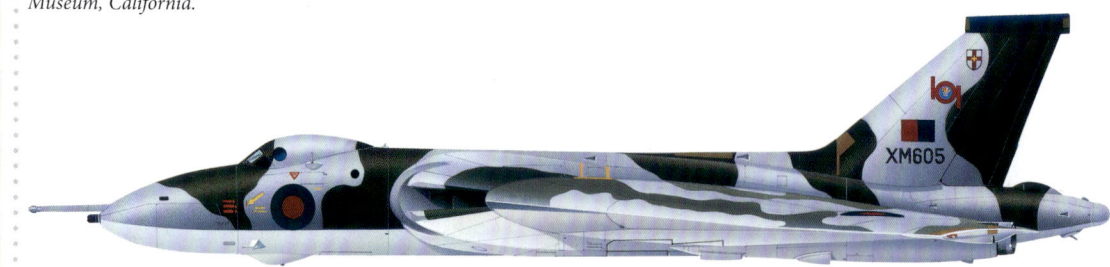

BOMBING RADAR
The Vulcan's bombing radar descended from the World War II H2S set. The 2m (61ft 6in) rotating antenna was housed in the underside of the nose.

DELTA WING
The delta wing gave a smooth ride at low level and was less prone to stress than the more conventional Vickers Valiant, which had to be withdrawn due to fatigue.

VULCAN B.MK 2
The definitive Vulcan B.Mk 2 arrived in 1960, with more powerful engines, in-flight refuelling capability and a modified wing of increased area. It could be armed with a Blue Steel missile, as seen on this example from 617 Sqn, which is painted in anti-flash white.

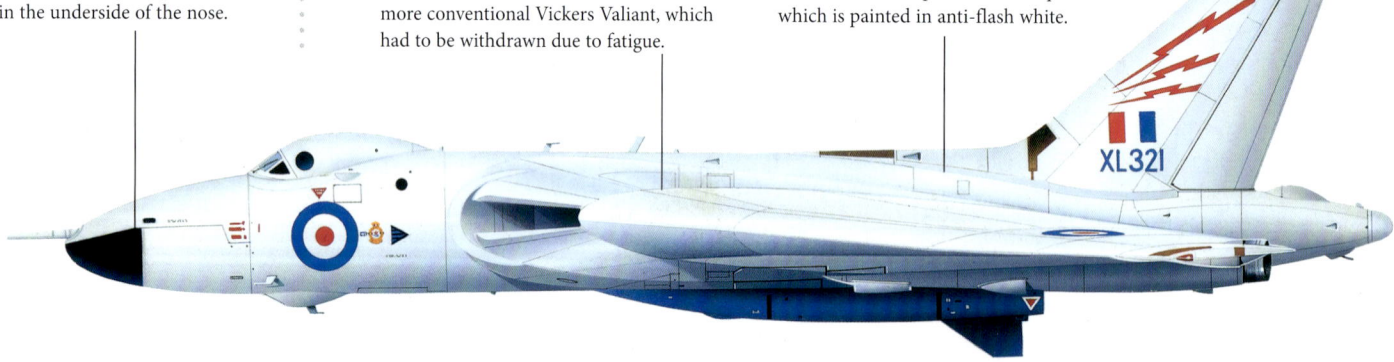

FALKLANDS RAID

The Vulcan force was midway through being wound down when, in April 1982, Argentine forces invaded the Falklands in the South Atlantic. In the course of round trips exceeding 12,870km (8000 miles), and under the codename 'Black Buck', the Vulcans bombed the occupied Port Stanley airfield and related radar installations in the Falklands, putting the runway out of action, albeit only temporarily. The last Vulcans to bow out of service were the K.Mk 2 tankers, finally retired in March 1984.

NUCLEAR BOMBER
A left underside view of a Vulcan B.Mk 2 jet bomber aircraft, donated by the British government for display at the Castle Air Museum, California. The Vulcan's original purpose was as a nuclear bomber, but it will be best remembered for its conventional bombing raids.

Grumman F9F Panther (1952)

TYPE • *Naval fighter* **COUNTRY** • *United States*

SPECIFICATIONS (F9F-2)	
Dimensions:	Length: 11.3m (37ft 5in); Wingspan: 11.6m (38ft); Height: 3.8m (11ft 4in)
Weight:	Empty 4220kg (9303lb); maximum 7462kg (16,450lb)
Powerplant:	One 26.5kN (5950lbf) thrust Pratt & Whitney J42-P-6/P-8 turbojet engine
Max Speed:	925km/h (575mph)
Range:	2100km (1300 miles)
Ceiling:	13,600m (44,600ft)
Crew:	1
Armament:	Four 20mm (0.79in) M2 cannon; up to 910kg (2000lb) bombs, six 127mm (5in) rockets

The F9F Panther brought the jet age to the US Navy. Although not the USN's first carrier jet, it was the first to reach widespread service and to win real popularity among the Navy and Marine pilots who flew in it.

Below: The Grumman Panther was the first truly successful American carrier-based jet. Nearly 1400 were built for the US Navy, Marines and Argentina. Panthers were the main naval strike aircraft in Korea.

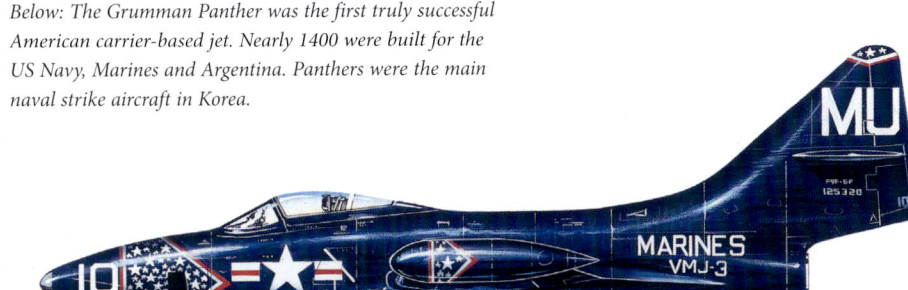

Nose cannon
Four 20mm (0.79in) cannon were mounted in the nose, which could be slid forward to gain access for reloading.

Control
The Panther suffered instability problems that were never entirely cured. With the hydraulic control boost inoperative, aileron stick forces were very high.

F9F-2 fuselage
The deep, sturdy fuselage was so shaped because the J42 engine was of centrifugal design. The shape was useful, however, as it gave a very large volume for internal fuel – twice as much as the British Hawker Sea Hawk.

GROUND-ATTACK
The Panther flew for the first time in 1947, and was the first carrier-based jet fighter to see combat. Extensively used on ground-attack duties in Korea, the Panther was a fine warplane in the hands of a trained pilot. Its structural strength, a trademark of the 'Grumman Iron Works', helped Marines enormously when they flew Panthers through gunfire to attack ground troops in the Korean War.

F9F CARRIER OVERFLIGHT
An F9F attached to the USS *Bon Homme Richard* (CVA31) flies over Task Force 77 engaged in carrier operations against North Korean targets. The last operational Panther was retired in October 1958.

Westland Whirlwind (1952)

TYPE • *Fighter* **COUNTRY** • *United Kingdom*

One of the world's most widely used helicopters, the Westland Whirlwind was a licence-built version of the Sikorsky S-55 that served throughout the Cold War. It first proved itself in the arduous conditions of the Korean War.

SPECIFICATIONS (HCC.Mk 12)

DIMENSIONS:	Length fuselage: 13.46m (44ft 2in); Rotor diameter: 16.2m (53ft); Height: 4.03m (13ft 2.5in)
WEIGHT:	Empty 2246kg (4952lb); maximum take-off 3629kg (8000lb)
POWERPLANT:	One 783kW (1050hp) Bristol Siddeley Gnome H.1000 shaft turbine engine mounted horizontally in nose compartment
MAX SPEED:	170km/h (106mph)
RANGE:	480km (300 miles)
CEILING:	5060m (16,600ft)
CREW:	2
ARMAMENT:	None

This page: A rectangular, light alloy semi-monocoque structure, the Whirlwind's fuselage had a rear cone-shaped extension carrying the tail rotor pylon.

INTERIOR
Room for eight passengers, 10 armed troops or six stretchers was available in the 9.63m² (340 cubic-foot) cabin interior.

QUEEN'S FLIGHT
In service with the Queen's Flight the two HCC.12s bore the distinctive Royal blue and scarlet livery.

POWERPLANT
A turboshaft engine replaced the radial piston engine in the Whirlwind from the Mk 9 onwards. Bristol Siddeley's 783-kW (1050-hp) Gnome H.1000 was installed in the Mks 9, 10 and 12, offering increased power but only one-third of the radial's total weight.

LICENCE BUILT
Early Whirlwinds were licence-built Sikorsky S-55s, with Pratt & Whitney Wasp or Wright Cyclone piston engines. Westland soon replaced the US engines with the more powerful Alvis Leonides Major, a two-row radial that gave improved performance. Then, in 1958, Westland decided to build a version using the Gnome, a licence-built version of the General Electric T58 turboshaft. The result was a much more capable machine.

BRITISH SERVICE
A licence-built Sikorsky design, the Whirlwind served the British armed forces for many years, mainly in transport, anti-submarine warfare and rescue roles. In RAF and Royal Navy service, search and rescue was a major Whirlwind role. The type often co-operated with RAF rescue motor launches.

Avro Canada CF-100 Canuck (1952)

TYPE • Fighter COUNTRY • Canada

The CF-100 Canuck was an excellent, all-weather warplane that was also Canada's first jet fighter. Avro Canada manufactured this two-man, twin-engined machine for the express purpose of defending North America from bomber attack.

SPECIFICATIONS (Mk 5)

DIMENSIONS: Length: 16.48m (54ft 1in); Wingspan: 17.68m (58ft); Height: 4.74m (15ft 7in)

WEIGHT: Empty 10,478kg (23,100lb); maximum take-off 16,783kg (37,000lb)

POWERPLANT: Two 32.36kN (7278lbf) Orenda 11 or 14 turbojet engines

MAX SPEED: 1046km/h (650mph) at 3050m (10,000ft)

RANGE: 3220km (2,000 miles)

CEILING: 16,460m (54,000ft)

CREW: 2

ARMAMENT: 29 x 70mm (2.75in) 'Mighty Mouse' folding-fin aircraft rockets (FFAR) in each wingtip pod (48 additional FFAR or eight machine guns on some versions)

Left: The definitive model of the CF-100 was the Mk 5, which introduced uprated powerplants and enhanced aerodynamics.

MK 5 IMPROVEMENTS
The definitive CF-100 Mk 5 model introduced a larger tailplane and increased wingspan, coupled with more powerful Orenda 14 turbojets, to improve high-altitude performance. Each non-afterburning engine could provide 32.36kN (7278lbf) thrust.

ALL-WEATHER RADAR
The Canuck Mk 5 shown here carried an all-weather APG-40 radar for the tracking of enemy bomber formations.

VENTRAL CAVITY
In the capacious ventral cavity under the forward fuselage, the Mk 5 could carry eight 12.7-mm (0.5-in) Colt-Browning machine guns or, alternatively, a tray containing 48 additional 70mm (2.75in) rockets.

EASY FLYING
Although it lacked the swept wings that became standard on most combat aircraft, the CF-100 was a fine all-weather fighter, its stability making it an excellent weapon platform and easy to fly on instruments. Pilots also found the Canuck (a slang term for a Canadian) an easy aircraft to handle at lower speeds when landing or manoeuvring in the airfield pattern. The Canuck flew alongside American interceptors such as the Northrop F-89 Scorpion and Lockheed F-94 Starfire.

CF-100 IN FLIGHT
Avro Canada began to design its first jet in 1946 and flew the CF-100 in 1950. This long-range interceptor became the backbone of Canada's contribution to the defence of North America and Western Europe.

Lockheed C-130 Hercules (1954)

TYPE • Transport **COUNTRY** • United States

SPECIFICATIONS (C-130E)

DIMENSIONS:	Length: 29.79m (97ft 9in); Wingspan: 40.41m (132ft 7in); Height: 11.66m (38ft 3in)
WEIGHT:	79,380kg (175,000lb) maximum take-off
POWERPLANT:	4 × Allison T56A-15 turboprops, 3362kW (4508hp)
MAX SPEED:	556km/h (345mph)
RANGE:	4002km (2487 miles) with maximum payload of 19,686kg (43,400lb)
CEILING:	10,060m (33,000ft)
CREW:	5
ARMAMENT:	None

Lockheed's C-130 Hercules has been in production for 70 years, with around 3000 aircraft built. It has become the standard military transport of the West, with few major nations not in possession of any 'Herks'.

Below: The third major production variant was the C-130E tailored towards Military Airlift Command operations, as opposed to the previous versions that had been tactical.

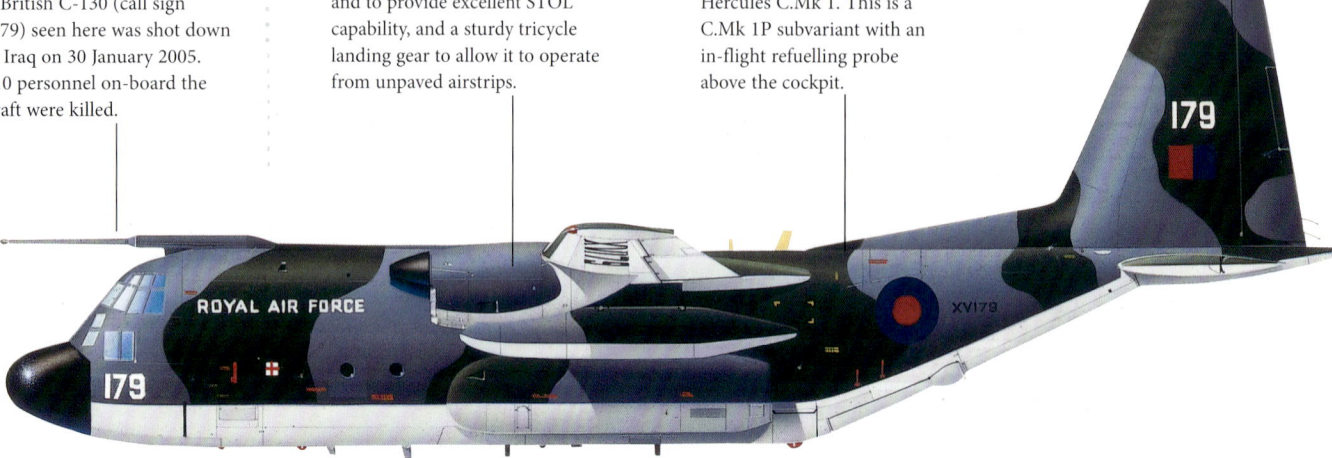

XV179
The British C-130 (call sign XV179) seen here was shot down over Iraq on 30 January 2005. All 10 personnel on-board the aircraft were killed.

LAYOUT
The Hercules uses turboprops for improved performance, a high-set wing to avoid encroaching on the cargo space and to provide excellent STOL capability, and a sturdy tricycle landing gear to allow it to operate from unpaved airstrips.

C-130K HERCULES
The C-130K version was built specifically for British use on the basis of the C-130H, and entered RAF service as the Hercules C.Mk 1. This is a C.Mk 1P subvariant with an in-flight refuelling probe above the cockpit.

VIETNAM OPS
As an airlifter, the C-130 proved its worth in Vietnam. In its basic form, the C-130 can carry 78 troops (or 92 if a high-density configuration is employed), 64 paratroopers or 74 litter patients. Later in the conflict in South East Asia, MC-130 Combat Talons were used to deliver special operations forces and for aerial refuelling. Once fitted with ground-target radar, 20mm (0.79in) Gatling guns, 40mm (1.57in) Bofors cannon and, later, a side-firing 105mm (4.1in) howitzer, the Hercules became the AC-130 gunship.

USAF C-130
Designed from 1951 for the US Air Force's Tactical Air Command, the C-130 set a new pattern for military transport aircraft. Its landing gear allowed it to operate from unpaved dirt surfaces.

Republic F-84F Thunderstreak (1954)

TYPE • *Fighter* COUNTRY • *United States*

Republic's F-84F Thunderstreak arrived when the straight-wing F-84 was rebuilt with swept wings, reconfigured air intake and canopy, and an improved engine. The resulting fighter-bomber served the United States and 12 other nations, mostly in NATO, for over a decade.

SPECIFICATIONS (F-84F)

DIMENSIONS:	Length: 13.23m (43ft 4.75in); Wingspan: 10.24m (33ft 7.25in); Height: 4.39m (14ft 4.75in)
WEIGHT:	12,701kg (28,000lb) loaded
POWERPLANT:	1 x 32kN (7220lbf) Wright J65-W-3 turbojet
MAX SPEED:	1118km/h (695mph)
RANGE:	2608km (1620 miles) with drop tanks
CEILING:	14,020kg (46,000ft)
CREW:	1
ARMAMENT:	6 x 12.7mm (0.5in) Browning M3 machine guns, external hardpoints with provision for up to 2722kg (6000lb) of stores

Below: Normally based at Rheims, this EC 3/3 F-84F fighter-bomber was one of those deployed to Akrotiri, Cyprus, for the Suez action. Also flying from Akrotiri were the French AF's EC 1/3 and EC 4/33, with the F-84F and the reconnaissance-configured RF-84F respectively. Further F-84Fs were based in Lydda (Lod), Israel.

CAMOUFLAGE
USAF F-84Fs were left in a natural polished metal finish, but most NATO air forces camouflaged their aircraft, especially when they 'went nuclear'.

WEST GERMAN F-84F
In total West Germany's post-war Luftwaffe adopted 480 F-84F Thunderstreaks. These were operational between 1957 and 1966.

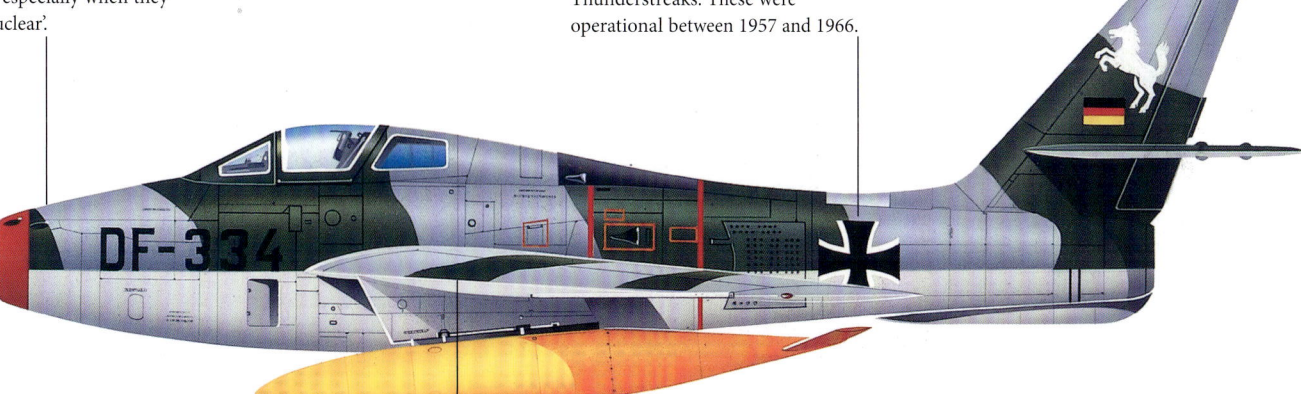

COMBAT ROLE
Early hopes that the Thunderstreak might prove to be a MiG-killer were forlorn: the F-84F introduced a new standard of precision as a fighter-bomber, but other performance shortfalls made it an adequate dogfighter at best. Although it was a little heavy on the controls, the F-84F gave its pilot a roomy cockpit with fair visibility. In some units pilots relentlessly practised a 'lob' technique to deliver tactical atomic bombs.

WING
The wing of the Thunderstreak was swept sharply back at 38.5 degrees. It was very broad in chord, and incorporated large leading-edge slats and large trailing-edge flaps.

F-84G THUNDERJET
The predecessor to the F-84F Thunderstreak was the F-84 Thunderjet, which was a noticeable straight-wing aircraft in comparison with the swept-wing Thunderstreak.

Gloster Meteor NF.14 (1954)

TYPE • *Fighter* **COUNTRY** • *United Kingdom*

SPECIFICATIONS

DIMENSIONS:	Length: 15.5m (51ft 4in); Wingspan: 13.10m (43ft); Height: 4.2m (13ft 9.5in)
WEIGHT:	Empty 5724kg (12,620lb) maximum take-off 9300kg (20,500lb)
POWERPLANT:	Two 16kN (3600lbf) thrust Rolls-Royce Derwent 8 turbojets
MAX SPEED:	930km/h (578mph) at 10,000m (33,000ft)
RANGE:	933km (580 miles)
CEILING:	12,192m (40,000ft)
CREW:	2
ARMAMENT:	Four 20mm (0.79in) cannon

The Gloster Meteor NF.14 was the last of the piloted Meteors (aside from some subsequent target drone variants). Aesthetically it was the most pleasing and it filled the gap until the deployment of the Gloster Javelin.

Below: Meteor NF.14s routinely carried three external fuel tanks, two under the wings and one belly tank, with a total capacity of 1705 litres (375 Imp gal).

VARIANTS
The NF.11, NF.12, NF.13 and NF.14 radar-equipped night fighters differed visually, mainly in fuselage length and canopy detail. Different models of engine and radar were the main internal differences.

GLOSTER METEOR NF.14
The NF.14 seen here, WS800, like all other Meteor night fighters, was built by W.G. Armstrong Whitworth. It is shown in the markings of the commanding officer of No. 60 Squadron, the last RAF Meteor squadron to form, in 1959.

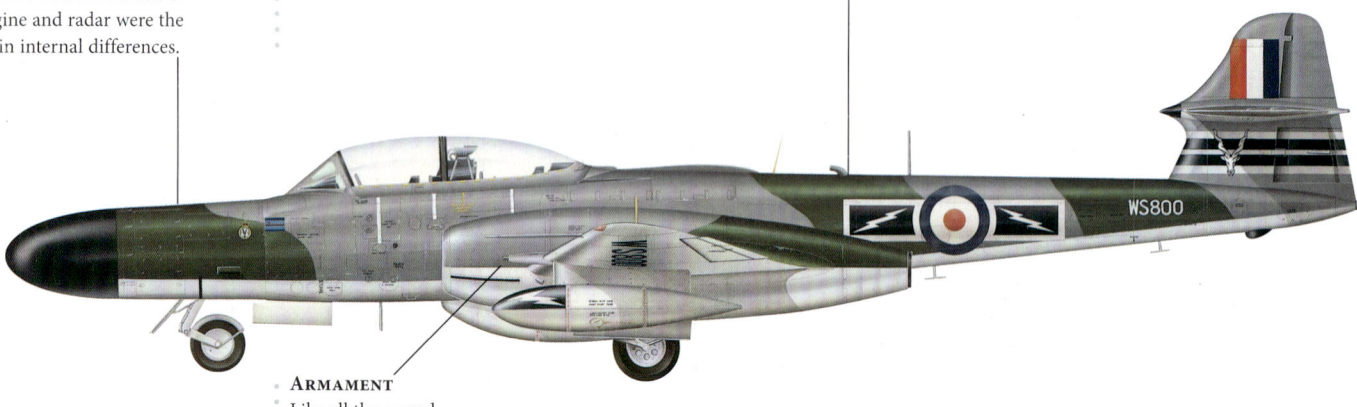

ARMAMENT
Like all the armed Meteor variants, the NF.14 was equipped with four Hispano 20mm (0.79in) cannon, which were wing-mounted in the

NIGHT FIGHTERS
Based on the Meteor T.7 airframe, the Meteor night fighter family, beginning with the NF.11, shared the trainer's wing, but featured a new tail unit derived from the F.Mk.8 and a lengthened nose to accommodate the AI radar equipment. Cabin pressurization was a new feature, as was the clear-view bubble canopy. The NF.14 was not fitted with ejection seats.

DANISH METEOR T.7
Denmark flew limited numbers of Meteors between 1949 and 1962. The aircraft seen here are the Meteor T.7, which was a two-seat training variant.

Hawker Hunter (1954)

TYPE • Fighter **COUNTRY** • United Kingdom

Hawker's classic Hunter first flew in July 1951, sharing a common ancestor with the Sea Hawk naval fighter in the P.1040 prototype of 1947. The Hunter became the first British-built swept-wing fighter and was the backbone of Royal Air Force (RAF) day-fighter squadrons from 1954.

SPECIFICATIONS (F.Mk 56)

DIMENSIONS:	Length: 14m (45ft 11in); Wingspan: 10.2m (33ft 8in); Height: 4.01m (13ft 2in)
WEIGHT:	5795kg (12,760lb) loaded
POWERPLANT:	One 44.48kN (10,000lbf) Rolls-Royce Avon 203 turbojet engine
MAX SPEED:	1150km/h (715mph)
RANGE:	689km (429 miles)
CEILING:	15,707m (51,000ft)
CREW:	1
ARMAMENT:	Four 30mm (1.18in) Aden cannon

Below: The Hunter T.Mk.8 seen here, XL598 of No. 738 Naval Air Squadron, was based at RNAS Brawdy, Wales, in 1965.

POWERPLANT
Thrust was delivered by a 44.48kN (10,000lbf) Rolls-Royce Avon 203 turbojet engine.

INDIAN AIR FORCE F.MK 6
India took more than 200 Hunters, which equipped seven squadrons and saw considerable action in the 1965 and 1971 wars with Pakistan. The bulk of India's Hunters were F.6s delivered as the F.Mk 56. Deliveries after the 1965 war were mostly the F.Mk 56A, which was optimized for ground attack.

30MM CANNON
Empty shell casings are ejected from beneath the fuselage (away from the jet intakes) as the pilot fires his four 30mm (1.18in) Aden cannon, the Hunter's principal armament.

HUNTER F.MK 6
The definitive RAF Hunter was the F.Mk 6, powered by the Avon 203 engine rated at 44.48kN (10,000lbf). The so-called 'bigbore' Hunter equipped 18 RAF squadrons and was the only pure fighter version in British use after 1963. Derivatives of the F.6 were exported to a number of countries, including Belgium, Denmark, India, Oman, Peru, Saudi Arabia, Sweden and Switzerland.

HUNTER PROTOTYPE
Designed to replace the Meteor, the prototype Hunter (P.1067) flew on 20 July 1951. The F.Mk.1, which entered service in 1954, suffered from engine problems in high-altitude gun-firing trials, resulting in some modifications to its Rolls-Royce Avon turbojet.

Tupolev Tu-16 Badger (1954)

TYPE • Bomber **COUNTRY** • Soviet Union

SPECIFICATIONS (Tu 16PM Badger-L)

DIMENSIONS: Length: 34.80m (114ft 2in); Wingspan: 32.99m (108ft 3in); Height: 10.36m (34ft)

WEIGHT: Empty 37,200kg (82,026lb); maximum take-off 75,800kg (167,139lb)

POWERPLANT: Two 85.3kN (19,200 lbf) thrust Mikulin RD-3M turbojets

MAX SPEED: 960km/h (597mph) at 6000m (19,685ft)

RANGE: 4800km (2983 miles)

CEILING: 15,000m (49,200ft)

CREW: 7

ARMAMENT: Two 23mm (0.906in) cannon in radar-controlled barbettes

The Tu-16 strategic jet bomber first flew in 1952 and became the most important bomber serving with the Soviet Air Force and Naval Air Arm. Deployed in the mid-1950s, Tupolev's Tu-16 was also the most effective of Russia's trio of new strategic bombers.

Below: Hardpoints for missile pylons were fitted on Tu-16 bombers, and similar hardpoints have been used on other variants for equipment pods. Most Badger-Ls carry such pods under the wing, those on this aircraft having cooling inlets in their noses.

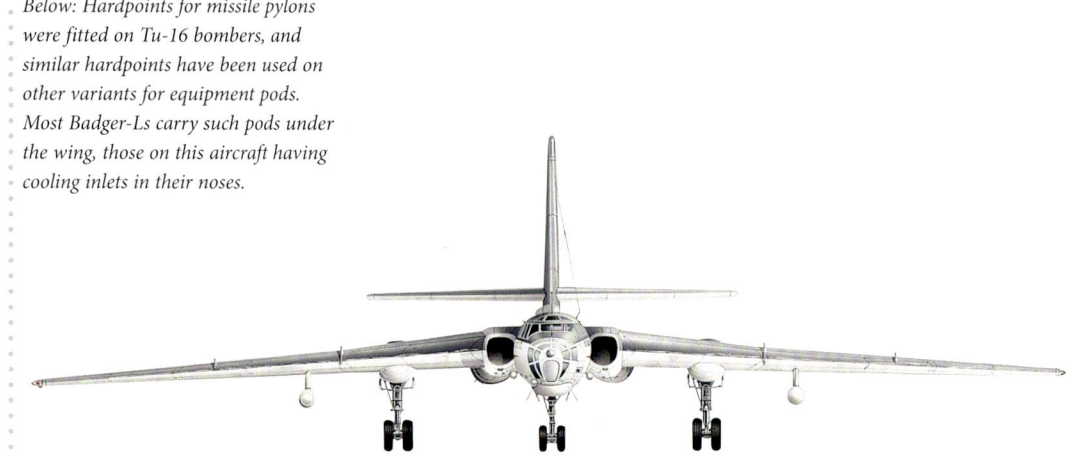

MARKINGS
This Badger-L has white undersides and displays an 'Excellence' award on the forward fuselage. The aircraft retains silver/natural metal upper surfaces. Regimental codes are applied in small numerals on the fin tip and nose wheel doors, and a construction number is painted below the cockpit.

DEFENCES
The Badger-L featured two 23mm (0.906in) cannon in radar controlled barbettes, one in a dorsal turret and one in a ventral turret. Other variants had heavier firepower, including two cannon in the tail and a fixed cannon in the nose.

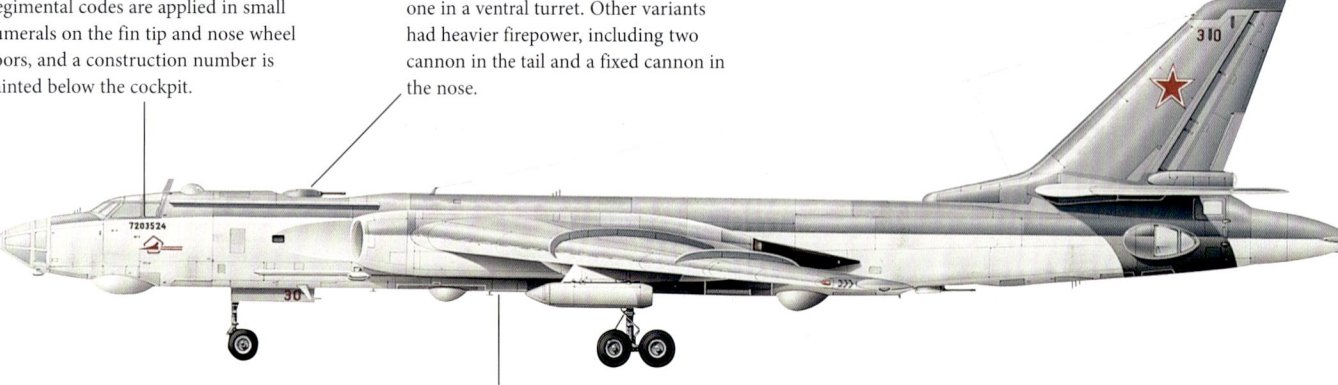

OPERATIONAL STATUS
The Tu-16 'Badger' was a vital part of Russia's Long Range Aviation, serving as a tanker, electronic jammer and reconnaissance aircraft. Despite being obsolete in its original role of nuclear free-fall bombing, the Tu-16 'Badger' continues to do sterling work in the hands of a small number of pilots, most notably in Russia and China. The Tu-16 airframe was highly adaptable, and was soon pressed into service in the first Soviet air-to-air refuelling squadrons.

WEAPONS BAY
The Tu-16's capacious weapons bay is the key to the aircraft's astonishing versatility, enabling bulky payloads to be carried with ease. Originally designed for early atomic bombs, which were bulky, the voluminous bomb bay is today used to accommodate electronic equipment, chaff dispensers and fuel.

TU-16 BADGER
The 'Badger' was also sold extensively overseas. Despite its age, the Tu-16 remains a very useful aircraft that will almost certainly be flying in the world for many years to come.

North American F-100 Super Sabre (1954)

TYPE • Fighter   COUNTRY • United States

Considering it suffered from so many inherent problems – including landing characteristics described by a pilot with 2000 hours on the type as 'a controlled crash' –the F-100 is remembered with respect and some affection by a generation of aviators.

SPECIFICATIONS (F-100D)

DIMENSIONS:	Length: 15.2m (50ft); Wingspan: 11.8m (38ft 9in); Height: 4.95m (16ft 3in)
WEIGHT:	13,085kg (28.847lb) loaded
POWERPLANT:	1 x 45kN (10,200lbf) Pratt & Whitney J57-P-21/21A turbojet
MAX SPEED:	1380km/h (864mph)
RANGE:	3210km (1995 miles)
CEILING:	15,000m (50,000ft)
CREW:	1
ARMAMENT:	4 x 20mm (0.79in) M39 cannon, 4 x AIM-9 Sidewinder missiles, provision for 3190kg (7040lb) nuclear bombs

Below: This F-100A-61-NA, named Jeanne Kay, was typical of the type serving in South East Asia. The 'Hun' was in action over Vietnam almost from the outset and proved a robust ground-attack platform, which bore the brunt of offensive operations in the early years of the conflict.

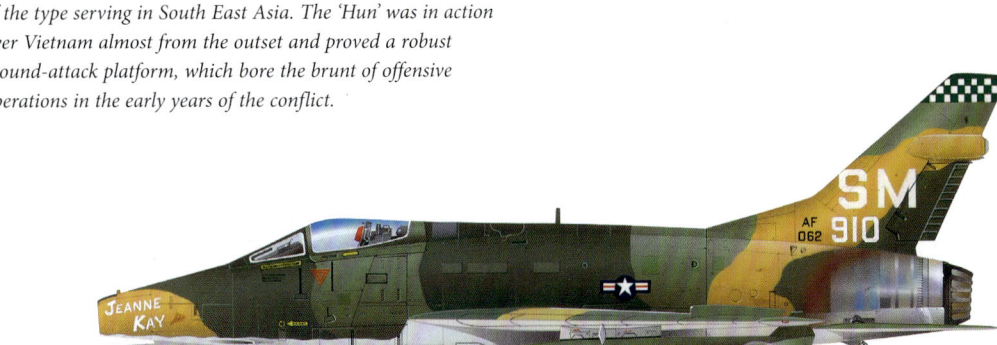

ARMAMENT
For ground-strafing and self-protection, four M39 20mm (0.79in) cannon were fitted in the nose of the F-100D. Two AIM-9 Sidewinder air-to-air missiles were also fitted on occasions.

F-100D SUPER SABRE
This F-100D strike fighter was based at Wethersfield, England, during the Berlin Crisis. At the time, USAFE operated two tactical F-100D wings in the UK, plus a further two wings in West Germany. Additional fighter-bombers based in Europe comprised one wing each of F-105D Thunderchiefs and F-101C Voodoos.

SUPERSONIC
Attempts to build a supersonic Sabre had been killed by the limitations of its wing and engine. The latter problem was solved by Pratt & Whitney's JT3 (J57) turbojet, so, with USAF agreement, a new aircraft was designed. After the first flight on 24 April 1953 the Super Sabre went into production for the USAF as the F-100A day fighter. However, a fatal crash caused by handling problems brought groundings and delays. A taller tail fin and longer wings provided the solution.

WING DESIGN
A key feature of the F-100D was the redesigned wing. This had a kinked trailing edge incorporating broad, slotted landing flaps. These were much needed and appreciated by pilots used to the high-speed landing run of the F-100C. Extra internal tankage was provided. D-models were built at US factories in Inglewood, California, and Columbus, Ohio.

ROCKET LAUNCHING
An F-100 launches a battery of rockets at a ground target during the Vietnam War. From 1954, the F-100C fighter-bomber was built, and it was this and the much-improved D-model that were built in the biggest numbers – 1750 in total.

Lockheed U-2 (1955)

TYPE • Reconnaissance **COUNTRY** • United States

The U-2 initially operated under the cover of a high-altitude research aircraft programme until CIA pilot Gary Powers was shot down on 1 May 1960 over Sverdlovsk. Later versions of the 'Dragon Lady' remain in service today.

SPECIFICATIONS (U-2R)

DIMENSIONS:	Length: 19.13m (62ft 9in); Wingspan: 31.39m (103ft); Height: 4.88m (16ft)
WEIGHT:	Empty 7031kg (15,500lb); maximum take-off weight 18,733kg (41,300lb)
POWERPLANT:	One 75.6kN (17,000lbf) Pratt & Whitney J75-P-13B turbojet
MAX SPEED:	Cruising: Mach 0.715 (412 kn; 470 mph; 760 km/h)
RANGE:	10,050km (6250 miles)
CEILING:	27,430m (90,000ft)
CREW:	1
ARMAMENT:	None

Below: The U-2 played a critical role during the Cuban Missile Crisis, with 102 USAF sorties being flown by the type in the period between 14 October and 6 December 1962. Painted light grey overall, this USAF U-2A was operated by the 4080th Strategic Reconnaissance Wing from Laughlin AFB, Texas, at the time of the crisis.

Q-BAY
The Q-bay behind the pilot housed the principal sensors and/or cameras. A smaller bay was built into the nose of the aircraft.

U-2R
The U-2R, first flown in 1967, was significantly larger and more capable than the original aircraft. A distinguishing feature is the addition of a large instrumentation 'superpod' under each wing. It was designed for stand-off tactical reconnaissance in Europe.

WINGS
The key to the U-2's altitude and range performance was its long-span wings. It was effectively a powered glider, with its high aspect ratio wing and lightweight structure.

DRAGON LADY
The 'Dragon Lady', as it became known, first flew in 1955 and entered service in the late 1950s, initially crewed by CIA personnel (the USAF later took over). The U (for utility) designation was used by the US Department of Defense to hide the true role of the aircraft. Flights over the Soviet Union were made until 1 May 1960, when one was shot down over Sverdlovsk. From that time on flights were restricted to spy sorties over non-Soviet territory.

U-2 IN FLIGHT
The U-2 was designed at the height of the Cold War to penetrate heavily defended airspace and bring back photos and data on the enemy's most secret installations. Although graceful in flight, the U-2 gained a reputation for being difficult to handle on landing.

Mikoyan-Gurevich MiG-21 (1955)

TYPE • *Fighter* **COUNTRY** • *Soviet Union*

Known by the Nato reporting name 'Fishbed,' the MiG-21 was a child of the Korean War, during which Soviet air combat experience had identified a need for a light, single-seat target defence interceptor with high supersonic manoeuvrability.

SPECIFICATIONS (MiG-21bis)

DIMENSIONS:	Length: 15.0m (49ft 2.5in) with pitot; Wingspan: 7.15m (23ft 5.66in); Height: 4.13m (13ft 6.4in)
WEIGHT:	8725kg (19,325lb) loaded
POWERPLANT:	1 x 69.6kN (15,650lbf) Tumansky R25-300 afterburning turbojet
MAX SPEED:	2237km/h (1468mph)
RANGE:	1210km (751 miles) on internal fuel
CEILING:	17,800m (58,400ft)
CREW:	1
ARMAMENT:	1 x 23mm (0.9in) cannon and 4 x air-to-air missiles or 2 x 500kg (1102lb) bombs

Below: India was one of the most important users of the MiG-21, and took the aircraft into combat against Pakistan. The specific variant pictured here is the MiG-21FL.

AIR DATA PROBE
A long boom mounted on the nose carries pitot-static heads for the airspeed system as well as pitch-yaw sensor vanes.

COCKPIT
Due to the poor rear visibility from the cockpit, the pilot is fitted with a rear-view mirror attached to the canopy. The canopy of the MiG-21M is sideways-opening.

MiG-21M
East Germany employed the MiG-21 for air defence, ground-attack and reconnaissance. It retained later versions of the type in frontline service until the fall of the Berlin Wall and German reunification. JG-7 operated this MiG-21M from Drewitz, near Cottbus.

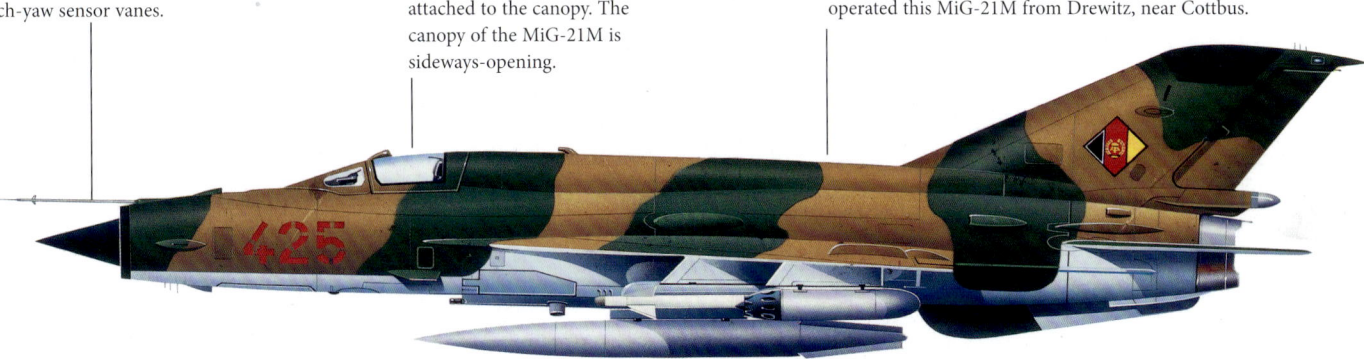

MODIFICATIONS

The MiG-21F was the first major production version; it entered service in 1960 and was progressively modified and updated over the years that followed. In the early 1970s, the MiG-21 was virtually redesigned, re-emerging as the MiG-21B (Fishbed-L) multi-role air superiority fighter and ground-attack version. In its several iterations, the MiG-21 became the most widely used jet fighter in the world, being licence-built in India, Czechoslovakia and China. In Vietnam, the MiG-21 was the Americans' deadliest opponent.

CZECH MiG-21S
Two Czech MiG-21s fly high over mountainous terrain. By 1975, a total of 391 MiG-21s were operational in Czechoslovakia, making it one of the largest Eastern Bloc users of the aircraft.

Vickers Valiant (1955)

TYPE • Bomber COUNTRY • United Kingdom

SPECIFICATIONS (B. Mk 1)

Dimensions:	Length: 32.99m (108ft 3in); Wingspan: 34.85m (114ft 4in); Height: 9.8m (32ft 2in)
Weight:	63,503kg (140,000lb) loaded
Powerplant:	4 x 44.7kN (10,054lbf) Rolls-Royce Avon RA.28 turbojets
Max Speed:	912km/h (576mph)
Range:	7242km (4500 miles)
Ceiling:	16,460m (54,000ft)
Crew:	5
Armament:	1 x 4536kg (10,000lb) nuclear bomb or up to 9525kg (21,000lb) conventional bombs

Three manufacturers responded to a 1946 requirement for a new British heavy jet bomber, with three designs – the Valiant, Vulcan and Victor – entering service in that order from 1954 to 1957. The Valiant was the most conservative but was put into production partly in case the others failed.

Below: This Vickers Valiant B.Mk served with 17 Squadron flying out of RAF Honington in 1957–56. Valiants gave exceptional service in their designed role, and also as strategic reconnaissance platforms and tankers.

Ejector seats
Only the pilot and co-pilot were provided with ejector seats, with the other three crew members escaping by parachute from the door below the cockpit. Two extra personnel were carried during the tanker role.

Anti-flash white
Experts thought that a bright, shiny paint scheme would reflect the radiation of a nuclear blast, protecting the bomber crew from its harmful effects.

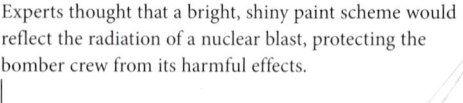

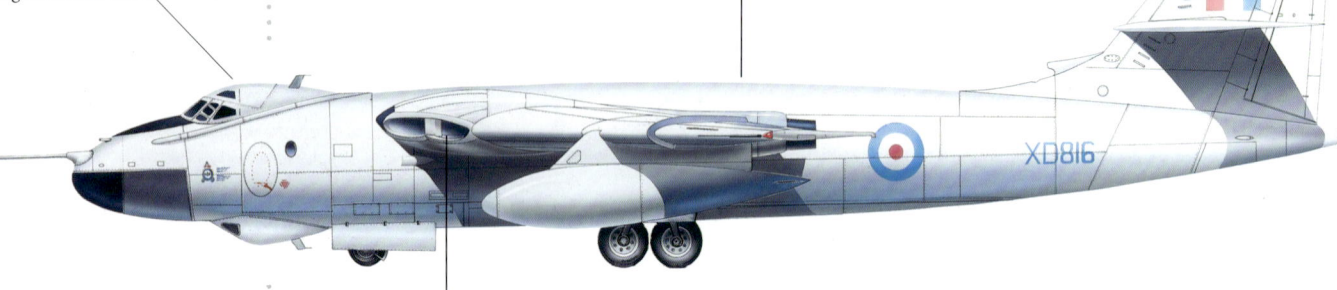

Engine placement
When the Valiant was designed, British thinking was that engines should be mounted in the wingroots. This was true of all the operational V-bombers. American designers preferred to fit podded engines, which were slung on pylons under the wings.

VALIANT FIRSTS
In October 1956 a Valiant dropped Britain's first air-launched nuclear weapon over southern Australia, and this was followed by a first British hydrogen bomb, dropped by another Valiant over Christmas Island in May 1957. The introduction of Soviet surface-to-air missiles (SAMs) saw a switch to low-level operations, which caused excessive fatigue, forcing its complete retirement by 1965.

VICKERS VALIANT B.2
The Vickers Valiant B.2 was a prototype aircraft intended to perform in the pathfinding role. It had a strengthened airframe, a lengthened fuselage and increased speed, but its development was cancelled in 1958.

Tupolev Tu-95 (1956)

TYPE • Bomber COUNTRY • Soviet Union

SPECIFICATIONS

DIMENSIONS: Length: 49.50m (162ft 5in); Wingspan: 51.10m (167ft 8in); Height: 12.12m (39ft 9in)
WEIGHT: 171,000kg (376,200lb) loaded
POWERPLANT: 4 x 11,000kW (14,800shp) Kuznetsov NK-12MV turboprops
MAX SPEED: 920km/h (575mph)
RANGE: 15,000km (9,400miles)
CEILING: 13,716m (45,000ft)
CREW: 7
ARMAMENT: 1 or 2 x 23mm (0.9in) AM-23 cannon in tail turret and up to 15,000kg (33,000lb) of bombs

The Tu-95 'Bear' was unique in being a swept-wing turboprop bomber. Although slower than its counterparts, it had enormous range. Entering service in 1956, it remains in Russian use. Later versions were equipped to carry a variety of cruise missiles.

Below: Named 'Veliky Novgorod', in honour of the city of the same name in the far west of Russia, RF- 94124 '16 Red' is one of the Tu-95MS bombers operated by the 184th Heavy Bomber Aviation Regiment at Engels air base. Many examples of the Tu-95 fleet have been individually named.

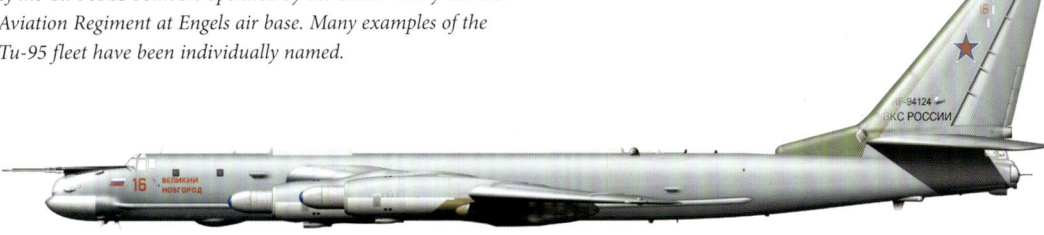

ESCAPE SYSTEM
The fuselage of the Tu-95 contains a unique aircrew escape system, in which a conveyer-belt style floor section carries the pilots to safety through an escape hatch. The tail gunner has a floor hatch.

DEFENSIVE ARMAMENT
The defensive armament of the 'Bear-C' included retractable dorsal and ventral turrets and a manned tail turret, each with two 23mm (0.9in) cannon. The 'Bear-D' retained only the rear gun turret.

ENGINE POWER
Huge engine power and a graceful swept wing push the 'Bear' faster than any other propeller-driven aircraft, achieving jet-like speeds in excess of 900km/h (560mph). Jet power was considered for the 'Bear' but was never adopted.

VARIANTS
The aircraft was first flown in prototype form on 12 November 1952, with series production being launched during 1955 in Samara and continuing until 1969 with a total of 173 built. In addition to the original Tu-95M bomber version, there were also Tu-95K missile carriers carrying the Kh-20 (AS-3) cruise missile that entered service in 1960 and the Tu-95RTs maritime reconnaissance aircraft, which was equipping naval aviation units from the mid-1960s.

COLD WAR INTRUDER
For decades, 'Bears' flew along the fringes of the West, shadowing NATO's fleets, prying out secrets and testing the reaction times of opposing air defence systems. As long as they were in the danger zone they were invariably intercepted and escorted by Allied fighters.

Boeing KC-135 Stratotanker (1956)

TYPE • *Refueller* COUNTRY • *United States*

SAC's fuel-thirsty jet bombers required an equally large fleet of tanker aircraft to support their patrols. The KC-135 was derived from the same prototype that parented the 707 airliner, and over 800 were built, the majority of them as KC-135As.

Below: A KC-135A of the Ohio Air National Guard. Upgraded aircraft, beginning with the KC-135E programme, have increased-span tailplanes that were taken from 707 airliners.

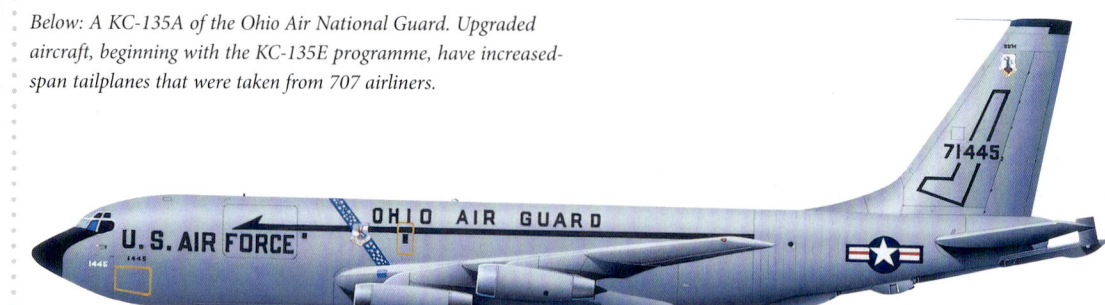

SPECIFICATIONS (KC-135A)

DIMENSIONS:	Length: 41.53m (136ft 3in); Wingspan: 39.88m (130ft 10in); Height: 12.7m (41ft 8in)
WEIGHT:	Empty 44,665kg (98,465lb); maximum 134,720kg (29,000lb)
POWERPLANT:	Four 244.7kN (55,000lbf) Pratt & Whitney J57-59W thrust turbojet engines
MAX SPEED:	982km/h (609mph)
RANGE:	4627km (2875 miles)
CEILING:	10,980m (36,000ft)
CREW:	3
ARMAMENT:	None

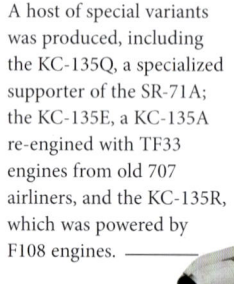

KC-135E STRATOTANKER
A host of special variants was produced, including the KC-135Q, a specialized supporter of the SR-71A; the KC-135E, a KC-135A re-engined with TF33 engines from old 707 airliners, and the KC-135R, which was powered by F108 engines.

CAPACITY
All of the Stratotanker's refuelling systems are mounted below floor level, which gives the aircraft great flexibility by allowing the carriage of cargo, or up to 80 passengers, or any combination of the two.

EARLY PROTOTYPE
Taking a huge risk, Boeing proposed, built and funded a military jet tanker/transport prototype in 1954. The aircraft was ordered into production as the KC-135, and one of the greatest success stories in military aviation had begun. Stratotankers have since served around the world, supporting all types of USAF missions, and have been involved in combat operations over Vietnam, the Persian Gulf, Iraq and Afghanistan.

LOADING
A large door on the left forward fuselage hinges upwards to allow cargo or passengers to be loaded. Up to 37,650kg (82,000lb) of palletized freight also may be accommodated.

REFUELLING EXERCISE
A US Navy F/A-18C Hornet aircraft assigned to the 'Knighthawks' of Strike Fighter Squadron One Three Six takes on fuel from a KC-135 Stratotanker aircraft on 7 October 2007 over Iraq during a close air support mission in support of US and coalition ground forces.

Gloster Javelin (1956)

TYPE • Interceptor **COUNTRY** • United Kingdom

SPECIFICATIONS (FAW.Mk. 9)

DIMENSIONS:	Length: 17.15m (56ft 3in); Wingspan: 15.85m (52ft); Height: 4.88m (16ft)
WEIGHT:	19,578kg (43,162lb) loaded
POWERPLANT:	2 x 48.94kN (11,007lbf) Armstrong Siddeley Sapphire 203 turbojets
MAX SPEED:	1130km/h (702mph)
RANGE:	1600km (994 miles)
CEILING:	16,000m (52,493ft)
CREW:	2
ARMAMENT:	2 x 30mm (1.18in) ADEN cannon in each wing; 4 x De Havilland Firestreak heat-seeking air-to-air missiles

Although plagued by serviceability and structural problems, the Javelin was a highly capable bomber-interceptor, equipped as it was with four 30mm (1.18in) ADEN cannon and, later, with the addition of four IR-seeking missiles.

This page: The Javelin served with 14 squadrons of the Royal Air Force, including No. 11 Squadron, which used various marks from 1960 to 1966.

RADAR
The all-important radar system was British in the FAW.Mk 1, but American in the FAW.Mks 2 and 6.

CREW
Both pilot and navigator sat on Martin-Baker ejection seats. The navigator operated the radar set.

TAILPLANE
The FAW.Mk 4 variant was the first to have a variable-incidence tailplane. The massive fin made the Javelin easy to recognize.

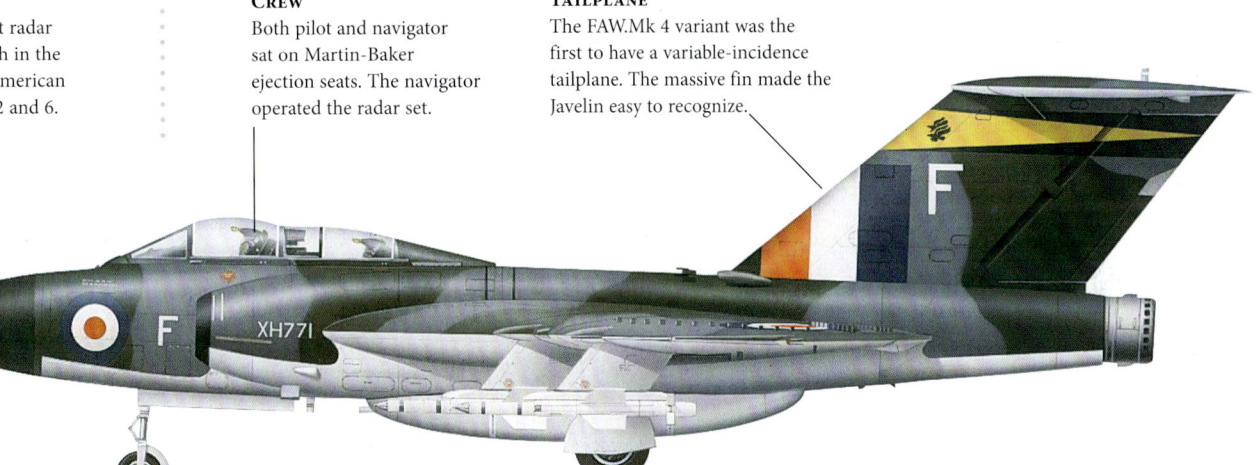

DELTA WING
The Cold War nuclear threat demanded that NATO air forces deploy radar-equipped interceptors able to halt attacking Soviet bombers far from the West's vulnerable cities. Britain's Gloster Javelin, the world's first twin-jet, delta-winged interceptor, was designed for the job. A two-man long-range interceptor, the Javelin was a little slow and difficult to fly, but its radar and missile combination made it one of the most advanced fighters of the 1950s.

GLOSTER GA.5 JAVELIN
This Javelin (WD804) was a prototype aircraft that performed development work for Gloster during the 1950s. It was lost in an air accident on 29 June 1952 after elevators broke off, although the pilot still managed to land the jet.

Douglas A-4 Skyhawk (1956)

TYPE • Naval fighter **COUNTRY** • United States

SPECIFICATIONS (A-4S)	
DIMENSIONS:	Length: 11.70m (38ft 5in); Wingspan: 8.38m (27ft 6in); Height: 4.27m (14ft 1in)
WEIGHT:	10,206kg (22,500lb) loaded
POWERPLANT:	One 37.37kN (8400lbf) Curtiss-Wright J65-W-20 turbojet engine
MAX SPEED:	1064km/h (661mph)
RANGE:	2680km (1665 miles)
CEILING:	13,716m (45,000ft)
CREW:	1
ARMAMENT:	Two 30mm (1.18in) ADEN cannon; ordnance of up to 2268kg (5000lb) including conventional bombs and rockets and AIM-9P Sidewinder missiles

One of the most versatile combat aircraft ever produced, the Douglas A-4 Skyhawk was designed to replace the piston-engined A4D Skyraider. Skyhawk pilots in Vietnam went into battle relying on their own ability and on the lightweight, nimble attack aircraft.

Below: A-4S number 681 was built for the US Navy as A-4B Bureau Number 145046 and is known to have flown with Reserve squadron VA-209 'Air Barons' before delivery to Singapore in 1972.

TACTICAL EQUIPMENT
The A-4E was the first Skyhawk variant optimized for air support and conventional bombing. New equipment included tactical air navigation (TACAN), Doppler navigation, a radio altimeter and new toss-bombing and low-altitude bombing systems.

A-4E SKYHAWK
BuNo. 149993 served with Attack Squadron 72 (VA-72) 'Blue Hawks' aboard the USS *Independence* in the South China Sea in May 1965. VA-72 was the first US Navy unit to operate Skyhawks, receiving A4D-1s in 1956.

TURBOJET
Originally intended for the cancelled A4D-3 all-weather variant, the Pratt & Whitney J52 turbojet made its debut in the A-4E (designated A4D-5 during development). Its lower fuel consumption increased range by 27 per cent.

'SCOOTER'
The A-4 Skyhawk (nicknamed the 'Scooter') was designed for a nuclear mission. Originally, it was intended to carry a single atomic bomb on a centreline rack beneath the fuselage, but the Navy and Marine Corps always saw the type with an additional role as a conventional bomber. From the start of its flight test programme in the early 1950s, the A-4 carried a variety of conventional bombs, rockets and missiles. When the United States launched its first air strike against North Vietnam in August 1964, the 'Scooter' was there.

ADVERSARY A-4S
These two A-4 Skyhawks were in USAF service as 'adversary aircraft', denoted by the red star on the tail. In this role they would act as enemy aircraft and participate in dogfighting and interception exercises.

Vought F-8 Crusader (1956)

TYPE • *Naval fighter* **COUNTRY** • *United States*

The first carrier-borne fighter capable of supersonic speed in level flight, the Crusader won a May 1953 US Navy competition for a new day fighter. It was developed in record time, with only 21 months elapsing between the prototype's first flight and the fighter's entry into service.

SPECIFICATIONS (F-8E)

DIMENSIONS:	Length: 16.61m (54ft 6in); Wingspan: 10.87m (35ft 8in); Height: 4.8m (15ft 9in)
WEIGHT:	Empty 9038kg (19,925lb); maximum take-off 15,420kg (34,000lb)
POWERPLANT:	One 80kN (18,000lbf) Pratt & Whitney J57-P-20A turbojet
MAX SPEED:	1827km/h (1135mph)
RANGE:	1932km (1200 miles)
CEILING:	17,680m (58,000ft)
CREW:	1
ARMAMENT:	4 x 20mm (0.79in) M39 cannon; provision up to 2268kg (5000lb) of stores, including two Matra R530 missiles or 8 x 127mm (5in) rockets

Left: This well-worn F-8E belongs to US Marine Corps all-weather fighter squadron VMF(AW)-235 and is depicted as it would have appeared during the time of the siege of Khe Sanh in 1968. The unit's 'Bozo Noses' nickname was derived from the red star-studded markings applied to the nose, ventral fins and tail.

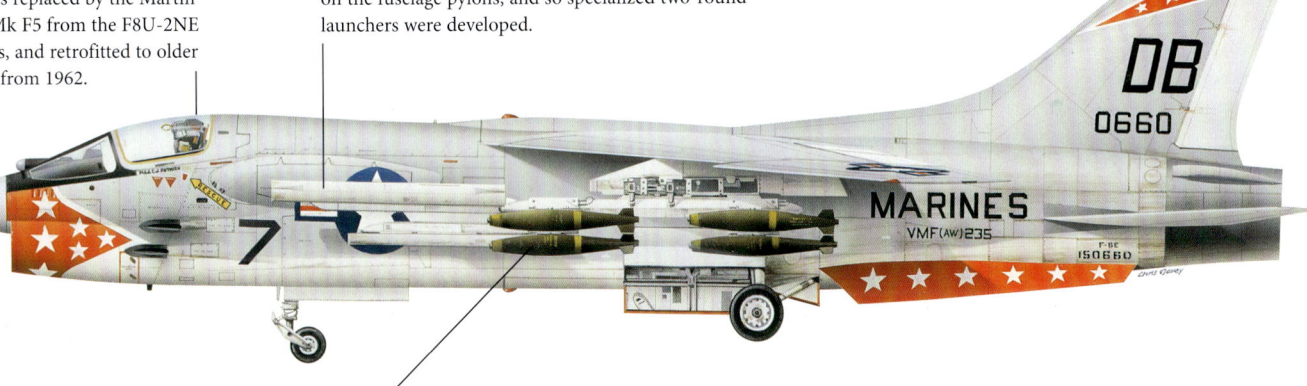

EJECTION SEATS
Production aircraft were initially fitted with a high-altitude ejection seat built by Vought. This was replaced by the Martin Baker Mk F5 from the F8U-2NE onwards, and retrofitted to older aircraft from 1962.

ZUNI ROCKET
The 127mm (5in) Zuni rocket was the most widely used of the F-8's attack weapons. The aircraft could not carry the standard four-round LAU-10 launcher on the fuselage pylons, and so specialized two-round launchers were developed.

BOMBLOAD
The F-8E was notable as the only fighter aircraft in Vietnam able to carry the 907kg (2000lb) Mk 84 bomb. The aircraft here is armed with eight 227kg (500lb) Mk 82 bombs carried in a 'slant-four' configuration on each rack.

DOGFIGHTER
Crusader jocks boasted that their manoeuvrable, cannon-armed jet was hotter than anything in the skies and scoffed when 'experts' said that missiles would make their fighters obsolete. In battle, the pilots were proved correct when their fighter was able to dogfight with Vietnamese MiGs on equal terms. The Crusader was the best-loved fighter in the post-war US Navy and remained in service in France until 2000.

TWO-SEAT TRAINER
A two-seat trainer variant of the F-8 Crusader in flight. One XF8U was converted to a two-seater, with only two cannon but a second set of controls. It first flew in 1962, but did not see active service.

Convair F-102 Delta Dagger (1956)

TYPE • Interceptor COUNTRY • Italy

SPECIFICATIONS	
DIMENSIONS:	Length: 20.84m (68ft 4in); Wingspan: 11.62m (38ft 1in); Height: 6.46m (21ft 2in)
WEIGHT:	Normal loaded 12,565kg (27,700lb); maximum 14,290kg (31,504lb)
POWERPLANT:	One 49.72kN (11,183lbf) Pratt & Whitney J57-P-23 turbojet, increasing to 76.51kN (17,208lbf) with afterburning
MAX SPEED:	1328km/h (825mph) at 12,190m (40,000ft)
RANGE:	2173km (1350 miles)
CEILING:	16,500m (54,134ft)
CREW:	1
ARMAMENT:	3 x Falcon heat-seeking missiles and one Nuclear Falcon, or three radar and three heatseeking air-to-air missiles; up to 24 unguided 70mm (2.75in) rockets

The Convair F-102 was the first supersonic warplane with a delta- or triangle-shaped wing. It was designed to defend North America against the nuclear-tipped Soviet bomber attack dreaded by so many during the Cold War.

Below: The needle-nosed 'Deuce' was the first interceptor to be developed as part of an overall air defence weapon system, known as WS201A. The F-102 was considered just the airframe portion of this system, to which was added the radar and missile sub-systems.

NOSE RADAR
The radar of the Hughes MG-10 fire control system could track several airborne targets simultaneously at ranges of up to 50km (31 miles) and lock-on to individual aircraft at circa 25km (16 miles).

FUSELAGE
The first F-102 design was disappointing in service, and was revised to incorporate 'area rule'. This produced a pinched waist, or 'Coke bottle' shape. To make the tail fatter in area to conform with the new design, large bulges were added each side of the rear fuselage. These were known, for obvious reasons, as 'Marilyns'.

ROCKET ARMED
Although it was a fighter, the F-102 carried no gun. It was armed with an almost unbelievable battery of rockets, including a Falcon missile with an atomic warhead intended to break up bomber formations. Pilots called this magnificent aircraft the 'Deuce'. It was huge and powerful, and a pleasure to fly. F-102s served briefly in Vietnam where, late in its career, this great plane was miscast in a limited war setting.

WEAPONS BAY
Putting the weapons in an internal bay was a vital factor in the F-102's speed. If they had been on outside pylons, the drag would have degraded performance.

HIGH PERFORMANCE
The F-102 combined a 'wasp waist' fuselage shape, technically called 'area rule', with blade-like wings and tail to become one of the fastest fighters of its era.

Grumman OV-1 Mohawk (1956)

TYPE • Reconnaissance COUNTRY • United States

SPECIFICATIONS (OV-1D)	
DIMENSIONS:	Length: 12.5m (41ft); Wingspan: 14.6m (48ft); Height: 3.86m (12ft 8in)
WEIGHT:	14.63m (48ft)
POWERPLANT:	2 × 1000kW (1400shp) Lycoming T53-L-701 turboprops
MAX SPEED:	491km/h (305mph)
RANGE:	1519km (944 miles)
CEILING:	7600m (25,000ft)
CREW:	2
ARMAMENT:	Stores include: 12.7mm (0.50in) machine-gun pods, FFARs and HVARs rocket pods; 114kg/227kg/454kg (250lb/500lb/1000lb) low-drag bombs, napalm bombs and Sidewinder air-to-air missiles

The Grumman Mohawk was introduced into US service in 1959, a quirky turboprop observation and attack aircraft that would not be retired from American operations until 1996, and even longer in export countries.

Below: An OV-1D Mohawk. The US Army lost 65 OV-1s operationally in the Vietnam War, to either ground fire or accidents.

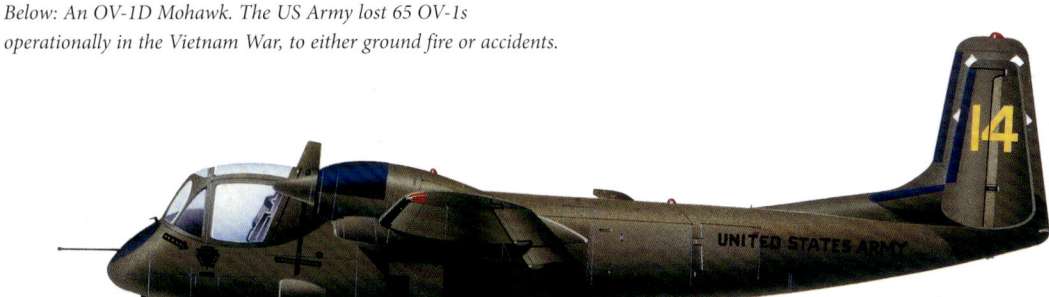

WINGS
To offset the weight and drag of the AN/APS-94B pod, the Mohawk was fitted with longer wings, reaching to 14.6m (48ft).

TRIPLE TAIL
The Mohawk's instant defining feature was its triple tail, although originally it had been designed with a T-shaped tail.

ARMED MOHAWKS
In order to evaluate the suitability of the Mohawk in the armed reconnaissance/ground support role, two OV-1As were modified as JOV-1As, equipped with an extra 227kg (500lb) stores station on each wing. A cockpit gunsight, gun firing and stores release equipment and armour plating were also added. Stores cleared for carriage by the aircraft included machine-gun pods, rockets, bombs and Sidewinder air-to-air missiles.

SLAR POD
The AN/APS-94B side-looking airborne radar (SLAR) was a sophisticated ground-mapping and target detection radar, fitted into a large pod beneath the Mohawk's fuselage.

RECOGNITION FEATURES
Apart from its triple tail, the OV-1 is also distinguished by its bulbous cockpit, which perfectly facilitates the crew's observation role, and its high-mounted turboprops on the wings.

Douglas C-133 Cargomaster (1957)

TYPE • Transport **COUNTRY** • United States

SPECIFICATIONS (C-133A)

DIMENSIONS:	Length: 48m (157ft 6in); Wingspan: 54.76m (179ft 8in); Height: 11.98m (39ft 3in)
WEIGHT:	Empty 50,662kg (111,700lb); maximum take-off 143,600kg (316,600lb)
POWERPLANT:	Four 4847kW (6500hp) Pratt & Whitney T34-P-7WA turboprops
MAX SPEED:	571km/h (355mph)
RANGE:	6395km (3975 miles) with a 23,556kg (52,000lb) load
CEILING:	5915m (19,400ft)
CREW:	5
ARMAMENT:	None

In the early 1950s, the Douglas Aircraft Company broke new design ground with the massive C-74 and C-124 Globemasters I and II airlifters. They were soon surpassed by the even mightier C-133 Cargomaster.

Below: The C-133's main landing gear consisted of two main units with four wheels each, in pods either side of the fuselage.

ENGINES
The Cargomaster's powerful engines drove 5.49m (18ft) diameter Curtiss-Wright turboelectric three-bladed, constant-speed reversible propellers.

HOLD
The C-133's enormous cargo hold was 27.4m (90ft) long and at least 3.6m (11ft 10in) wide. With a large loading door and a floor at truck bed height, large loads such as ballistic missiles could be easily loaded and carried. Alternative loads included 16 fully loaded jeeps or up to 200 passengers.

CLAMSHELL DOORS
Clamshell doors were fitted to the last two C-133As and all 15 C-133Bs to improve the usable load capacity of the aircraft's cargo hold. Adding 0.91m (3ft) to the length of the hold, this modification allowed a Titan ICBM to be carried in one piece.

NO PROTOTYPES
First flown on 23 April 1956, the giant C-133 was originally conceived as a means of transporting the American intercontinental ballistic missiles (ICBMs) and intermediate range ballistic missiles that were under development. There were no prototypes, and the Cargomaster went straight into production as the C-133A. It entered service with the USAF Military Air Transport Service in 1957 and equipped the 1st, 39th and 84th Air Transport Squadrons.

CARGO CARRIER
The great size of the C-133 is apparent in this photograph. The aircraft featured a high-mounted wing, blister fairings on either side of the fuselage to accommodate the landing gear, as well as both rear-loading and side-loading cargo doors.

Lockheed T-2V1 SeaStar (1957)

TYPE • Trainer COUNTRY • United States

SPECIFICATIONS

DIMENSIONS:	Length: 11.74m (38ft 6.5in); Wingspan: 13.05m (42ft 10in); Height: 4.08m (13ft 4in)
WEIGHT:	Empty 5209kg (11,500lb); maximum take-off 7610kg (16,800lb)
POWERPLANT:	One 27kN (6100lbf) thrust Allison J33-A-24 turbojet
MAX SPEED:	933km/h (580mph) at 7620m (25,000ft)
RANGE:	1448km (900 miles)
CEILING:	12,200m (40,000ft)
CREW:	2
ARMAMENT:	None

The Lockheed T2V-1 SeaStar was developed from the most widely used advanced trainer in the world, the Lockheed T-33, which first flew in 1948 and was developed from the F-80C Shooting Star airframe.

Left: This T2V-1, the sixth production aircraft, carries the markings of the Naval Air Test Center, and is seen as it appeared during qualification trials aboard the carrier USS Antietam.

COCKPIT REDESIGN
A substantial redesign of the TV-2 (T-33B) was necessary to produce an aircraft more suited to the naval training role. Among the changes was the incorporation of a stepped cockpit to improve the instructor's view from the rear seat.

TIP TANKS
To provide a range of 1448km (900 miles), the SeaStar had increased fuel capacity over that of the T-33. It also had 1741 litres (460 Imp gal) in the tip tanks, which were non-jettisonable.

ARRESTOR HOOK
An obvious modification for this carrier-capable aircraft was the addition of a tailhook, hinged in line with the wing trailing edge.

TRAINING TYPE

The SeaStar was designated T-33A in USAF service and T-33B (T-2V) with the US Navy and Marine Corps. It is estimated that 90 per cent of the free world's military jet pilots trained on the T-33 during the 1950s and 1960s. T-33 production totalled 5691 in the USA alone, and many others were built under licence in Canada and Japan. A version adapted to carry underwing weapons was offered to small air arms in the counter-insurgency role.

LOCKHEED T-2V
The T-2V started as a private venture under the designation L-245; it featured a raised instructor's seat, a dorsal fairing, one-piece windscreen, single-point refuelling, strengthened undercarriage (for deck landing operations), larger tail surfaces, boundary layer control flaps and leading-edge slats.

McDonnell F-101 Voodoo (1957)

TYPE • *Fighter* **COUNTRY** • *United States*

SPECIFICATIONS (F-101B)

DIMENSIONS:	Length: 20.54m (67ft 4.67in); Wingspan: 12.09m (39ft 8in); Height: 5.49m (18ft)
WEIGHT:	Empty 13,141kg (28,970lb); maximum take-off 23,768kg (52,400lb)
POWERPLANT:	Two 53.3kN (11,990lbf) thrust Pratt & Whitney J57-P-55 turbojets
MAX SPEED:	1965km/h (1221mph) at 12,190m (40,000ft)
RANGE:	2494km (1550 miles)
CEILING:	16,705m (54,800ft)
CREW:	2
ARMAMENT:	Two nuclear-tipped and four conventional AAMs

To pilots, the McDonnell F-101 Voodoo was a friendly monster. It was not a forgiving mount; in fact, it was one of the hardest-to-fly aircraft ever to serve in the US Air Force. However, once a pilot learned its quirks, the Voodoo was an extravagant performer.

Below: The aircraft intended to replace the Avro Canada CF-100 was the powerful and very advanced CF-105 Arrow, which was cancelled late in its development, with a prototype already flying. Ex-USAF Voodoos, designated CF-101B in Canadian service, were bought instead. The example seen here was a USAF F-101B.

WINGS
With its distinctive boundary layer fences, the Voodoo's wing, swept at 45 degrees, was very small, but not quite as thin as it appeared.

HIGH TAIL
The distinctive 5.4m (18ft) high tail of the Voodoo combined with the low wing to produce a tendency to pitch up, which could make it tricky to fly.

FLIGHT CHALLENGES
The Voodoo was possibly the first warplane to exceed supersonic speed on its first flight (on 29 September 1954), but it never became easy to fly. Tucking in the Voodoo's nosewheel was a challenge and the aircraft had a tendency to 'pitch up', for which various cures were attempted, never with success. It killed test pilots and challenged service pilots. It remained totally unforgiving throughout – but when used properly it was a world-beater.

WHEEL TRACK
With its main wheel track of 6.06m (19.9ft) the Voodoo had a light 'footprint'. On ice-covered airfield aprons, pilots knew that they had a firm grip and could taxi at speeds of up to 80km/h (50mph) without fear of mishap.

VOODOOS ON EXERCISE
An air-to-air left side view of two F-101 Voodoo aircraft near Niagara Falls during exercise Sentry Castle in 1981. The aircraft are assigned to the 107th Fighter Interceptor Group, 136th Fighter Interceptor Squadron, New York Air National Guard.

Supermarine Scimitar (1957)

TYPE • Naval fighter   COUNTRY • United Kingdom

With the arrival of new fleet carriers like HMS *Hermes* and *Ark Royal*, and supersonic fighters and attack aircraft like the Scimitar, the Fleet Air Arm was able to project airpower and deploy nuclear weapons over large parts of the world.

SPECIFICATIONS (Scimitar F.1)

DIMENSIONS:	Length: 16.87m (55ft 4in); Wingspan: 11.33m (37ft 2in); Height: 4.65m (15ft 3in)
WEIGHT:	Empty 9525kg (21,000lb); maximum take-off weight 15,513kg (34,200lb)
POWERPLANT:	Two 50kN (11,250lbf) Rolls-Royce Avon 202 turbojets
MAX SPEED:	1143km/h (710mph)
RANGE:	966km (600 miles)
CEILING:	15,240m (50,000ft)
CREW:	1
ARMAMENT:	Four 30mm (1.18in) Aden cannon, four 454kg (1000lb) bombs or four Bullpup air-to-ground missiles, or four Sidewinder air-to-air missiles or drop tanks

Right: A Supermarine Scimitar F.1 makes a carrier landing, always a perilous time for the fast and heavy jet.

TAIL
In the original design phase, the Scimitar was to be fitted with a V-shaped tail and a straight wing, but it eventually received a butterfly tail and a swept wing.

POWERPLANT
The Scimitar was given thrust from two 50kN (11,250lbf) Rolls-Royce Avon 202 turbojets, which could take aircraft to a maximum speed of 1143km/h (710mph).

HARDPOINTS
Each of the underwing hardpoints could carry 454kg (1000lb) of ordnance or stores; the aircraft was also capable of transporting a 'Red Beard' nuclear bomb.

DEVELOPMENT

The Scimitar had an extremely protracted gestation period. The first prototype, the Supermarine 508, was a thin, straight-winged design with a butterfly tail. Production aircraft were delivered from August 1957. A total of 76 were built, providing the Fleet Air Arm with a capable low-level supersonic attacker until the Scimitar was superseded by the Buccaneer in 1969.

SCIMITAR F.1
Supermarine Scimitar F.1 XD328 in flight. The attrition rate for Scimitars during carrier operations in the 1950s and 1960s was horrendous: of 76 Scimitars produced, a total of 39 were lost in accidents and mechanical failure.

McDonnell Douglas F-4 Phantom II (1958)

TYPE • Fighter **COUNTRY** • United States

Arguably the greatest warplane of its era, the McDonnell Douglas F-4 Phantom II stemmed from a 1954 project for an advanced naval fighter. Its superb flight and combat characteristics meant that it spread beyond carrier decks to land-based air units across the world.

SPECIFICATIONS (F-4E Phantom II)

DIMENSIONS:	Length: 17.76m (58ft 3in); Wingspan: 11.70m (38ft 5in); Height: 4.96m (16ft 3in)
WEIGHT:	Empty 12,700kg (28,000lb); maximum take-off 26,309kg (58,000lb)
POWERPLANT:	Two 79.38kN (17,845lbf) thrust General Electric J79-GE-17 turbojets
MAX SPEED:	2390km/h (1485mph) at altitude
RANGE:	2817km (1750 miles)
CEILING:	18,975m (62,250ft)
CREW:	2
ARMAMENT:	One 20mm (0.79in) cannon and four AAMs; up to 5888kg (12,980lb) of ordnance

Left: The F-4E Phantom was used in large numbers by the Japanese Air Self-Defence Force, the F-4EJ being the licence-built version. Of the 140 F-4EJs procured, about two-thirds were upgraded to F-4EJ Kai standard with improved weapons and avionics systems, the Kai suffix denoting 'modified'.

KAI MODIFICATIONS
F-4EJ Kai modifications centre around the adaptation of the Northrop Grumman (Westinghouse) APG-66J radar, based on that fitted to the F-16, in place of the old Westinghouse APQ-120. The new radar was lighter and smaller, with better performance and reliability.

CREW
All Phantoms had two crew and many had dual controls. USAF aircraft carried a pilot and weapons system officer (WSO, or 'Wizzo' or 'Bear' in the F-4G). Initially the USAF wanted to equip their F-4s with two qualified pilots, an expensive option that was later dropped.

EXTERNAL MODIFICATIONS
External additions to the Kai-modified aircraft include twin aft-facing radar warning receivers on the tip of the tailfin, similar forward-facing antennae on the wingtips and a large blade aerial mounted mid-spine for the new UHF radio.

VIETNAM SERVICE
In 1965, F-4s began their campaign against MiG fighters in the skies over Hanoi. MiG-17s and MiG-21s were smaller, simpler and more nimble than the robust and ungraceful Phantom – but the F-4 entered each dogfight with brute force and power. The Vietnam War proved the merit of the Phantom, not just in air-to-air combat, but also in nearly every military task a jet aircraft could perform, including close support, bombing and reconnaissance.

REFUELLING
A German Air Force F-4F Phantom aircraft flies a refuelling mission near Royal Air Force (RAF) Mildenhall, United Kingdom.

Fiat/Aeritalia G.91 (1958)

TYPE • *Fighter-bomber* COUNTRY • *Italy*

In the 1950s, NATO air commanders were keen to develop a lightweight tactical jet fighter-bomber that could operate from makeshift airstrips if conventional runways were destroyed. The result was the G.91.

SPECIFICATIONS (G-91Y)

DIMENSIONS:	Length: 11.67m (38ft 3.5in); Wingspan: 9.01m (29ft 6.5in); Height: 4.43m (14ft 6in)
WEIGHT:	Empty 3900kg (8600lb); maximum take-off 8700kg (19,184lb)
POWERPLANT:	Two General Electric J85-GE-13A turbojets, delivering 18.1kN (4080lbf) thrust
MAX SPEED:	1110km/h (690mph)
RANGE:	1500km (932 miles)
CEILING:	12,500m (41,000ft)
CREW:	1
ARMAMENT:	Two 30mm (1.19in) cannon; four pylons for various ordnance; three camera installations

This page: The G-91Y pictured bears insignia of the Gruppo 13 of Stormo 32, which replaced its G-91Rs with the Y model from August 1973. All three of Gruppo 13's squadrons were re-equipped by September 1974 and declared combat-ready in 1975.

FUSELAGE
The Fiat G.91Y used the basic fuselage of the G-91T trainer, with the rear cockpit removed and replaced by extra fuel tankage. Fuel capacity was twice that of the earlier single-seat variants.

UNDERCARRIAGE
The G.91 was designed to operate from rough, unprepared airstrips, and was fitted with a tall undercarriage to avoid damage and to give ground clearance for factors such as underwing munition.

UNDERWING PYLONS
Each of the G-91Y's four underwing pylons was stressed for the carriage of a 454kg (1000lb) bomb. This was a substantial increase over the load that could be carried by the earlier G.91R. Combat radius was also increased.

NATO REQUIREMENT
Produced by Fiat, the Aeritalia G.91 lightweight ground-attack fighter was designed in response to a NATO requirement issued in 1954. The first prototype flew on 9 August 1956 and proved easily capable of meeting all the demands of the NATO specification. Total production of all G.91 variants from 1956 to 1977 was 756. The last G.91 variant was the G.91Y, which was substantially redesigned and had two General Electric J85 turbojets to allow a greater tactical load.

AIRSHOW G.91
An Italian Fiat G.91 is parked during Airshow for the 100th anniversary of the Italian Air Force in Pratica di Mare, Rome, June 2023. This example is the only one of the type airworthy in the world.

Beriev Be-10 'Mallow' (1958)

TYPE • Seaplane   COUNTRY • Soviet Union

SPECIFICATIONS

DIMENSIONS:	Length: 31.45m (103ft 2in); Wingspan: 28.6m (93ft 10in); Height: 10.7m (35ft 1in)
WEIGHT:	Empty 27,600kg (60,900lb); maximum 48,500kg (106,900lb)
POWERPLANT:	Two 71.2kN (16,000lbf) thrust Lyulka AL-7PB turbojet engines
MAX SPEED:	910km/h (565mph)
RANGE:	3150km (1957 miles)
CEILING:	12,500m (41,010ft)
CREW:	3
ARMAMENT:	Four 23mm (0.9in) AM-23 cannon; up to 1360kg (3000lb) of bombs, torpedoes or mines

Given the NATO reporting name 'Mallow', the Be-10 was one of the few jet-engined flying-boat types that were built in small numbers in the 1940s and 1950s. The Be-10 was displayed in public in 1961 and broke several records in its class, but did not enter large-scale service.

Right: A Soviet-era Beriev Be-10 sits patiently on the water. The aircraft was able to operate on waters with waves of up to 1.2m (4ft).

SPRAY FENCES
The spray fences fitted to the bow of the aircraft served to deflect sea spray and waves away from the air intakes for the engines and the wings.

AIR INTAKES
To minimize the seawater being ingested into the Lyulka engines during take-off, the air intakes sit close to fuselage, just beneath the wing roots.

HULL
The Be-10 was designed for good handling on the water, with a single-step hull, a high length-to-beam ratio (optimal in rough seas) and a sea rudder fitted under the rear fuselage.

IN SERVICE
Despite breaking several records, the Beriev Be-10 did not have a trouble-free career. Designed as a flying-boat patrol bomber, it entered service in 1958 with the 977th Independent Naval Long-range Reconnaissance Air Regiment (977th OMDRAP). In service use, it was subsequently found to suffer from metal fatigue and corrosion, so was removed from the Soviet arsenal by the late 1960s.

BE-10 POSTAGE STAMP
A Soviet postage stamp celebrates the Be-10 in artistic fashion, depicting it roaring off choppy seas. A specially modifed version of the aircraft won 21 international records for flying boats in 1961 (under the designation M-10), including speed records and altitude records.

Lockheed F-104 Starfighter (1958)

TYPE • Fighter COUNTRY • United States

Development of the F-104 began in 1951, when the lessons of the Korean air war were starting to bring about profound changes in combat aircraft design. A contract for two XF-104 prototypes was placed in 1953 and the first of these flew on 7 February 1954, only 11 months later.

SPECIFICATIONS (F-104G)

DIMENSIONS:	Length: 16.66m (54ft 8in); Wingspan: 6.68m (21ft 11in); Height: 4.09m (13ft 5in)
WEIGHT:	Empty 6348kg (13,995lb); maximum take-off 13,170kg (29,035lb)
POWERPLANT:	One 69kN (15,600lbf) thrust General Electric J79-GE-11A turbojet
MAX SPEED:	1845km/h (1146mph) at 15,240m (50,000ft)
RANGE:	1740km (1081 miles)
CEILING:	15,240m (50,000ft)
CREW:	1
ARMAMENT:	One 20mm (0.79in) cannon; AAMs; up to 1814kg (4000lb) of ordnance

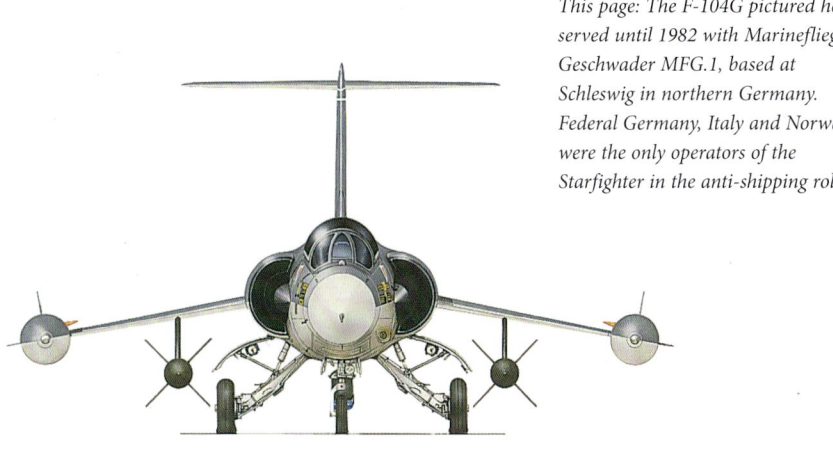

This page: The F-104G pictured here served until 1982 with Marineflieger Geschwader MFG.1, based at Schleswig in northern Germany. Federal Germany, Italy and Norway were the only operators of the Starfighter in the anti-shipping role.

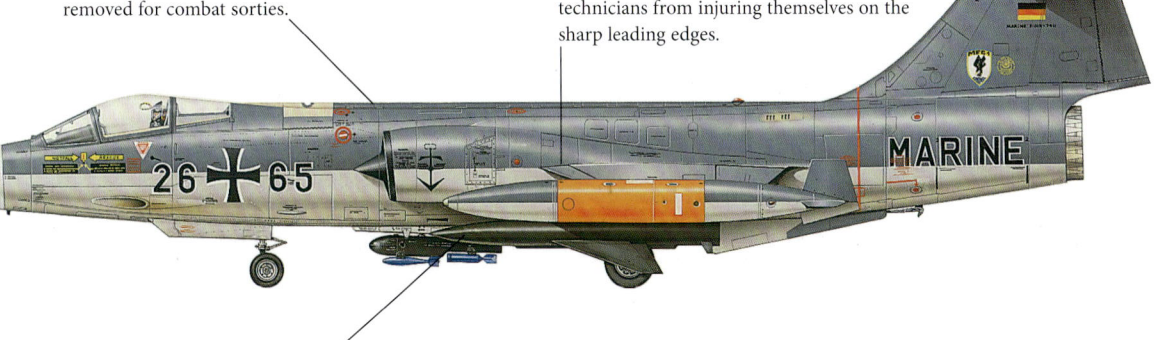

CAMOUFLAGE
German Navy Starfighters received a dark grey upper surface camouflage, rendering them almost invisible against the Baltic at low level. The orange patches on the tanks were a safety measure, and would have been removed for combat sorties.

WINGS
The Starfighter featured an ultra-thin, short-span wing with a leading-edge sweep of 26 degrees. Felt strips could be added to the wing on the ground to prevent technicians from injuring themselves on the sharp leading edges.

ANTI-SHIP MISSILES
German Navy Starfighters carried a pair of Kormoran anti-ship missiles from 1977 onwards, replacing the French AS30 weapon. These gave the aircraft the ability to fire at enemy shipping from a range of 30km (18.6 miles).

'ZIPPER'

The Lockheed F-104A Starfighter was known as the 'Zipper' to pilots who flew the aircraft during its brief, unsuccessful US military career. But the sleek, futuristic fighter was not a failure. Designed with speed as its primary feature, the F-104A was a fast performer. It extended the boundaries of fighter performance in the 1950s, and brought Mach 2 capability to more than two dozen air forces around the world. Many fighter-bomber and multi-mission variants were produced.

F-104 WITH SIDEWINDERS
An air-to-air right side view of an F-104 Starfighter aircraft carrying two AIM-9J Sidewinder missiles. The aircraft flew with the 69th Tactical Fighter Training Squadron, 58th Tactical Training Wing, 12th Air Force.

Republic F-105 Thunderchief (1958)

TYPE • *Fighter*   COUNTRY • *United States*

Conceived as a tactical nuclear strike aircraft, the Republic F-105 Thunderchief suffered heavy losses over North Vietnam, carrying out a role for which it was never intended. Mimicking the sound the aircraft made when it hit the ground, pilots nicknamed it the 'Thud'.

SPECIFICATIONS (F-105G)

DIMENSIONS:	Length: 19.58m (64ft 3in); Wingspan: 10.65m (34ft 11.25in); Height: 5.99m (19ft 8in)
WEIGHT:	Empty 12,474kg (27,500lb); maximum take-off 23,967kg (52,838lb)
POWERPLANT:	One 117.88 kN (26,500 lbf) thrust Pratt & Whitney J75-P-19W turbojet
MAX SPEED:	2382km/h (1480mph)
RANGE:	2975 km (1849 miles)
CEILING:	15,850m (52,000ft)
CREW:	2
ARMAMENT:	One 20mm (0.79in) cannon; provision for up to 2722kg (6000lb) of munitions or ECM pods

Right: A Republic F-105D in flight with a full bombload. Almost one-third of all F-105s came to the end of their lives in the fiercely defended skies over North Vietnam.

IN-FLIGHT REFUELLING
The F-105 was equipped with the two types of fuel receiving equipment. A standard receptacle for a tanker boom was housed in the upper nose, while a probe (for use with drogue equipment) could be extended from the upper forward fuselage.

TACTICAL MARKINGS
This F-105G, 63-8320, bears the markings of the 333rd Tactical Fighter Squadron, 355th Tactical Fighter Wing, which was based at Takhli AFB, Thailand, in 1969.

CONTESTED AIRSPACE
In the Vietnam War, an F-105 heading north from Thailand with a typical load of eight 340kg (750lb) bombs would be able to attack its target successfully only by outwitting the enemy's missiles, MiGs and anti-aircraft fire. Typically, the F-105 pilot refuelled from a tanker, communicated with a command and control aircraft or a forward air controller (FAC), and then plunged into the hell of enemy airspace. Many aircraft fell and a host of pilots died fighting the most comprehensive anti-aircraft defences assembled up to that time.

SHRIKE MISSILES
Seen here on the F-105's outer wing pylons, the Shrike missile was derived from the AIM-7 Sparrow AAM and was the original US anti-radiation missile. It was replaced by the AGM-88 HARM in the 1980s.

F-105 REFUELLING
Three Air Force F-105 Thunderchief pilots en route to bomb military targets in Vietnam pull up to a flying Air Force 'gas station', in this case an Air Force KC-135 Stratotanker in January 1966.

Handley Page Victor (1958)

TYPE • Bomber COUNTRY • United Kingdom

The crescent-winged Victor – one of the three British Cold War 'V-bombers' – was futuristic in appearance, but still basically conventional, unlike the tailless delta-wing Vulcan. It saw action as a conventional bomber and later use as an aerial tanker.

SPECIFICATIONS (Victor B.2)

- **Dimensions:** Length: 35.05m (114ft 11in); Wingspan: 36.58m (120ft); Height: 9.2m (30ft 1.5in)
- **Weight:** Empty 41,277kg (91,000lb); maximum take-off 105,687kg (233,000lb)
- **Powerplant:** Four 91.6kN (20,600lbf) Rolls-Royce Conway Mk 201 turbofans
- **Max Speed:** 1030km/h (640mph)
- **Range:** 7400km (4600 miles) with internal fuel
- **Ceiling:** 18,290m (60,000ft)
- **Crew:** 5
- **Armament:** One Avro Blue Steel Mk 1 stand-off missile or or 35 to 48 x 454kg (1000lb) bombs

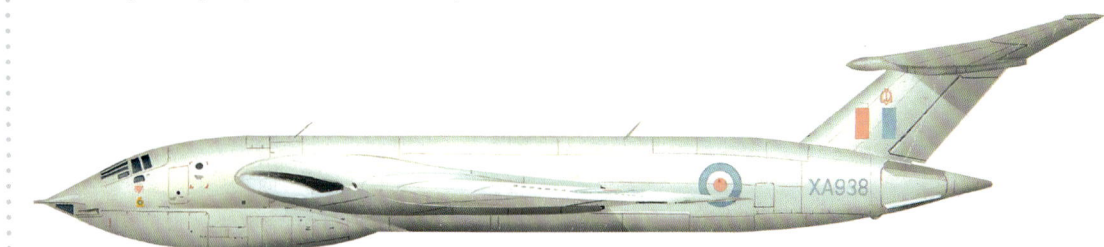

Below: A Victor B.2. The Victor's all-white paint scheme was intended to reflect the flash from its own nuclear weapons.

Tail
Only a small portion of the Victor's high-mounted tailplane was fixed, the majority of it moving to act as a giant elevator. The bullet fairing that covered the fin/tailplane joint housed an antenna at either end for the ARI 18228 radar warning receiver. At the base of the fin was a large intake for a heat exchanger, added to the B.Mk.2 to cool the ECM equipment.

Crew
The Victor was operated by the standard V-bomber crew of five (two pilots, two navigators and an air electronics operator). The pilots were strapped into ejection seats, while the three rear crew were expected to exit via a door that rotated out from the fuselage side to create a bail-out chute protected from the airstream.

Refuelling unit
Mounted under each wing was a FR.20B hose-drogue unit, with capacity for 659 litres (145 Imp gal) of fuel. The Mk.20 was equipped with a 15m (49ft) hose that deployed by air resistance. It was winched in by power from the ram air turbine on the front of the pod.

BOMBER ROLE
Designed to carry nuclear bombs to the Soviet Union, the Victor suffered from a protracted development compared to its rivals, only entering service in 1957, 12 years after design had started. The Victor had a larger bomb-carrying capacity than the other V-bombers and would have been at the vanguard of a strategic strike if war had come. It was equipped with the Blue Steel nuclear-tipped stand-off missile, but the weapon was never used outside trials.

B.1 IN FLIGHT
The last of the V-bombers to enter service, the Victor prototype finally flew in 1952. The nose was slightly lengthened and the tailplane shortened in the production aircraft.

Convair F-106 Delta Dart (1959)

TYPE • Interceptor COUNTRY • United States

The development line of the Convair F-106 Delta Dart began in 1950, when the US Air Force formulated a requirement for a night and all-weather interceptor incorporating the latest fire control system.

SPECIFICATIONS (F-106A)

DIMENSIONS:	Length: 21.58m (70ft 9in); Wingspan: 11.66m (38ft 3in); Height: 6.18m (20ft 3in)
WEIGHT:	Empty 10,712kg (23,616lb); maximum take-off 18,949kg (41,775lb)
POWERPLANT:	One 109kN (24,500lbf) thrust Pratt & Whitney J75-P-17 turbojet
MAX SPEED:	1328km/h (825mph) at 10,970m (36,000ft)
RANGE:	2172km (1350 miles)
CEILING:	16,460m (54,000ft)
CREW:	1
ARMAMENT:	One unguided AAM with a 1.5kT yield nuclear warhead; four infrared homing AAMs; unguided AAM later replaced by 20mm (0.79in) cannon

This page: The F-106 was the primary air defence aircraft of the USAF between 1959 and the late 1970s, when the F-15C Eagle began entering service in numbers. The Delta Dart carried its weaponry in an internal bay, a most unusual feature. A semi-retractable M61 cannon, also fitted in the bay, was introduced in 1973 following the experience of fighter pilots with other aircraft in Vietnam.

WINDSCREEN
The F-106 had optically flat windscreen panels that met at their forward edges. A blade-like metal 'vision splitter' prevented internal reflections without obstructing the pilot's view.

POWERPLANT
With its bigger engine, the F-106A had a top speed twice that of the F-102A. Yet its acceleration left a lot to be desired. Early models took 4.5 minutes to go from Mach 1.0 to Mach 1.7 at 17,373m (57,000ft), and a further 2.5 minutes to reach Mach 1.8.

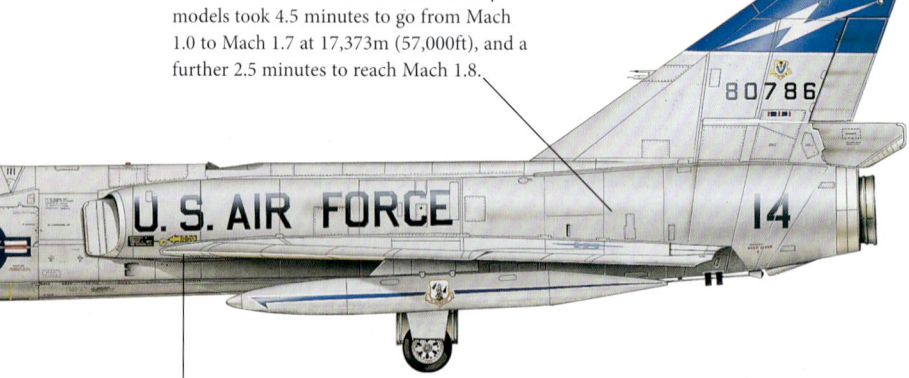

WINGS
YF-106s had the same boundary-layer fences as the F-102, but production aircraft had a slot in the leading edge, performing the same function more efficiently. Apart from this, the wing of the Delta Dart was nearly identical to that of the Delta Dagger.

AMERICA'S GUARDIAN
Though it never fired a shot in combat, the F-106 Delta Dart is one of the most fondly remembered fighters to serve in the US Air Force. Building on experience from its predecessor, the F-102A Delta Dagger, the F-106 carried the flame as the guardian of North America at the Cold War's height. Its job was simple: as an all-weather interceptor it was to detect, identify and destroy Russian bombers carrying atomic weapons to American cities.

F-106A OVER MOJAVE DESERT
An F-106A Delta Dart aircraft passes over the Mojave Desert while en route to Davis-Monthan Air Force Base, Arizona, where it was to be used in the QF-106 drone programme. The second-to-last F-106 in active service, it had been used as a safety chase aircraft in the B-1B production acceptance flight test programme.

Sukhoi Su-7 (1959)

TYPE • Fighter **COUNTRY** • Soviet Union

SPECIFICATIONS
(Su-7BMK 'Fitter-A')

DIMENSIONS:	Length: 17.37m (57ft); Wingspan: 8.93m (29ft 3.5in); Height 4.7m (15ft 5in)
WEIGHT:	Empty 8620kg (19,000lb); maximum take-off weight 13,500kg (29,750lb)
POWERPLANT:	One 88.2kN (19,842lbf) Lyulka AL-7F turbojet
MAX SPEED:	1700km/h (1056mph)
RANGE:	675km (42mph)
CEILING:	15,150m (49,700ft)
CREW:	1
ARMAMENT:	Two 30mm (1.18in) NR-30 cannon; four external pylons for two 750kg (1653lb) and two 500kg (1102lb) bombs

The Sukhoi Design Bureau produced the widely exported 'Fitter' series. The simple Su-7 with a fixed, swept wing was followed by the Su-17 (designated Su-20 for export) with variable-geometry (swing) wings and the Su-22M with improved avionics and more fuel.

Below: The Su-7B was ordered into production in 1958 and became the standard Soviet Bloc attack aircraft. This aircraft is an Egyptian Su-7BM. Thousands were supplied to all Warsaw Pact nations, among other countries.

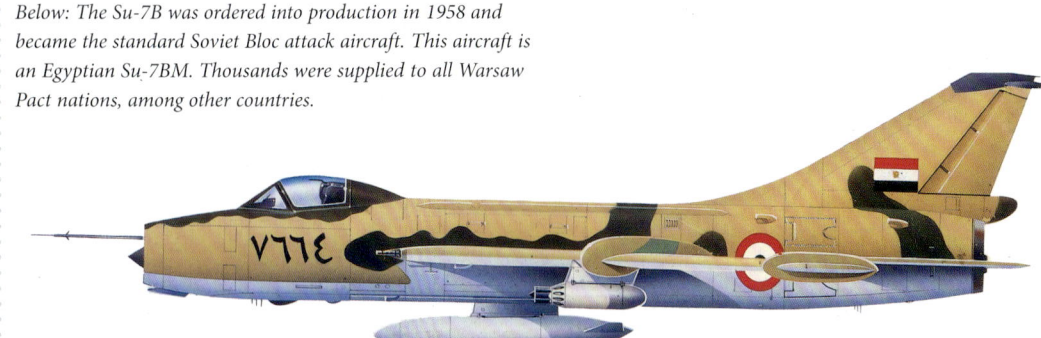

DURABILITY
The Su-7 was a physically tough aircraft, able to keep flying in adverse and sparse environments. This was one of the reasons why the aircraft enjoyed reasonable export success.

WINGS
The Su-7's wings were thin and highly swept. This was favourable for speed, but manoeuvrability in dogfighting was poor and the landing speed was precariously high.

SU-7BMK
The Su-7BMK was a version built largely for export between 1968 and 1971. Some served with the Soviet Air Force's Frontal Aviation arm, including the one illustrated, which was on the strength of a unit in the Trans-Baikal Military District in 1978.

LESSONS FROM KOREA
The air war over Korea provided the Soviet forces with many lessons in modern jet combat. Always tactically minded, they fielded the Sukhoi Su-7 'Fitter', which replaced ground-attack versions of the MiG-15 and MiG-17 – neither of which was suited to the role – from 1956. The Fitter, which was also widely exported, became the standard Soviet tactical support fighter-bomber, and remained so for two decades in a number of developed versions.

SOVIET 'FITTER'
The original Sukhoi Su-7 'Fitter' was a swept-wing contemporary of the MiG-21. Entering service with Frontal Aviation in the late 1960s, swing-wing 'Fitters' were upgraded several times to carry more fuel, weapons and better avionics.

De Havilland Sea Vixen (1959)

TYPE • *Naval fighter* **COUNTRY** • *United Kingdom*

SPECIFICATIONS (FAW.2)

DIMENSIONS:	Length: 17.02m (55ft 7in); Wingspan: 15.54m (51ft); Height: 3.28m (10ft 9in)
WEIGHT:	Maximum take-off 18,858kg (41,575lb)
POWERPLANT:	Two 49.9kN (11,230lbf) Rolls-Royce Avon 208 turbojets
MAX SPEED:	1110km/h (690mph)
RANGE:	1287.5km (800 miles)
CEILING:	21,790m (48,000ft)
CREW:	2
ARMAMENT:	Four Red Top air-to-air missiles (FAW 2); on outer pylons 454kg (1000lb) bombs, Bullpup air-to-surface missiles or equivalent stores

The Sea Vixen, like many of the aircraft operated by the Royal Navy, was originally designed to a 1946 RAF requirement for a land-based, all-weather interceptor.

Right: A Sea Vixen FAW.1 lifts off from the deck of an aircraft carrier. Mk 1s featured a hinged and pointed radome, power-folding wings and hydraulically steerable nosewheel. The FAW.2 had increased fuel capacity and provision for four Red Top missiles.

RADOME
The Sea Vixen's main sensor was the AI.18 radar in the nose. The radome hinged to starboard for maintenance.

TWIN BOOM
The distinctive twin-boom layout was inherited from the earlier Vampire and Venom fighters, and provided good stability in flight.

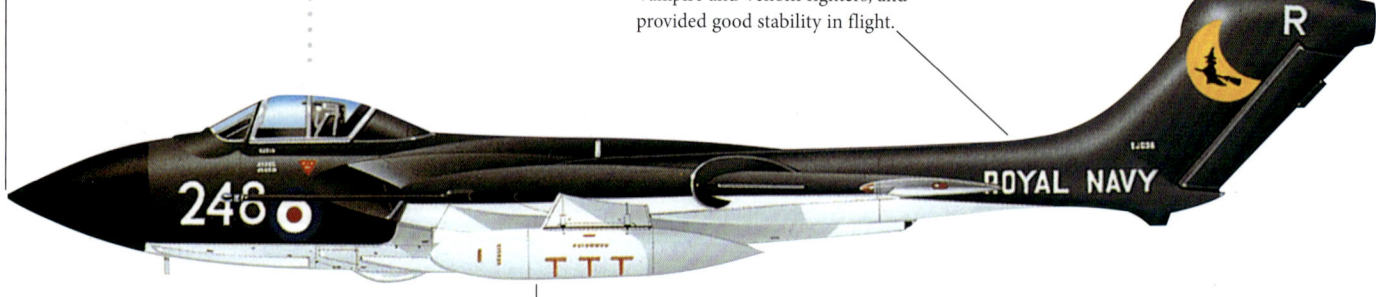

EXTERNAL FUEL
The outer two of the six wing stores hardpoints were almost always used for the carriage of external fuel tanks to improve the Sea Vixen's already impressive range.

MODERN FIGHTER
One feature that marked the Sea Vixen as a modern fighter for its day was its armament of four Firestreak infra-red homing missiles. Another was the large nose radome. The radar was responsible for one of the most unusual characteristics of the Sea Vixen: the lack of a canopy for the observer. His compartment was on the right side of the fuselage and was enclosed to make the most of the dim screen on the radar display.

ROCKET RUN
A Sea Vixen FAW.1 launches a battery of rockets. Sea Vixens could carry a wide range of defensive and attack stores on the wing pylons, typically Red Top or Firestreak air-to-air missiles, 454kg (1000lb) bombs, Bullpup air-to-surface missiles or 96 51mm (2in) rockets in four Microcell packs (as seen here).

Bell UH-1 Iroquois (Huey) (1959)

TYPE • Helicopter COUNTRY • United States

SPECIFICATIONS

DIMENSIONS:	Length: 12.31m (40ft 5in); Main rotor diameter: 14.63m (48ft); Height: 3.77m (12ft 5in)
WEIGHT:	Empty 2177kg (4789lb); maximum take-off 3856kg (8483lb)
POWERPLANT:	One Lycoming T53-L-11 turboshaft, 820kW (1100hp)
MAX SPEED:	217km/h (135mph)
RANGE:	383km (237 miles) at sea level
CEILING:	5790m (19,000ft)
CREW:	1–4
ARMAMENT:	Up to 3800kg (8300lb) of guns, rockets, missiles and grenades

Bell's Model 204 formed the basis for one of the most successful series of helicopters ever built. Flown for the first time in October 1956, it was designated XH-40, then HU-1, by the US Army (who called it the 'Huey'), before a designation change to HU-1A Iroquois.

Below: The D (seen here) and H models of the UH-1 Iroquois offered greater lifting capability, had a longer fuselage and a larger loading door than earlier helicopters in the series.

UH-1B IROQUOIS
Pilots always checked the glass-fibre rotor blades before a flight to ensure that the surface was not delaminating. They also inspected the so-called 'Jesus nut', which held the main rotor blades securely.

VIETNAM ICON
The UH-1 'Huey' became an icon of the Vietnam conflict. The original UH-1B was joined in 1963 by the UH-1D with provision for 14 rather than seven troops. In 1967 the further improved UH-1H became available. Air Cavalry missions involved UH-1s, aided by USAF close-support assets, such as the A-37B. UH-1s were also backed by observation helicopters, like the OH-6A, which provided support around the landing zone.

CREW COMPARTMENT
The main compartment could carry up to 10 troops in combat gear or six stretchers. The crew chief supervised loading of cargo and a gunner operated the M-60 machine gun.

SIDE DOOR
The main cabin door slid backwards to allow the troops to dismount or the gunner to fire. Gunships carried their armament on side-mounted sponsons.

UH-1 IN FLIGHT
A UH-1 Huey helicopter flies in the skies over a forward area refuelling point as it prepares to land and take on fuel in Yuma, Arizona, June 2007.

Saab J-35 Draken (1960)

TYPE • Fighter **COUNTRY** • Sweden

SPECIFICATIONS

DIMENSIONS:	Length: 15.40m (50ft 6.5in); Wingspan: 9.40m (30ft 10in); Height: 3.90m (12ft 9in)
WEIGHT:	Empty 7425kg (16,372lb); maximum take-off 16,000kg (35,280lb)
POWERPLANT:	One 78.4kN (17,600lbf) Svenska Flygmotor RM6C (licence-built Rolls-Royce Avon 300 series) turbojet
MAX SPEED:	2125km/h (1320mph)
RANGE:	3250km (2020 miles)
CEILING:	20,000m (65,615ft)
CREW:	1
ARMAMENT:	One 30mm (1.19in) cannon; four AAMs of various types; up to 4082kg (9000lb) of bombs

In the 1950s, the Swedish Government, determined to defend its neutrality against the growing threat from the east, built up one of the world's best-equipped air forces. Its SAAB J-35 Draken (Dragon) was Western Europe's first supersonic interceptor.

This page: This J-35J bears the original disruptive camouflage applied to J-35s, which was replaced by overall two-tone grey in the 1990s.

ADEN CANNON
The J-35F and J-35J had only a single internal M/55 ADEN 30mm (1.19in) belt-fed cannon with 90 rounds, fitted well aft on the starboard wing root. Earlier variants had twin ADEN guns.

DELTA WING
The Saab J-35's distinctive 'double delta' wing planform was first tested on a small research aircraft, the Saab 210, in the early 1950s. It still looks futuristic today, and its aerodynamic advantage was that it combined the low drag of the delta wing with astonishing low-speed manoeuvrability.

ORDNANCE
Even after upgrade to J-35J standards, Swedish Drakens continued to use the licence-built variants of the Hughes Falcon AAM, the Rb27 semi-active radar homing version and the Rb28 infrared homing version, as well as the AIM-9 Sidewinder.

J-35J DRAKEN
The J-35J Draken was an upgraded version of the J-35F for the Swedish Air Force. The programme was undertaken to keep two squadrons of Drakens operational during a period of delays with Saab's JAS 39 Gripen. To create the J 35J, the nose and cockpit were removed and rebuilt by Saab, while FFV refurbished the rest of the airframe. External changes included an infrared search-and-track sensor unit under the nose and two extra fuselage pylons, shown here carrying AIM-9J Sidewinders. The wing pylons have Rb28 (AIM-4C Falcon) missiles.

INNOVATIVE DESIGN
Two Drakens perform manoeuvres. Sweden's Saab company has become famous for its innovative fighter designs. One of its more unusual designs, the Draken (Dragon) was also the first Saab fighter to be exported in reasonable numbers.

English Electric Lightning (1960)

TYPE • *Interceptor* **COUNTRY •** *United Kingdom*

Out of all the world's air forces, only the RAF made the jump from subsonic to Mach 2 fighter with no Mach 1-plus intermediary, replacing the Hawker Hunter day fighter and the Gloster Javelin all-weather fighter with the Mach 2 English Electric (later BAC) Lightning.

SPECIFICATIONS (F.2A)

- **DIMENSIONS:** Length: 16.84m (5ft 3in); Wingspan: 10.61m (34ft 10in); Height: 5.97m (19ft 7in)
- **WEIGHT:** Empty 12,700kg (28,000lb); maximum take-off 22,680kg (50,000lb)
- **POWERPLANT:** Two 7112kg (15,682lb) thrust Rolls-Royce Avon 211R turbojets
- **MAX SPEED:** 2415km/h (1500mph)
- **RANGE:** 1287km (800 miles)
- **CEILING:** 18,920m (62,000ft)
- **CREW:** 1
- **ARMAMENT:** Two nose-mounted 30mm (1.19in) guns; two AAMs

Left: This Lightning F.Mk.2A, was essentially a hybrid, combining the fine handling of its F.Mk.2A origins with the extra features of the F.Mk.6.

INSIGNIA
The Lightning F.Mk.3 seen here wears the insignia of No. 111 (Treble One) Squadron, which was based at RAF Wattisham in Suffolk. The unit re-equipped with the Lightning F.Mk.1A in 1961, subsequently converting to the F.Mk.3 in 1964. Treble One continued to operate the Lightning until 1974, when it received the Phantom FG.1/FGR.2.

NOSE SECTION
The nose section was built in two halves to allow easy access for the installation of wiring, pneumatic and hydraulic lines. The windscreen and canopy frames were manufactured from heavy forgings, with stretched acrylic Perspex mouldings and an optically flat armoured glass windscreen.

ENDURANCE
With internal fuel alone and no in-flight refuelling, the Lightning had only about 50 minutes endurance.

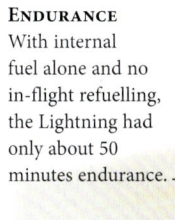

SPEED AND POWER
Flying the English Electric Lightning was like being saddled to a skyrocket. This great interceptor, with its sharply swept-back wings and tail, guarded Britain against air attack for many years. The Lightning was a creature of raw power and brute force – it had the incredible ability to climb to 10,000m (32,808ft) in two minutes – and was regarded with great affection by the pilots who flew it.

T.5 PROTOTYPE
An English Electric Lightning T.5 prototype (XM967) goes into a steep bank. The Lightning was a very fast-reacting interceptor. Even as the big jet reached old age in the 1990s, only fighters such as the F-15 and Su-27 could match its superb climb performance.

Convair B-58 Hustler (1960)

TYPE • Bomber COUNTRY • United States

SPECIFICATIONS (B-58A)	
Dimensions:	Length: 29.49m (96ft 9in); Wingspan: 17.32m (56ft 10in); Height: 9.58m (31ft 5in)
Weight:	Empty 25,200kg (55,560lb); maximum take-off 72,576kg (160,000lb)
Powerplant:	Four 67kN (15,600lbf) thrust General Electric J79-GE-5 turbojets
Max Speed:	2228km/h (1385mph)
Range:	8248km (5125 miles)
Ceiling:	19,500m (64,000ft)
Crew:	3
Armament:	One 20mm (0.79in) Vulcan six-barrel cannon; 8820kg (19,450lb) of nuclear or conventional bombs

The world's first operational supersonic bomber, the Convair B-58 Hustler was a bold departure from conventional design, having a delta wing with a conical-cambered leading edge and an area-ruled fuselage.

This page: This B-58 was assigned to the 305th Bomb Wing based at Bunker Hill AFB, Indiana.

Crew
The crew sat in individual cockpits, with the navigator behind the pilot and the defensive systems operator in the rear cockpit.

Rear cannon
Chasing a B-58 was a dangerous business because a Vulcan M61 20mm (0.79in) cannon was mounted in the tail.

Fuselage pod
The B-58's under-fuselage pod carried both fuel and weapons. Most of the fuel was carried in the lower component, which was jettisoned when the fuel was consumed; more fuel was contained in the upper component, which also housed a Mk 48 variable-yield nuclear bomb. Four more weapons could be carried on underwing pylons.

PRODUCTION
Convair built 116 B-58s – 30 test and pre-production aircraft and 86 for operational service – flying as part of the Strategic Air Command between 1960 and 1970. B-58s also won five different aviation trophies, setting 19 world speed and altitude records.

B-58A HUSTLER
A Convair B-58A Hustler, bulging with fuel and weapons. The Hustler was designed to cruise at Mach 2.2 at 21336m (70,000ft), but improved Soviet SAM defences forced the USAF to switch its B-58s to the low-level role.

Westland Wessex (1960)

TYPE • Helicopter **COUNTRY** • United Kingdom

Westland's Wessex performed solidly in every task a helicopter could undertake. The popular British-built version of the Sikorsky S-58 entered service in 1961 and from then it appeared on ship decks and battlefields in many different countries and climates.

Below: This HAS Mk 3 Wessex (nicknamed 'Humphrey') flew from the British destroyer HMS Antrim *in South Georgia during the Falklands War in 1982 and rescued an SAS patrol.*

SPECIFICATIONS (HC.Mk 2)

DIMENSIONS: Length: 20.04m (65ft 9in) rotors turning; Main rotor diameter: 17.07m (56ft); Height: 4.93m (15ft 3in)

WEIGHT: Empty 3767kg (8287lb); maximum take-off 6123kg (13,470lb)

POWERPLANT: Two coupled 1007kW (1350hp) Bristol Siddeley Gnome Mk 110 and Mk 111 turboshaft engines

MAX SPEED: 212km/h (132mph)

RANGE: 769km (477 miles)

CEILING: 3658m (12,000ft)

CREW: 2

ARMAMENT: Provision for 7.62-mm GPMGs, rocket pods or SS.11 or AS.12 missiles; naval versions carried two Mk 44 torpedoes or Mk 11 depth charges

ROTOR HEAD
The Wessex rotor head was an old-style conventional type with dragging and flapping hinges. The blade pitch was altered by control rods attached to rotating and non-rotating swash plates.

REAR FUSELAGE
Royal Navy Wessexes featured hinged rear fuselages for ease of stowage on ships.

ROTOR BLADES
The Wessex Mk 3 had four all-metal main rotor blades driven by a Napier Gazelle rather than the twin Gnomes of the Wessex Mk 2.

SIKORSKY ORIGINS
The Wessex was based on the Sikorsky H-34. It was intended to have a traditional piston engine, but Westland introduced the gas turbine Gazelle engine, which enhanced both performance and economy. With its stalky fuselage and jutting nose, the turbine-powered Wessex was easy to recognize. This ageing machine remained in operation as a tactical support and search-and-rescue helicopter until it retired from RAF service in 2004.

RAF SEARCH AND RESCUE
The bright yellow search-and-rescue colour scheme became one of the more familiar signatures of the Wessex in later years of British service.

Nanchang CJ-6 (1960)

TYPE • Trainer COUNTRY • China

First flown in 1958, the Nanchang CJ-6 is a Chinese development of the Yak-18 that has been in service since 1960. It remains the PLAAF's standard basic trainer over 60 years later; in 2023, at least 400 examples were on strength.

SPECIFICATIONS

DIMENSIONS:	Length 8.46m (27ft 9in); Wingspan: 10.22m (33ft 6in); Height: 3.3m (10ft 10in)
WEIGHT:	Maximum take-off 1400kg (3086lb)
POWERPLANT:	One 213kW (286hp) Zhuzhou HS-6A (Ivchenko AI-14) 9-cylinder, air-cooled radial piston engine
MAX SPEED:	370km/h (230mph)
RANGE:	700km (430 miles)
CEILING:	6250m (20,510ft)
CREW:	2
ARMAMENT:	Usually none but provision for two 7.62mm (0.3in) machine guns fixed forward-firing in nose

Right: The primary virtues of the CJ-6 are its simplicity and ease of flight and its rugged reliability.

POWERPLANT
The standard powerplant of the Nanchang CJ-6 is the 213kW (286hp) Zhuzhou HS-6A (Ivchenko AI-14) 9-cylinder, air-cooled radial piston engine.

PLAAF CJ-6
Here we see the CJ-6 in the standard colour scheme of the Chinese air force. Variants have also been built for agricultural top-dressing, aerial spraying and firefighting.

WIDE EXPORT
More than 3000 CJ-6s are believed to have been built, most of the improved CJ-6A models that substituted a 213kW (286hp) Zhuzhou Huosai radial engine in place of the 194kW (260hp) unit originally fitted and the aircraft was exported to several friendly nations such as Albania, Bangladesh, Cambodia, North Korea, Tanzania and Sri Lanka.
In its original role with the PLAAF, however, the CJ-6 has for some time been expected to be replaced by the Hongdu/Yakovlev CJ-7 primary trainer.

UNDERCARRIAGE
The aircraft is regularly confused with the Soviet Yak-18A, which also features a tricycle undercarriage, although the CJ-6 can easily be distinguished by its angular tail and rudder and pronounced dihedral on the outer wing panels.

CIVILIAN CJ-6
Its comparative cheapness, easy availability, reliability and impeccable flying characteristics have seen the CJ-6 become a widespread 'warbird', with several hundred civilian-owned examples airworthy worldwide.

Dassault Mirage III (1961)

TYPE • Fighter COUNTRY • France

Undertaken as a private venture, the delta-wing Mirage III became an enormous export success as well as arming the French Air Force. The aircraft fought in several conflicts in the mid- to late-twentieth century.

SPECIFICATIONS (Mirage IIICJ)

DIMENSIONS:	Length: 14.75m (48ft 5in) Wingspan: 8.22m (27ft) Height: 4.5m (14ft 9in)
WEIGHT:	12,700kg (27,998lb)
POWERPLANT:	SNECMA Atar 9C afterburning turbojet, 41.97kN (944 lbf)
MAX SPEED:	2112km/h (1320mph)
RANGE:	3335km (2072 miles)
CEILING:	17,000m (55,770ft)
CREW:	1
ARMAMENT:	2 × 30mm (1.1in) cannon plus 4000kg (8818lb) of external stores, typically 2 × Sidewinder or Magic heat-seeking missile plus 1 × Matra R.530 radar-guided missile

Below: Additional new equipment in service with the Pakistan AF at the time of the 1971 war were French Mirage IIIs, a single squadron of which became operational in June 1969. As well as missile-armed Mirage IIIEP interceptor versions (seen here), the 1971 war saw Pakistan make use of the Mirage IIIRP version for reconnaissance.

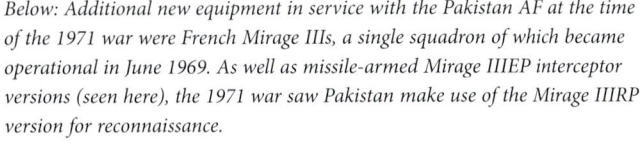

PILOT VISIBILITY
Pilot visibility ahead and to the side was good, although the rear view was limited especially when compared with fighters such as the F-16 and MiG-29.

DELTA WING
The high sweep of the Mirage III's delta wing resulted in reduced drag at high speeds, while enabling the aircraft to dispense with a conventional tail surface.

ENGINE
The Atar turbojet engine of the Mirage III was used in a variety of Mirage variants and also in the Etendard and Super Etendard naval strike aircraft.

EFFECTIVE SERVICE
The Mirage III was bought by a number of export clients and built under licence in several nations. A major user was Israel, whose pilots contributed to the Mirage's export success by demonstrating its effectiveness in the Six-Day War of 1967. More than 80 per cent of Arab aircraft downed in the conflict were at the hands of Mirages, although ground-attack performance was less impressive.

RAAF MIRAGES
Two Mirage III aircraft of the Royal Australian Air Force take off on a mission during the joint Australian, New Zealand and US (ANZUS) Exercise TRIAD '84.

North American A-5A Vigilante (1961)

TYPE • *Fighter* COUNTRY • *United States*

Designed to deliver a nuclear weapon in a low-level Mach 2 dash over the target, the A-5 Vigilante was a potent weapon in the US Navy's arsenal, however, it was as a reconnaissance aircraft that it proved its true worth.

SPECIFICATIONS	
DIMENSIONS:	Length: 23.11m (75ft 10in); Wingspan: 16.15m (53ft); Height: 5.92m (19ft 5in)
WEIGHT:	Empty 17,010kg (37,500lb); maximum take-off 36,094kg (79,588lb)
POWERPLANT:	Two 79.4kN (17,860lbf) thrust General Electric J79-GE-10 turbojets
MAX SPEED:	2230km/h (1385mph)
RANGE:	2415km (1500 miles)
CEILING:	14,752m (48,400ft)
CREW:	2
ARMAMENT:	Nuclear or conventional weapons in internal bomb bay

This page: The A3J-1 was redesignated A-5A in 1962. The aircraft seen here, 149283, was one of the last A-5As built and is depicted as it appeared on the first cruise by Heavy Attack Squadron 7 (VAH-7) aboard the USS Enterprise to the Mediterranean in 1962.

COCKPIT
In mock-up, the A3J-1's rear cockpit had a fully transparent canopy, but a blacked-out version proved better for viewing the radar display and gave better protection from a nuclear flash, so a solid canopy was proposed instead. Unhappy with this arrangement, navigator/bombardiers pressed for a compromise on production aircraft, and two small glazed panels were provided.

HANDLING
The A-5A's high approach speed made it tricky to handle around a carrier, and several aircraft and crews were lost in accidents. Handling was improved in the A-5B by fitting blown leading-edge flaps in place of blown trailing-edge surfaces.

WEAPON TUNNEL
First flown as the YA3J-1 on 31 August 1958, the Vigilante completed its carrier trials in July 1960. The aircraft was designed to carry either conventional or nuclear weapons in a tunnel inside the fuselage, the bombs being ejected rearwards between the two jet pipes. Fifty-seven A-5As were built, followed by 20 A-5Bs, an interim long-range variant.

POWERPLANT
Initial production A-5As were equipped with a pair of J79-GE-2 turbojets producing 71.62kN (16,100lbf) of thrust, but the later GE-10 upgrade increased this to 79.4kN (17,860lbf).

XA3J-1 VIGILANTE PROTOTYPE
The Vigilante's career as an attack bomber was relatively short-lived and the majority of A-5A and A-5B airframes were later converted to the RA-5C reconnaissance configuration.

Boeing RC-135 (1961)

TYPE • Reconnaissance COUNTRY • United States

SPECIFICATIONS (RC-135V)

DIMENSIONS:	Length: 41.53m (136ft 3in); Wingspan: 39.88m (130ft 10in); Height: 12.7m (41ft 8in)
WEIGHT:	124,965kg (275,500lb) loaded
POWERPLANT:	4 x 80kN (18,000lbf) Pratt & Whitney TF33-P-9 turbojets
MAX SPEED:	991km/h (616mph)
RANGE:	4305km (2675 miles)
CEILING:	12,375m (40,600ft)
CREW:	27
ARMAMENT:	6 x 7.92mm (0.3in) MGs; 1000kg (2205lb) bomb load

For decades the Boeing RC-135 – the aerial espionage cousin of the KC-135 tanker – has been in the vanguard of the secret world of reconnaissance, giving its crews hours of boring routine interrupted by seconds of sheer terror.

Right: A US Air Force RC-135 Rivet Joint reconnaissance aircraft, 343rd Reconnaissance Squadron, Offutt Air Force Base, Nebraska, moves into position for an aerial refuelling over Manas Air Base, Kyrgyzstan on 14 March 2006.

RADAR
The elongated nose contains a side-looking radar that provides an accurate picture of the coastline for precise navigation. This is very important when snooping close to a hostile country's airspace.

CHEEK FAIRINGS
Large cheek fairings on either side of the fuselage contain flat antennas. These 'listen out' across a wide range of frequencies for signals, which are analyzed by the onboard crew.

FUSELAGE BULGE
This bulge is inherited from the KC-135 tanker from which the RC-135 is derived. In the tanker it is used to mount the refuelling boom, but in the RC-135 it incorporates yet more antennas. Some RC-135s also have a downward-facing camera in the bulge.

SIGINT PLATFORM
The Boeing RC-135 strategic reconnaissance aircraft is the offspring of the Boeing KC-135 Stratotanker and is closely related to the spectacularly successful Boeing 707 airliner. Designed as an electronic eavesdropper, the RC-135 collects SIGINT (signals intelligence), including an enemy's radar emissions, radio communications or missile telemetry. The closely related EC-135 was an airborne command post for Strategic Air Command.

USAF RC-135
A US Air Force (USAF) RC-135 Rivet Joint, 343rd Reconnaissance Squadron, Offutt Air Force Base, Nebraska, taxies to the runway prior to take off, in support of an operational readiness exercise (ORE).

Boeing CH-47 Chinook (1962)

TYPE • Helicopter COUNTRY • United States

From Vietnam to Afghanistan, the Boeing Chinook has been the most successful Western tactical medium/heavy-lift helicopter. Fast, massively powerful and with a large rear loading ramp and long interior, the Chinook takes loads that other helicopters cannot.

SPECIFICATIONS (CH-47F)

DIMENSIONS:	Length: 30m (90ft); Rotor diameter: 18m (60ft); Height: 5.77m (18ft 11in)
WEIGHT:	22,680kg (50,000lb)
POWERPLANT:	Two 3529kW (4733hp) Lycoming T55-GA-714A turboshaft engines
MAX SPEED:	315km/h (196mph)
RANGE:	740km (460 miles)
CEILING:	6100m (20,000ft)
CREW:	3
ARMAMENT:	Optionally up to three pintle-mounted MGs, usually 7.62mm (0.3in) M240/FN MAG but can be armed with the 7.62mm (0.3in) M134 Minigun rotary MG

Left: A CH-37C of the Royal Moroccan Air Force, which received 12 CH-47Cs built by Meridionali in Italy. The Chinook was also operated in North Africa by Libya and Egypt.

MH-47 CHINOOK
The huge refuelling probe of the special operations MH-47 makes it easily distinguishable from standard Chinooks. This all-black MH-47E is fitted with advanced terrain-following radar and a FLIR (forward-looking infra-red) sensor, essential for night operations.

CABIN
Fifty-five equipped troops or 24 litters can be accommodated in the main cabin. Small vehicles can also be carried inside the main fuselage.

REAR PYLON
The rear pylon carries both engines and the gearbox synchronization unit. Chaff and flare dispensers and infra-red jammers can be mounted on the pylon.

TWIN ROTOR
Designed to meet a US Army requirement for a heavy-lift helicopter, the CH-47 Chinook first flew in 1962. It remains one of the few helicopters to successfully use the 'twin-rotor' layout. Each engine can drive both rotors if one fails, and a synchronization unit keeps the intermeshing rotors clear of each other. Vietnam proved that the Chinook was a superb performer. It could lift artillery pieces, trucks, fuel bladders and even shot-down UH-1 Hueys.

AFGHANISTAN OPERATIONS
A US Army CH-47 Chinook helicopter lands to refuel near the town of Khowst, Afghanistan, before transporting soldiers to Narizah to conduct a cordon search in that area during Operation Mountain Sweep, 2002.

Lockheed SR-71 'Blackbird' (1962)

TYPE • Reconnaissance COUNTRY • United States

SPECIFICATIONS

DIMENSIONS:	Length: 32.74m (107ft 5in); Wingspan: 16.94m (55ft 7in); Height: 5.64m (18ft 6in)
WEIGHT:	78,017kg (172,000lb) maximum take-off
POWERPLANT:	2 × 144.57kN (32,500lbf) Pratt & Whitney J58 afterburning turbojets
MAX SPEED:	Mach 3.35 at 24,385m (80,000ft)
RANGE:	5230km (2250 miles) at Mach 3, unrefuelled
CEILING:	24,385m (85,000ft)
CREW:	2
ARMAMENT:	None

First revealed in 1964, the 'Blackbird' remains the world's fastest air-breathing manned vehicle. During the Cold War years, in which it served as a strategic reconnaissance platform for the USAF, the Mach 3+ capable SR-71 was effectively immune to interception.

This page: The 64-17978, an SR-71A of the 9th Strategic Reconnaissance Wing, was the first of three 'Blackbirds' to be sent to the Far East during the initial deployment phase, departing Beale for Okinawa in March 1968.

CHINES
Aerodynamic sharp edges leading aft from either side of the nose along the fuselage provided extra lift, reduced drag and improved directional stability.

SENSOR BAYS
Four compartments each accommodated an interchangeable pallet carrying sensors including cameras, side-looking radar, infrared linescan and electronic intelligence receivers.

POWERPLANT
The Pratt & Whitney J58 engine was a scaled-down version of the JT9 developed for the XB-70 bomber. Uniquely, it was designed for sustained Mach 3 operations.

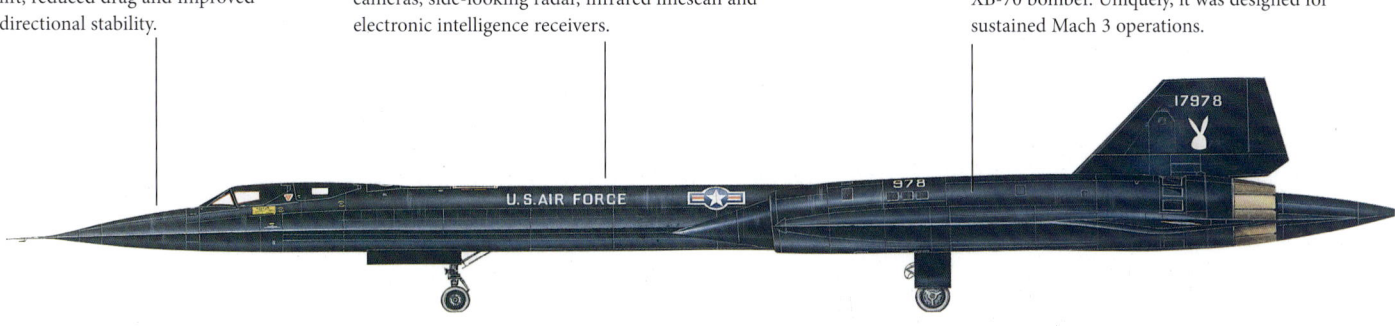

DESIGN FEATURES
The SR-71 had numerous innovative design features. First, its largely titanium construction allowed it to withstand kinetic heating and retain structural integrity at speeds beyond Mach 3. Integral lifting surfaces, known as chines, were built into the forward fuselage, preventing the nose from pitching down at high speed. At Mach 3, the engines were designed to provide just 18 per cent of total thrust, the remainder being derived from suction in the engine intakes and via special exhaust nozzles at the rear of the multiple-flow nacelles.

9TH SRW BLACKBIRD
An SR-71 Blackbird of the 9th Strategic Reconnaissance Wing flies through scattered clouds on a mission out of Beale Air Force Base, California. In order to cope with the aerodynamic drag found at Mach 3, the design employed a slim fuselage and very thin delta wings.

Lockheed P-3 Orion (1962)

TYPE • *Reconnaissance* COUNTRY • *United States*

A development of the Lockheed Electra airliner, the P-3 (formerly P3V-1) Orion was Lockheed's winning submission in a 1958 US Navy contest for a new off-the-shelf ASW aircraft that could be brought into service rapidly by modifying an existing type.

SPECIFICATIONS (P-3C)

DIMENSIONS:	Length: 35.61m (116ft 10in); Wingspan: 30.37m (99ft 8in); Height: 10.29m (33ft 8in)
WEIGHT:	Empty 27,982kg (61,700lb); maximum take-off 64,410kg (142,000lb)
POWERPLANT:	Four 3661kW (4910hp) Allison T56-A-14 turboprops
MAX SPEED:	761km/h (473mph) at 4570m (15,000ft)
RANGE:	3835km (2383 miles)
CEILING:	8625m (28,300ft)
CREW:	10
ARMAMENT:	Up to 8735kg (19,250lb) of ASW munitions

This page: The P-3C seen here served with VP-4 'Skinny Dragons' at Kaneohe Bay, Hawaii. VP-4 was active during Operation Desert Storm, flying from Masirah, Oman.

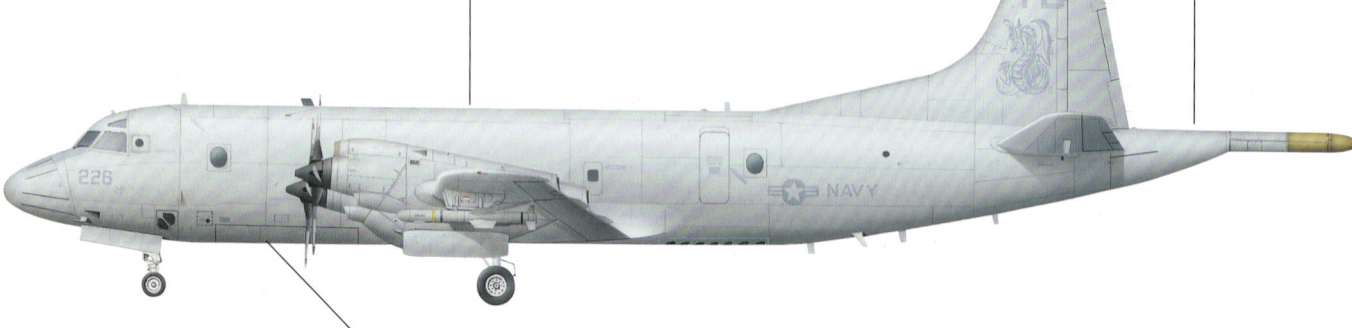

CREW
The standard Orion crew is ten: two pilots and a flight engineer on the flight deck; the tactical co-ordinator (Tacco), navigator/ communications operator (nav/comm) and three sensor operators in the forward part of the 'tube' area; the ordnanceman and in-flight technician in the rear.

TAILCONE
The tailcone has been adapted to house electronic equipment, namely the AN/ASQ-81 magnetic anomaly detector (MAD) for detecting and tracking submerged submarines.

WEAPONS BAY
The internal bay can carry up to eight 227kg (500lb) bombs, depth charges or mines, eight Mk 46 torpedoes or six Mk 50 Barracudas.

AIRLINER ADAPTATION
Adapting airliner designs to the maritime reconnaissance role has long been an inexpensive solution adopted by nations requiring an airborne sea-search capability. The Lockheed Orion, based on the civil Electra, offered four-engined reliability and the necessary range and 'loiter' time for hours of patrol duty. Capable of considerable modification and updating, the Orion has served the US Navy and other air arms for more than six decades.

PATROL AIRCRAFT
The land-based Lockheed P-3 Orion has been the premier maritime patrol and anti-submarine warfare aircraft with the US Navy and many other nations since its introduction in 1962.

Tupolev Tu-22 Blinder (1962)

TYPE • Bomber COUNTRY • Soviet Union

The 1961 annual Soviet air display came as a shock to Western aviation experts. For the first time the Russians revealed combat aircraft as advanced as, if not better than, anything that existed in the West. Among the revelations was a hitherto unknown supersonic bomber.

SPECIFICATIONS (Tu-22P Blinder E)

Dimensions: Length: 40.53m (132ft 11.67in); Wingspan: 23.75m (77ft 11in); Height: 10.67m (35ft)

Weight: Empty 40,000kg (88,200lb); maximum take-off 84,000kg (185,200lb)

Powerplant: Two 16,000kg (35,280lb) thrust Koliesov VD-7M turbojets

Max Speed: 1487km/h (924mph) at 6000m (19,685ft)

Range: 3100km (1926 miles)

Ceiling: 18,300m (60,040ft)

Crew: 3–4

Armament: One 23mm (0.906in) cannon

This page: The Tu-22P was the final production version of the Blinder and fulfilled much the same role as the GD/Grumman EF-111E, as an electronic support aircraft.

Aerial refuelling
This was one of the most difficult operations for an inexperienced crew. The Tu-22 flight controls were not delicate, and many inexperienced pilots found themselves closing in on the refuelling drogue too quickly, forcing them to abandon the link-up.

Handling
The Tu-22 did not have a reputation as a pilot's aircraft and could be very difficult and dangerous to fly. The aircraft could not be allowed to slow to less than 290km/h (180mph) on the approach, as this risked an uncontrollable pitch-up and stall.

Fuselage
The Tu-22's 'waisted' fuselage came about as a result of the drag-reducing 'area ruling' parameters built into its design. A supersonic (Mach 1.5) dash capability was essential for the Tu-22B bomber's original attack profile, and the ability to launch missiles at supersonic speed was important for the Tu-22K.

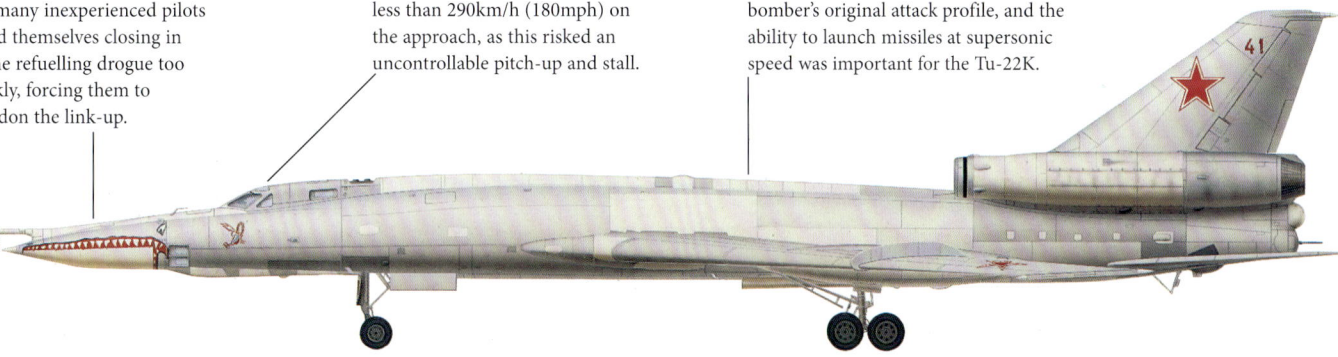

ADVANCED AIRFRAME
The Tupolev Tu-22 'Blinder' is now an obsolete design, but its sleek airframe was highly advanced when it entered service. One of the fastest bombers of its time, the 'Blinder' was designed with fighter-like aerodynamics, and could dash through hostile defences to deliver a heavy load of conventional or nuclear weapons. Export 'Blinders' saw action in Africa and in the Iran-Iraq War, on long-range strikes.

SOVIET-ERA TU-22
An impressive design, the Tu-22 'Blinder' remains an almost unique example of a tail-engined jet bomber. It was designed with an area-ruled fuselage and sharply swept wing to reduce supersonic drag.

Shenyang J-6/F-6 (1962)

TYPE • Fighter COUNTRY • China

Chinese production of the Soviet MiG-19 began at Shenyang and Nanchang in the late 1950s. The first aircraft actually to fly were assembled from Soviet parts. The first true Chinese-built J-6 followed in September 1959, with series production beginning soon after.

SPECIFICATIONS (F-6)

DIMENSIONS: Length: 14.9m (48ft 10.5in); Wingspan: 9.2m (30ft 2.25in); Height: 3.88m (12ft 8.75in)

WEIGHT: 10,000kg (22,046lb) maximum

POWERPLANT: 2 x 31.9kN (7165lbf) Shenyang WP-6 turbojets

MAX SPEED: 1540km/h (957mph)

RANGE: 1390km (864 miles)

CEILING: 17,900m (58,725ft)

CREW: 1

ARMAMENT: 3 x 30mm NR-30 cannon; four external hardpoints for up to 500kg (1102lb) of stores, including air-to-air missiles, 250kg (551lb) bombs, 55mm (2.1in) rocket-launcher pods, 212mm (8.34in) rockets or drop tanks

Below: The Shenyang F-6 (MiG-19SF 'Farmer') was the export variant of the J-6. This Pakistani F-6, one of 135 delivered, was reportedly one of the first to be received by the Pakistan AF, in 1966.

AIR INTAKE
In common with early Mikoyan-Gurevich jet fighters, the J-6 features a large splitter plate at the mouth of the intake. This is contoured for more effective air flow to the twin turbojet engines.

CANOPY
A flat, almost razor-straight canopy profile identifies the two-seat 'Farmer'. The twin Shenyang ejection seats were mounted high up, requiring tall pilots to wear old-style leather flying helmets instead of modern 'bone domes'.

J-6 PRCAF
Entering service in the early 1960s, PLA J-6 fighters (licence-built versions of the MiG-19) were involved in a number of Cold War skirmishes against Nationalist Chinese and other air arms.

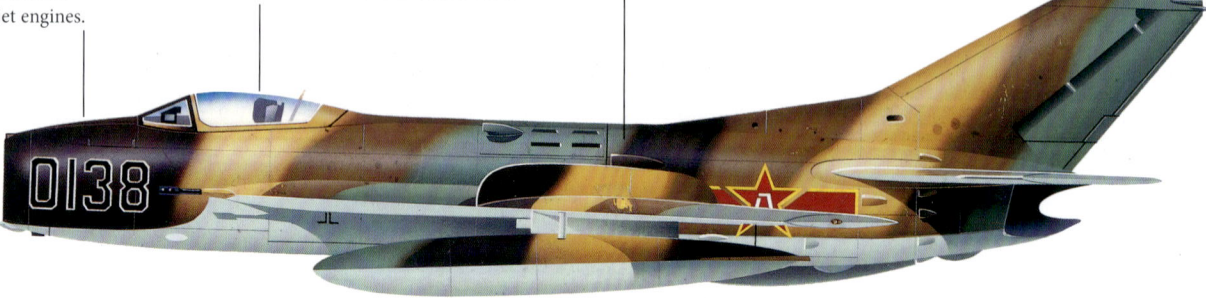

VARIANTS

Although the J-6 became the standard fighter of the Chinese air force from 1962, the Cultural Revolution of the late 1960s badly hampered production. As a result, it was 1973 before the first new model was delivered. Designated J-6III, it was a high-speed day fighter with cropped wings and more powerful engines. The J-6C was an improved version of the basic J-6 and featured, among other things, a relocated brake parachute. China's Guizhou factory used this variant as the basis for the all-weather J-6A.

PAF F-6 ON DISPLAY
A retired Shenyang F-6 fighter aircraft of Pakistan Air Force is displayed in the city of Lahore, Pakistan. F-6s were exported to Pakistan, Egypt, Vietnam, Bangladesh, Cuba, Tanzania and Somalia, among others.

Grumman A-6 Intruder (1963)

TYPE • Ground-attack **COUNTRY** • United States

One of the primary weapon systems of the 1991 Gulf conflict, the 1960s-vintage Grumman A-6 Intruder provided tactical bombing and close support airpower from American aircraft carriers and land bases.

SPECIFICATIONS (A-6A)

DIMENSIONS:	Length: 16.64m (54ft 7in); Wingspan: 16.15m (53ft); Height: 4.93m (16ft 2in)
WEIGHT:	18,918kg (41,715lb) loaded
POWERPLANT:	Two 41.3kN (9300lbf) Pratt & Whitney J52-P-8A turbojets
MAX SPEED:	1016km/h (631mph)
RANGE:	3254km (2021 miles)
CEILING:	14,480m (47,500ft)
CREW:	2
ARMAMENT:	Five external hardpoints for up to 8165kg (18,000lb) of ordnance

Below: The aircraft here is the KA-6D tanker version of the Intruder. These conversions of the A-6A had most of their bombing and weapon systems removed, including the radar, although in theory they retained a daylight bombing capability.

TRAM
The target recognition and attack multi-sensor (TRAM) turret contained a forward-looking infra-red (FLIR) sensor and a laser designator.

CREW
The two-man crew consisted of a pilot and a bombardier/navigator. The main part of the avionics suite was the Norden AN/APQ-148 multi-mode radar.

FUEL CAPACITY
Maximum internal fuel capacity is 8873 litres (2344 Imp gal), to which can be added up to five tanks of 1135 litres (300 Imp gal) or 1514 litres (400 Imp gal) capacity each. An air-to-air refuelling probe is fitted forward of the cockpit.

BOMBLOAD
The A-6 is capable of carrying all US and NATO air-to-ground weapons within its five external store stations – a total payload of 8165kg (18,000lb), including 28 × Mk 82 227kg (500lb) general purpose bombs (illustrated above).

WINNING DESIGN
Designed as a carrier-based low-level attack bomber with the ability to deliver both nuclear and conventional warloads with pinpoint accuracy in all weathers, the Grumman A-6 was the winner among 11 competitors in a US Navy design contest in 1957.

General Dynamics F-111 (1964)

TYPE • Bomber COUNTRY • United States

A controversial aircraft, after a troubled development the F-111 suffered high-profile combat losses early on. Once refined, however, it became the most capable precision attack aircraft of its generation, seeing combat over Vietnam, Libya and Iraq.

SPECIFICATIONS (F-111F)

DIMENSIONS:	Length: 22.4m (73ft 6in); Wingspan: 19.2m (63ft) wings spread; Height: 5.22m (17ft 1in)
WEIGHT:	45,359kg (100,000lb) maximum take-off
POWERPLANT:	2 × 111.65kN (25,100lbf) Pratt & Whitney TF30-100 afterburning turbofans
MAX SPEED:	2655km/h (1650mph)
RANGE:	4707km (2925 miles)
CEILING:	18,290m (60,000ft)
CREW:	2
ARMAMENT:	Up to 14,228kg (31,500lb) of disposable stores carried in a lower-fuselage weapons bay and on 6 underwing pylons

Below: This General Dynamics F-111F Aardvark was seen at San Diego, California, in the early 1990s.

FLAPS
The entire trailing edge of the variable-geometry wing was fitted with a powerful double-slotted flap, used to increase lift and, at high angles, drag.

RADOME
This housed the primary attack radar, a multimode General Electric AN/APQ-114, as well as the Texas Instruments AN/APQ-146 terrain-following radar.

INTERNAL BAY
The internal weapons bay was used mainly for carriage of additional fuel or equipment, including the Pave Tack electro-optical targeting pod in the case of the F-111F.

REFUELLING
An air-to-air front view of an F-111 aircraft during a refuelling mission over the North Sea off the coast of Britain. Above the fuselage was a receptacle for an in-flight refuelling 'flying boom'. Provision was not made for the probe and drogue method of aerial refuelling.

CONTROVERSY

The early career of the F-111 was hugely controversial due to its protracted development and the failure of the F-111B version, which was intended for the US Navy's fleet defence role. The F-111A of the US Air Force (USAF) first flew in December 1964 and made its combat debut in Vietnam in 1968. The initial deployment was notable for its 50 per cent loss rate, but the F-111, nicknamed the 'Aardvark', soon proved the most effective long-range strike platform, using its terrain-following radar to navigate to and attack important targets at night and in bad weather.

Grumman E-2 Hawkeye (1964)

TYPE • Reconnaissance COUNTRY • United States

SPECIFICATIONS (Lockheed E-2C)

DIMENSIONS:	Length: 17.56m (57ft 7in); Wingspan: 24.58m (80ft 7in); Height: 5.58m (18ft 4in)
WEIGHT:	23,391kg (55,000lb) loaded
POWERPLANT:	2 x 3800kW (5096hp) Allison T56-A-425 or -427 turboprop
MAX SPEED:	604km/h (375mph)
RANGE:	2583km (1605 miles)
CEILING:	9300m (30,800ft)
CREW:	5
ARMAMENT:	None

The US Navy's principal electronic surveillance aircraft in the Gulf War, and the mainstay of the US Navy's early warning capability for many years, was the Grumman E-2 Hawkeye, the prototype of which first flew on 20 October 1960.

Below: The E-2 fleet lagged behind the rest of the US Navy in adopting low-visibility schemes. The example shown here is from VAW-126 'Seahawks', and is illustrated as it appeared when operating from USS Kennedy.

LAYOUT
The E-2's airframe needs to be large to contain the equipment, operators and fuel for the long-range AEW mission, yet the aircraft must still fit a carrier hangar deck. The wings are designed to fold and rotate through 90 degrees to lie parallel to the rear fuselage.

RADAR DOME
The E-2's APS-145 radar sweeps a 3-million cubic mile envelope of air space, while simultaneously plotting the positions of surface vessels. Up to 2000 targets can be tracked at one time, thanks to high-speed processing and long-range automatic track initiation, and 40 separate intercepts handled.

TAIL FINS
The unusual fin arrangement of the Hawkeye stems from the requirement to provide sufficient keel area yet still fit the aircraft into the cramped confines of the carrier hangar. The port inner fin is the only one not fitted with a double-hinged rudder.

AWACS
Designed as a flying radar station, the Hawkeye became the US Navy's eye in the sky. Sometimes called 'the affordable AWACS' (airborne warning and control system), it was just what the US Navy requires to guard its aircraft-carrier battle groups and to direct friendly warplanes when the action begins. This twin-engined aircraft, with its long, slender wing, huge tail and saucer-shaped rotodome, is still a familiar sight in the US Navy.

US NAVY HAWKEYE
An early E-2A Hawkeye onboard USS *Oriskany*'s (CV-34) starboard catapult, October 1964.

Tupolev Tu-128 (1964)

TYPE • *Interceptor* COUNTRY • *Soviet Union*

The largest interceptor to enter service anywhere in the world, the Tupolev Tu-128 (NATO codename 'Fiddler') was tailored for long-range patrol missions in defence of the USSR's vast and featureless northern and Arctic frontiers.

SPECIFICATIONS

DIMENSIONS:	Length: 27.20m (89ft 4in); Wingspan: 18.10m (59ft 5in); Height: 7m (23ft)
WEIGHT:	40,000kg (88,185lb) loaded
POWERPLANT:	2 x 107.9kN (24,300lbf) Lyulka AL-7F-2 turbojets
MAX SPEED:	1740km/h (1089mph)
RANGE:	3200km (2000 miles)
CEILING:	18,000m (59,055ft)
CREW:	2
ARMAMENT:	4 x Bisnovat R-4 air-to-air missiles

Below: A powerful radar scanner in the nose enabled the Tu-128 to locate intruders at long ranges. Twin afterburning engines enabled it to reach speeds of up to 2085 km/h (1296 mph) and it could take off with enough fuel for patrols lasting up to six hours.

AVIONICS
Avionics included an AP-7P autopilot, an NVU-B1 navigational complex and a Put-4 flight control system giving semi-automatic level flight guidance, airfield homing, altitude/heading hold, auto runway approach and auto-return to a pre-programmed position.

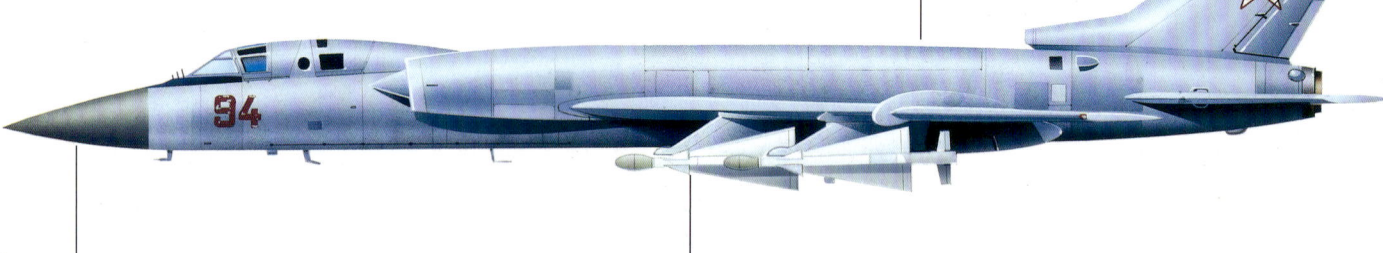

RADOME
The large radome contained an RP-5 radar as part of the Smerch fire-control system. The prototype aircraft had a large ventral fairing containing a receiver antenna for target echo-reception.

MISSILE ARMAMENT
Four wing pylons carried R-4T (later R-4TM) infra-red and R-4R (later R-4PM) radar-guided air-to-air missiles. Generally, the IR weapons were carried inboard, with the radar-guided weapons outboard. The weapons had to be loaded using a specially built powered trolley because they were so heavy.

FULLY LOADED
The aircraft's huge size enabled it to carry all the equipment needed to intercept bombers before they could penetrate to within their missiles' effective range.

BOMBER INTERCEPTOR
The Tu-128 was developed in the late 1950s to defend the Soviet Union against the threat of B-52 bombers armed with long-range stand-off missiles. This enormous interceptor was based on an unsuccessful bomber design, the Tu-98 'Backfin', and carried long-range missiles designed specifically for it. The Tu-128's highly specialized role meant that only limited numbers were built, and it was replaced by the MiG-31 and Su-27.

Lockheed C-141 StarLifter (1965)

TYPE • Transport COUNTRY • United States

Designed in the early 1960s, the Lockheed C-141 StarLifter was the most numerous of Military Airlift Command's strategic transport aircraft. The aircraft were delivered to the US Air Force between April 1965 and February 1968.

SPECIFICATIONS (C-141B)

DIMENSIONS:	Length: 51.29m (168ft 3.5in); Wingspan: 48.74m (159ft 11in); Height: 11.96m (39ft 3in)
WEIGHT:	Empty 67,186kg (148,120lb); maximum take-off weight 155,582kg (343,000lb)
POWERPLANT:	Four 93.4kN (21,000lbf) Pratt & Whitney TF33-7 turbofans
MAX SPEED:	912km/h (567mph)
RANGE:	4723km (2935 miles) with maximum payload
CEILING:	12,879m (42,250ft)
CREW:	5–6
ARMAMENT:	None

Below: All 270 surviving C-141As were converted in the late 1970s to C-141B standard by stretching the fuselage by 7.11m (23ft 4in). The aircraft saw service in Vietnam, Grenada and in the 1991 Gulf War, before being retired in 2006.

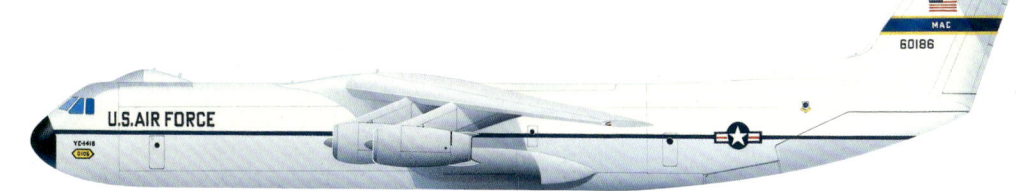

IN-FLIGHT REFUELLING
The raised area above the forward fuselage housed the in-flight-refuelling receptacle, into which the tanker's boom is connected to pass fuel to the StarLifter.

C-141B STARLIFTER
Modification of the C-141A to the C-141B involved the insertion of newly fabricated fuselage sections ahead of and behind the wing, resulting in a stretch of 7.11m (23ft.). This gave an increase in the volume of cargo that could be carried. The C-141B also had a new, more streamlined wingroot fairing and in-flight-refuelling capability.

REAR DOORS
The large rear ramp doors could be opened fully in flight for aerial load dropping, while a built-in loading ramp could be extended and lowered for vehicle access when on the ground.

HEAVY LIFT
Lockheed's C-141 StarLifter was the heavy muscle of the American military air transport fleet. For more than 40 years the StarLifter supplied US military and sometimes civilian installations around the world with vital supplies and reinforcements. In times of crisis the C-141 could, in partnership with the massive C-5, fully equip an entire army in a matter of days anywhere in the world.

TROOP DEPLOYMENT
Marines from Camp Pendleton wait in line to board a C-141 bound for Saudi Arabia. The aircraft was the backbone of the 1991 Desert Shield airlift, the entire force of 85 Starlifters being used to carry 42 per cent of all air-delivered cargo.

Sukhoi Su-15 (1965)

TYPE • Interceptor **COUNTRY** • Soviet Union

SPECIFICATIONS	
DIMENSIONS:	Length: 21.33m (70ft); Wingspan: 8.61m (28ft 3in); Height: 5.1m (16ft 8in)
WEIGHT:	19,000kg (41,890lb) loaded
POWERPLANT:	2 x 66.8kN (13,669lbf) Tumanskii R-11 turbojets
MAX SPEED:	1180km/h (733mph)
RANGE:	1380km (860 miles)
CEILING:	20,000m (65,615ft)
CREW:	1
ARMAMENT:	4 x external pylons for two R-8M medium-range air-to-air missiles outboard and 2 x AA-8 'Aphid' short-range AAMs inboard, plus 2 x pylons for 23mm (0.9in) UPK-23 cannon pods or drop tank

The Su-15 'Flagon' gained notoriety for its role in the destruction of a Korean Airlines 747 airliner in September 1983. The twin-engined, all-weather interceptor was part of a new wave of equipment introduced by the IA-PVO in the 1970s.

Right: The original conical radar nose radome was replaced in later versions by one of ogival shape and housed the improved 'Taifun-M' interception radar. The entire avionics system could be linked to ground control stations for automated interceptions.

FORWARD FUSELAGE
Forward fuselage is circular in section, with a bubble cockpit canopy. The side ram-type air intakes are rectangular with blow-in auxiliary intakes ahead of the wing.

TURBOJETS
The two afterburning Tumanskii turbojets were mounted side by side in the rear fuselage and had protruding variable-area nozzles below the rudder. Each engine drove separate electrical and hydraulic systems.

UNDERWING STORES
Underwing external stores were carried by a single pylon on early versions and two pylons on later 'Flagons'. To supplement the AA-3 missile, four infrared close-range R-60 AA-8 'Aphid' air-to-air missiles could also be fitted on twin PD-62 pylons. There is provision for UPK gun pods under the fuselage.

DESIGN REQUIREMENT
Designed to fully meet the demanding requirement for an all-weather interceptor that would be able to remain operational in the most adverse conditions, the Sukhoi Su-15 was a conventional interceptor. The Sukhoi Design Bureau decided not to adopt a variable geometry wing, which was internationally favoured at that time. Instead, the Su-15 emerged as a needle-nosed single-seater incorporating some 'tried and tested' systems and equipment, including Tumanskii engines.

SU-15 WITH R-98M AA-3 MISSILE
The twin-turbojet, delta-winged Su-15 was first flown on 30 May 1962. It took nearly 10 years of development before it was fully operational as an all-weather air-defence fighter. More than 1500 had been built when production ended in 1979.

Transsall C-160 (1965)

TYPE • *Transport* **COUNTRY** • *France/Germany*

SPECIFICATIONS (C-160NG)

DIMENSIONS:	Length: 32.40m (106ft 3in); Wingspan: 40m (131ft 3in); Height: 11.65m (38ft 5in)
WEIGHT:	16,000kg (35,270lb) loaded
POWERPLANT:	Two 4548kW (6100hp) Rolls-Royce Tyne RTy.20 Mk 22 turboprops
MAX SPEED:	536km/h (333mph)
RANGE:	4558km (2832 miles)
CEILING:	8500m (27,900ft)
CREW:	4
ARMAMENT:	None

Built by the Franco-German Transporter Allianz consortium over a 20-year period, the Transall C-160 became the main airlifter for L'Armée de l'Air and the Luftwaffe. Other military users were Turkey and South Africa, and total production numbered more than 200 aircraft.

Below: The Transall C-160 tactical transport was designed and produced as a joint venture between France and Federal Germany, Transall being an abbreviation of the specially formed consortium Transporter Allianz, comprising the companies of MBB, Aerospatiale and VFW-Fokker.

FLIGHT DECK
The flight-deck crew of the C-160 consists of a pilot, co-pilot and flight engineer. The aircraft is fully pressurized and air-conditioned in flight and on the ground.

TURBOPROPS
Two Rolls-Royce Tyne turboprops power all Transall C-160s. This powerplant was also used in the Atlantic maritime patrol aircraft. Two four-bladed British Aerospace Dynamics propellers are fitted.

MAIN CABIN
The main cabin of the Transall will hold up to 93 troops, 61 to 88 fully equipped paratroops or 62 stretchers and four attendants. Other typical loads include armoured vehicles up to 16,000kg (35,200lb) in weight.

VARIANTS

The principal variants were the C-160A, consisting of six pre-series aircraft; the C-160D for the Luftwaffe (90 built); the C-160F for France (60 built); the C-160T (20 built for export to Turkey); and the C-160Z (nine built for South Africa). Production of a second series was authorized in 1977. The new version was designated the C-160NG (Nouvelle Generation) and was fitted with improved avionics and a reinforced wing with additional fuel tanks.

FRENCH C-160
A French Air Force Transall C-160 taxis in to pick up paratroopers for the D-Day observation jump near Djibouti-Ambouli International Airport, Djibouti on 6 June 2018.

Chengdu J-7/F-7 (1965)

TYPE • *Fighter* COUNTRY • *China*

For many years, the backbone of the PLAAF fighter fleet was provided by the J-7, a locally produced version of the ubiquitous Soviet-designed Mikoyan-Gurevich MiG-21. Today, more advanced versions of the J-7 remain in PLAAF service in only very limited numbers.

SPECIFICATIONS

DIMENSIONS:	Length: 14.88m (48ft 10in); Wingspan: 8.32m (27ft 4in); Height: 4.11m (13ft 6in)
WEIGHT:	Maximum take-off 7540kg (16,623lb)
POWERPLANT:	One Liyang Wopen-13F afterburning turbojet, 44.1kN (9,900lbf) thrust dry
MAX SPEED:	2200km/h (1400mph)
RANGE:	850km (530 miles)
CEILING:	17,500m (57,400ft)
CREW:	1
ARMAMENT:	Two 30mm (1.18in) cannon; five hardpoints: four underwing, up to 500kg (1100lb) each, one centreline underfuselage 2000kg (4400lb) maximum; 55mm (2.1in) rocket pod (12 rounds), 90mm (3.5in) rocket pod (7 rounds)

Left: Because of its simplicity and low unit cost, the F-7 has proved an ideal aircraft for small developing world air forces such as that of Sri Lanka. This is an F-7 in use by Pakistan, the main export customer for the aircraft.

CANOPY
Many export F-7s have been fitted with Martin-Baker Mk 10 zero-zero ejection seats. Whereas early J-7I aircraft had single-piece canopies (as on the MiG-21F-13), later machines have more conventional two-piece units.

CHENGDU J-7L
This PLAAF J-7, serial number 21002, was seen at Dalian Zhoushuizi International Airport, China, in 2016.

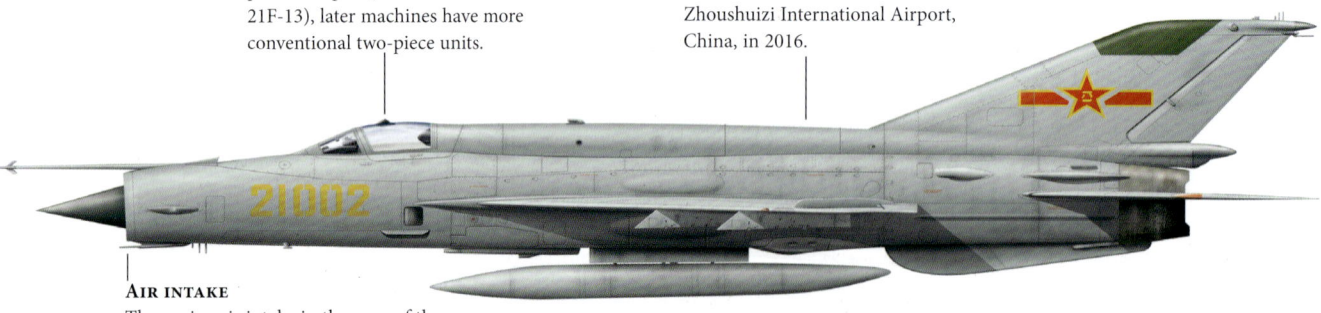

AIR INTAKE
The engine air intake in the nose of the aircraft features a variable shockcone, as in the MiG-21. Computer controlled, this is fully variable and houses a basic radar set.

VARIANTS

Flown for the first time in January 1966, the Shenyang J-7 was a Chinese version of the Soviet MiG-21. By then Shenyang was working on the J-8, so development was transferred to Chengdu, where derivatives have been in production ever since. The J-7I, designated F-7B for export, was produced in small numbers before giving way to the J-7II, with an improved WP-7B engine. With Western avionics the J-7II became the F-7M Airguard. Development continued with the F-7MG, aimed at the export market.

PRODUCTION J-7II
After its maiden flight in December 1978, the J-7II was approved for service in September 1979. The demand for the new fighter was such that a new production line was opened by Guizhou. Between them, Chengdu and Guizhou produced around 475 J-7IIs up until 1986.

Sikorsky S-65/CH-53 Sea Stallion (1966)

TYPE • Helicopter COUNTRY • United States

The US Marines provided the inspiration for the largest and most powerful helicopter in the world outside the Soviet Union. One of the rotary-wing marvels of the Vietnam era, the Sikorsky S-65 first flew in October 1965.

SPECIFICATIONS (CH-53G)

Dimensions:	Length: 26.97m (88ft 6in); Rotor diameter: 22.01m (72ft 3in); Height 7.6m (24ft 11in)
Weight:	15,227kg (33,500lb) loaded
Powerplant:	2 x 2927kW (3925shp) General Electric T64-GE-413 turboshaft
Max Speed:	395km/h (196mph)
Range:	1000km (620 miles)
Ceiling:	5106m (16,750ft)
Crew:	3
Armament:	2 x 7.62mm (0.3in) MG3 MGs

Below: The CH-53 Sea Stallion was designed primarily for the United States Marine Corps, which uses the type as its primary heavy-lift assault helicopter. The CH-53D is illustrated here.

GE TURBOSHAFTS
Twin General Electric T64 turboshafts are mounted in pods on each side of the central gearbox.

GERMAN CH-53G
The German Army and Luftwaffe provided a significant transport force for NATO's battlefield commanders. Heavy-lift CH-53Gs served with frontline Heeresflieger regiments at Rheine-Bentlage, Lauphiem and Niedermendig.

LIFT CAPACITY
The latest version of the CH-53, the CH-53K King Stallion, has a lift capability of 12,247kg (12,000lb), which it can transport over a mission radius of 203km (126 miles)

ASSAULT TRANSPORT
The Sikorsky S-65's dynamic parts (rotor, gearboxes and control system) were developed from those of the earlier S-64 SkyCrane and made extensive use of titanium. Fitted with folding rotor blades for shipboard stowage and given the designation CH-53 Sea Stallion by the US Marines, the S-65 emerged as the world's most capable assault transport.

CH-53E SUPER STALLION
US Marines assigned to Marine Heavy Helicopter Squadron 465 exit a CH-53E Super Stallion helicopter at Naval Air Station Lemoore, California, 30 June 2008. Marines with the unit were conducting firefighting support operations as part of the 302nd Air Expeditionary Group.

Bell OH-58 Kiowa (1966)

TYPE • Helicopter COUNTRY • United States

SPECIFICATIONS (OH-58D)

DIMENSIONS:	Length: 12.85m (42ft 2in), Rotor diameter: 10.67m (35ft); Height: 3.93m (12ft 10in)
WEIGHT:	2495kg (5500lb)
POWERPLANT:	485kW (650hp) Allison T703-AD-700A turboshaft
MAX SPEED:	240km/h (149mph)
RANGE:	556km (161 miles)
CEILING:	4575m (15,000ft)
CREW:	2
ARMAMENT:	One 12.7mm (0.5in) machine-gun; seven-tube 70mm (2.75in) rocket pods, plus provision for Stinger air-to-air missiles and Hellfire anti-armour missiles

The Kiowa is the military version of Bell's extremely successful Jet Ranger. First flown in 1964, the D-250, the helicopter that would become the OH-58, was developed in response to a US Army request in 1960 for a light observation helicopter.

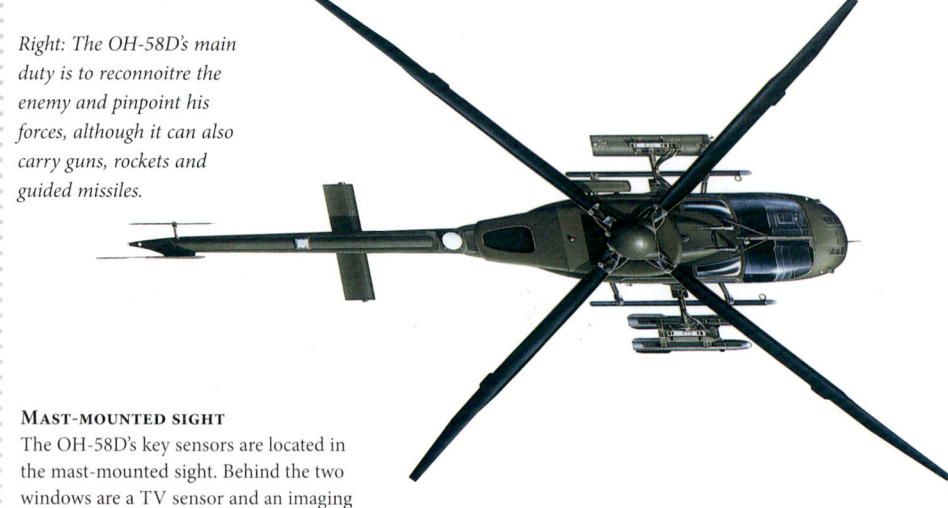

Right: The OH-58D's main duty is to reconnoitre the enemy and pinpoint his forces, although it can also carry guns, rockets and guided missiles.

MAST-MOUNTED SIGHT
The OH-58D's key sensors are located in the mast-mounted sight. Behind the two windows are a TV sensor and an imaging infra-red, which provide targeting and navigational information in all light and weather conditions.

WIRE CUTTERS
High-tension wires can cause a helicopter to crash if its rotors become entangled. Wire-cutters are fitted above and below the cockpit to deal with these hazards.

DEFENSIVE TURRET
The ALQ-144 IRCM turret, placed just behind the cabin and the powerplant, provides protection against heat-seeking missiles.

OH-58D
The OH-58 has enjoyed a long and successful career as a special forces' asset. By the mid-1980s, the Kiowa had been updated to OH-58D Kiowa Warrior standard, an armed variant of the original helicopter, capable of carrying combinations of Hellfire missiles, air-to-air Stinger missiles, seven-shot 70mm (2.75in) Hydra-70 rocket pods and an M296 12.7mm (0.50in) calibre machine gun.

OH-58D IN IRAQ
An OH-58D Kiowa Warrior armed reconnaissance helicopter from 2nd Squadron, 17th Cavalry Regiment, lands on Forward Operating Base Warrior near Kirkuk in Iraq.

Cessna O-2 Skymaster (1966)

TYPE • Utility   COUNTRY • United States

SPECIFICATIONS

DIMENSIONS:	Length: 9.07m (29ft 8in); Wingspan: 11.63m (38ft 2in); Height 2.79m (9ft 2in)
WEIGHT:	Empty 1292kg (2848lb); loaded 2448kg (5400lb)
POWERPLANT:	Two 157kW (210hp) Continental IO-360C flat-6 piston engines
MAX SPEED:	322km/h (200mph)
RANGE:	2132km (1325 miles)
CEILING:	5940m (18,000ft)
CREW:	1–2
ARMAMENT:	Pylons for various munitions, including rockets, flares and 7.62mm (0.3in) Minigun pack

Over the years many civilian light aircraft have found military applications – the famous Piper Cub of pre-World War II vintage was a classic example. In another war, the military version of the Cessna 337 also found its uses.

This page: The black paintwork of this O-2A shows that it was engaged in night operations along the Ho Chi Minh Trail.

CREW
O-2A FAC aircraft were flown both solo and with an observer/artillery spotter, depending on the mission. Visibility was relatively poor compared to the O-1, which had a narrow fuselage and tandem seating. A cut-out window in the starboard door improved the view for a solo pilot, who flew from the left-hand seat.

COMMUNICATIONS
Pilots referred to the O-2 as a 'flying radio set', equipped as it was with a transceiver for communication with USAF and USN strike aircraft and with friendly ground forces. This equipment included VHF (FM-662 and Wilcox 807) radio sets.

O-2A SKYMASTER
The Cessna Model 337's unusual push-and-pull engine configuration gave it twin-engined power with single-engined drag and handling characteristics. As the O-2A, it made an effective FAC platform. This one was based at Nakhom Phanom, Thailand, in 1970.

VIETNAM SURVIVOR
The Cessna O-2 was a military version of the Cessna Model 337 Super Skymaster – in 1966 the USAF selected the O-2 military variant. Its twin tail booms and tractor-pusher (pullpush) propeller arrangement were distinctive. In Vietnam, the twin engines enabled the O-2 to absorb ground fire and still get home safely, making it well liked by its crews.

O-2A SKYMASTER
A Cessna O-2A Skymaster in original US Air Force livery. Two O2 models were produced. The O-2A was equipped with wing pylons to carry rockets, flares and other ordnance, and was used for marking enemy targets and co-ordinating air strikes. The O-2B was a psychological warfare aircraft equipped with loudspeakers and leaflet dispensers.

Cessna A-37B Dragonfly (1966)

TYPE • Attack aircraft COUNTRY • United States

Derived from the T-37 'Tweety Bird' trainer, the A-37 Dragonfly (or 'Super Tweet') was a cheap, reliable and hard-hitting light attack platform that saw action in numerous conflicts in battlefields from Vietnam to Latin America.

SPECIFICATIONS (A-37B)

Dimensions:	Length: 8.93m (29ft 2.5in) excluding probe; Wingspan: 10.93m (35ft 10.5in)
Weight:	Maximum take-off 6350kg (14,000lb)
Powerplant:	Two 12.68kW (2850lb) General Electric J85-GE-17A turbojets
Max speed:	816km/g (507mph)
Range:	1628km (1012 miles)
Ceiling:	12,730m (41,765ft)
Crew:	2
Armament:	Up to 1860kg (4100lb) of ordnance on eight underwing pylons

Right: A US-registered Cessna A-37 Dragonfly performs during an air show. Several Latin American countries still operate the A-37, including Colombia, El Salvador, Guatemala, Honduras, Peru and Uruguay.

Tip tanks
The prototype, designated YAT-37D, flew in October 1963 and was followed by a batch of T-37Bs converted to A-37As, with an internal 20mm (0.79in) minigun, wing pylons and tip tanks.

Airframe
The definitive A-37B had a significantly strengthened airframe and consequently weighed more than twice as much as a T-37.

Cessna A-27B Dragonfly
Depicted carrying Mk 81 bombs with 'daisy cutter' fuse extenders, this is one of 19 A-37Bs delivered to El Salvador from surplus USAF stocks from 1982.

US SERVICE

US and South Vietnamese squadrons used the Dragonfly to great effect in Vietnam, where it proved a more accurate bomber than the much more sophisticated F-4 Phantom, as well as being easier and cheaper to maintain. After Vietnam, the United States continued to operate a small number of A-37s, some of them designated as OA-37s for the forward air control (FAC) role. The USAF finally retired the A-37 in 1992.

COIN AIRCRAFT

A-37s have been supplied to numerous nations for the counter-insurgency role, particularly in Latin America. Many of these have seen action against or on behalf of rebels in various conflicts, notably El Salvador and Nicaragua in the 1980s.

Antonov An-22 (1967)

TYPE • Transport COUNTRY • Soviet Union

SPECIFICATIONS

DIMENSIONS:	Length: 57.9m (190ft); Wingspan: 64.4m (211ft 3in); Height: 12.53m (41ft 1in)
WEIGHT:	250,000kg (551,000lb) loaded
POWERPLANT:	4 x 11,030kW (15,000hp) Kuznetsov NK-12MA turboprops
MAX SPEED:	740km/h (460mph)
RANGE:	5000km (3100 miles)
CEILING:	8000m (26,240ft)
CREW:	5–6
ARMAMENT:	None

Antonov's design bureau, based at Kiev in the Ukraine, is renowned for producing some of the world's biggest aeroplanes. The An-22 – with its maximum take-off weight of 250 tonnes (275 tons) including an 80-tonne (88-ton) payload – was the largest of its day.

Below: Flying into Kabul and Shindand, the An-22 was the most capable strategic airlifter available to the Soviets at the time of the Afghanistan invasion. The initial combat force of 6000 troops was sent there in around 300 sorties flown by transport aircraft between 24 and 26 December 1979.

CABIN
When designing the An-22, Antonov was unwilling to take on the complexities of building a complete pressurized fuselage. As a result, only the forward section of the cabin, with seating for passengers, is pressurized.

WINGS
Even in profile the pronounced droop of the Antei's outer wings is evident. The main wing box forms a fuel tank with a capacity of 55,800 litres (14,740 Imp gal) or 43,000kg (94,800lb) of fuel. The tank runs from almost one wingtip to the other. Pressure refuelling points are mounted in the undercarriage fairings.

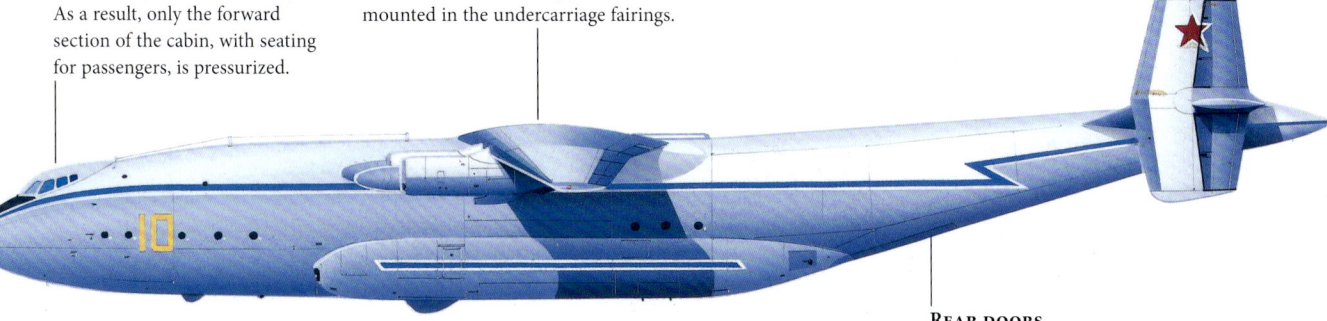

REAR DOORS
There are two doors at the rear of the cabin. The lower one opens to ground level for easy loading and also contains tracks that allow the cabin cranes to move out onto the opened ramp.

CARGO CAPACITY
The An-22's titanium-floored main hold has two travelling cranes and two winches to assist with loading and unloading. With 33m (108ft) of usable length and a 4.40m (14ft 5in) square cross-section, the cabin can accommodate main battle tanks and other similar heavy loads. With a 45-tonne (50-ton) payload the Antei is capable of travelling more than 10,000km (6,200 miles). The pressurized forward fuselage can seat 29 passengers, but the 'Cock' was built to carry freight rather than people.

LANDING-GEAR FAIRINGS
Large fairings on either side of the lower fuselage contain the 12-wheeled main landing gear. Six wheels are used on each side, mounted in rows of three twin-wheeled bogies. Auxiliary power units and systems controlling cabin air are also contained in the fairings.

Mil Mi-8 (1967)

TYPE • Helicopter COUNTRY • Soviet Union

The world's most-produced helicopter type, the Mil Mi-8 (NATO reporting name 'Hip') and its export variant, the Mil Mi-17, remain in production and thousands are in military and civilian service worldwide.

SPECIFICATIONS (Mil Mi-8MTV-5-1)

DIMENSIONS: Length: 18.65m (61ft 2in); Rotor diameter: 21.29m (69ft 10in); Height: 5.54m (18ft 2in)

WEIGHT: Maximum take-off 13,000kg (28,660lb)

POWERPLANT: Two Klimov/St Petersburg TV3-117VM turboshaft engines, each rated at 1491kW (2000shp)

MAX SPEED: 250km/h (155mph)

RANGE: 690km (429 miles)

CEILING: 6000m (19,685ft)

CREW: 3

ARMAMENT: Up to 4000kg (8818lb) weapons and stores

Left: A long-lived aircraft, thousands of Mi-8s have served dozens of countries since the mid-1960s. This Soviet Mi-8T was built in 1975 and operated by Aeroflot until it was put into storage in 1993.

ROTOR BLADES
Although slightly redesigned, the Mi-8's large five-bladed main rotor was also used by the later Mi-24 gunship. Like all Mil designs, it rotates clockwise when viewed from above. The rotors have an automatic ice detection and thermal de-icing system, essential for operations in Russian conditions.

ENGINES
The Isotov TV-2 engines of the Mi-8 are very similar to the TV-3 engines in the Mi-24 and Mi-17. The TV-3 proved more reliable and economical and dramatically improved performance in 'hot-and-high' conditions.

CLAMSHELL DOORS
Loading a Mi-8 is easy, thanks to the rear clamshell doors, which can accommodate wide cargoes and allow infantry to exit swiftly in an assault.

'HIP' DESIGN
Design of the Mi-8 'Hip' began in 1960. Unlike the earlier Mi-4 'Hound', the Mi-8 has a more efficient shape with the turboshaft powerplant above the fuselage leaving maximum space for payload. Except on specialized models, large clamshell doors swing open at the rear fuselage. Nearly a dozen versions of the Mi-8 and its upgraded Mi-17 derivative were used by Soviet forces and exported to Moscow's allies.

'HIP-C', KAZAKHSTAN
Kazakhstan paratroopers fly into a drop zone area in a Mil Mi-8 'Hip-C' during an international mass jump in September 2000. Military Mi-8s are often equipped to a high specification, including additional cockpit armour, infra-red jammers, chaff and flare dispensers and exhaust gas diffusers.

Bell AH-1 HueyCobra (1967)

TYPE • Helicopter   COUNTRY • United States

SPECIFICATIONS (AH-1T)

Dimensions:	Length: 13.4m (44ft 5in); Rotor diameter: 13.4m (44ft); Height: 4.1m (13ft 6in)
Weight:	4500kg (10,000lb) loaded
Powerplant:	1 x 820kW (1100shp) Lycoming T53-L-13 turboshaft
Max Speed:	352km/h (219mph)
Range:	574km (357 miles)
Ceiling:	3475m (11,400ft)
Crew:	2
Armament:	2 x 7.62mm MGs; 70mm (2.75in) rockets; M18 7.62mm Minigun pod

The HueyCobra was the first helicopter designed for armed battlefield duties. Although it had a number of features in common with the famous UH-1 'Huey', the AH-1 was the first of the real anti-tank helicopter gunships.

Above: Four AH-1Ts were used in Grenada in 1983 in support of the US Marine Corps transport helicopters that delivered US troops from the assault ship USS Guam. Two SeaCobras were shot down during the battle for Fort Frederick. The USMC helicopter component also included UH-1Ns for command and control duties.

Slender design
The tail and fuselage are very slender. This enables the helicopter to fly tight and low at tree-top level to help mask its presence.

Powerplant
Early Cobras had a single Textron Lycoming T53 turboshaft rated at 994kW (1332hp). Later models had two.

Tandem seat
The now common tandem seating of the gunner and pilot was first introduced into combat on a helicopter by the AH-1.

VIETNAM NEED
In 1965 the US Army finalized its requirement for the world's first armed battlefield helicopter, the Bell AH-1 Cobra, often called the HueyCobra. The idea had arisen before Vietnam, but the Cobra arrived on the scene just when it was required in the South East Asia conflict. The AH-1 featured a streamlined, narrow-width fuselage that accommodated a two-man crew in tandem seats with the pilot above and behind the co-pilot/gunner.

VIETNAM WAR
A US Army AH-1G HueyCobra 'Patricia Ann' makes a combat sweep over a hamlet in Vietnam. The Cobra has also proved well-suited to fighting in an urban theatre.

Lockheed AC-130 Spectre (1967)

TYPE • Gunship COUNTRY • United States

The AC-130 is the definitive gunship, introduced into Vietnam in September 1967 and still in use in modernized form today. The AC-130A Spectre was used to hunt for night movements on the Ho Chi Minh Trail supply routes into the south. It also provided fire support for US bases.

SPECIFICATIONS (AC-130A)

DIMENSIONS:	Length: 29.8m (97ft 9in); Wingspan: 40.4m (132ft 7in); Height: 11.7m (38ft 6in)
WEIGHT:	Maximum 69,750kg (155,000lb)
POWERPLANT:	Four 3661kW (4910hp) Allison T56-A-15 turboprop engines
MAX SPEED:	480km/h (300mph)
RANGE:	4070km (2530 miles)
CEILING:	10,060m (33,000ft)
CREW:	7
ARMAMENT:	Four 7.62mm GAU-2/A Miniguns and four 20mm (0.79in) Vulcan cannon; later variants included 105mm (4.13in) howitzer and 40mm (1.57in) Bofors

Above: An AC-130A Spectre. The latest version of the aircraft in service in the United States is the AC-130J Ghostrider.

CREW SURVIVAL
Dramatic improvements in the survivability for the crew were made since the AC-130A was first developed. Crews became surrounded by Spectra ceramic armour, with the fuel carried in explosion-suppressing fuel tanks.

AC-130H SPECTRE
The AC-130H saw action in Grenada (1983), Panama (1989), over Iraq during the 1990–91 Gulf War and following the 2003 invasion, and over Afghanistan.

HEAVY WEAPONS
The heavy punch of the AC-130H is provided by the 105mm (4.13in) howitzer and the 40mm (1.57in) Bofors cannon. They are positioned in the rear of the fuselage. Situated opposite these guns are racks for the vast amount of ammunition required.

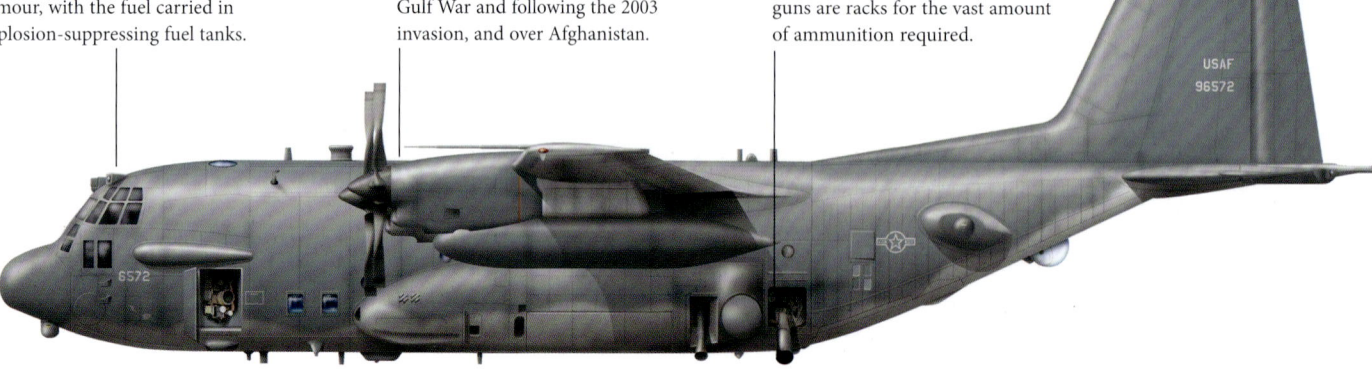

HOWITZER ADDITION
After its successful combat debut in Vietnam, the AC-130 Spectre became the standard USAF night attack gunship. The early 'Hercules' gunships were subsequently upgraded, up-gunned and replaced by newer airframes. For example, the AC-130U carried fewer guns than its predecessor but, with its computer-controlled 105mm (4.13in) howitzer, it could bring down devastating aerial firepower on the enemy.

GUNSHIP IN FLIGHT
This front view of an AC-130 in-flight near Hurlburt Field, Florida, clearly shows the firepower bristling from the side of the aircraft. When the heavy weapons are used, the pilot puts the aircraft into a constant shallow bank to present the barrels to the target.

Vought A-7 Corsair II (1967)

TYPE • *Attack aircraft* **COUNTRY** • *United States*

SPECIFICATIONS (A-7E)

DIMENSIONS:	Length: 14.06m (46ft 1.5in); Wingspan: 11.8m (38ft 9in); Height: 4.9m (16ft 0.67in)
WEIGHT:	Empty 8973kg (19,781lb); maximum take-off 19,050kg (42,000lb)
POWERPLANT:	One 64.5kN (14,255lb) thrust Allison TF41-1 (Rolls-Royce Spey) turbofan
MAX SPEED:	1123km/h (698mph)
RANGE:	1127km (700 miles)
CEILING:	15,545m (51,000ft)
CREW:	1
ARMAMENT:	One 20mm (0.79in) cannon; up to 6804kg (15,000lb) of munitions

Although it could hardly be described as pleasing to the eye, the Vought A-7 Corsair II's ability to lift a considerably heavy war-load made it as successful an attack aircraft as its illustrious World War II namesake.

Right: Mounted on LAU-7/A launch rails, the A-7E's pair of AIM-9L Sidewinder AAMs gave the aircraft a respectable self-defence capability.

EJECTION SEAT
The A-7D was fitted with a McDonnell Douglas ESCAPAC Model 1C2/3 ejection seat. A miniature detonating cord shatters the cockpit canopy with initiation of the ejection sequence.

CANNON AMMUNITION
Ammunition for the A-7D's Vulcan cannon was housed in a drum in the upper fuselage, behind the cockpit, connected to the weapon via a double flexible feed chute, one side carrying live ammunition to the gun, the other returning linkages to the drum. A total of 1000 20mm (0.79in) rounds could be carried.

A-7D
The A-7D pictured here was assigned to the 125th TFS, 138th TFG, of the Oklahoma Air National Guard, based at Tulsa. Before converting to the A-7D in July 1978, the Oklahoma ANG flew the F-100D/F Super Sabre, and began converting to the F-16C/D in 1993.

SERVICE ISSUES
To naval pilots, the A-7 presented them with an ideal attack platform from which to deliver a wide range of bombs and air-to-air missiles. Despite its operational success, however, the Corsair II encountered problems in service. Pilots found the aircraft had a tendency to suck up catapult steam during launches, which resulted in more than a few accidents. Yet the Corsair II achieved the reputation of being one of the most able attack aircraft ever.

GULF WAR A-7E
An A-7E of Attack Squadron 72 (VA-72) en route to targets during Operation Desert Storm. The aircraft is carrying Mark 20 Rockeye II anti-tank cluster bombs on their outboard wing pylons and AIM-9 Sidewinder missiles on their fuselage pylons.

Agusta-Bell 212 (1968)

TYPE • Helicopter COUNTRY • Italy

Developed as a civil version of the UH-1N, the twin-turboshaft Agusta-Bell 212 helicopter also found military operators, and was successfully followed by the more powerful single-engined Model 214.

SPECIFICATIONS (AB.212ASW)

DIMENSIONS:	Length: 17.4m (57ft); Main rotor diameter: 14.63m (48ft); Height: 4.53m (14ft 10in)
WEIGHT:	Empty 3420kg (7,524lb); maximum take-off 5070kg (11,154lb)
POWERPLANT:	One 1398kW (1875hp) Pratt & Whitney Canada PT6T-6 Turbo Twin-Pac
MAX SPEED:	238km/h (149mph) at sea level
RANGE:	439km (272 miles)
CEILING:	3200m (10,500ft)
CREW:	4
ARMAMENT:	Two Mk 44, Mk 46 or MQ 44 torpedoes or two air-to-surface missiles

Above: AB.212ASW '7-20' of Italy's 5° Gruppo Elicotteri wears the standard overall medium sea-grey camouflage scheme. Day-Glo orange patches are also applied as an aid to visibility.

CABIN
Depending on the internal equipment fit and the type of mission to be flown, the cabin can hold up to two crew, in addition to the pilots, who operate the radar and dipping sonar. Alternatively, the cabin can hold seven passengers or four stretcher cases.

SEARCH RADAR
All but 12 of the Italian navy's AB.212ASWs are fitted with an APS-705 or -706 search radar. These have a range of 0.9 to 148km (0.5 to 92 miles).

TAIL ROTOR
The twin-bladed tail rotor is a typical Bell design feature. The tip of each blade is painted in Italy's national colours, making it easy for ground crew to avoid the spinning rotor.

BELL CONTRACT

The product of a contract between Bell Helicopters and the Canadian government, the successful 212/214/412 series began when the first twin-turbine development of the Model 205 took to the air. The USAF took 79 examples (as the UH-1N), and the series was adopted by more than 40 different military services and numerous other civil operators. Licence production around the world continues today.

AUSTRIAN AB.212
An AB.212 of the Austrian Army demonstrates its performance at an air show. Bell installed a Pratt & Whitney Turbo Twin-Pac engine into its 205 airframe to produce the AB.212 – the equivalent of a twin-engined helicopter.

Aérospatiale Puma (1968)

TYPE • *Helicopter* COUNTRY • *France*

SPECIFICATIONS (SA 330L)

DIMENSIONS: Length: 18.15m (59ft 6in); Main rotor diameter: 15m (4 ft 3in); Height: 5.14m (16ft 10in)

WEIGHT: Empty 3615kg (7,953lb); maximum take-off 7400kg (16,280lb)

POWERPLANT: Two 1175kW (1575hp) Turboméca Turmo IVC turboshafts.

MAX SPEED: 294km/h (182mph)

RANGE: 572km (355 miles)

CEILING: 6000m (19,700ft)

CREW: 2–3

ARMAMENT: Optional provision for weapons combinations including cannons, MGs, rockets and missiles

Building on its experience with the earlier, larger Super Frelon, Sud Aviation (later Aérospatiale) answered the French army's call for an all-weather medium transport helicopter with the Puma. France and later Britain ordered sizeable fleets.

Above: An SA 330 Puma HC.1 SA 330Js and Ls were the first Western helicopters certified for all-weather flight including operations in Arctic conditions.

FLIGHT CONTROLS
Dual flight controls are standard on the Puma, which flies with two flight deck crew.

MAIN ROTOR
The fully articulated main rotor with four aluminum blades on the initial production SA 330s were replaced by a new rotor with composite blades in the SA 330J and L.

TIGER STRIPES
Royal Air Force SA 330E (Puma HC.Mk 1) XW229 was painted in tiger stripes for a NATO Tiger Meet while serving with No. 230 Squadron based in West Germany in the 1980s.

PERFORMANCE

The Puma was designed to be capable of high speeds, possess good manoeuvrability and have decent hot-and-high performance. The helicopter also possesses a relatively small 'footprint' for its size, which makes it ideal for operations in urban areas. The engines have a high reserve of power that enables a Puma to continue flying at maximum weight with only one functioning engine if necessary.

RAF PUMA
During the Gulf War of 1990–91, an RAF Puma embarks British soldiers of the Staffordshire Regiment. In British service, as the HC Mk 1, the Puma also saw action in the Falklands (it was also used by Argentinian forces) and became a regular transport vehicle for British special forces.

Lockheed C-5 Galaxy (1968)

TYPE • Transport **COUNTRY** • United States

SPECIFICATIONS (C-5B)

DIMENSIONS:	Length: 75.54m (247ft 10in); Wingspan: 67.88m (222ft 8in); Height: 19.85m (65ft 1in)
WEIGHT:	348,810kg (769,000lb) maximum take-off
POWERPLANT:	4 × General Electric TF39-GE-1C turbofans, 191.27kN (43,000lbf)
MAX SPEED:	890km/h (553mph) at 7620m (25,000ft)
RANGE:	6035km (3750 miles) maximum payload of 100,228kg (220,967lb)
CEILING:	10,360m (34,000ft)
CREW:	5
ARMAMENT:	None

For many years the world's biggest military cargo aircraft, the C-5 Galaxy remains the largest and only strategic airlifter in the US Air Force inventory and can carry more cargo farther distances than any other aircraft. It is now modified to C-5M Super Galaxy standard.

Above: A C-5B Galaxy of the USAF's 436th Military Airlift Wing in the European One 'lizard' camouflage scheme worn during the 1980s.

WINGS
The Galaxy's huge wing is equipped with track-mounted, slotted flaps on the trailing edge. These consist of three inboard and three outboard sections. The wing leading edge is fitted with four sealed inboard slat sections, and three slotted outboard sections.

TROOP TRANSPORT
As an alternative to the typical cargo load (up to a maximum of 100,228kg/220,967lb), the C-5 can also be adapted for the transport of up to 350 fully equipped troops.

AIR MOBILITY
The C-5 is the largest and most capable asset in Air Mobility Command (AMC) service. Led by a general, the AMC serves as the USAF component of US Transportation Command (TRANSCOM), and is responsible for a single numbered air force, the 18th Air Force/Air Forces Transportation headquartered at Scott Air Force Base, Illinois. The C-5 can carry almost any item in the US military inventory, from Abrams main battle tanks to more than 360 fully equipped paratroops.

POWERPLANT
The four General Electric TF39-GE-1C engines are twin-shaft high-bypass turbofans, with a two-stage fan and a 16-stage axial flow compressor. The rear section of the cowling serves as a thrust reverser.

CARGO LOADING
A US Air Force C-5 Galaxy sits with its nose up before loading a US Army Stryker Infantry Carrier Vehicle at Joint Base Balad, Iraq, 25 June 2008. The massive hold can accommodate a wide variety of outsized loads, from helicopters and tanks to trucks and cargo containers.

BAC 167 Strikemaster (1968)

TYPE • *Attack aircraft* COUNTRY • *United Kingdom*

In 1967, BAC flew a strengthened version of the Jet Provost Mk 5 as the Strikemaster, intended for the export market as a light attack aircraft and weapons trainer. With a more powerful Viper and extra wing hardpoints, the Strikemaster could carry a useful weapons load.

SPECIFICATIONS

DIMENSIONS:	Length: 10.36m (34ft); Wingspan: 11.23m (36ft 10in); Height: 3.10m (10ft 2in)
WEIGHT:	5216kg (11,500lb) loaded
POWERPLANT:	One 15.17kN (3410lbf) Rolls-Royce Viper 20 Mk 525 turbojet engine
MAX SPEED:	724km/h (450mph)
RANGE:	2224km (1382 miles)
CEILING:	12,190m (40,000ft)
CREW:	2
ARMAMENT:	Up to 1361kg (3000lb) of ordnance, including bombs, rockets or gun pods

Below: A Royal Saudi Air Force Strikemaster. More than 100 Strikemasters were sold to Ecuador, Kenya, Kuwait, New Zealand, Oman, Saudi Arabia, Singapore, Sudan and South Yemen. Botswana later acquired some ex-Kuwaiti aircraft.

FATIGUE ISSUES
In the late 1980s, the RNZAF Strikemasters suffered from fatigue problems and had been phased out in favour of the Aermacchi MB 339 by 1993.

MK 88 STRIKEMASTER
This is one of 16 Strikemaster Mk 88s acquired by the Royal New Zealand Air Force (RNZAF) in the 1970s to replace the two-seat Vampire in the advanced training, jet conversion and weapons training roles.

JET PROVOST
The basic design of the British Aircraft Corporation (BAC) Strikemaster can be traced to the piston-engined Percival Provost trainer of 1950. Hunting Aviation took the wings and tail, and created a jet version with the Bristol Siddeley (later Rolls-Royce) Viper engine in 1954. The Jet Provost became the Royal Air Force's most numerous advanced trainer and served in this role into the 1990s, the later versions being pressurized with wingtip fuel tanks.

HARDPOINTS
Part of the conversion of the Jet Provost to the Strikemaster involved the addition of hardpoints under the wings for up to 1361kg (3000lb) of ordnance.

STRIKEMASTER MK 82A
BAC Strikemaster Mk 82A is put through its paces during a flying display at RAF Fairford, Gloucestershire, UK during the Royal International Air Tattoo in 2016.

Sikorsky MH-53 Pave Low (1968)

TYPE • Helicopter COUNTRY • United States

The infiltration and extraction of special forces to and from enemy territory by air dates back to World War II. It was for this task that the US Air Force's 'Pave Low' programme was initiated, using helicopters.

SPECIFICATIONS

DIMENSIONS:	Length: 30.19m (99ft 0.5in); Main rotor diameter: 22.02m (72ft); Height: 8.64m (28ft 4in)
WEIGHT:	Empty 10,691kg (23,569lb); take-off 17,344kg (38,238lb)
POWERPLANT:	Three 3266kW (4380hp) General Electric T64-GE-416 turboshaft engines
MAX SPEED:	315km/h (196mph)
RANGE:	2075km (1288 miles)
CEILING:	5640m (18,500ft)
CREW:	6
ARMAMENT:	Three 7.62mm (0.30in) Miniguns or 12.7mm (0.50in) machine guns

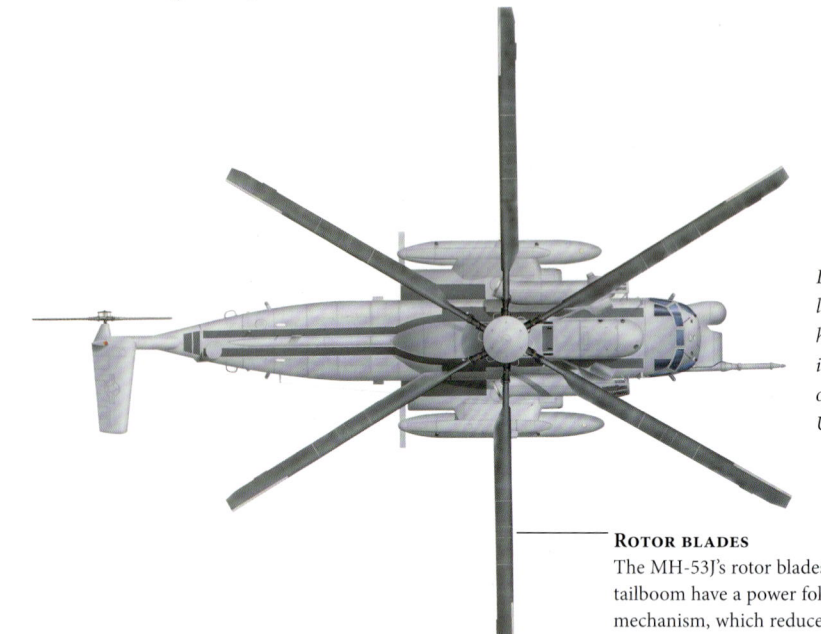

Left: The US Air Force has long used the H-53 as a rescue helicopter. The MH-53J is in service with the special operations squadrons of the US Air Force.

ROTOR BLADES
The MH-53J's rotor blades and tailboom have a power folding mechanism, which reduces the time needed to prepare the helicopter for air transport aboard the C-5 Galaxy.

TAIL BUMPER
The tail bumper is fully retractable and the four-bladed tail rotor is slightly canted to port.

REFUELLING PROBE
The long in-flight-refuelling probe extends forward, well clear of the rotor blades, when in use.

SIKORSKY S-35

The foundation of the MH-53 is the Sikorsky S-65, which during the Vietnam era was the largest helicopter built outside the USSR. Its dynamic parts (rotor, gearboxes and control system) were developed from those of the earlier S-64 SkyCrane and made extensive use of titanium. Fitted with folding rotor blades for shipboard stowage and given the designation CH-53 Sea Stallion by the US Marines, the S-65 emerged as the world's most capable assault transport.

MH-53J FLY-BY
A US Navy MH-53J Pave Low performs a fly-by over the Ohio River. An extensive avionics fit includes terrain following radar and forward-looking infra-red sensors.

Ilyushin Il-38 (1968)

TYPE • *Maritime patrol* **COUNTRY** • *Soviet Union*

The Il-38 had its origins as an airliner. The fuselage of the Il-18 'Coot' was stretched and ASW equipment fitted, including a magnetic anomaly detector (MAD) in a tail boom to create the Il-38 in 1968.

SPECIFICATIONS

DIMENSIONS:	Length 40.14m (131ft 8in); Wingspan: 37.42m (122ft 9in); Height: 10.17m (33ft 4in)
WEIGHT:	Maximum take-off: 66,000kg (145,505lb)
POWERPLANT:	Four 3169kW (4250hp) Ivchenko-Progress AI-20M turboprops
MAX SPEED:	722km/h (448mph)
RANGE:	6500km (4039 miles)
CEILING:	11,000m (36,089ft)
CREW:	7–8
ARMAMENT:	Up to 8400kg (18,519lb) weapons and disposable stores, including depth charges, mines, torpedoes and bombs

Right: A massive radome dominated the Il-38's forward fuselage. It housed a search radar, NATO codename 'Wet Eye', which was used for detecting submarine periscopes and surface vessels.

FLIGHT CREW
The Il-38 carries three flight crew. Separated from the flight deck by a pressure bulkhead, the main cabin houses the equipment and operating consoles for nine mission specialists.

COLOUR SCHEME
Il-38s were originally painted in Aeroflot colours in an attempt to hide their true nature. Later they were finished more conventionally in overall grey. The Novella radar is housed in the ventral radome, with MAD equipment (used to detect submarines) housed in the tail 'sting'.

CREW ROLES
In addition to the two pilots and flight engineer on the flight deck, the aircraft carries a crew of nine systems operators in the main cabin. Their job is to monitor the displays showing targets detected by the radar and MAD sensors, and to track submarines using sonobuoys dispensed from the stores bays. Contacts may be destroyed using depth charges, torpedoes or missiles.

ENGINES
Four powerful and efficient turboprops power the Il-38. Identical to those fitted on the Il-18D 'Coot' airliner, they give the 'May' a respectable top speed of 722km/h (448mph) and a patrol endurance of 12 hours.

IL-38 WITH RADOME
An Il-38 taxis to position. This aircraft displays the prominent search radar radome fitted to many Il-38s just above the flight crew compartment.

North American Rockwell OV-10 Bronco (1969)

TYPE • Attack aircraft COUNTRY • United States

Rockwell's OV-10 Bronco was a product from early lessons learned in the Vietnam War. Designed as a COIN (counterinsurgency) aircraft, it evolved into an armed, agile forward air control (FAC) machine directing fighter-bombers to their targets.

SPECIFICATIONS (OV-10D)

Dimensions:	Length 13.41 (44ft); Wingspan: 12.19m (40ft) Height: 4.62m (15ft 2in)
Weight:	Maximum take-off 4,494kg (9,908lb)
Powerplant:	Two Garrett T76-G-420/421 turboprop engines, 1,040 shp (780 kW) each equivalent
Max Speed:	460km/h (290mph)
Range:	367km (198 miles)
Ceiling:	9,100m (30,000ft)
Crew:	2
Armament:	1 × 20mm (0.79in) M197 electric cannon; 5 fuselage and 2 underwing, with provisions to carry combinations of 7- or 19-tube launchers for 70mm (2.75in) FFARs/WAFARs; up to 227kg (500lb) bombs

Above: Here we see an OV-10A Bronco of the US AF 20 Tactical Air Support Squadron (TASS), painted with a distinctive 'sharkmouth' nose.

Canopy
The long bulged canopy and short nose allowed an excellent view of the target area. This OV-10A was assigned to the USMC's VMO-4 squadron at Quang Tri, Republic of Vietnam.

Tailplane
The tailplane is a fixed-incidence unit with an inset elevator, providing excellent control at low speeds. The aircraft's broad high-lift wing meant it could fly slowly enough to escort helicopters.

Fuel tanks
For extended range and loiter time, a 568-litre (150-gal.) drop-tank could be carried, in addition to the 976 litres (250 gal.) of internal fuel.

BRONCO IN SERVICE

The OV-10 was ordered in 1964 and reached Vietnam in 1969. Bronco variants served in Germany (which used a turbojet-boosted model for target towing), Indonesia, Morocco and Venezuela. The most advanced version was the Marine Corps OV-10D-Plus, which incorporated night observation capability and forward-looking infra-red sensors and was employed on covert special forces insertion missions as well as forward air control.

EXPORT MODELS
A Rockwell OV-10B Bronco light attack and observation aircraft in German Air Force colours lands at Leeuwarden airbase, Netherlands, 2019. This aircraft is part of a demonstration team, but the Bronco can still be found in limited service internationally.

Antonov An-30 (1969)

TYPE • Transport **COUNTRY** • Soviet Union

Developed from the An-24, the An-30 was intended to serve as a reconnaissance platform in addition to its transport role. It gained a new forward fuselage and specialist features, such as a glazed nose, but retained its cargo handling equipment.

SPECIFICATIONS

DIMENSIONS:	Length: 24.26m (79ft 7in); Wingspan: 29.20m (95ft 10in); Height: 8.32m (27ft 4in)
WEIGHT:	Maximum take-off: 23,000kg (50,706lb)
POWERPLANT:	Two Ivchenko AI-24TVT turboprop engines, 2090kW (2803hp) each
CRUISE SPEED:	430km/h (270mph)
RANGE:	2630km (1630 miles)
CEILING:	8,300m (27,200ft)
CREW:	7
ARMAMENT:	None

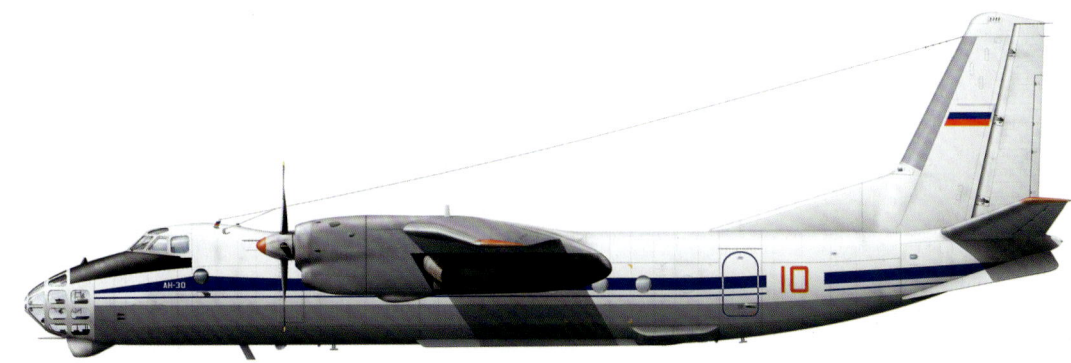

Above: A Russian Antonov An-30 'Red 10'. In Cold War years, the An-30 was given the supremely awkward NATO codename 'Clank'.

ANTONOV AN-30 UKRAINE AF
The glazed nose of the An-30 is intended to facilitate its reconnaissance and aerial photography roles. The aircraft are still also capable of serving as cargo haulers.

CREW POSITION
An-30 pilots have a far better field of view than their An-24 colleagues. The raised cockpit was primarily installed to provide more space for the navigator and survey equipment.

SURVEY EQUIPMENT
A computer aboard the An-30 is programmed with the route and controls the aircraft during surveys. In basic fit the An-30 cabin has four camera apertures and a light meter.

AN-24 UPGRADE
Developed from the An-24RT, the An-30 is a specialized aerial survey aircraft used in small numbers by Russia and a handful of former Soviet allies. Externally, the most obvious difference between the An-30 and An-24RT is the glazed nose and raised cockpit canopy of the 'Clank'. Internally, though, the later model carries survey cameras along with a darkroom and map-making equipment or other geographical survey equipment.

UKRAINIAN AN-30
A Ukrainian Antonov AN-30, seen in 2019. The date of the photograph illustrates the fundamentally sound design principles behind the aircraft, which is still in service more than 50 years after its first flight.

Ilyushin Il-20 and Il-22 (1969)

TYPE • Transport **COUNTRY** • Soviet Union

SPECIFICATIONS (Il-20)

DIMENSIONS:	Length: 35.9 m (117ft 9in); Wingspan 37.42m (122ft 9in); Height: 10.17m (33ft 4in)
WEIGHT:	Maximum take-off 64,000kg (141,056lb)
POWERPLANT:	4 x 3170kW (4250hp) Ivchenko Progress AI-20M series 4 turboprops
MAX SPEED:	660km/h (410mph)
RANGE:	6000km (3728 miles)
CEILING:	10,000m (32,808ft)
CREW:	5–11
ARMAMENT:	None

Both aircraft based on the Il-18D airliner, the Il-20 (NATO reporting name 'Coot-A') is a radar reconnaissance and electronic intelligence (ELINT) platform, and the Il-22 ('Coot-B') is an airborne command post.

Above: This 'Coot' is shown as it appeared on 2 June 2020 when it was intercepted by Lithuania-based RAF Typhoon FGR.4s tasked with NATO's Baltic Air Policing mission. Il-20s regularly fly intelligence gathering flights between mainland Russia and the Baltic enclave of Kaliningrad.

FLIGHT CREW
Like civil Il-18Ds, the 'Coot-A' has a flight crew of four or five, including two pilots, a navigator, radio operator and flight engineer. A mission crew of about 20 is carried in the main cabin to operate the aircraft's systems.

POWERPLANT
Powerplants on the Il-20 are four standard Ivchenko AI-20Ms rated at 3170kW (4250hp), which drive AV-68I four-bladed reversible propellers. These engines, like similar Western designs, date from the 1950s.

IL-22M-11
One of the Russian Air Force's Il-22M-11 'Zebra' radio-relay aircraft in its Aeroflot-derived colour scheme. These aircraft were still active in the 2022 Russian invasion of Ukraine but the failure to neutralize Ukrainian air defences has required them to operate at a significant stand-off distance, limiting their effectiveness.

CREW COMPOSITION
The prototype for the Il-20, converted by Ilyushin from an Il-18D, made its maiden flight on 21 March 1968. Tests with the side-looking radar fitted started in 1968. Work on an airborne command post began in September 1965, with Ilyushin preparing a draft design for the Il-22 Bison in 1967 and the first of two prototypes making its maiden flight in 1970. Both aircraft have a flight crew of five (pilot, co-pilot, navigator, flight engineer and communication operator) plus tactical crew of six.

RUSSIAN IL-20
A Russian Air Force Ilyushin IL-20 lands at Kubinka air base in 2011. The first Il-20 series aircraft were produced by the Znamya Truda plant in Moscow in late 1969. Production continued in small batches until 1976, for a total of 18 aircraft.

McDonnell Douglas F-4 Phantom (exports) (1969)

TYPE • Fighter COUNTRY • United States

The excellence of the F-4 Phantom meant that it achieved major export success. The major production version was the F-4E, 913 of which were delivered to the USAF between October 1967 and December 1976, in addition to export orders of 558 aircraft.

SPECIFICATIONS (FGR.2)

- **DIMENSIONS:** Length: 17.55m (57ft 7in); Wingspan: 11.7m (38ft 5in); Height: 4.96m (16ft 3in)
- **WEIGHT:** 26,308kg (58,000lb) loaded
- **POWERPLANT:** 2 x 91.2kN (20,515lbf) Rolls-Royce Spey 202 turbofans
- **MAX SPEED:** 2230km/h (1386mph)
- **RANGE:** 2817km (1750 miles)
- **CEILING:** 18,300m (60,000ft)
- **CREW:** 2
- **ARMAMENT:** 4 x AIM-7 Sparrow missiles; two wing pylons for 2 x AIM-7, or 4 x AIM-9 Sidewinders, provision for 20mm cannon; plus stores up to 7257kg (16,000lb)

Below: An RF-4EJ of the Japanese Air Self-Defence Force, which equipped five squadrons with 140 Phantom F-4EJs, most of which were built under licence.

RADAR
The British Phantoms incorporated the Westinghouse AN/AWG-10 radar carried by the F-4J, although this was built under licence by Ferranti.

AIRFRAME CHANGES
Fitting the Rolls-Royce Spey engines to the Phantom required larger air intakes and other significant airframe changes, particularly a thickening of the rear fuselage.

PHANTOM FG.1
The first British Phantoms were powered by Rolls-Royce RB.168-25R Spey 201 engines, the FGR.2 receiving the uprated 91.2kN (20,515lbf) Rolls-Royce Spey 202 turbofans.

BRITISH PHANTOMS
The first foreign nation to order the Phantom was the UK. Versions for the Royal Navy and the RAF were designated F-4K and F-4M respectively. Fifty-two F-4Ks were delivered to the RN in 1968/69 and these were progressively handed over to the RAF with the run-down of the RAF's fixed wing units, becoming the Phantom FG.1 in RAF service. The RAF's own version, the F-4M Phantom FGR.2, equipped 13 air defence, strike and reconnaissance squadrons.

RAF FGR.2
The British Phantoms served a variety of tactical roles in the Royal Navy and the Royal Air Force, including bomber interception, close air support and carrierborne fleet protection.

Hawker Siddeley/BAE Nimrod (1969)

TYPE • *Maritime patrol* COUNTRY • *United Kingdom*

SPECIFICATIONS (MR.Mk 2P)

DIMENSIONS:	Length: 38.65m (126ft 9in); Wingspan: 35m (114ft 10in); Height: 9.14m (31ft)
WEIGHT:	87,090kg (192,000lb) loaded
POWERPLANT:	4 x 53.98kN (12,140lbf) Rolls-Royce Spey Mk 250 turbofans
MAX SPEED:	925km/h (575mph)
RANGE:	9265km (5755 miles)
CEILING:	12,800m (41,995ft)
CREW:	12
ARMAMENT:	Provision for 6123kg (13,500lb) of ordnance

Descended from the Comet jetliner, but in such modified form that it first appeared to be a totally new design, the Nimrod was the world's only four-jet maritime patroller, able to detect and sink submarines, or mount anti-shipping strikes with Harpoon missiles.

Left: The Nimrod fleet was hastily modified with air-to-air refuelling probes during the Falklands War in 1982. This allowed the Nimrod to make long-surface search patrols into the South Atlantic, with missions often lasting well over 12 hours.

NIMROD MR.MK 2P
The MR.Mk 2P version of the Nimrod appeared during the Falklands campaign, and introduced an aerial refuelling capability with a forward fuselage probe.

REAR CABIN
Two teams of systems operators were housed in the rear cabin. The 'wet' team was responsible for anti-submarine engagements, while the 'dry' team handled surface searches and actions.

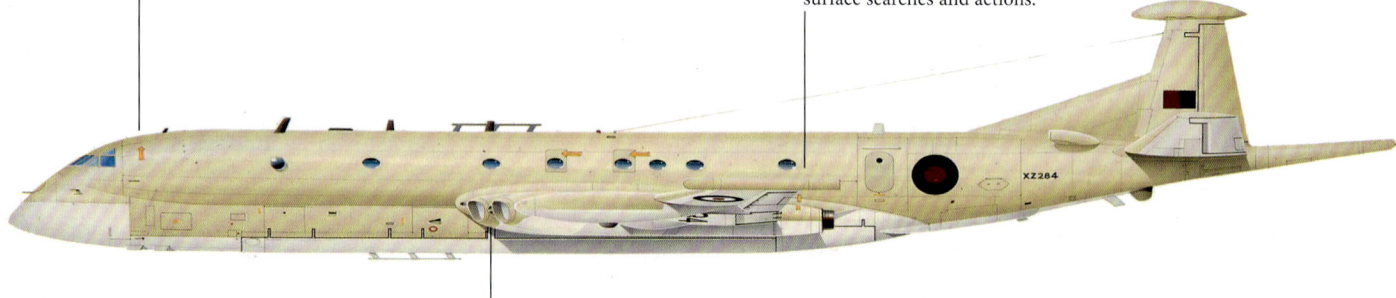

TURBOFANS
The four Rolls-Royce Spey turbofans were similar to the engines formerly used in the RAF's Buccaneers and Phantoms.

ELINT NIMRODS
A trio of Nimrods went to the RAF for the very different job of ELINT (electronics intelligence) gathering, snooping on an enemy's activity with hi-tech 'black boxes'. One of the three was lost in an accident in 1995. A planned airborne early warning version of the Nimrod – in effect, a flying radar station – never overcame technical difficulties, and Britain purchased Boeing's E-3 Sentry AWACS instead.

NIMROD MRA4
A Nimrod MRA4 at RAF Fairford, Gloucestershire, UK, in 2009. Prowling the oceans of the world, the Nimrod was uniquely qualified to oversee surface shipping and to hunt down hostile submarines. It was, however, finally retired in 2011.

Xi'an H-6 (1969)

TYPE • Bomber COUNTRY • China

The early Cold War-era Tupolev Tu-16 'Badger' was the template for the H-6, which remains the primary long-range bomber in the Chinese inventory. However, local development means the latest versions of this aircraft are far removed from the original design.

SPECIFICATIONS (H-6K)

DIMENSIONS:	Length 34.8m (114ft 2in); Wingspan: 33m (108ft 3in); Height: 10.36m (34ft)
WEIGHT:	Maximum take-off 95,000kg (209,439lb)
POWERPLANT:	Two Soloviev D-30KP-2 turbofan engines each rated at 118kN (27,000lbf)
MAX SPEED:	1050km/h (650 mph)
RANGE:	6000km (3700 miles)
CEILING:	12,800m (42,000ft)
CREW:	4
ARMAMENT:	Six underwing pylons for air-launched or land-attack cruise missiles

Below: An H-6N assigned to the 106th Air Brigade, another component of the 36th Bomber Division, based at Nanyang-Neixing air base in Henan province.

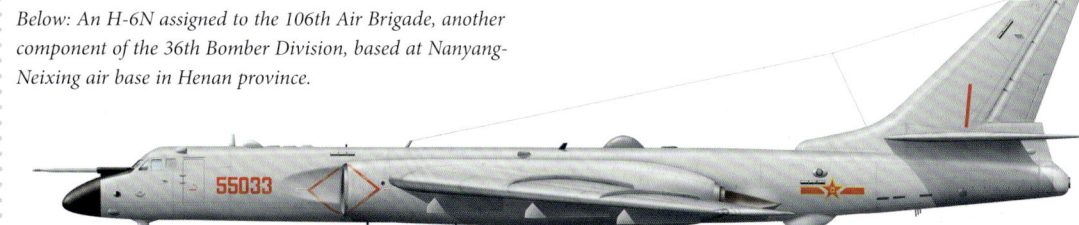

COCKPIT FEATURES
Modernized variants of the Xi'an H-6 feature an advanced cockpit layout, including an LCD multi-function display replacing earlier generations of analogue and digital controls.

H-6K
This H-6K is operated by the People's Liberation Army Air Force's 108th Air Regiment, part of the 36th Bomber Division, which has operated successive variants of the H-6 since the mid-1960s. The 108th Air Regiment flies both the H-6K and the H-6M from Wugong in Shaanxi province.

DEVELOPMENT
In 1956 China came to an agreement with the Soviet Union to establish an assembly line for the Tu-16 at Harbin in China, where the aircraft would be known as H-6. Moscow provided a pair of Tu-16s as pattern aircraft in early 1959, followed by another in kit form. At the same time, work began to produce the Mikulin RD-3 turbojet engine under licence as the WP-8. The first Chinese-assembled H-6 flew in September 1959. Two years later, production of the Chinese 'Badger' switched to Xi'an, but it wasn't until December 1968 that an H-6 built at the new production centre finally took flight.

FREEFALL BOMBS
While the H-6 is chiefly a missile carrier, it can also carry 12,000kg (26,400lb) of freefall bombs, including precision-guided munitions.

H-6 BOMBER
An H-6 roars into the air from its air base in China. While the H-6E/F added new navigation and electronic countermeasures equipment, they were still freefall bombers, while the H-6H appeared in the late 1990s with a new long-range missile capability.

Grumman F-14 Tomcat (1970)

TYPE • *Naval fighter* **COUNTRY** • *United States*

SPECIFICATIONS (F-14A)

DIMENSIONS:	Length: 19.1m (62ft 8in); Wingspan: 19.45m (64ft 1in); Height: 4.88m (16ft)
WEIGHT:	33,724kg (74,348lb) maximum take-off
POWERPLANT:	2 × Pratt & Whitney TF30-P-412 afterburning turbofans, 92.97kN (20,900lbf)
MAX SPEED:	2485km/h (1544mph)
RANGE:	1994km (1239 miles)
CEILING:	17,070m (56,000ft)
CREW:	2
ARMAMENT:	One 20mm (0.79in) Vulcan rotary cannon, plus a variety of AAMs

The F-14 entered service as a dedicated carrier-based interceptor. Following the end of the Cold War, the Tomcat was reborn as a multi-role fighter-bomber. In this form, it saw out its career with combat duty over Afghanistan and Iraq.

This page: F-14As were originally delivered in a two-tone camouflage of white undersides and light gull-grey topsides with white flying surfaces.

VULCAN CANNON
The General Electric M61A1 Vulcan cannon in the nose was provided with 675 rounds of ammunition and could be selected to fire at 4000 or 6000 rounds per minute.

REAR FUSELAGE
This includes upper and lower speed brakes in the tapered rear fuselage decking between the engine nozzles. An arrester hook is provided below this decking, lying flush when not required.

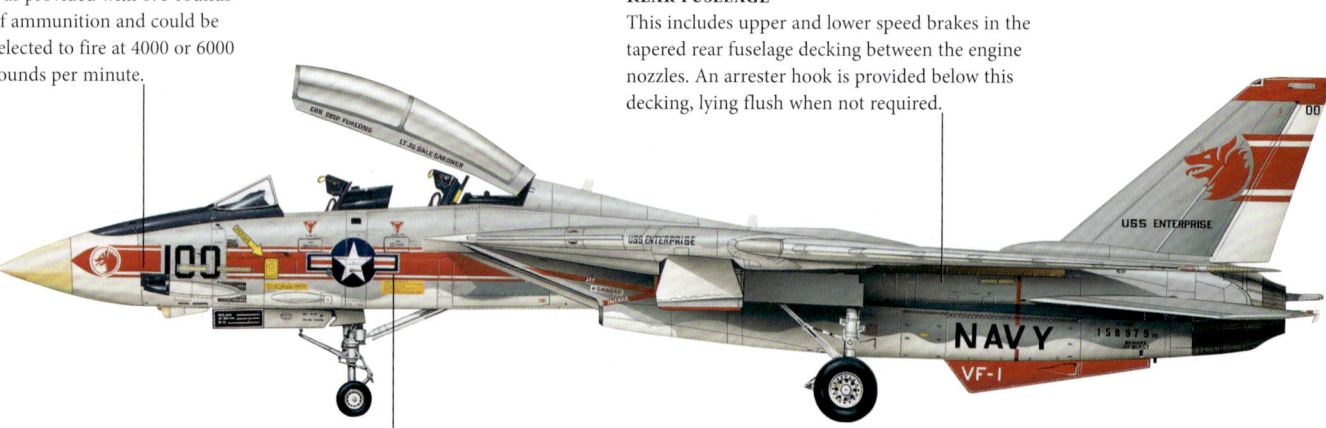

MARKINGS
The early days of US Navy Tomcat operations saw units wear flamboyant markings. Typical was this F-14A of VF-1 'Wolfpack', on board USS *Enterprise* (CVN 65) in the mid-1970s.

FIRST KILLS
The initial F-14A version entered service in 1972 and saw its first combat in US Navy hands in August 1981, when examples shot down a pair of Libyan Su-22 fighters over the Mediterranean. In another action in January 1989, US Navy Tomcats downed a pair of Libyan MiG-23s over the Gulf of Sidra. Meanwhile, the only export operator, Iran, put its Tomcat fleet to good use during the Iran–Iraq War, in the course of which the type was credited with as many as 64 Iraqi aircraft confirmed destroyed.

GULF WAR F-14
A Fighter Squadron 114 (VF-114) F-14 Tomcat aircraft flies over oil well fires still burning in the aftermath of Operation Desert Storm in 1991. The F-14 remained one of the world's most formidable warplanes well into the 1990s.

Antonov An-26 (1970)

TYPE • *Transport* **COUNTRY** • *Soviet Union*

SPECIFICATIONS

DIMENSIONS:	Length: 23.8m (78ft 1in); Wingspan: 29.3m (96ft 2in); Height: 8.58m (28ft 2in)
WEIGHT:	Maximum take-off 24,000kg (52,911lb)
POWERPLANT:	Two Progress AI-24VT Turboprop engines, 2103kW (2820hp) each
CRUISE SPEED:	440km/h (270mph)
RANGE:	2500km (1,600 miles)
CEILING:	7500m (24,600ft)
CREW:	5
ARMAMENT:	None

Operated by both Ukraine and Russia in the current conflict in Ukraine, the Antonov An-26 is a simple and robust aircraft developed from the An-24 as a general purpose transport for personnel and cargo.

Above: This Antonov An-26 'Blue 30' serves with the Russian Navy. The An-26 represents a trade-off between cargo capacity and the ability to operate from short or rough fields. It bears similarities to Western aircraft serving in the same role.

TROOP CARRIER
As well as being used to carry cargo, the An-26 is also applied to long-haul troops deployments. Properly configured, its internal bay can carry 40 fully equipped soldiers.

CONVENTIONAL LAYOUT
The aircraft that emerged from the An-26's development process was conventional for the type, with a high wing and twin turboprop engines.

REAR DOOR
Ukrainian Air Force Antonov An-26 'Yellow 05'. The powered rear door of the An-26 is one of its most important features. Loading and unloading are greatly facilitated, reducing turnaround time at forward airstrips, which may be vulnerable to attack.

LOGISTICS
While logistics is not the most glamorous aspect of the conflict in Ukraine, it is vital to the military campaign. Air transport is relatively fast and can bypass most obstacles or enemy-created impediments, such as destroyed bridges, although there is always a risk of interception. The An-26's ability to operate from very basic airstrips and to handle a harsh climate was a design requirement proven in the conflict zone.

CZECH AIR FORCE AN-26
The AN-26 has been widely exported. One of its most convincing features is its powered rear door. The preceding An-24 was satisfactory in most ways, but had a ventral loading hatch that prevented the airdropping of troops and supplies. Additionally, vehicles could not be loaded aboard. The An-26 remedied this defect and was adopted by several militaries worldwide.

Hawker Siddeley/BAe Harrier (1970)

TYPE • Naval fighter **COUNTRY** • United Kingdom

The Hawker Siddeley Harrier was a great aviation breakthrough. Designed in the late 1950s by Sydney Camm, the Harrier became the world's first V/STOL (vertical/short take-off and landing) combat aircraft.

SPECIFICATIONS (GR.7)

DIMENSIONS:	Length: 14.36m (47ft 1.5in); Wingspan: 9.25m (30ft 4in); Height: 3.55m (11ft 8in)
WEIGHT:	Empty 7050kg (15,542lb); maximum take-off 14,061kg (31,000lb)
POWERPLANT:	One 96.7kN (21,750lbf) Rolls-Royce Pegasus vectored-thrust turbofan
MAX SPEED:	1065km/h (661mph)
RANGE:	5382km (3310miles) with drop tanks
CEILING:	15,240m (50,000ft)
CREW:	1
ARMAMENT:	Provision for up to 4082kg (9000lb) (short take-off) or 3175kg (7000lb) (vertical take-off) of weapons stores

Below: The Harrier's astonishing ability to take off and land vertically liberated its users from the tyranny of the runway, basing aircraft anywhere from car parks to forest clearings.

NOSE EQUIPMENT
The extended nose of the Harrier GR.Mk 3 housed the Ferranti laser rangefinder and marked target seeker, which can search for and detect energy reflected from a target by a ground-based designator.

POWERPLANT
A single Rolls-Royce Pegasus engine powered the Harrier. This was one of the most powerful jet engines then used in combat aircraft. In the GR.Mk 3 it delivered nearly 10 tonnes (11 tons) of thrust without afterburning.

HOVERING
At hovering speeds, when aerodynamic surfaces have no effect, the Harrier was controlled by 'puffer' jets in the nose, tail and wingtips. These operated on high-pressure air bled from the engine.

V/STOL AIRCRAFT
The notion of a V/STOL combat plane, proven during the years when GR.Mk 1s served in Germany, had wide appeal. The US Marine Corps picked up the idea and had enough confidence in the concept to sponsor advanced new versions of the Harrier. Pilots did not master the Harrier easily, but, once they invested the effort, they commanded an aircraft with unique and exciting performance. It was also agile: in a dogfight the Harrier could hold its own with any fighter in the sky.

ROCKET ATTACK
An RAF Harrier GR.3 fires a stream of unguided rockets. The Matra SNEB rocket pod contained 19 unguided rocket projectiles with 'flip-out' stabilizing fins. Available with both HEAT and fragmentation warheads, the 68mm (2.68in) rockets were effective against most armoured targets.

Mikoyan-Gurevich MiG-23 (1970)

TYPE • Fighter COUNTRY • Soviet Union

SPECIFICATIONS (MiG-23MF 'Flogger-B')

DIMENSIONS: Length: 16.71m (54ft 10in); Wingspan: 13.97m (45ft 10in) spread and 7.78m (25ft 6.25in) swept; Height: 4.82m (15ft 9.75in)

WEIGHT: Empty 10,400kg (22,932lb); maximum loaded 18,145kg (40,000lb)

POWERPLANT: One 98kN (22,046lbf) Rumanskii R-27F2M-300 turbojet

MAX SPEED: 2445km/h (1520mph)

RANGE: 1900 km (1200 miles)

CEILING: 18,290m (60,000ft)

CREW: 1

ARMAMENT: One 23mm (0.9in) GSh-23L cannon, underwing pylons for AA-3 Anab, AA-7 Apex, and/or AA-8 Aphid air-to-air missiles

The MiG-23, which entered service with the Frontal Aviation's attack units of the 16th Air Army in East Germany in 1973, was a variable-geometry fighter-bomber with wings sweeping from 23 to 71 degrees, and was the Soviet Air Force's first true multi-role combat aircraft.

Below: Developed as a ground-attack counterpart to the MiG-23 fighter, the MiG-23BN was optimized for export, and saw service with Bulgaria, Czechoslovakia (as here) and East Germany within Eastern Europe. In addition to free-fall bombs and unguided rockets, the MiG-23BN could launch Kh-23 air-to-surface missiles.

COCKPIT
The fact that the MiG-23 was not designed to dogfight can be seen in its cockpit. Rearward visibility is very poor, even with mirrors fitted on the canopy arch.

MiG-23BN
The variable-geometry MiG-23BN was the most advanced offensive asset available to the Syrian AF at the time of the 1982 conflict in Lebanon. In fact, Syria had deployed its MiG-23BNs over Lebanon before the Israeli invasion, with two examples being claimed destroyed over Bekaa during an April 1982 raid.

RADAR
The MiG-23 was always constrained by the performance of its radar, which lacked a true 'lookdown' capability until the final 'ML' variant was introduced.

MIG-23 VARIANTS
The MiG-23M Flogger-B was the first series production version and equipped all the major Warsaw Pact air forces; a simplified version for export to Libya and other Middle East air forces was designated MiG-23MS Flogger-E. The MiG-23UB Flogger-C was a two-seat trainer, retaining the combat capability of the single-seat variants, while the MiG-23BN/BM Flogger-F and Flogger-H were fighter-bomber versions for export.

SOVIET MIG-23
A Soviet MiG-23, flying in May 1989. The MiG-23 equipped nearly all allies of the former Soviet Union, and was still being refined and improved three decades after entering service.

Saab SF-37 Viggen (1971)

TYPE • Fighter COUNTRY • Sweden

Until the debut of the Panavia Tornado, the Saab Viggen was probably the most advanced combat aircraft ever produced in Europe, with a more advanced radar, greater speed range and more comprehensive avionics than its contemporaries.

SPECIFICATIONS (AJ-37)

DIMENSIONS:	Length: 16.40m (53ft 9.75in); Wingspan: 10.60m (34ft 9.25in); Height: 5.6m (18ft 3in)
WEIGHT:	17,000kg (37,478lb) loaded
POWERPLANT:	One 125.04kN (28,110lbf) Volvo Flygmotor RM8B afterburning turbofan
MAX SPEED:	2126km/h (1321mph)
RANGE:	2000km (1243 miles)
CEILING:	18,000m (59,055ft)
CREW:	1
ARMAMENT:	Up to 5897kg (13,000lb) of bombs, rockets or air-to-surface missiles

Right: The Viggen's wraparound single-piece windscreen gave the pilot an excellent view forward, and was strengthened to withstand bird strikes at high speed.

RADAR
The Ericsson PS-37 radar was capable of ground mapping, air-to-ground ranging and ground proximity warning.

FOLDING FIN
The fin could be folded down to port, reducing aircraft height and facilitating storage in Sweden's network of underground hangars.

WINGS
The wing incorporated hydraulically activated two-section elevons on the trailing edge; the leading edge has compound sweep and is extended forward on the outer sections, outboard of the prominent bullet fairings, which accommodate RW antennae.

MOTORWAY OPS
The Saab 37 Viggen (Thunderbolt) was designed to carry out the four roles of attack, interception, reconnaissance and training. Powered by a Swedish version of the Pratt & Whitney JT8D turbofan engine, with a powerful Swedish-developed afterburner, the aircraft had excellent acceleration and climb performance. Part of the requirement was that it should be capable of operating from Swedish motorways.

VINTAGE VIGGEN
A vintage Saab Viggen fighter performs at the Sanicole Sunset Airshow in Belgium, 2019. The Viggen incorporated many advanced features that have become standard, including a head-up display and a navigation/attack computer for accurate delivery of ordnance.

Tupolev Tu-134 (1971)

TYPE • *Trainer* COUNTRY • *Soviet Union*

A trainer for bomber pilots and navigators, the Tupolev Tu-134 (NATO reporting name 'Crusty') was based on the Tu-134 regional commercial aircraft and had a flight crew of four – two pilots, navigator and flight engineer.

SPECIFICATIONS (Tu-134Sh)

DIMENSIONS:	Length: 37.05m (121ft 7in); Wingspan: 29.01m (95ft 2in); Height: 9.14m (30ft)
WEIGHT:	Maximum take-off 47,000kg (103,617lb)
POWERPLANT:	Two Aviadvigatel D-30 series II turbofans each rated at 68.0kN (15,287lbf)
MAX SPEED:	860km/h (534mph)
RANGE:	3400km (2113 miles)
CEILING:	12,100m (39,700ft)
CREW:	4 (plus 13 trainees)
ARMAMENT:	Up to 600kg (1323lb) of practice bombs

Right: The first Tu-134Sh (Shturmanskiy, Russian word for navigators) navigator/bombardier trainer flew in Kharkiv on 12 February 1971; 90 aircraft of this variant were built and used for training navigators of bomber aircraft and for bombing practice.

CABIN
The cabin had 13 workstations for trainee navigators, the two at the front having a bombsight and a radar sight respectively.

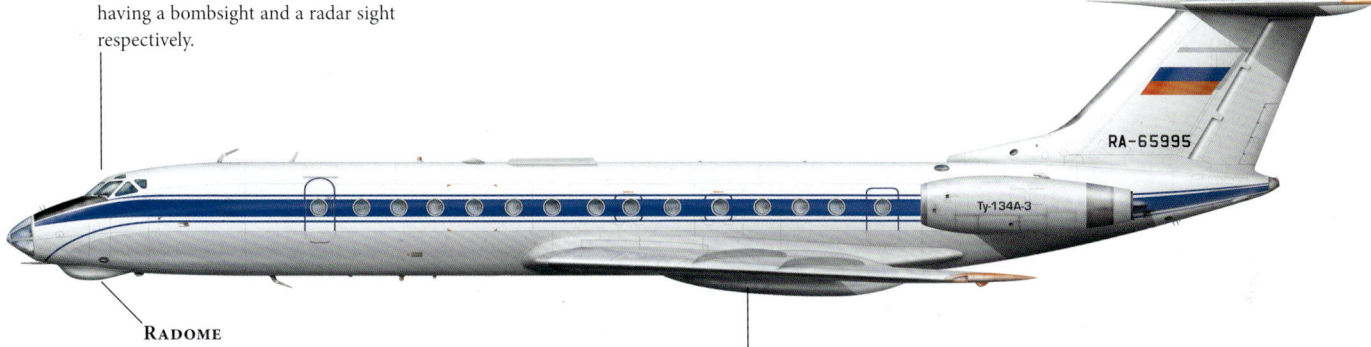

RADOME
The Tu-134Sh was similar in overall appearance to the Tu-134A passenger version, but it featured an enlarged radome under the glazed nose.

BOMB RACKS
Two bomb racks, each for four 120kg (264lb) practice bombs were fitted, later replaced with two bomb racks for six 50kg (110lb) practice bombs each.

CIVIL VARIANTS

Design of the Tu-134 was inspired by the visit of Soviet leader Nikita Khrushchev to France in 1960, when he flew in a Caravelle airliner. He then ordered Andrei Tupolev to relocate the engines of the Tu-124 airliner to the rear fuselage. On 29 July 1963 the prototype Tu-134 (Tu-124A) made its maiden flight in Kharkiv. The aircraft entered scheduled service with Aeroflot in September 1967. The major passenger versions comprised the Tu-134 with 72 passengers, Tu-134A with fuselage stretch and 76 passengers, and Tu-134B for 80 passengers.

TRAINER AIRCRAFT
Today around a dozen examples of the Tu-134Sh, Tu-134UB-L and Tu-134UB-KM trainer aircraft are operated by the navigator school in Chelyabinsk, a pilot school in Balashov, plus operational training centres in Ryazan and Yeysk.

Grumman EA-6B Prowler (1971)

TYPE • Ground-attack COUNTRY • United States

SPECIFICATIONS

DIMENSIONS:	Length: 17.98m (59ft); Wingspan: 16.15m (53ft); Height: 4.57m (15ft)
WEIGHT:	Empty 15,237kg (33,600lb); maximum take-off 27,890kg (61,500lb)
POWERPLANT:	Two 5080kg (11,200lb) thrust Pratt & Whitney J52-P-408 turbojets
MAX SPEED:	982km/h (610mph)
RANGE:	1769km (1099 miles)
CEILING:	11,580m (38,000ft)
CREW:	4
ARMAMENT:	Provision for four anti-radar missiles on external hardpoints

Based on the Grumman A-6 Intruder, the EA-6B Prowler electronically 'cleaned' combat zones so that friendly warplanes could attack in safety. The Prowler took a pilot and three operators into action with a powerhouse of 'black boxes', intent on jamming an enemy's radar.

Right: The EA-6B is equipped with nosewheel steering that becomes active once the arrester hook is deployed. The steering is controlled by the rudder pedals and will allow turns of up to 60 degrees.

JAMMING
Self-protection jamming to decoy enemy radar-guided missiles is provided by a deception jamming suite. The antenna for this is located next to the refuelling probe.

CREW
The pilot sits in the front port cockpit, surrounded by three electronic countermeasures officers (ECMOs). ECMO one sits by his side and operates the navigation, radar and communications equipment, with ECMOs two and three operating the tactical jamming suite.

RECEIVER
The large pod on top of the tailfin houses the system integration receiver, which detects hostile radar emissions and sends them to a central computer for threat analysis.

JAMMING PODS
The heart of the Prowler was the AN/ALQ-99 Tactical Jamming System. The Prowler could carry five jamming pods, one belly-mounted and the others on the wings; each pod was integrally powered and housed two jamming transmitters that covered seven frequency bands. The EA-6B was upgraded several times and was finally retired from US military service in 2019.

PROWLER ON STATIC DISPLAY
Although the Prowler is now retired, it acquired a prestigious combat record over its service career. The Prowler fought in Vietnam (1972) and in many wars since – Grenada (1983), Libya (1986), both Gulf Wars, Bosnia (1995) and Afghanistan.

Bell AH-1J SeaCobra (1971)

TYPE • Helicopter COUNTRY • United States

Developed specifically to meet USMC close-support requirements, the AH-1J introduced a twin-turboshaft powerplant to the AH-1 airframe, and its chin turret mounted a three-barrel M197 20mm (0.79in) cannon.

SPECIFICATIONS

DIMENSIONS:	Length: 13.5m (44ft 3in); Main rotor diameter: 13.4m (43ft 11in); Height: 4.1m (13ft 5in)
WEIGHT:	4525kg (9979lb) loaded
POWERPLANT:	1 x 1342kW (1800shp) Pratt & Whitney Canada T400-CP-400 turboshaft
MAX SPEED:	352km/h (218mph)
RANGE:	571km (355 miles)
CEILING:	3475m (11,398ft)
CREW:	2
ARMAMENT:	1 x 20mm (0.8in) M197 cannon; 14 x 70mm (2.75in) Mk 40 rockets; 8 x 127mm (5in) Zuni rockets; 2 AIM-9 anti-aircraft missiles

Above: The AH-1J SeaCobra was a formidable attack platform. USMC SeaCobras operated from ships and were used to support ground operations, particularly beach assaults.

UPGRADE
The AH-1T 'Improved SeaCobra' that followed the AH-1J had a lengthened fuselage and tailboom.

M197 CANNON
The M197 cannon in a three-barrel, electrically powered rotary design, with a rate of fire between 750 and 1500rpm.

STUB WINGS
The stub wings on the side of the fuselage not only provided extra lift in forward flight, but they also offered a weapon mounting platform, with two hardpoints per wing.

ATTACK HELICOPTER

Development of a dedicated attack helicopter for the US Marine Corps was inspired by the US Army's application of the AH-1G Cobra in Vietnam. In 1968, the Bell was contracted to produce 49 SeaCobras (although the Marines also used AH-1Gs). The AH-1J was a twin-engine variety of Cobra and it came fitted with a Pratt & Whitney Canada T400-CP-400 turboshaft and the potent three-barrel 20mm (0.79in) M197 cannon under the nose.

USMC AH-1J
The AH-1J SeaCobra was upgraded for the Marines during the 1980s as the AH-1T variant, capable of handling Hellfire and Sidewinder missiles. This was in turn superseded by the AH-1W SuperCobra, then the AH-1Z Viper, both with more potent weaponry and fire control.

Tupolev Tu-22M (1972)

TYPE • Bomber **COUNTRY** • Soviet Union

SPECIFICATIONS

DIMENSIONS:	Length 42.46m (139ft 4in); Wingspan: 23.3m (76ft 5in) fully swept or 34.28m (112ft 6in) fully spread; Height: 11.05m (36ft 3in)
WEIGHT:	Maximum take-off 124,000kg (273,373lb)
POWERPLANT:	Two Kuznetsov NK-25 (izdeliye Ye) afterburning turbofans each rated at 140.2kN (31,526lbf) dry and 245.18kN (55,115lbf) with afterburning
MAX SPEED:	2300km/h (1429mph)
RANGE:	6800km (4225 miles)
CEILING:	14,000m (45,932ft)
CREW:	4
ARMAMENT:	One 23mm (0.906in) Gryazev-Shipunov GSh-23 twin-barrel Gast auto-cannon in remotely controlled tail turret; up to 24,000kg (53,000lb) of bombs, missiles or mines

A supersonic intermediate-range bomber and missile carrier intended to deliver both nuclear and conventional weapons, the Tupolev Tu-22M (NATO reporting name 'Backfire') has been in service for over 50 years.

This page: This Tupolev Tu-22M-3 'Backfire-C' is a Russian navy aircraft configured for a long-range anti-shipping mission.

TU-22M-3
The Tu-22M-3 introduced not only more powerful turbofans, but reconfigured high-performance ramp-type intakes similar in shape to those on the Russian MiG-25 'Foxbat'.

CREW
The 'Backfire-C' carries a crew of four: a pilot and co-pilot side-by-side in front and two further crew behind. All four have ejection seats under gull-wing doors.

NEW REQUIREMENT
After the disappointing performance of the original Tu-22 (NATO reporting name Blinder), the Soviet government called upon Tupolev to develop the Tu-22M in November 1967. Although maintaining the Tu-22 designation, the Tu-22M was effectively a completely new design and was to prove vastly more successful than its predecessor. Requirements included a maximum speed of 2300–2500km/h (1429–1553mph) and subsonic range of 7000km (4349 miles) while armed with one Kh-22 missile.

ORDNANCE
For conventional missions 12,000kg (26,400lb) of bombs can be carried in the weapons bay and another 12,000kg (26,400lb) on underwing racks. The heaviest weapon is the FAB-3000 3000kg (6600lb) bomb.

RUSSIAN TU-22M
Around 60 Tu-22Ms remain in service in Russia today (Ukraine's were scrapped by 2006), with more in storage. Main armament is the Raduga Kh-22 (AS-4 Kitchen) supersonic heavy missile that dives onto its target at a fearsome speed of Mach 4.15.

Fairchild Republic A-10 Thunderbolt II (1972)

TYPE • Ground-attack **COUNTRY** • United States

SPECIFICATIONS

DIMENSIONS:	Length: 16.26m (53ft 4in); Wingspan: 17.53m (57ft 6in); Height: 4.47m (14ft 8in)
WEIGHT:	22,680kg (50,000lb) maximum take-off
POWERPLANT:	2 × General Electric TF34-GE-100 high-bypass turbofans, 40.32kN (9065lbf)
MAX SPEED:	706km/h (439mph)
RANGE:	3949km (2454 miles) with drop tanks
CEILING:	13,716m (45,000ft)
CREW:	1
ARMAMENT:	1 × 30mm (1.18in) rotary cannon plus up to 7260kg (16,000lb) of external ordnance on 8 wing and 3 fuselage pylons

In December 1970 Fairchild Republic and Northrop were each selected to build a prototype of a new close support aircraft for evaluation under the USAF's A-X programme. The winner was Fairchild Republic's formidable Thunderbolt II, affectionately nicknamed the 'Warthog'.

This page: An A-10A of the 706th Tactical Fighter Squadron, 926th Tactical Fighter Group, as it appeared during Operation Desert Storm wearing the 1980s-era European One camouflage scheme.

AMMUNITION
The Avenger cannon can fire three different rounds: high-explosive incendiary, armour-piercing incendiary and training practice.

STORES STATIONS
Each wing supports four stores stations, with another three under the fuselage.

TAIL
The A-10 is intended to be able to fly with either half of the twin-finned tail shot away. All parts of the left and right tails are interchangeable.

WARTHOG AT WAR
At the end of the Cold War, the USAF considered withdrawing the A-10, however, the 'Warthog' received a new lease of life with the 1990 Iraqi invasion of Kuwait. A total of 152 OA/A-10s from bases in the United States and UK conducted missions during the Gulf War, with a notable success rate during Operation Desert Storm. As well as two air-to-air victories against Iraqi helicopters, it averaged a kill rate of more than 25 tanks a day. By the end of the war, confirmed A-10 tank kills approached 1000 tanks destroyed.

MAVERICK MISSILE LAUNCH
A US pilot fires an AGM-65 Maverick missile from an A-10 Thunderbolt II over the Pacific Alaska Range Complex during live-fire training. The Maverick is an air-to-ground missile suited to targets ranging from armour to storage facilities.

Aero L-39 Albatros (1972)

TYPE • *Trainer* COUNTRY • *Czechoslovakia*

An excellent design, the Aero L-39 Albatros became the standard jet trainer in the air forces of the Warsaw Pact, and was exported as a light attack aircraft. In its most modern variants, it remains in widespread production and service today.

SPECIFICATIONS (L-39C)

DIMENSIONS:	Length: 12.32m (40ft 5in); Wingspan: 9.46m (31ft); Height 4.72m (15ft 5.5in)
WEIGHT:	Empty 3330kg (7341lb); maximum take-off 5270kg (11,618lb)
POWERPLANT:	One 1720kg (3793lb) Ivchenko AI-25TL turbofan
MAX SPEED:	630km/h (391mph)
RANGE:	1750km (1087 miles)
CEILING:	9000m (29,525ft)
CREW:	1-2
ARMAMENT:	One 23mm (0.9in) twin-barrel cannon; 2 AAMs; 57 or 130mm (2.24 or 5.12in) rocket pods; bombs up to 500kg (1102lb)

Above: The L-39's Ivchenko turbofan is fed by two high-set air intakes behind the cockpit. Their position was chosen to minimize the FOD (foreign object damage) ingestion.

TRAINING CONFIGURATION
The L-39C's student and instructor sat in tandem in separate, well-appointed (if outdated by western standards) cockpits.

TAILPLANE
A feature of the L-39, except for the L-39MS/L-59, is an electrically operated variable-incidence tailplane. Elevators are also fitted, although these are manually actuated with a small trim tab.

UNDERCARRIAGE
The L-39 was fitted with a simple and robust undercarriage incorporating many features that made it suitable for operation from semi-prepared strips.

MODERNIZATION

The L-39 Albatros, successor to Aero's earlier L-29, was until recent years Russia's standard trainer. More than 2000 L-39s were delivered to the former Soviet Union between 1973 and 1989. The type sold well outside the former Eastern Bloc. Serving with the air forces of at least 30 other countries, the Albatros has been progressively modernized, and resulted in the more capable L-59 and a number of other variants.

CZECH AIR FORCE L-39
Czechoslovakia was a major customer for the L-38. For the Czech air force, Aero also produced the L-159, a single-seat attack version with a 28kN (6,300lbf-thrust) Garrett F124 engine.

SEPECAT Jaguar (1972)

TYPE • Bomber COUNTRY • France/United Kingdom

Designed as a tactical support aircraft, the Anglo-French SEPECAT Jaguar saw considerable action during the Gulf War of 1991 and proved its worth beyond all doubt in precision attacks on Iraqi positions.

SPECIFICATIONS (GR.Mk.1A)

DIMENSIONS:	Length: 16.83m (55ft 2.5in); Wingspan: 8.69m (28ft 6in); Height: 4.89m (16ft 0.5in)
WEIGHT:	Empty 7000kg (15,435lb); maximum take-off 15,500kg (34,178lb)
POWERPLANT:	Two 3313kg (7305lb) thrust Rolls-Royce/Turbomeca Adour Mk 102 turbofans
MAX SPEED:	1593km/h (990mph)
RANGE:	1600km (994 miles)
CEILING:	15,240m (50,000ft)
CREW:	1
ARMAMENT:	Two 30mm (1.19in) cannon; five external hardpoints for 4536kg (10,000lb) of munitions; two AAMs

Left: As the Jaguar was a strike aircraft, its single-seat cockpit did not require the all-round visibility of an air superiority fighter. Internally, the cockpit was old-fashioned, having been designed long before the era of digital displays.

JAGUAR GR.1A
This Jaguar GR.1A, XZ364 'Sadman' of the Coltishall Wing, deployed to Muharraq in Bahrain for combat operations during Operation Desert Storm, in which it flew 47 sorties. Note the desert camouflage.

ARMAMENT
The Jaguar could carry a maximum warload of eight 454kg (1000lb) bombs, but four was a more typical load on Gulf War missions. This example is carrying two laser-guided bombs, with Sidewinders on overwing mounts for self-defence.

CANNON
A hard-hitting 30mm (1.19in) ADEN Mk.4 cannon was mounted on either side of the fuselage, below the engine air intakes. Each gun carried 150 rounds.

ANGLO-FRENCH
A Franco/British project, the result of collaboration between the British Aircraft Corporation (now British Aerospace) and Dassault-Breguet, the SEPECAT Jaguar was first flown on 8 September 1968 as a single-seat attack aircraft with limited all-weather capability. It was intended to serve both the Armée de l'Air and the RAF; the French Jaguar A entered service first, in May 1972. The RAF took delivery of its first GR.Mk 1 in May 1973.

FRENCH JAGUAR A/E
A French Air Force Jaguar A/E flies a refuelling mission over the Adriatic Sea in support of Operation Joint Forge, a stabilization operation conducted over the former Yugoslavia between 1998 and 2004.

Mi-24 'Hind' (1972)

TYPE *Helicopter* **COUNTRY** • *Soviet Union*

SPECIFICATIONS (Mi24P)

DIMENSIONS:	Fuselage length: 17.51m (57ft 5in); Main rotor diameter: 17.3m (56ft 9in); Height: 5.47m (17ft 11in)
WEIGHT:	Maximum take-off 11,500kg (25,353lb)
POWERPLANT:	Two Klimov-St Petersburg TV3-117VM or TV3-117VMA turboshaft engines each rated at 1641kW (2200shp)
MAX SPEED:	320km/h (199mph)
RANGE:	450km (280 miles)
CEILING:	4500m (14,764ft)
CREW:	2 or 3
ARMAMENT:	One 30mm (1.2in) Gryazev-Shipunov GSh-30-2 twin-barrel Gast autocannon flexibly mounted under nose; up to 1500kg (3307lb) of weapons and stores

The archetypal combat helicopter of the Soviet Union, the Mi24 'Hind' remains in service with Russia and over 50 other nations. In modernized form it remains in production at Rostvertol in Rostov-on-Don.

Above: The Mil Mi-35M is to date the most advanced version of the 'Hind' series to have entered serial production, featuring updated avionics, targeting, and communication systems. As well as serving with the Russian Aerospace Forces, the Mi-35M has been exported; '03 Red' serves with the Air Defence Forces of Kazakhstan.

POWERPLANT
Twin 1434-kW Isotov TV-3 turboshafts powered most Mi-24s, although early production batches were fitted with smaller TV-2s as installed in the Mi-8, on which the Mi-24 design was based.

MI-24 HIND-A
Libya was among four Soviet allies to receive the 'Hind-A', the others being Algeria, Afghanistan and Vietnam. It is believed that few, if any, 'Hind-As' remain in service.

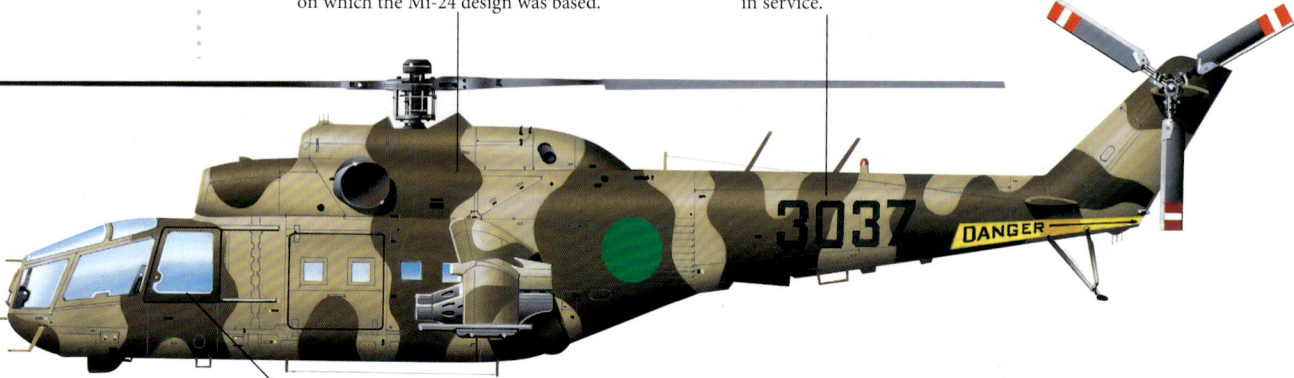

GUNSHIP
Most 'Hinds' are gunships, with a stepped tandem canopy housing a weapons operator in front and a pilot higher to the rear. Either can aim the gun with a magnifying sight in a bulge under the nose, which also contains a laser tracker for missiles. After combat experience in Afghanistan, Mil introduced an improved 'Hind' with a twin-barrelled GSh-23L 30mm (1.18in) cannon. This, together with its rockets and missiles, makes the 'Hind' very much a close-support weapon.

CREW POSITIONS HIND-A
The cockpit had three seats. The crew consisted of a gunner in the centre front, a flight engineer/co-pilot behind him and the pilot offset to the left rear.

MI-24, SLOVAKIA
The Mi-24's distinctive stub wings support pylons for up to 1500kg (3307lb) of weapons. These include unguided rockets and anti-tank guided missiles as well as versions of man-portable anti-air missiles capable of engaging slow airborne targets (such as other helicopters).

Tupolev Tu-142 (1972)

TYPE • Bomber COUNTRY • Soviet Union

In 1963 the Soviet government called for Tupolev to develop a long-range ASW version of the Tu-95 strategic bomber able to patrol and destroy enemy submarines at 4000km (2485 miles) from its base. The result was the Tu-142 (NATO reporting name 'Bear-F').

SPECIFICATIONS (Tu-142MK)

- **Dimensions:** Length: 53.09m (174ft 2in); Wingspan: 50.04m (164ft 2in); Height: 13.62m (44ft 8in)
- **Weight:** Maximum take-off: 185,000kg (407,885lb)
- **Powerplant:** Four 11,185ekW (15,000ehp) Kuznetsov NK-12MV turboprops
- **Max Speed:** 855km/h (531mph)
- **Range:** 12,000km (7456 miles)
- **Ceiling:** 12,000m (39,370ft)
- **Crew:** 10–13
- **Armament:** Up to 9000kg (19842lb) of weapons and stores

Right: An Indian Navy Tu-142MK-E. The Indian Navy flew the Tu-142 for 29 years, INAS (Indian Naval Air Squadron) 312 operating eight ex-Soviet aircraft between 1988 and 2017.

Crew
The Tu-142 was typically operated by a normal crew of 10, consisting of two pilots, two navigators, navigator/weapon system operator, two radio sonobuoy system operators, communications operator, onboard technician and rear gunner.

Communications
The Tu-142MR variant carries the Oryol communication relay system, with a suite of HF radios and satellite communications to connect ground-based or airborne command posts, very-low frequency (VLF) radio to connect submerged submarines, a data encrypting system and relay system.

Payload
Both the Tu-142MK and MZ can carry up to 9000kg (19,841lb) of stores in two fuselage weapon bays. A typical payload includes three torpedoes and 135 sonobuoys of different types.

DEVELOPMENT
Originally designated Tu-95PLO (Protivo-Lodochnoy Oborony, anti-submarine defence), by the time the aircraft made its maiden flight in Kuibyshev (present-day Samara) on 18 June 1968 the designation had changed to Tu-142 (izdeliye VP), known to NATO as Bear-F Mod 1. It was operated by a normal crew of 10. Series production began in Samara the same year, the first 18 aircraft being fitted with the Berkut-95 mission system derived from the Berkut system used in the Il-38.

SOVIET TU-142
Over 100 Tu-142s of all versions were completed. Currently, the Russian Navy has two Tu-142 squadrons, with Tu-142MK and Tu-142MZ versions, based at Kipelovo-Fedotovo with the Northern Fleet, and at Mongokhto with the Pacific Fleet, for a total of around 16 airworthy aircraft.

Lockheed Martin F-16 Fighting Falcon A/B (1974)

TYPE • Fighter COUNTRY • United States

The F-16 is proof that one aircraft can move back the boundaries of aviation. This marvellous warplane introduced lightweight computers, 'fly-by-wire' electronic controls and a breathtaking arsenal of hi-tech weaponry.

SPECIFICATIONS
(F-16B Block 20 MLU)

DIMENSIONS:	Length: 15.14m (49ft 8in); Wingspan: 9.97m (32ft 8in); Height: 4.9m (16ft 1in)
WEIGHT:	Maximum takeoff 17,010kg (37,500lb)
POWERPLANT:	One Pratt & Whitney F100-PW-220 turbofan
MAX SPEED:	1473km/h (915mph)
RANGE:	1166.77km (725 miles)
CEILING:	6599m (21,650ft)
CREW:	2
ARMAMENT:	One M61A1 20mm (0.79in) cannon; AIM-9 Sidewinder, AIM-7 Sparrow, AIM-120 AMRAAM air-to-air missiles; AGM-65 Maverick missile, GBU-31 (JDAM) and GBU-38 (JDAM) guided bombs

Above: Nearly all F-16s are painted in grey – the colour found to be the most difficult to see across different weather conditions.

RADAR
The radar of the F-16 is as versatile as the aircraft. With a flick of a switch the pilot can change from air-to-air operation to air-to-ground. When dogfighting, the radar automatically follows the enemy and gives the pilot a steering cue on the display in front of him.

CONFIGURATION
With its curved surfaces blending the fuselage and wing together, and its fly-by-wire electric flight control system, the F-16 ushered in a new era of fighter design. The radical shape had far better aerodynamics than earlier designs.

DANISH SERVICE
Built in the Belgian SABCA factory, this F-16B Block 20 MLU is part of the Royal Danish Air Force, Eskadrille 727. Esk 727 was the first Danish squadron to receive the F-16 fighter.

MANOEUVRABILITY
Designed as a multi-role fighter, the Fighting Falcon could turn and fight with unbridled fury when provoked and was one of the first operational 'fly-by-wire' aircraft, its controls being electronically operated and computer-controlled. The pilot sits in a seat which reclines at 30 degrees to withstand high-g manoeuvres, allowing the gut-wrenching turns that give the F-16 an advantage over its rivals.

MISSILE LAUNCH
An F-16 launches an AGM-88 HARM (high-speed anti-radiation missile). The F-16's underwing hardpoints are stressed for manoeuvres up to 9g, enabling the aircraft to dogfight while still carrying weaponry.

Panavia Tornado (1974)

TYPE • *Strike aircraft* **COUNTRY** • *Multi-national*

SPECIFICATIONS (Tornado ADV)

DIMENSIONS:	Length: 18.68m (61ft 3in); Wingspan: 8.6m (28ft 2in) maximum sweep; Height: 5.95m (19ft 6in)
WEIGHT:	27,987kg (61,700lb) maximum take-off
POWERPLANT:	2 × Turbo-Union RB.199 Mk 104 afterburning turbofans, 71.50kN (16,075lbf)
MAX SPEED:	2338km/h (1453mph) at 10,975m (36,000ft)
RANGE:	3890km (2417 miles)
CEILING:	15,240m (50,000ft)
CREW:	2
ARMAMENT:	Two 27mm (1.063in) cannon; up to 5806kg (12,800lb) of AAMs

Developed as an all-weather strike aircraft and interdictor, by the mid-1980s the Tornado had established itself as arguably the most important combat aircraft in Western Europe. It also served as the basis for a long-range interceptor and a dedicated defence suppression aircraft.

Right: A Tornado ADV of the Royal Saudi Air Force. RSAF Tornado ADVs flew 451 sorties during Operation Desert Storm, protecting Saudi airspace against any Iraqi intrusion.

COCKPIT
As an interceptor, the ADV does not need the all-round visibility of a high-agility dogfighter from its comfortable, modern cockpit.

STRUCTURE
Structural changes for the ADV involved stretching the fuselage by 136cm (53in). The wing root glove was also given increased sweep, moving the centre of pressure forward to compensate for the resultant change of centre of gravity and reduce wave drag.

AIR DEFENCE VARIANT
With an outstanding RAF requirement for a long-range all-weather interceptor, Panavia developed the Air Defence Version (ADV), which was first flown in October 1979 as the Tornado F.Mk 2. The aircraft accommodated Foxhunter radar and tandem pairs of Sky Flash missiles semi-recessed under a lengthened fuselage. Internal fuel was increased to extend unrefuelled endurance to around four hours 30 minutes.

ARMAMENT CHANGE
A further change involved the deletion of one of the ADV's two planned Mauser 27mm (1.063in) cannon, giving more space for avionics.

WEAPONS LOAD
The original Tornado ADV study envisaged four Sky Flash missiles under the wings, long-range tanks under the fuselage and a modified nose to accommodate the AI radar. Early aerodynamic trials showed that with pylon-mounted missiles the ADV gave little or no advantage over the Phantom it was intended to replace. The answer was to carry the AAMs semi-submerged under the fuselage (seen here), reducing drag.

Dassault Mirage F.1 (1974)

TYPE • Fighter COUNTRY • France

France's Dassault company has always been prepared to risk privately funded ventures, and the risk has been amply justified by its export record. The Mirage F.1 was one of its major success stories of the 1970s and 1980s.

SPECIFICATIONS (F.1AZ)

Dimensions: Length: 15.0m (49ft 2.25in); Wingspan: 8.40m (27ft 6.67in); Height: 4.50m (14ft 9in)

Weight: Maximum take-off 15,200kg (33,510lb)

Powerplant: One 7200kg (15,876lb) thrust SNECMA Atar 9K-50 turbojet

Max Speed: 2350km/h (1460mph)

Range: 900km (560 miles)

Ceiling: 20,000m (65,615ft)

Crew: 1

Armament: Two 30mm (1.19in) cannon; up to 6300kg (13,892lb) of external munitions

Right: This Mirage F.1AZ (Z for Zuid Afrika) served with No. 1 Squadron, South African Air Force, at Hoedspruit AB. The SAAF's last Mirage F.1s were retired in 1997.

Radar warning receiver
The fin-mounted forward- and rearward-facing antennae for the Thomson-CSF BF radar warning receiver. Sideways cover was provided by disc antennae flush with the fin sides.

Ranging radar
The F.1A carried a small EMD Aida 2 ranging radar in the extreme nose. The radar had a fixed antenna and provided automatic search, acquisition, ranging and tracking for targets within its 16 degree field of view. Data was presented to the pilot in his gyro gunsight.

Armament
The F.1AZ's principal armament comprised two internal cannon, with most stores carried on multiple dispensers on the centreline. As shown here, the F.1AZ could be fitted with wingtip launch rails for the V3B Kukri or V3C Darter indigenous air-to-air missile.

F.1 WING DESIGN
The Mirage F.1's wing, a departure from the traditional Dassault delta format, was fitted with elaborate high-lift devices, which permitted the aircraft to take off and land within 500–800m (1600–2600ft) at average combat mission weight. The primary role of the Mirage F.1 was all-weather interception at any altitude, and the original production version used the same weapon systems as the Mirage III.

EXPORT F.1
Aircraft Mirage F.1 taking part in an exhibition at the first airshow of Cadiz in September 2008. The Mirage F.1 was the subject of large overseas export orders, notably to countries in the Middle East.

Lockheed S-3 Viking (1974)

TYPE • *Maritime patrol* COUNTRY • *United States*

Although it began life as a dedicated anti-submarine warfare (ASW) platform, the extremely versatile Lockheed S-3 Viking went on to undertake a multitude of other roles, including tanking, mining and limited electronic surveillance.

SPECIFICATIONS (S-3B)

DIMENSIONS:	Length: 16.26m (53ft 4in); Wingspan: 20.93m (68ft 8in); Height: 6.93m (22ft 9in)
WEIGHT:	Maximum take-off 19,278kg (42,500lb)
POWERPLANT:	Two 4207kg (9275lb) thrust General Electric TF34-GE-400B turbofans
MAX SPEED:	828km/h (518mph)
RANGE:	3705km (2302 miles)
CEILING:	12,200m (40,000ft)
CREW:	4
ARMAMENT:	Provision for up to 1794kg (3958lb) of munitions

Above: When fulfilling its tanker mission, the S-3B used an ARS 31-301 'buddy' pod mounted on the left wing station. Almost all Viking missions were flown with the 'buddy' pod affixed.

S-3B VIKING
This S-3B of VS-245 'Scouts' is seen as it appeared during its 1997 cruise aboard the carrier USS *John F. Kennedy* (CV-67) as part of Carrier Air Wing Eight. VS-24 was the first unit to deploy the S-3B in combat, during the 1991 Operation Desert Storm.

CREW
For most of its career, the Viking flew with a crew of four in the ASW role, comprising two pilots, a tactical co-ordinator (Tacco) and an enlisted sensor operator (Senso). Aircraft converted as permanent tankers had a two-man crew – the pilot and a naval flight officer.

S-3B STANDARD
The Viking fleet was substantially updated to S-3B standard in the early 1990s, some aircraft being converted to the electronic warfare role as ES-3As. The S-3B's high-speed computer system processed information from the acoustic and non-acoustic target sensor systems; these included new inverse synthetic aperture radar (ISAR) and ESM systems suites.

MISSION LOAD
A typical flexible mission load consisted of one AGM-84D Harpoon ASM and a refuelling pod, with two Mk 82 bombs and two Mk 46 torpedoes in the aircraft's weapons bays.

NASA S-3B
The last operational S-3B Vikings were actually flown by the National Aeronautics and Space Administration (NASA), using the aircraft to support various research programmes. They were retired in July 2021.

Ilyushin Il-76 and Il-78 (1974)

TYPE • Transport COUNTRY • Soviet Union

The Il-76 (NATO reporting name 'Candid') was first seen in the West at the 1971 Paris Air Salon. With a high cruising speed and intercontinental range, it was designed as a capable freighter that could operate from relatively poor and partially prepared airstrips.

SPECIFICATIONS (IL-76M)

DIMENSIONS:	Length: 46.59m (152ft 10.25in); Wingspan: 50.5m (165ft 8.2in); Height: 14.76m (48ft 5in)
WEIGHT:	Maximum take-off 170,000kg (374,786lb)
POWERPLANT:	Four 117.6kN (26,455lbf) Soloviev D-30KP-1 turbofans
MAX SPEED:	850km/h (528mph)
RANGE:	5000km (3107 miles)
CEILING:	15,500m (50,850ft)
CREW:	7
ARMAMENT:	Optional two 23-mm (0.9in) GSh-23L twin-barrelled cannon in tail turret

Above: In July 1993 Il-38 '34 Blue' arrived in dramatic style at the Royal International Air Tattoo at RAF Fairford in the UK, by appearing over the airfield while simulating refuelling a Tu-95MS 'Bear-S'. The pods containing the refuelling drogues are apparent under the wings and on the rear fuselage.

COCKPIT
The cockpit seats a crew of seven, including two freight handlers. The glazed nose houses a navigator for negotiating combat landings without the use of the chin radar. All systems are designed for all-weather, day or night operations.

IL-76MD
A Ukrainian Air Force Ilyushin Il-76MD. Although the Il-76MD has more powerful engines and advanced electronic systems than its predecessor, its heavy lift capability is of limited use to the Ukrainian forces.

AWACS RESPONSE
As the effectiveness of Boeing's E-3 Sentry AWACS (airborne warning and control system) aircraft became apparent, Soviet designers began working on an equivalent. The Ilyushin Il-76 'Candid' formed the basis of Beriev's A-50 response to the requirement, while Ilyushin itself had been working on a further modification of the Il-76, this time to tanker configuration. The definitive Il-78M 'Midas' emerged as a useful three-point in-flight refuelling tanker.

HOLD EQUIPMENT
The 'Candid' has a hold with reinforced titanium flooring and folding roller conveyors. In the roof are two travelling lifter cranes each with two hoists of 2500kg (5500lb) capacity. At the front of the hold are twin winches for loading cargo.

VICTORY DAY OVERFLIGHT
Two Il-76Ms taking part in the Victory Day Parade on 7 May 2021 over Moscow. Russian jet engines tend to be smokier than their Western counterparts and the Il-76's D-30s are no exception to this general rule.

Northrop F-5E/F Tiger II (1974)

TYPE • Fighter COUNTRY • Unietd States

The F-5E/F Tiger II was one of the most successful US export fighters of the Cold War and it continues in service today with a diminishing number of operators, its frontline status a testament to its suitability for a range of avionics, aerodynamics and weapons upgrades.

SPECIFICATIONS (F-5N)

DIMENSIONS:	Length: 14.4m (47ft 4.7in); Wingspan: 8.1m (26ft 8in); Height: 4.1m (13ft 4.25in)
WEIGHT:	Maximum take-off 11,214kg (24,722lb)
POWERPLANT:	Two 22.24kN (5,000lbf) General Electric J85-GE-21C turbojet engines
MAX SPEED:	Mach 1.64 (2025km/h)
RANGE:	1328km (825 miles)
CEILING:	15,240m (50,000ft)
CREW:	1
ARMAMENT:	None

Below: In the Middle East, Bahrain and Iran are the remaining F-5E/F operators, the Islamic Republic of Iran Air Force (seen here) retaining the survivors from what was once the world's second largest F-5 fleet.

F-5E Tiger II
This F-5E Tiger II is in the markings of the Brazilian Air Force's 1 Escadron, 1 Grupo de Aviação Caza; the 'fighting ostrich' insignia was originally used when the unit flew P-47D Thunderbolts with the USAAF's 350th Fighter Group in Italy during World War II.

Refuelling probe
The original Brazilian F-5Es are almost invariably fitted with the detachable flight refuelling probe beneath the starboard canopy rail, allowing the aircraft to refuel at a rate of up to 907kg (2000lb) per minute.

Cannon
The F-5E is armed with a pair of 20mm (0.79in) Pontiac (Colt-Browning) M39A-2 cannon, with up to 280 rounds per gun stored in ammunition boxes below the barrels. The cannon each have a rate of fire of 1500 rounds per minute.

ADVERSARY AIRCRAFT

Developed as an improved version of the F-5A/B Freedom Fighter, the first single-seat F-5E took to the air on 11 August 1972. In total, Northrop completed over 790 F-5Es and 140 two-seat F-5Fs, plus 12 RF-5E reconnaissance variants. Another 90 examples were built under licence in Switzerland, which continues to use the type for air defence. Today, the US Navy and Marine Corps employ 44 former Swiss F-5Es (redesignated as F-5Ns) and F-5Fs as adversary aircraft, ideally suited for simulating types such as the MiG-21.

F-5E AGGRESSORS
Pilots of the 26th Aggressor Squadron aboard five F-5E Tiger II aircraft wait for final preflight checks during Exercise Cope Thunder, conducted in 1984–87.

ShinMaywa SS-2 (1975)

TYPE • *Seaplane* COUNTRY • *Japan*

Introduced in 1967, the ShinMaywa SS-2 is one of the last great flying boats. This aircraft served Japan's Maritime Self-Defence Force so well that in 1992 it was returned to production, designated as the US-1.

SPECIFICATIONS (US-1A)

DIMENSIONS:	Length: 33.46m (109ft 9in); Wingspan: 33.15m (108ft 9in); Height: 9.82m (32ft 3in)
WEIGHT:	Maximum take-off 45,000kg (99,225lb)
POWERPLANT:	Four 2602kW (3490hp) Ishikawajima-built General Electric T64-1H1-10J turboprops; one LHTEC T800 turboshaft, 1,017kW (1,364hp)
MAX SPEED:	495km/h (310mph)
RANGE:	4200km (2610 miles)
CEILING:	8200m (27,000ft)
CREW:	9
ARMAMENT:	None

Right: The SS-2 is almost unique in having a hidden fifth engine. The T-58, installed in the top of the fuselage, blows hot air through ducts to the wing and tail surfaces to help lift performance.

NOSE RADAR
The nose houses a powerful search radar, although the sonar equipment fitted to the older anti-submarine aircraft has been deleted.

ENGINES
The first seven ShinMaywa US-1s were delivered with 2282kW (3060hp) T64-IHI-10E turboprops, but were later retrofitted with the 2602kW (3490hp) 10J engines used on later aircraft.

RESCUE CAPACITY
The US-1 could carry up to 20 seated survivors and 12 stretcher cases in its standard fit, although alternative configurations allowed greater numbers to be carried.

SEAFARER
The seafaring capability of this aircraft makes it a 'guardian angel' to those in peril, and it is credited with several hundred rescues. Its comprehensive rescue equipment includes flares, float lights, droppable life raft containers, maritime markers and even a lifeboat with an outboard motor. The SS-2 with amphibious gear and more powerful engines was almost certainly the most hi-tech seaplane ever built. Today its latest incarnation is the ShinMaywa US-2, introduced in 2007.

SHINMAYWA US-2
A ShinMaywa US-2 of the Japan Maritime Self-Defence Force. Upgrades in this variant compared to the US-1A include more powerful Rolls-Royce engines, a pressurized fuselage and an electronic cockpit.

McDonnell Douglas F-15 Eagle (1975)

TYPE • *Fighter* **COUNTRY** • *United States*

In the 1960s, a new generation of fast and agile Soviet jet fighters presented a challenge to NATO air superiority, and created a pressing need for a combat aircraft to restore the balance. The result was the F-15 Eagle.

SPECIFICATIONS (F-15C)

DIMENSIONS:	Length: 19.43m (63ft 9in); Wingspan: 13.05m (42ft 9.67in); Height 5.63m (18ft 5in)
WEIGHT:	Maximum take-off 30,845kg (68,000lb)
POWERPLANT:	Two 10,800kg (23,810lb) Pratt & Whitney F100-PW-220 turbofans
MAX SPEED:	2655km/h (1650mph)
RANGE:	1930km (1200 miles) on internal fuel
CEILING:	30,500m (100,000ft)
CREW:	1
ARMAMENT:	One 20mm (0.79in) cannon; eight AAMs; up to 7620kg (16,800lb) of munitions

Right: Even when CFTs were fitted, the F-15C could carry Sparrow or AMRAAM missiles on the lower corners of the fuselage. When CFTs were fitted, the Sparrow missiles were mounted on their corners.

RADAR
The AN/APG-63 radar of the F-15C was equipped with a programmable signal processor (PSP), a high-speed, special-purpose computer. It allowed much more rapid switching of the radar between different modes for maximum operational flexibility.

F-15C EAGLE
The F-15C Eagle pictured here bears the markings of the 58th Tactical Fighter Squadron of the 33rd Fighter Wing. It accounted for four of the 58th TFS's 16 kills in the 1991 Gulf War.

DEVELOPMENT
In 1969 it was announced that McDonnell Douglas had been selected as prime airframe contractor for a new air superiority fighter, designated FX. As the F-15A Eagle, it flew for the first time on 27 July 1972, and first deliveries of operational aircraft were made to the USAF in 1975. The F-15 Eagle was designed to outfly and outfight any opponent it might encounter. The tandem-seat F-15B was developed alongside the F-15A, and the main production version was the F-15C (built under licence in Japan as the F-15J).

FAST PACKS
The F-15C can be fitted with FAST (fuel and sensor tactical) packs, now called conformal fuel tanks (CFTs), attached to the side of the fuselage outside each air intake. CFTs carry extra fuel, reconnaissance cameras, infrared equipment, radar warning receivers and jammers.

F-15D EAGLE
The F-15 had many innovations to improve performance. Its wing, for example, was given a conical camber and an airfoil section optimized to reduce wave drag at high speed. The last 20 per cent of the cord was thickened to delay boundary layer separation and so reduce drag.

Mil Mi-14 (1975)

TYPE • Helicopter **COUNTRY** • Soviet Union

Based on the Mi-8 'Hip', the Mi-14 was developed as a land-based anti-submarine helicopter in the early 1970s. A boat hull and retractable landing gear were used to make it suitable for amphibious operations, with more powerful engines compensating for the additional weight.

SPECIFICATIONS (Mil Mi-14PL)

DIMENSIONS:	Length: 18.37m (60ft 3in); Main rotor diameter: 21.29m (69ft 10in); Height: 6.94m (22ft 9in)
WEIGHT:	Maximum take-off 14,000kg (30,865lb)
POWERPLANT:	Two Klimov TV3-117M turboshafts (M for Morskoi, each rated at 1454kW (1950shp)
MAX SPEED:	240km/h (149mph)
RANGE:	1135km (705 miles)
CEILING:	4000m (13,125ft)
CREW:	4
ARMAMENT:	Up to 2000kg (4409lb) of weapons and stores

Right: In common with the Mi-8 'Hip' from which it was developed, the Mi-14 has a five-blade main rotor. A three-blade tail rotor is fitted on the left side of the tail boom, as on the improved Mi-17.

SEARCH RADAR
The under-fuselage radome contains a Type 12-M search radar. A watertight weapons bay on the centreline can carry depth charges and torpedoes.

WEAPONS BAY
Weapons and stores are carried in a weapons bay below the cabin floor. In 'hunter' configuration, the Mi-14 carries up to 36 sonobuoys; the 'killer' variant is armed with one torpedo or eight depth charges, or 12 depth charges, or a single nuclear depth charge.

MIL MI-14PL
Poland is one of the current operators of the 'Haze'. This example operated from Darlowo Airbase during the 1990s.

HAZE-A
The first production helicopter from the Kazan factory flew in January 1974 and in September 1975, and the first Mi-14s were delivered to an operational Soviet Navy unit at Donskoye, near Kaliningrad. The first variant in service was the Mi-14PL Haze-A, the standard anti-submarine version with the Kalmar search/attack system comprising analogue mission computer, Initsiativa series radar, dipping sonar and datalink.

EXPORT MI-14 HAZE
Mi-14s were exported during the Soviet era to Bulgaria, Cuba, East Germany, Libya, Poland, Syria and Yugoslavia. Post-Soviet exports of second-hand helicopters went to Ethiopia, Pakistan and Yemen.

Sukhoi Su-24 (1975)

TYPE • Bomber COUNTRY • Soviet Union

The Sukhoi Su-24 (NATO reporting name 'Fencer') was developed during the Cold War as a high-speed nuclear and conventional tactical bomber, with missions including air interdiction and defence suppression.

SPECIFICATIONS (Su-24M)

DIMENSIONS: Length: 24.53m (80ft 6in); Wingspan: 10.366m (34ft) fully swept or 17.638m (57ft 10in) fully spread; Height: 6.19m (20ft 3in)

WEIGHT: Maximum take-off 39,700kg (87,523lb)

POWERPLANT: Two Lyulka-Saturn AL-21F3 (or AL-21F3A or AL-21F3AT), izdeliye 89 turbojets, each rated at 108.36kN (24361lbf) with afterburning

MAX SPEED: 1450km/h (901mph)

RANGE: 2775km (1724 miles)

CEILING: 11,500m (37,730ft)

CREW: 2

ARMAMENT: One 23mm (0.9in) Gryazev-Shipunov GSh-6-23M rotary cannon in starboard lower fuselage; up to 7500kg (16,535lb) of weapons and stores

Below: The Fencer-C variant seen here featured a new radar warning receiver, long-range navigation system and identification friend or foe (IFF).

VARIABLE GEOMETRY WING
The Fencer's wing has four angle of sweep settings: 16 degrees for take-off and landing, 35 and 45 degrees for cruise at different altitudes, and 69 degrees for high-speed low-level dashes.

SUKHOI SU-24M2
Operated by the 277th Bomber Aviation Regiment, RF-95108 77 is one of a relatively small number of 'Fencers' upgraded to Su-24M2 standard. The aircraft is launching a Kh-25ML (AS-10 'Karen') air-to-surface missile.

SU-24M
The Su-24M originated in a 1975 requirement for a much-improved second-generation bomber, or T-6M, with a wider array of precision-guided weapons. This required a new fire control system and the aircraft also added a more comprehensive self-defence system and in-flight refuelling capability. The first production Su-24M appeared in 1979 and could be distinguished from the basic Su-24 by its nose, which was lengthened by 750mm (29in).

EXTERNAL FUEL
The Su-24 suffers from a limited range. With 3000kg (6600lb) of ordnance and external tanks the combat radius is 615km (382 miles) at low level.

SU-24M, RUSSIA, 2016
This Russian Su-24M served as a combat aircraft over Syria. The Su-24M provided the basis for the all-weather, day and night, multi-sensor tactical reconnaissance Su-24MR, first flown in May 1980.

Mikoyan-Gurevich MiG-27 (1975)

TYPE • Ground-attack COUNTRY • Soviet Union

Based on the MiG-23BM, the MiG-27 (NATO reporting name 'Flogger-D') is a dedicated ground-attack aircraft and is equipped with a six-barrelled 30mm (1.18in) cannon for use against ground targets.

SPECIFICATIONS (MiG-27L 'Flogger-J')

DIMENSIONS:	Length: 16.71m (54ft 10in); Wingspan: 13.97m (45ft 10in) spread and 7.78m (25ft 6in) swept; Height: 4.82m (15ft 9in)
WEIGHT:	18,145kg (40,000lb) loaded
POWERPLANT:	One 10,000kg (22,046lb) Tumanskii R-27F2M-300 turbojet
MAX SPEED:	2445km/h (1520mph)
RANGE:	2500 km (1600 miles)
CEILING:	18,290m (60,000ft)
CREW:	0
ARMAMENT:	One 23mm (0.91in) GSh-23L cannon; underwing pylons for various combinations of air-to-air missiles and offensive stores

Below: A MiG-27 serving with the Soviet Frontal Aviation in the mid-1980s.

MiG-27L Flogger-J
The Indian Air Force's MiG-27s were designated MiG-27L. The first aircraft were assembled from Soviet-supplied kits by Hindustan Aeronautics Ltd, but at a later stage in the programme major Indian sub-assemblies were incorporated. In Indian Air Force service, the type is known as *Bahadur* (Valiant).

Nose sensors
The MiG-27's broad, flat nose contains a small ranging radar and a laser rangefinder, which are capable of locking on to laser energy from a marked target. The nose also holds air data probes and other antennas. The MiG-27 is known to the Russians as *utkanos* (ducknose).

AFGHANISTAN
Soviet forces used MiG-27s during the later stages of the war in Afghanistan in 1987–89. They were fitted with dispensers for chaff and flares to help protect them against surface-to-air missiles used by the mujahideen guerillas. The subsequent break-up of the Soviet Union left MiG-27s in the hands of several of the newly independent states, including Ukraine.

Stabilizing fin
A stabilizing ventral fin automatically folds up when the undercarriage is lowered.

MIG-27 IN STORAGE
NATO's shift to the doctrine of 'flexible response' saw the increasing primacy of the fighter-bomber with the Soviet air arms, and introduction of ever more capable strike/attack assets. The MiG-27 was the backbone of the 16th Air Army strike force by the end of the Cold War, with four regiments in East Germany.

FMA Pucará (1975)

TYPE • *Fighter* COUNTRY • *Argentina*

SPECIFICATIONS (IA-58 Pucará)

Dimensions:	Length: 14.25m (46ft 9in); Wingspan: 14.5m (47ft 6in); Height: 5.36m (17ft 7in)
Weight:	6800kg (14,991lb) loaded
Powerplant:	2 x 729kW (978hp) Turbomeca Astazou XVIG turboprop engines
Max speed:	500km/h (310mph)
Range:	3710km (2305 miles)
Ceiling:	10,000m (31,800ft)
Crew:	2
Armament:	2 x 20mm (0.79in) Hispano-Suiza HS.804 cannon and four 7.62mm (0.3in)FM M2-20 MGs; up to 1500kg (3300lb) of bombs or rockets

The Fábrica Militar de Aviones (FMA) Pucará was an indigenous Argentine counter-insurgency (COIN) aircraft. Most of those deployed to the islands were destroyed in a Special Air Services raid during the Falklands War of 1982.

Above: The Argentine Air Force Pucará was named after the stone forts built by indigenous South American tribes. It was a manoeuvrable and rugged aircraft able to operate from small, rough airstrips – as short as 80m (262ft) when boosted by JATO (jet-assisted take-off) bottles.

Front fuselage
Two 20-mm (0.79in) Hispano-Suiza DCA-804 cannon were carried in the nose, each equipped with up to 270 rounds. The crew were protected by armoured plate glass in the canopy.

FMA Pucará
The first Pucára COIN aircraft arrived in the Falklands on 2 April 1982, the day of the Argentine invasion. By late April 24 Pucáras were operating from Stanley and Goose Green, which were known by the Argentines as Base Aérea Militar (BAM) Malvinas and BAM Condor respectively.

DESIGN
The Pucará dated back to the early 1960s and was a unique design. A tall, retractable tricycle undercarriage provided ample space for weapons and the generous propeller ground clearance needed for flights from unpaved ground. The two crew members were strapped into Martin-Baker Mk 6 ejection seats, with the rear seat positioned 25cm (10in) higher. In practice, many missions were flown by a single pilot.

Stores
Fuselage attachments could carry tandem pylons, and stores pylons were permanently attached under the wings at the junction between the rectangular centre section and the outer panels.

ARGENTINE PUCARÁ
Argentina developed this twin-turboprop anticipating that an aircraft would be required to undertake anti-guerrilla and counter-insurgency (COIN) operations. Repeatedly upgraded and improved, the Pucará had limited success in the Falklands War. The COIN aircraft was better suited for counter-drug work in Colombia.

Israeli Aircraft Industries (IAI) Kfir (1976)

TYPE • *Multi-role fighter* COUNTRY • *Israel*

The Kfir (Lion Cub) cemented a worldwide reputation for Israel Aircraft Industries, an aggressive and innovative builder of hi-tech warplanes in a very small country. The Kfir, based on France's remarkable Mirage III, appeared in the mid-1970s and became a key weapon in the Israeli arsenal.

SPECIFICATIONS (Kfir C.2)

DIMENSIONS:	Length: 15.65m (51ft 4in); Wingspan: 8.22m (26ft 11.5in); Height: 4.55m (14ft 11in)
WEIGHT:	Maximum take-off 16,200kg (35,721lb)
POWERPLANT:	One 8119kg (17,900lb) General Electric J79-J1E turbojet
MAX SPEED:	2445km/h (1520mph)
RANGE:	Combat radius 346km (215 miles)
CEILING:	17,680m (58,000ft)
CREW:	1
ARMAMENT:	One 30mm (1.19in) cannon; nine external hardpoints with provision for up to 5775kg (12,732lb) of ordnance

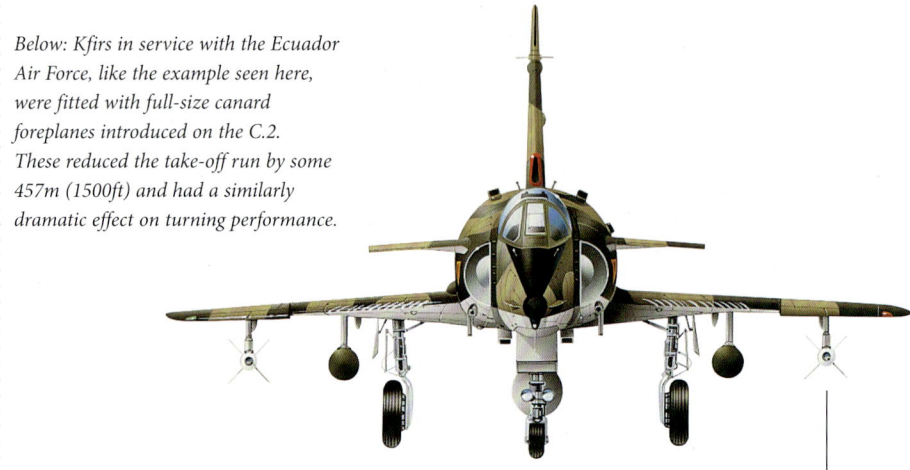

Below: Kfirs in service with the Ecuador Air Force, like the example seen here, were fitted with full-size canard foreplanes introduced on the C.2. These reduced the take-off run by some 457m (1500ft) and had a similarly dramatic effect on turning performance.

ARMAMENT
The Kfir C.2 carried a fixed armament of two 30mm (1.19in) DEFA 553 cannon just ahead of the wing roots, each with 125 rounds of ammunition. The aircraft also carried two Rafael Shafrir air-to-air infra-red homing missiles.

WINGS
The Kfir's wings lack the sawcut leading edge of the Mirage III, and instead have extended outboard leading edges, giving a pronounced saw-tooth leading-edge discontinuity.

KFIR-C7
Although the Kfir originally resembled the Mirage 5, a version first seen in 1976 introduced small, swept-back foreplanes that improved handling, agility and low-speed performance. From 1983 Israel began operating the Kfir-C7, which is an upgraded aircraft fitted with advanced radar and avionics and a boosted engine. A highly capable interceptor, it was also as good a ground-attack aircraft as could be found anywhere in the world.

POWERPLANT
The Kfir was powered by the most powerful production variant of the General Electric J79 engine, the J79-J1E. Due to its greater mass flow than the original Mirage III Atar engine, installation of the J79 necessitated larger intakes.

COLOMBIAN KFIRS
Colombian Air Force Kfir fighter jets fly over Arizona in July 2018, training at the Davis-Monthan AFB in preparation for a US-Colombian tactical flying exercise. Israel successfully exported the Kfir to Ecuador and Colombia.

Yakovlev Yak-38 (1976)

TYPE • *Naval VTOL* COUNTRY • *Soviet Union*

SPECIFICATIONS (Yak-38 Forger-A)	
DIMENSIONS:	Length: 15.5m (50ft 10in); Wingspan: 7.32m (24 ft); Height: 4.37m (14ft 4in)
WEIGHT:	Maximum take-off 11,700kg (25,795lb)
POWERPLANT:	Two 29.9kN (6724lbf) Rybinsk RD-36-35VFR lift turbojets; one 6950kg (15,322lb) Tumanskii R-27V-300 vectored-thrust turbojet
MAX SPEED:	1009km/h (627mph)
RANGE:	1,00 km (810 miles)
CEILING:	12,000m (39,370ft)
CREW:	1
ARMAMENT:	Four external hardpoints with provision for 2000kg (4409lb) of stores, including missiles, bombs, pods and drop tanks

Apart from the Harrier, the Yakovlev Yak-38 (NATO reporting name 'Forger') was the only operational jet VTOL aircraft in the world, albeit a far less capable one. Operational service began in 1976.

Left: The Yak-36 could not carry a major warload and was no match for land-based fighters, but it was more than capable of destroying enemy bombers and anti-submarine aircraft.

ENGINE VECTORING
Unlike the Harrier, the Yak-38 used two fixed turbojets mounted in tandem behind the cockpit for lift, with auxiliary inlets on the top of the fuselage. These were augmented by a third vectoring thrust unit in the rear fuselage.

FOLDING WINGS
The short-span wing folds for shipboard stowage, with a hinge between the flap and ailerons. A 600-litre (160-Imp gal) fuel tank could be fitted under each wing.

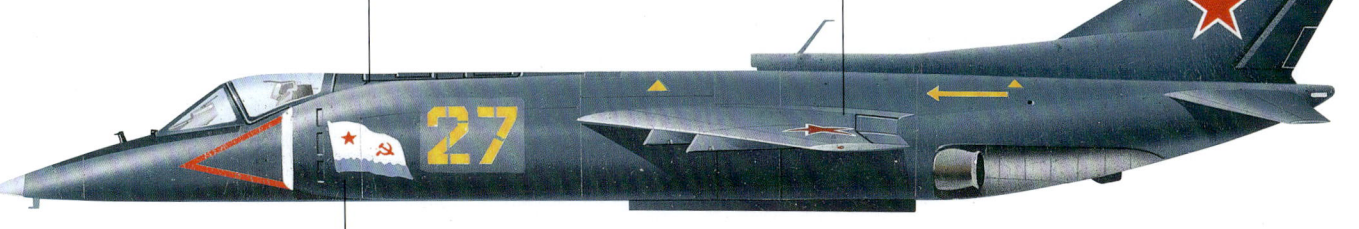

YAK-38 'FORGER-A'
The Yak-38 served aboard the *Kiev*-class aircraft carriers *Kiev*, *Minsk*, *Novorossiysk* and *Baku* (later renamed *Admiral Gorshkov*). This Yak-38 carries the badge of the Red Banner Northern Fleet.

TACTICAL LIMITATIONS

The Yak-38 was never designed to be in the class of conventional naval fighters; it was aimed at warding off NATO maritime patrol and strike aircraft such as the P-3 Orion and BAe Nimrod. Its radar had only a very limited range, and it carried only the short-range infra-red 'Atoll' missile and cannon. It had limited strike capability, with armament including the unguided UV-32 rocket pod and the short-range AS-7 'Kerry' tactical missile.

NAVAL JETS
The Yak-38 gave the Soviet navy experience with high-performance jets at sea, and was a useful stepping stone towards the fixed-wing naval fighters that later came into service with the Russian navy.

British Aerospace Hawk (1976)

TYPE • Trainer COUNTRY • United Kingdom

SPECIFICATIONS (T.Mk 1A)

DIMENSIONS: Length: 11.17m (36ft 7.75in); Wingspan: 9.39m (30ft 9.75in); Height: 3.99m (13ft 1.75in)

WEIGHT: Maximum take-off 7750kg (17,085lb)

POWERPLANT: One 23.1kN (5200lbf) Rolls-Royce/Turbomeca Adour Mk 151 turbofan

MAX SPEED: 1038km/h (645mph)

RANGE: 2500km (1800 miles) with two drop tanks

CEILING: 15,240m (50,000ft)

CREW: 2

ARMAMENT: Underfuselage/wing hardpoints with provision for up to 2567kg (5660lb) of stores, wingtip mounted air-to-air missiles

The first British Aerospace (originally Hawker Siddeley) Hawk flew in August 1974, the culmination of a lengthy process to create a subsonic aircraft to replace the Gnat, Hunter T.7 and some two-seat Jaguars for advanced and weapons training.

Above: The Royal Air Force aerobatic team, the 'Red Arrows', flew its first season with the Hawk T.Mk 1 in 1980, replacing the Folland Gnat. The team uses nine Hawks, modified with tanks and piping able to make trails of red, white and blue smoke to enhance the display.

COCKPIT
The Hawk cockpit has two Martin Baker ejection seats. The raised rear seat gives an instructor a good view of a student in action.

TURBOFAN ENGINE
The Hawk is powered by a non-afterburning Rolls-Royce Adour turbofan engine, which has proved reliable and economical in service.

BRITISH SUCCESS

The Hawk has been one of the truly outstanding successes of the modern British aerospace industry. Much of this success is due to the exceptional service life of the airframe, low maintenance requirements, the relatively inexpensive purchase price when originally offered for export, large optional payload and its ability to operate in the medium-range attack and air-superiority role for a fraction of the cost of more powerful types. The first operational aircraft were delivered in 1976.

HAWK T1A
The RAF operates the T.Mk 1 for weapons instruction. The T.Mk 1A has three pylons; the central one is normally occupied by a 30mm (1.2in) Aden cannon, the two underwing pylons can be fitted with a mixture of weapons, including Matra rocket pods.

HAWK T.MK 1
A British Aerospace Hawk T.Mk 1 flies at the Royal International Air Tattoo at RAF Fairford, Gloucestershire in 2010. RAF Hawks now use a new gloss-black colour scheme to make them more visible.

Sukhoi Su-27 'Flanker' (1977)

TYPE • *Multi-role* COUNTRY • *Soviet Union*

Designed as a response to the American F-15 Eagle, the Sukhoi Su-27 (NATO reporting name 'Flanker') was the most potent fighter developed by the Soviet Union and remains in widespread service today, with many combat missions logged.

SPECIFICATIONS (Su-27P)

DIMENSIONS:	Length: 21.94m (72ft); Wingspan :14.7m (48ft 3in); Height: 5.93m (19ft 6in)
WEIGHT:	Maximum take-off 28,300kg (62,391lb)
POWERPLANT:	Two Lyulka-Saturn/ Moscow AL-31F (izdeliye 99V) turbofans each rated at 75.22kN (16,909lbf) dry and 122.58kN (27,558lbf) with afterburning
MAX SPEED:	2400km/h (1491mph)
RANGE:	3720km (2312 miles)
CEILING:	18,500m (60,700ft)
CREW:	1
ARMAMENT:	One 30mm (1.2in) Gryazev-Shipunov GSh-30-1 autocannon; up to 4430kg (9770lb) of weapons and stores

Above: The Su-27 was developed as a heavy air-superiority fighter capable of escorting bombers. The Su-27P is the standard variant of the type (shown here in Russian markings), but without air-to-ground weapons control system.

AIRCRAFT DETECTION
A glazed ball mounted in front of the windscreen houses a laser rangefinder and an infra-red search and track system, which can detect enemy aircraft at extended distances.

FIN PROFILE
Pre-series Flanker-Bs sported a square-cut flat top to the fins, with a large anti-flutter mass protruding forward. Early series production Flanker-Bs retained the antiflutter mass, but displayed pointed fins. This aircraft has the latter configuration.

SU-27S
The Su-27S was the initial production single-seater with an improved AL-31F engine. 'Blue 15' serves with the Ukrainian Air Force.

COUNTER RESPONSE
The aircraft that became the Su-27 was conceived as a counter to the US F-15 Eagle, which also started as a purely air-to-air platform but developed into a potent strike aircraft. Its first flight was in 1977. Up to 1999, 645 single-seat Su-27s were built, with an additional 190 of the two-seat version. These were given the NATO reporting names Flanker-B and Flanker-C respectively.

SU-27, AIR SHOW, 2011
More recently, Russia has procured three more new-build 'Flanker' versions: the two-seat Su-30M2 and Su-30SM, as well as the thrust-vectoring single-seat Su-35S. The Su-30M2 and Su-30SM are both domestic versions of the significantly improved two-seat derivatives of the 'Flanker'.

Mikoyan MiG-29 'Fulcrum' (1977)

TYPE • *Fighter* **COUNTRY** • *Soviet Union*

The appearance in the early 1980s of the MiG-29, with its superb agility and apparent ability to perform combat manoeuvres that could not be matched by any aircraft in the West, came as an unpleasant surprise to NATO.

SPECIFICATIONS (MiG-29 'Fulcrum-A')

DIMENSIONS:	Length: 17.32m (56ft 10in); Wingspan: 11.36m (37ft 3.25in); Height: 4.73m (15ft 6.2in)
WEIGHT:	18,500kg (40,785lb) loade
POWERPLANT:	Two 81.39kN (18,298lbf) Klimov/Leningrad RD-33 afterburning turbofan engines
MAX SPEED:	2445km/h (1519mph)
RANGE:	1200km (745 miles)
CEILING:	17,000m (55,775ft)
CREW:	1
ARMAMENT:	One GSh-301 30mm (1.18in) cannon; maximum stores of 3000kg (6614lb)

Below: A product of the ARZ-558 aircraft repair plant in Baranovichi, Belarus, the MiG-29BM upgrade is similar to the MiG-29SM, with N019P, R-27ER and R-77 AAMs, plus precision air-to-ground munitions, cockpit tweaks and in-flight refuelling probe. Eight have been upgraded.

TARGETING
The original fire control system was based around an N019 pulse-Doppler radar with lookdown/shootdown capability, coupled with an infrared search and track (IRST) system.

WING DESIGN
The MiG-29's wing and leading-edge extensions were designed at TsAGI, the Soviet equivalent of NASA. They are extremely efficient aerodynamically, and contribute to the 'Fulcrum's' superb performance.

ADVANCED FIGHTER

The first prototype of the MiG-29 flew in October 1977, and deliveries to Soviet Frontal Aviation began in 1983. It showed advanced technologies. The powerful pulse-Doppler radar was backed up with a passive infrared search-and-track (IRST) system. This could detect, track and engage a target while leaving the radar in a non-emitting mode. For close-in engagements, a helmet-mounted sight could be used to cue infrared-homing missiles onto an off-boresight target.

SLOVAKIAN MIG-29
In the immediate aftermath of the end of the Soviet Union, MiG made several efforts to offer more advanced versions of the MiG-29 for export. These included the MiG-29SD and MiG-29SE, both with export-optimized N019ME radar, R-77 AAMs and some Western avionics.

Dassault/Dornier Alpha Jet (1977)

TYPE • *Attack aircraft* COUNTRY • *Germany/France*

In 1968, France and Germany, both of whom had projects for an advanced jet trainer under study, decided to pool their resources and expertise. The result was the excellent Alpha Jet, which offered near-fighter performance and attack capability in an economical package.

SPECIFICATIONS
(Alpha Jet E)

DIMENSIONS:	Length: 13.23m (43ft 5in); Wingspan: 9.11m (29ft 10.67in); Height: 4.19m (13ft 9in)
WEIGHT:	Maximum take-off 8000kg (17,640lb)
POWERPLANT:	Two 1350kg (2977lb) Turbomeca Larzac 04 turbofans
MAX SPEED:	927km/h (576mph)
RANGE:	583km (363 miles)
CEILING:	14,000m (45,930ft)
CREW:	2
ARMAMENT:	One 30mm (1.19in) cannon (optional)

Left: Two SNECMA/Turbomeca Larzac 04-C6 turbofans are installed in the Alpha Jet in nacelles on the fuselage sides.

COCKPIT
The two crew members are accommodated in tandem under individual transparencies on Martin-Baker Mk 10 ejection seats, the forward pilot being provided with a simplified head-up display. Instruments are duplicated in front and rear cockpits.

CAMOUFLAGE
This Alpha Jet E displays the original training colours worn by the aircraft on its entry into service with the Belgian Air Force. Aircraft were later repainted in a two-tone grey camouflage as they returned from overhaul, the orange training bands being retained.

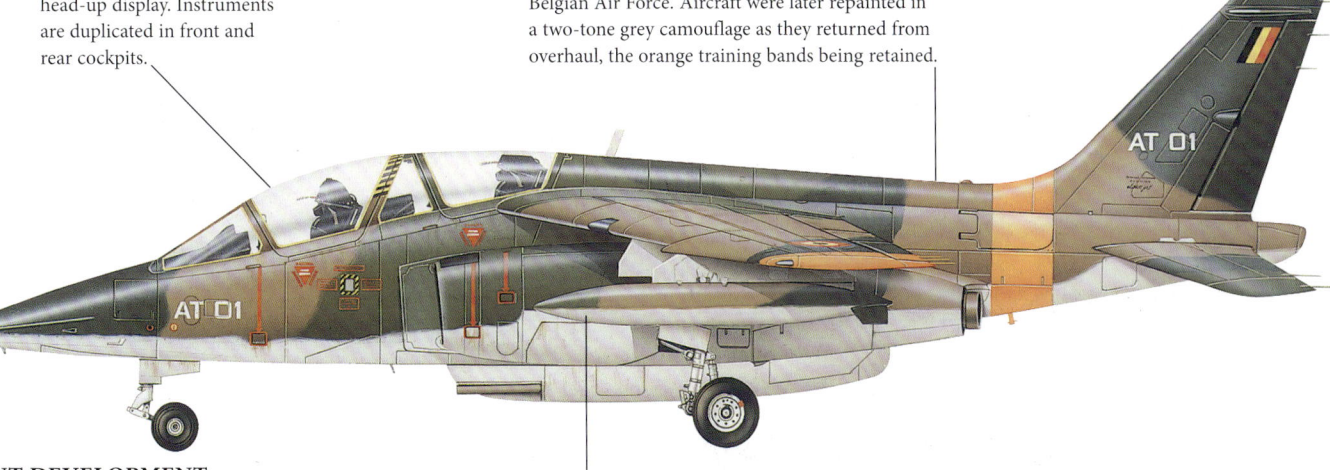

JOINT DEVELOPMENT
Early in the joint development programme the Germans decided that they had no requirement for a training version of the Alpha Jet, but a need was identified for a light attack aircraft to replace the Fiat G.91R fleet. In February 1972 two prototypes each were ordered by France and Germany, and the type (the French-built version) flew for the first time on 26 October 1973. Production began some time later, the first trials aircraft being delivered to the French Air Force late in 1977.

EXTERNAL STORES
Although Alpha Jets are capable of carrying a wide variety of rockets and bombs, this Belgian Air Force aircraft is fitted with the standard 310-litre (68.2-Imp gal) drop tanks.

ALPHA JET E
The Alpha Jet A exhibited several differences from the French E model seen here, most notably in the cockpit. The pilot sat on a Stencel S-III-S3AJ ejection seat, licence-built by Messerschmitt-Bölkow-Blohm (MBB), whereas French standard Alpha Jets were equipped with the Martin-Baker Mk 4 seat.

Mitsubishi F-1 (1978)

TYPE • *Fighter* **COUNTRY** • *Japan*

SPECIFICATIONS

DIMENSIONS:	Length: 17.86m (58ft 7in); Wingspan: 7.88m (25ft 10in); Height: 4.39m (14ft 4.67in)
WEIGHT:	Maximum take-off 13,700kg (30,203lb)
POWERPLANT:	Two 3315kg (7308lb) Ishikawajima-Harima TF40 IHI- 801A turbojets
MAX SPEED:	1708km/h (1061mph)
RANGE:	2870 km (1780 miles)
CEILING:	15,240m (50,000ft)
CREW:	1
ARMAMENT:	One 20mm (0.79in) six-barrel cannon; five hardpoints for up to 2722kg (6000lb) of munitions

Mitsubishi's F-1 drew its main features from the earlier T-2 supersonic trainer. The F-1 was developed for close air support and anti-ship operations, performing these tasks admirably in the Japan Air Self-Defence Force (JASDF).

This page: The F-1 depicted here bears the three-tone upper surface camouflage and light grey undersides that was the standard scheme for most of these aircraft.

FIRE CONTROL
The Mitsubishi/Denki J/AWG-12 fire control system fitted to the F-1 was a Japanese-built version of the AWG-12 installed in the RAF's F-4M Phantoms.

REAR COCKPIT
In the F-1, the redundant rear cockpit was used as an avionics bay, the rear canopy having been replaced with a metal access hatch.

HOME DEFENCE

The Mitsubishi F-1 was created because of the threatening nature of Soviet naval activity around the Japanese home islands in the 1970s and 1980s. Following the success of the T-2 twin-turbofan trainer, Mitsubishi went ahead with this close-support fighter version, intended primarily to fly from land bases on defence missions against enemy shipping. The F-1 had a comprehensive avionics fit, its principal weapon being the Japanese-developed Mitsubishi ASM-1 (Type 80).

ANTI-SHIP WEAPON
For most of its career the F-1's primary anti-ship weapon was the locally designed Type 80 ASM-1. This used inertial guidance and active radar terminal homing, had a range of about 50km (31 miles) and was fitted with a 150kg (331lb) semi-armour-piercing warhead.

F-1, TSUIKI AIRBASE
A Japanese F-1 at rainy Tsuiki Airbase. The F-1 remained in service until 2006, by which time it had been replaced by the Mitsubishi F-2 – an adaptation of the American F-16 Fighting Falcon.

Boeing E-3 Sentry (1978)

TYPE • *AEW&C* COUNTRY • *United States*

SPECIFICATIONS (E-3A)

DIMENSIONS:	Length: 46.61m (152ft 11in); Wingspan: 44.42m (145ft 9in); Height: 12.6m (41ft 4in)
WEIGHT:	Maximum take-off M47,400kg (325,000lb)
POWERPLANT:	Four 93kN (21,000lbf) Pratt & Whitney TF33-PW-100A turbofan engines
MAX SPEED:	855km/h (530mph)
RANGE:	7400km (4598 miles)
CEILING:	12,500m (41,000ft)
CREW:	Up to 20
ARMAMENT:	None

The Boeing E-3 Sentry introduced the term AEW&C to common use, and became the platform of choice for the United States, Saudi Arabia, the United Kingdom, France and NATO. The airframe is based on that of the Boeing 707.

Right: An RAD E-3A Sentry. The RAF had intended to phase these aircraft out of service, but they have made a reappearance during the Russo-Ukrainian conflict.

CREW
Seventeen personnel constitute the standard crew of an RAF Sentry, including a flight deck crew of two pilots, navigator and flight engineer. In the main cabin there are typically 13 mission operatives.

WINGTIP PODS
RAF E-3A Sentries – all eight of which are deployed with No. 8 Squadron – are fitted with wingtip pods containing Loral 1017 'Yellow Gate' ESM pods for passive radar detection. USAF E-3s carry this equipment in fairings on either side of the forward fuselage.

ROTODOME
The Westinghouse APY-1 radar in its giant rotodome can detect targets up to 650km (400 miles) away, although the range drops for low-flying aircraft.

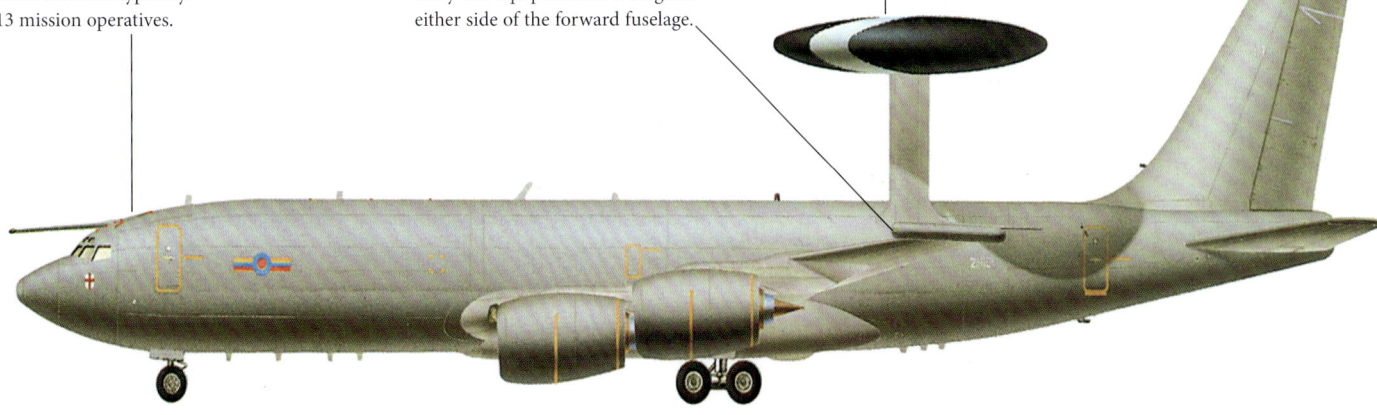

EYE IN THE SKY
To the technicians who sit out back, the E-3 AEW&C is the eyes and ears of the battlefield commander, watching, analyzing and directing. During Operation Desert Storm, 30 air-to-air victories were scored by Allied fighters who were guided into action by AEW&C crews. With its long range and endurance, the E-3 Sentry can spy on an entire battlefield or, if necessary, an entire nation.

E-3 SENTRY, THAILAND
A US Air Force E-3 Sentry of the 962nd Airborne Air Control Squadron (AACS) prepares to receive airborne refuelling from a KC-135E Stratotanker of the 108th Air Refuelling Squadron (ARS) in January 2002, over Thailand.

Dassault Mirage 2000 (1978)

TYPE • *Multi-role fighter* **COUNTRY** • *France*

SPECIFICATIONS

DIMENSIONS:	Length: 14.36m (47ft 1in); Wingspan: 9.13m (29ft 11in); Height: 5.2m (17ft)
WEIGHT:	15,000kg (33,069lb) maximum take-off
POWERPLANT:	98.06kN (22,046lbf) SNECMA M53-P20 afterburning turbofan
MAX SPEED:	2335km/h (1451mph)
RANGE:	3333km (2071 miles) with drop tanks
CEILING:	16,460m (54,000ft)
CREW:	2
ARMAMENT:	2 × 30mm (1.18in) cannon and up to 6300kg (13,889lb) of disposable stores carried on 5 under-fuselage and 4 underwing hardpoints

Continuing the tradition of delta-winged Dassault fighters, the Mirage 2000 brought the family up to date and established itself not only as the backbone of the French Air Force but also as a genuine success on the export market.

Right: The 2000C was adopted by the French government in December 1975 as the primary combat aircraft of the French air force. It was developed initially under contract as an interceptor and air-superiority fighter.

Mirage 2000C
A Mirage 2000C of the 5e Escadre de Chasse, as it appeared during Operation Daguet, France's contribution to the 1991 Gulf War. The aircraft was one of 14 based in Al Ahsa, Saudi Arabia.

Wing
Large and lightly loaded, the wing of the Mirage 2000 is fitted with automatic leading-edge slats for manoeuvrability and two-piece elevons on the trailing edge.

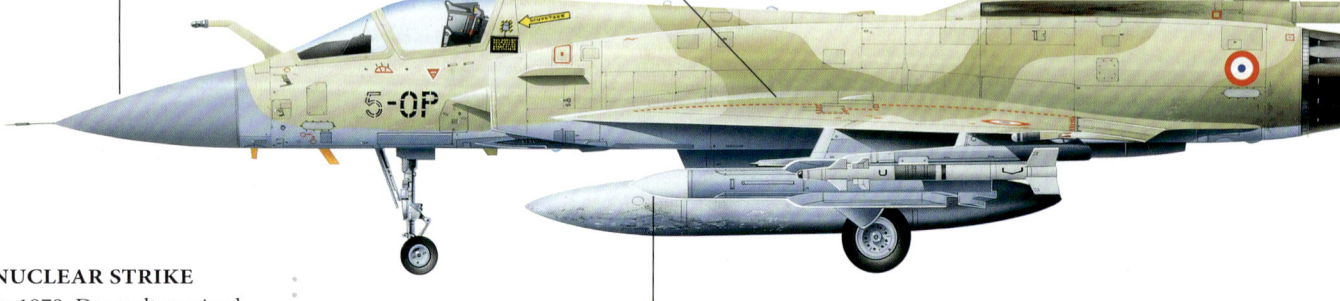

Nuclear Strike
In 1979, Dassault received a contract to produce two prototypes of a nuclear strike version, which became the Mirage 2000N. Based on the two-seat Mirage 2000B airframe, this features an airframe strengthened for low-level operations, and attack avionics based around the Antilope 5 radar. The primary weapon is the ASMP stand-off nuclear missile. A total of 75 Mirage 2000Ns were built for the French Air Force; the type achieved initial operational capability in 1988.

Matra missile
For the medium-range air interception role, the Mirage 2000 carries the Matra Super 530D missile. This is the latest in a development line stretching back to the original R530, which entered French service in 1963, with an interchangeable IR or semi-active radar homing seeker.

MIRAGE 2000 IN AFGHANISTAN
An Emirates Mirage 2000 multi-role combat aircraft conducts a flight over Afghanistan in November 2008, in support of exercise Iron Falcon.

McDonnell Douglas F/A-18 Hornet (1978)

TYPE • *Naval multi-role* COUNTRY • *United States*

Replacing the F-4 Phantom II in the fleet air defence role, and the A-7 Corsair II attack aircraft, the Hornet gave the US Navy's carrier air wing a genuine multi-role carrier fighter. The F/A-18 continues to serve with the US Marine Corps and a number of export operators.

SPECIFICATIONS (F/A-18A)

DIMENSIONS:	Length: 17.07m (56ft); Wingspan: 11.43m (37ft 6in) without wingtip missiles; Height: 4.66m (15ft 3in)
WEIGHT:	21,888kg (48,253lb) maximum take-off
POWERPLANT:	2 × 78.73kN (17,700lbf) General Electric F404-GE-402 afterburning turbofans
MAX SPEED:	1915km/h (1190mph)+
RANGE:	3336km (2073 miles) ferry with drop tanks
CEILING:	15,240m (50,000ft)
CREW:	1
ARMAMENT:	1 × 20mm (0.79in) rotary cannon and up to 6200kg (13,700lb) of disposable stores

Right: An F/A-18A of Marine Fighter Attack Squadron (VMFA) 314 'Black Knights', aboard the USS Coral Sea *in the Mediterranean in 1986. The unit was engaged in attacks on Libyan radar sites.*

RADAR
In the original F/A-18A, two-seat F/A-18B and early C-model aircraft the radar was the Hughes AN/APG-65 multi-mode system, which later gave way to the APG-73.

FUEL
This is carried in four main fuel tanks in the aircraft's spine. The internal total of 5300 litres (1400 Imp gal) can be supplemented by up to three drop tanks each of 1249-litre (330-Imp gal) capacity.

CF-18 HORNET
The Canadian CF-18 Hornet was introduced into service with the Royal Canadian Air Force in 1982. The aircraft were largely denavalized for Canadian service, but wing folding and the arrestor hook were retained.

EXPORT HORNETS
The first export customer for the type was Canada, which took delivery of 98 single-seat CF-188As and 40 two-seat CF-188Bs between 1982 and 1988. Australia was next, taking 57 F/A-18As and 18 F/A-18Bs. Spain acquired 60 EF-18As and 12 EF-18Bs. Thereafter, export aircraft were all completed to F/A-18C/D standard, comprising 32 F/A-18Cs and eight F/A-18Ds for Kuwait; 26 F/A-18Cs and eight F/A-18Ds for Switzerland; 57 F-18Cs and seven F-18Ds for Finland and eight F/A-18Ds for Malaysia.

F/A-18 HORNET LAUNCH
An F/A-18 Hornet assigned to the 'Blue Blasters' of Strike Fighter Squadron (VFA) 34, lands on the flight deck of the aircraft carrier USS *Ronald Reagan* (CVN 76) in the Pacific Ocean in 2012.

McDonnell Douglas AV-8B Harrier II (1978)

TYPE • *Naval VTOL* COUNTRY • *United States*

SPECIFICATIONS
(AV-8B Harrier II)

DIMENSIONS:	Length: 14.12m (46ft 4in); Wingspan: 9.25m (30ft 4in); Height: 3.55m (11ft 7.67in)
WEIGHT:	Maximum take-off 14,061kg (31,000lb)
POWERPLANT:	One 10,796kg (23,800lb) Rolls-Royce F402-RR-408 vectored thrust turbofan
MAX SPEED:	1065km/h (661mph)
RANGE:	1001km (684 miles)
CEILING:	15,240m (50,000ft)+
CREW:	1
ARMAMENT:	1 × 25mm (0.98in) rotary cannon and up to 6003kg (13,235lb) of disposable stores, after STO

A key element in modern battlefield support is the short take-off, vertical landing (STOVL) aircraft, epitomized by the McDonnell Douglas AV-8B Harrier, a machine that revolutionized military aircraft technology.

Above: A recent Boeing AV-8B(R) Harrier II used in combat operations by the US Marine Corps' VMA-223.

RADAR
The compact AN/APG-65 radar includes air-to-air and air-to-ground modes and all-digital performance. Terrain-avoidance information is provided for low-level flying.

REFUELLING PROBE
Normally a standard feature, the bolt-on refuelling probe is retractable and is housed in a streamlined fairing when not in use.

BOEING AV-8B(R) HARRIER II+
This Boeing AV-8B(R) Harrier II+ was flown by the Spanish Navy, one of many export customers for the aircraft.

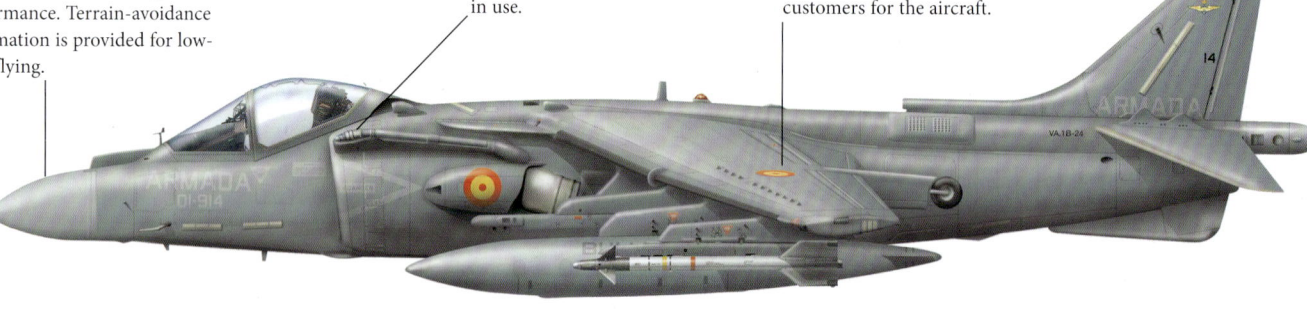

ENHANCEMENTS
USMC AV-8Bs saw extensive use in Iraq and Afghanistan; the aircraft is now subject to life-extension programmes, addressing the service life of the wing, the vertical tail and some of the sub-systems. Life-extension work has focused on the F402-RR-408B-powered aircraft, and the TAV-8Bs have also now received the uprated Dash 408B engine. In a bid to increase the AV-8B's utility in its primary close air support mission, the Litening targeting pod has been integrated; this can be used to pass target coordinates to the aircraft's Joint Direct Attack Munitions (JDAM).

VERTICAL TAKE-OFF
A US Marine Corps aviator makes a vertical take-off in an AV-8B Harrier from the US Navy Wasp Class Amphibious Assault Ship USS *Kearsage* (LHD 3) en route to King Faisal Air Base, Jordan, for the bilateral training mission Infinite Moonlight in 2005.

Westland Lynx (1978)

TYPE • Helicopter **COUNTRY** • United Kingdom

SPECIFICATIONS (AH.1)

DIMENSIONS:	Length: 15.24m (49ft 12in); Main rotor diameter: 12.8m (41ft 12in); Height: 3.73m (12ft 3in)
WEIGHT:	3030kg (6666lb) empty
POWERPLANT:	2 × Rolls-Royce Gem 42-1 turboshaft engines, each 835kW (1119hp)
MAX SPEED:	270km/h (168mph)
RANGE:	685km (426 miles)
CEILING:	5790m (19,000ft)
CREW:	2–3
ARMAMENT:	Multiple machine gun, rocket, missile, gun pod and mine options

Westland and Aérospatiale produced the Lynx, together with the Puma and Gazelle, under the Anglo-French helicopter agreement of 1967. The first of 13 prototypes flew in March 1971; subsequent production included both army and navy versions.

Above: Seen in service with No. 1 Wing, British Army of the Rhine, XZ669 was converted to AH.7 standard. It was based at Wattisham, Suffolk, with No. 669 Squadron of No. 4 Regiment, Army Air Corps.

ROOF-MOUNTED SIGHT
British Aerospace, under licence from Hughes, built the Lynx roof-mounted sight. It allows guidance for the TOW missiles.

POWERPLANT
Originally fitted with Gem 41-1 engines, from 1987 the Lynx's powerplant was upgraded to Gem 42-1 standard in the AH.7. These aircraft also have composite rotor blades.

ARMY LYNX
Designated AH.1, the first British Army Lynxes were delivered in 1977. They could carry nine troops, more than 1350kg (3000lb) of external cargo or eight TOW (tube-launched, optically-tracked, wire-guided) missiles, aimed using a sight on the cabin roof. More powerful Gem 41 engines were introduced in the AH.7, in addition to improved avionics and a more powerful tail rotor.

CABIN
Up to nine soldiers could be carried in the main cabin of the AH.1, or six with full combat equipment.

AW159 WILDCAT
The Wildcat shares little with its Lynx predecessor except its design layout and rotor gearbox. In Royal Navy service since 2015, the AW159 has been highly active in anti-drugs operations in the Arabian Sea.

Dassault Super Etendard (1978)

TYPE • *Naval fighter* COUNTRY • *France*

SPECIFICATIONS

DIMENSIONS:	Length: 14.31m (46ft 11in); Wingspan: 9.60m (31ft 6in); Height: 3.86m (12ft 8in)
WEIGHT:	Maximum take-off 12,000kg (26,455lb)
POWERPLANT:	One 5000kg (11025lb) SNECMA Atar 8K-50 turbojet
MAX SPEED:	1180km/h (733mph) at low level
RANGE:	Combat radius 850km (528 miles)
CEILING:	13,700m (44,947ft)
CREW:	1
ARMAMENT:	Two 30mm (1.19in) cannon; provision for up to 2100kg (4630lb) of external munitions

The Dassault Super Etendard achieved brief notoriety during the Falklands War of 1982, when aircraft armed with Exocet missiles sank the British destroyer HMS *Sheffield* and the container ship *Atlantic Conveyor*.

Right: The Super Etendard was originally configured to carry the AN52 tactical free-fall nuclear weapon (seen here), which had a yield of 15kT and which was replaced in service by the medium-range ASMP (Air-Sol Moyen Portée) nuclear missile.

POWERPLANT
Improvements in performance over the Etendard were made possible by the extra 336kg (740lb) of thrust obtained by using the 8K50 version of SNECMA's Atar turbojet.

HEAD-UP DISPLAY
Another improvement to the Etendard was the Thomson-CSF head-up display (HUD) with TV or infrared imaging. Threats received from the radar-warning receiver could be displayed on the HUD.

RADAR
The Super Etendard's Thomson-CSF/ESD Agave radar is a simple lightweight set, able to detect an object the size of a patrol boat at about 40km (25 miles) and a fighter at 19km (12 miles). It is controlled by a sidestick on the left side of the cockpit.

ANTI-SHIP AIRCRAFT
Dassault's Super Etendard was a sleek, arrow-like aircraft that achieved a superb military record with French and Argentine naval forces. The carrier-based Super Etendard could meet many tactical challenges, including reconnaissance in its unarmed photo version. Most importantly, it could deliver the famous Exocet anti-ship missile with devastating results, using the aircraft's Thomson-CSF/EMD Agave multi-mode radar.

US CARRIER OPERATIONS
A French Super Etendard from the nuclear-powered aircraft carrier *Charles de Gaulle* (R 91) completes a touch-and-go on the flight deck of the Nimitz-class aircraft carrier USS *John C. Stennis* (CVN 74) while operating in the Arabian Sea. The Super Etendard was retired from French service in 2016.

Eurocopter AS332 Super Puma (1978)

TYPE • Helicopter COUNTRY • France

Proving itself a highly capable transport helicopter, the Super Puma fast became the first choice for operators who required a dependable and safe helicopter. The exceptional power available coupled with long range ensure sales success.

SPECIFICATIONS (AS 332L-1)

DIMENSIONS:	Length: 16.29m (53ft 5in); Main rotor diameter: 15.60m (51ft 2in); Height: 4.92m (16ft 2in)
WEIGHT:	Maximum take-off 8600kg (18920ft)
POWERPLANT:	Two 1184kW (1590hp) (continuous rating) Turbomeca Makila 1A1 turboshafts
CRUISE SPEED:	266km/h (165mph)
RANGE:	870km (539 miles
CEILING:	4600m (15,100ft)
CREW:	2
ARMAMENT:	None

Above: This Puma was one of three examples used by the Belgian Gendarmerie. They were based at Brasschaat in the north of the country and employed for patrol and VIP duties.

CAPACITY
Puma can accommodate 24 seated passengers or a sizeable cargo load. This particular version operates in the passenger configuration.

ROTOR BLADES
Aérospatiale was able to equip the Super Puma with the latest technology. Apart from the improved avionics and radar, the helicopter is fitted with glass-fibre rotor blades to increase its performance.

ADAPTABILITY
An established favourite with helicopter companies specializing in the support of offshore oil exploration and production, the Super Puma, along with its Cougar military variant, has also won orders from many other companies and agencies for a wide variety of applications. They range from VIP transport to the support of UN peace-keeping forces in global trouble spots. The type is readily adaptable for a whole host of other tasks.

LANDING GEAR
Positioned either side of the fuselage in streamlined fairings is the large single-wheel high-energy absorption landing gear. The landing gear can be retracted.

AS332 SUPER PUMA IN FLIGHT
Super Pumas and Cougars have been used to transport no fewer than 25 heads of state, and more than 50 VIP versions are in service. Apart from the 8m (26ft) long cabin, the type's attractions include low noise and vibration levels and large windows along with high performance and long range.

British Aerospace Sea Harrier (1978)

TYPE • *Naval VTOL* COUNTRY • *United Kingdom*

The British Aerospace Sea Harrier will forever be remembered for its role in the Falklands War of 1982, where it provided an effective defence for the British task force against the full strength of the Argentine Air Force.

SPECIFICATIONS (FA.2)

- **Dimensions:** Length: 14.17m (46ft 6in); Wingspan: 7.70m (25ft 3in); Height: 3.71m (12ft 2in)
- **Weight:** Maximum take-off 11,884kg (26,200lb)
- **Powerplant:** One 9750kg (21,500lb) Rolls-Royce Pegasus Mk 106 vectored thrust turbofan
- **Max speed:** 1185km/h (736mph)
- **Range:** 740km (460 miles) combat
- **Ceiling:** 15,545m (51,000ft)
- **Crew:** 1
- **Armament:** Two 25mm (0.3in) cannon; five pylons for various AAMs or anti-ship missiles, up to a total of 3629kg (8000lb)

Below: This Sea Harrier FRS.Mk 1 served aboard HMS Invincible *during the early 1980s. Typical armaments included a pair of Sidewinder missiles and 30mm cannon pods.*

FA.2 UPGRADE
Although tailored more closely to the air defence mission than the Sea Harrier FRS.1, the FA.2 can carry numerous underwing munitions, including CRV-7 rocket pods, 454kg (1000lb) bombs, Lepus flares and other ordnance.

TRAILING EDGE
An extra 0.35m (1ft 2in) plug was inserted aft of the wing trailing edge of the FA.2 to provide greater internal capacity for avionics equipment.

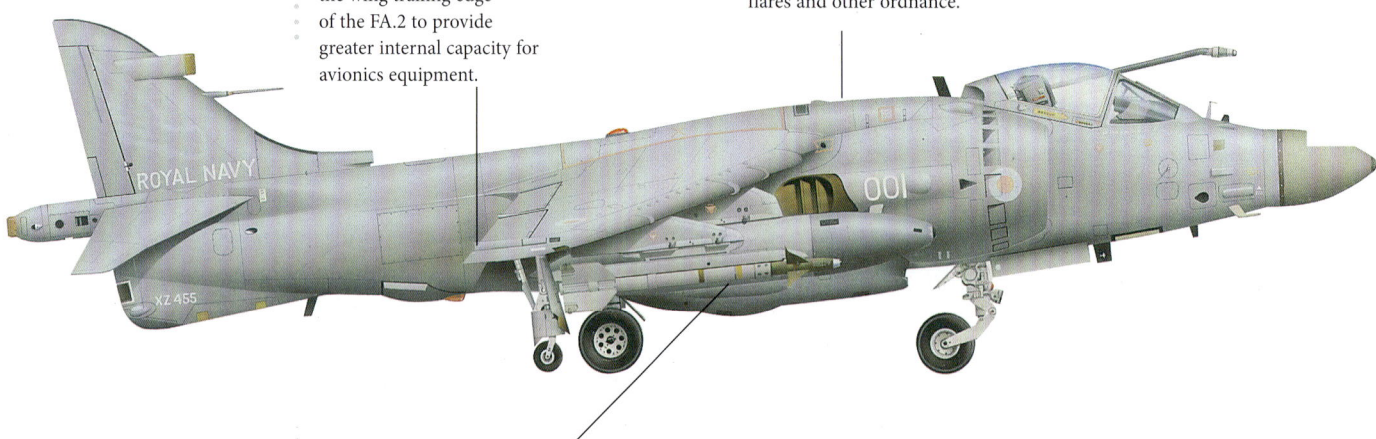

AMRAAM
Designed from the outset to be fully compatible with AMRAAM, the radar allowed the ripple-firing of all four missiles carried by the aircraft.

EARLY DEVELOPMENT

Developed from the RAF's Harrier GR.Mk 3, the Sea Harrier FRS.Mk 1 introduced a redesigned forward fuselage and nose fitted with a Ferranti Blue Fox radar, a new canopy and raised cockpit for improved view and a more powerful Pegasus Mk 104 engine. An initial order was placed in 1975 for 24 FRS.Mk 1s. Later orders brought the total to 57 for the Royal Navy, and another 24 for India. A two-seat trainer without radar was developed as the Sea Harrier T.4.

SEA HARRIER FA.2
A Sea Harrier FA.2 taking off from the British aircraft carrier HMS *Invincible* during Exercise Magic Carpet 05. The Sea Harrier was ordered to equip the Royal Navy's three *Invincible*-class aircraft carriers.

Aermacchi MB.339A (1979)

TYPE • Attack aircraft **COUNTRY** • Italy

SPECIFICATIONS

DIMENSIONS:	Length: 10.97m (36ft); Wingspan: 10.86m (35ft 7in); Height: 3.60m (11ft 9in)
WEIGHT:	4400kg (9700lb) loaded
POWERPLANT:	1 x 17.8kN (4000lbf) Rolls-Royce Viper Mk. 632 turbojet
MAX SPEED:	898km/h (558mph)
RANGE:	1760km (1093 miles)
CEILING:	14,630m (48,000ft)
CREW:	2
ARMAMENT:	Up to 1800kg (3968lb) of weapons

A beautiful aircraft and a delight to fly, the Aermacchi MB.339 is the standard Italian air force trainer, which has also been developed into a potent light-attack warplane. It is familiar on the air show circuit thanks to its appearances with the Italian national aerobatic team Frecce Tricolori.

Below: The MB.339K Veltro 2 was a single-seat, dedicated attack variant that first flew in May 1980.

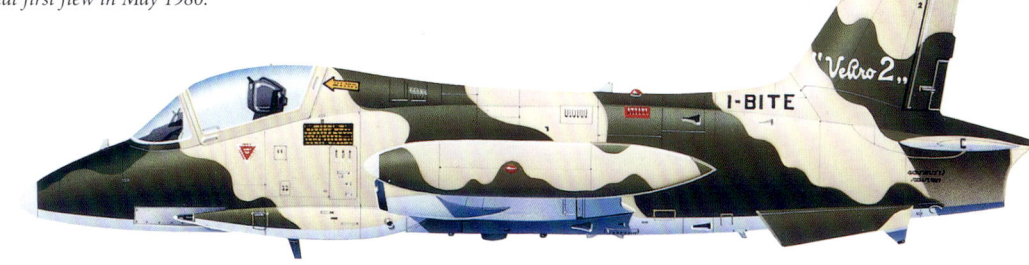

ADVANCED SYSTEMS
Systems fitted to the MB.339C version include laser rangefinder, Kaiser head-up display and weapon aiming computer.

COCKPIT
Compared to the MB.326, forward vision for the instructor in the rear seat is greatly improved. Both cockpits have full pressurization and dual controls. A gunsight can also be fitted to the rear cockpit to allow an instructor to monitor a student's shooting.

EASY HANDLING
Based upon Aermacchi's earlier MB.326 used by 12 nations, the MB.339 looks 'hot' but has very docile handling qualities. This makes it ideal for flight instruction yet highly adaptable for combat duties. The MB.326 is first and foremost a trainer that has taught thousands of fast-jet students how to fly. In service with seven air forces, it was an unsuccessful competitor in the Pentagon's JPATS competition for a new primary trainer for the US Air Force and Navy in the 1990s.

UNDERCARRIAGE
The main undercarriage retracts into the wing. It has an anti-skid braking system and is capable of operation from semi-prepared surfaces.

MB.339 IN THE UAE
An MB.339 of the Al Fursan aerobatics team from the United Arab Emirates. The two-seat 'lead-in fighter trainer' variant is designed to carry a wide variety of weapons to teach future fighter pilots how to fire them, and can be used as an effective light-attack and anti-shipping strike aircraft.

Sikorsky UH-60 Black Hawk/SH-60 Sea Hawk (1979)

TYPE • Helicopter   COUNTRY • United States

SPECIFICATIONS (UH-60A)

DIMENSIONS:	Length: 19.76m (65ft); Main rotor diameter: 16.36m (54ft); Height: 5.13m (17ft)
WEIGHT:	Empty 4819kg (10,600lb)
POWERPLANT:	2 × General Electric T700-GE-701C turboshaft, 1410kW (1890hp) each
MAX SPEED:	294km/h (183mph)
RANGE:	600km (370 miles)
CEILING:	5790m (19,000ft)
CREW:	2
ARMAMENT:	Usually 2 × 7.62mm (0.3in) door guns, but with a variety of rocket, missile, gun pod and mine options

Sikorsky's UH-60 Black Hawk is one of the most important combat helicopters in service, proven in conventional and special forces operations. The naval equivalent in the SH-60 Sea Hawk, with an equally illustrious operational career.

Below: The UH-60A, the first of many versions of the Black Hawk family, saw action during the invasion of Grenada in 1981. It has since been in action in Lebanon, Somalia and both Gulf Wars.

FOLDING COMPONENTS
To take up the minimum amount of space on the aircraft carrier, the SH-60F has a folding tail, tailplane and rotors. The main blades fold back to lie above the rear fuselage.

SH-60F SEAHAWK
This Seahawk carries the markings of Helicopter Anti-submarine Squadron 3 'Tridents', the first SH-60F unit in the US Navy's Atlantic Fleet, when it was deployed aboard USS *Theodore Roosevelt*.

FUSELAGE
The fuselage plan is noticeably broad and long, giving a generous internal capacity while allowing a very flat profile.

SURVIVABILITY

The UH-60 was designed with all the years of experience of battle in Vietnam in mind. The low profile of the airframe makes it a difficult target, and safer if it crashes. In an assault landing, the UH-60 comes in fast. Its undercarriage is designed to absorb vertical impacts of up to 45km/h (28mph), dramatically reducing the risk of injury among the helicopter's crew and occupants.

BLACK HAWKS IN IRAQ
US Army soldiers walk towards the passenger terminal after arriving at Sather Air Base, Iraq, from a UH-60 Black Hawk helicopter, 2008. Although the Black Hawk can carry armament, it is essentially a troop carrier. Its cabin and hatches are designed to allow a squad of infantry to get into action fast.

Mil Mi-26 (1980)

TYPE • Helicopter **COUNTRY** • United States

SPECIFICATIONS

DIMENSIONS:	Length 33.73m (110ft 8in); Rotor diameter: 32m (105ft); Height: 8.15m (26ft 9in)
WEIGHT:	Maximum take-off 56,000 kg (123,459lb)
POWERPLANT:	Two Ivchenko-Progress/Zaporizhzhya D-136 turboshaft engines each rated at 7457kW (10,000shp)
MAX SPEED:	295km/h (183mph)
RANGE:	800km (497 miles)
CEILING:	4600m (15,090ft)
CREW:	5
ARMAMENT:	None

A very substantial transport helicopter, and also capable of air assault missions, the Mil Mi-26 (NATO reporting name 'Halo') remains one of the world's largest and heaviest helicopters to ever enter production.

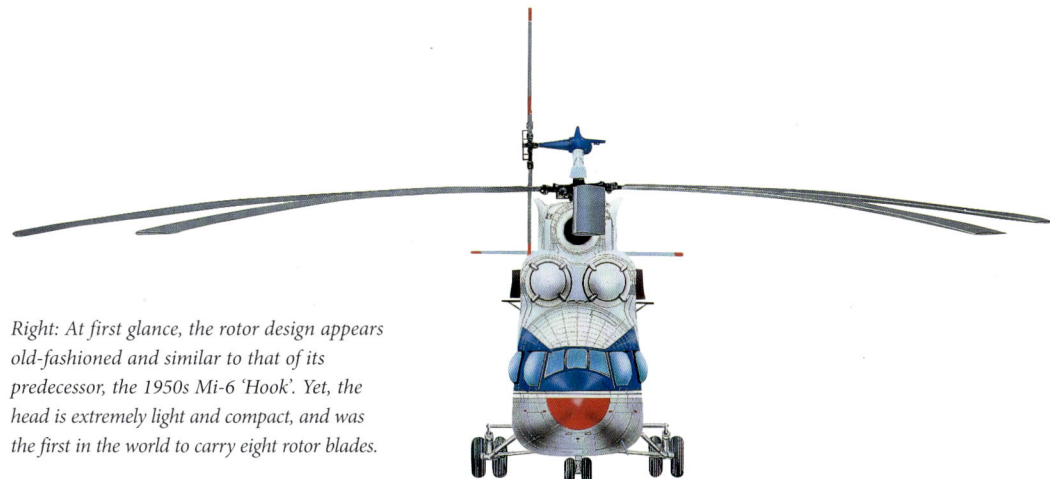

Right: At first glance, the rotor design appears old-fashioned and similar to that of its predecessor, the 1950s Mi-6 'Hook'. Yet, the head is extremely light and compact, and was the first in the world to carry eight rotor blades.

MI26 H-351
The huge Mi-26 offers a capability unmatched by any other helicopter. Bearing the Paris Airshow number H-351, CCCP-06141 is shown as it was displayed in August 1981. It subsequently visited Farnborough Airshow, England, in 1984.

WEIGHT BEARING
At rest, the 'Halo' adopts a distinctive tail-heavy stance, similar to that of other Mil-designed helicopters. To protect the underside from damage when the helicopter is operating in 'hot-and-high' conditions, a heavy-duty tail skid is fitted.

TAIL UNIT
Unlike the main unit, the tail rotor unit was a completely new design and features five composite blades. The fin to which it is fitted incorporates a low-speed aerofoil section, enabling better stability at cruising speed.

DEVELOPMENT
Developed by the Mil design bureau in Moscow as a successor to the Mi-6, first flown in 1957, the Mi-26 adopted the same conventional layout but added a new powerplant and eight-blade main rotor. After three prototypes were built by Mil, series production was launched in 1980 by Rostvertol in Rostov-on-Don. The first production Mi-26 (izdeliye 90) was flown in October 1980 and the helicopter has been in series production ever since, although after the demise of the Soviet Union production was greatly reduced.

Mi-26 RUSSIA
Mil set about designing its Mi-26 'Halo' in the early 1970s, aiming to produce a thoroughly conventional helicopter on a huge scale, with a payload up to one and a half times greater than that of any previous rotary-winged type.

Shenyang J-8 (1980)

TYPE • Fighter **COUNTRY** • China

Alongside the MiG-21-derived J-7, successive versions of the significantly larger Shenyang J-8 – which received the Western reporting name 'Finback' — served the PLAAF for many years, although the type is in the twilight of its career, with only a handful of units still flying the type.

SPECIFICATIONS (J-8F)

DIMENSIONS:	Length: 21.39m (70ft 2in); Wingspan: 9.34m (30ft 8in); Height: 5.41m (17ft 9in)
WEIGHT:	Maximum take-off 15,288kg (33,704lb)
POWERPLANT:	Two Guizhou WP-13B afterburning turbojet engines, 47.1kN (10,580lbf) thrust each dry, 68.6kN (15,430lbf) with afterburner
MAX SPEED:	Mach 1.8 (2222.64m/h)
RANGE:	1000km (620 miles) with drop tanks
CEILING:	18,000m (59,000ft)
CREW:	1
ARMAMENT:	One 23mm (0.9in) Type 23-III cannon; one centreline and six under-wing hardpoints with a capacity of three drop tanks; two PL-11 missiles SARH AAM

Right: The JZ-8F was a reconnaissance version of the J-8F with the Type 23-III cannon replaced by an internal camera.

ADVANCED RADAR
First flown in 2000, the J-8F includes a new Type 1492 pulse-Doppler radar, which is compatible with PL-12 AAM, an advanced weapon with active radar homing, providing a 'fire and forget' capability.

SHENYANG J-8F
This J-8F, serial number 72200, was seen at Changchun Dafangshen Airport in October 2019. The J-8F was a development of the J-8C, but equipped with JL-10 (Type 1473) radar and PL-11 medium-range semi-active radar homing (SARH) air-to-air missile (AAM).

UPGRADES
Other improvements built into the J-8F included a revised cockpit featuring digital displays and a head-up display, while the powerplant comprises two WP-13BII engines.

DESIGN OPTIONS
By 1964, work on a successor to the J-7 had been launched. Shenyang examined two different development paths. One was an all-new single-engine fighter powered by a new turbofan engine, known as the J-9. Somewhat less ambitious was a further development of the basic J-7 configuration, with two engines, and optimized for high-altitude interception — this was the J-8. Shenyang opted for the J-8, essentially an enlarged J-7, with large delta wings, a nose intake containing a Type 204 fire control radar, and a wider fuselage containing a pair of WP-7A turbojets.

SHENYANG J-8I
Improvements introduced by the J-8I or J-8A included new avionics, among them a Type 204 radar, while a new HTY-2 ejection seat was installed below a revised cockpit canopy. A new twin-barrel Type 23-III cannon was fitted, with missile armament comprising up to four PL-2B or PL-5B air-to-air missiles (AAMs).

Lockheed F-117 Nighthawk (1981)

TYPE • *Bomber* COUNTRY • *United States*

Lockheed's 'stealth' attack aircraft was developed under conditions of utmost secrecy before being propelled to legendary status during Operation *Desert Storm*. The aircraft fulfilled a niche role with the US Air Force until it was finally retired from service in 2008.

SPECIFICATIONS

DIMENSIONS:	Length: 20.08m (65ft 11in); Wingspan: 13.2m (43ft 4in); Height: 3.78m (12ft 5in)
WEIGHT:	23,814kg (52,500lb) maximum take-off
POWERPLANT:	2 × 48.04kN (10,800lbf) General Electric F404-GE-F1D2 turbofans
MAX SPEED:	1040km/h (646mph)
RANGE:	862km (535 miles) radius with 1814kg (4000lb) warload
CEILING:	11,765m (38,600ft)
CREW:	1
ARMAMENT:	Up to 2268kg (5000lb) of disposable stores

Above: This F-117A was flown by the commander of the 37th Tactical Fighter Wing. The wing consisted of three units: the 415th TFS, 416 TFS and the 417th Tactical Fighter Training Squadron.

COCKPIT
In original form, this was equipped with Texas Instruments monochrome displays and an array of off-the-shelf instruments taken from other aircraft, including the F/A-18.

WING
This forms a simple aerofoil, with flat surfaces underwing that blend into underfuselage surfaces to create a single unified lifting surface.

STEALTH COATING
Almost every surface of the F-117 was covered with radar absorbent material (RAM). This provided a critical defence against radar detection, but proved laborious and costly to maintain. After each mission, maintenance specialists had to closely examine the aircraft's special coating to identify whether any repairs were needed. If required, the special coatings were reapplied, allowed to cure and then reinspected.

INTERNAL WEAPONS
This F-117A is seen releasing a typical load of two 907kg (2000lb) GBU-27A/B laser-guided bombs fitted with hardened BLU-109/B penetrator warheads.

RETIREMENT
A total of 59 F-117As were built between 1981 and 1990, six of which were lost in non-combat accidents. Once expected to serve until at least 2018, the USAF retired the F-117 fleet in April 2008 after 27 years of service, in order to free up funds for modernization.

Mikoyan MiG-31 (1981)

TYPE • Interceptor **COUNTRY** • Soviet Union

One of the world's fastest combat aircraft, the MiG-31 (NATO reporting name 'Foxhound') was designed in the Soviet era by the Mikoyan Design Bureau in Moscow as a long-range interceptor capable of flying at more than twice the speed of sound.

SPECIFICATIONS (MiG-31B/BS Foxhound)

DIMENSIONS:	Length: 20.62m (67ft 8in) without probe; Wingspan: 13.46m (44ft 2in); Height: 6.45m (21ft 2in)
WEIGHT:	Maximum take-off 46,200kg (101,853lb)
POWERPLANT:	Two Aviadvigatel/Perm D-30F-6 (izdeliye 48) turbofans, each rated at 93.17kN (20,944lbf) dry and 152kN (34,172lbf) with afterburning
MAX SPEED:	Mach 2.35 (2901.78m/h)
RANGE:	3000km (1860 miles) with underwing tanks
CEILING:	20,600m (67,585ft)
CREW:	2
ARMAMENT:	One six-barrel 23mm (0.9in) GSh-6-23 cannon in the starboard fuselage; Up to 9000kg (19,840lb) of weapons and stores

Right: Viewed from the front, the MiG-31 airframe is dominated by the truly enormous intakes for its two Aviadvigatel/Perm D-30F-6 turbofans.

MiG-31M
The first MiG-31 seen in the West, '057 Blue' was the seventh MiG-31M to be built and included the improved Zaslon-M radar system, six K-37, four K-77 AAMs.

Cancellation
The advanced MiG-31M Foxhound-B interceptor and the anti-satellite MiG-31D were developed in the 1980s, but were cancelled in the early 1990s.

MIG-25 VARIANT
Initially considered a relatively straightforward two-seat MiG-25 variant featuring multi-target capability and extended endurance designated as the MiG-25MP, the programme to develop what would become the extensively redesigned MiG-31 was launched in May 1968. The first flight of the MiG-25MP occurred on 16 September 1975 and the initial production MiG-31 appeared in late 1976, with series production undertaken by the Sokol factory in Nizhny Novgorod (formerly Gorky).

Airframe and engines
The MiG-31M Air featured a modified airframe with around two tons of additional fuel and more powerful D-30D-6M engines.

PRODUCTION MiG-31
Total production between 1976 and 1994 amounted to 519, of which 349 were of the initial MiG-31 version followed by 101 MiG-31DZ (izdeliye 01DZ) that introduced an in-flight refuelling probe and entered production in 1989. These were followed by 69 MiG-31B versions featuring improved Zaslon-A radar and nuclear-tipped R-33S missiles.

Sukhoi Su-25 (1981)

TYPE • Ground-attack **COUNTRY** • Soviet Union

SPECIFICATIONS (Su-25SM)

DIMENSIONS: Length: 15.53m (50ft 11.5in); Wingspan: 14.36m (47ft 1in); Height: 4.80m (15ft 9in)

WEIGHT: Maximum take-off 19,000kg (41,888lb)

POWERPLANT: Two non-afterburning Soyuz/Moscow R-95Sh turbojets each developing 40.21kN (9,039lbf) of thrust

MAX SPEED: Mach 0.82

RANGE: 1850km (1450 miles) ferry range with drop tanks

CEILING: 7000m (22,966ft)

CREW: 1

ARMAMENT: One 30mm (1.2in) Gryazev-Shipunov GSh-30-2 twin-barrel Gast autocannon in port forward fuselage; up to 4400kg (9700lb) of weapons and stores

The Su-25 (NATO reporting name 'Frogfoot') single-seat subsonic close air support aircraft features heavy armour protection and a good degree of manoeuvrability, conferring upon it the ability to deliver accurate attacks against small ground targets.

Above: A Ukrainian Air Force Su-25UBK, the upgraded export model of the Su-25K. In addition to bombs and missiles, the Su-25's pylons can carry gun pods containing twin-barrel GSh-23 23mm (0.9in) guns.

AVIONICS
A mid-life upgrade launched in 1999 was based around new PrNK-25SM mission avionics, including Kopyo radar and a modernized cockpit with two digital displays. Eventually Russia opted for a more modest option, without radar, the Su-25SM.

SUKHOI SU-25
This Ukrainian Air Force Su-25 'Frogfoot' is part of the 299 Tactical Aviation Brigade, flying low-level missions over the Donbas region in eastern Ukraine.

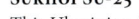

TWO-SEATERS
Single-seat combat versions of the Su-25 were manufactured at the Tbilisi aircraft plant between 1979 and 1992, while two-seaters came from the Ulan-Ude (Russia) factory between 1987 and 1992. Initial two-seaters were the Su-25UB Frogfoot-B, tandem two-seat combat trainer in which the weapons systems were retained and which featured an enlarged vertical fin compared to the single seaters. The export equivalent of the initial production two-seater was designated the Su-25UBK.

UPGRADES
The upgraded Russian Su-25SM can carry up to 5000kg (11,023lb) on the eight main hardpoints. New weapons include TV-guided KAB-500Kr bombs, S-25 laser-guided rockets and R-73 (AA-11 Archer) air-to-air missiles.

MODERN SU-25S
About 1320 Su-25s were built but the Ulan-Ude plant retains the infrastructure to build more if ordered. Russia currently has some 200 Su-25s including 20 Su-25UB trainers in service. Other post-Soviet operators are Armenia, Azerbaijan, Belarus, Georgia, Kazakhstan, Turkmenistan, Ukraine and Uzbekistan.

Shaanxi Y-8 (1981)

TYPE • Transport **COUNTRY** • China

SPECIFICATIONS	
DIMENSIONS:	Length: 34m (111ft 6in); Wingspan: 38m (124ft 7in); Height: 11.16m (36ft 7in)
WEIGHT:	Maximum take-off 61,000kg (134,100lb)
POWERPLANT:	Four Zhuzhou WoJiang WJ-6 turboprops, developing 3170kW (4250hp)
MAX SPEED:	660km/h (410mph)
RANGE:	5615km (3489 miles)
CEILING:	10,400m (34,120ft)
CREW:	5
ARMAMENT:	Two 23mm (0.9in) cannon

The Soviet-designed Antonov An-12, Western reporting name 'Cub', was the pre-eminent Eastern Bloc four-engine tactical transport during the Cold War. China embarked on domestic development of the aircraft, as the Shaanxi Y-8, which has had an illustrious career.

Below: This Y-8F-100 (serial number B-4155), seen in Hanzhong, China, in June 2023, is fitted with an electronic flight instrument system (EFIS), colour weather radar, traffic collision avoidance system (TCAS) and GPS.

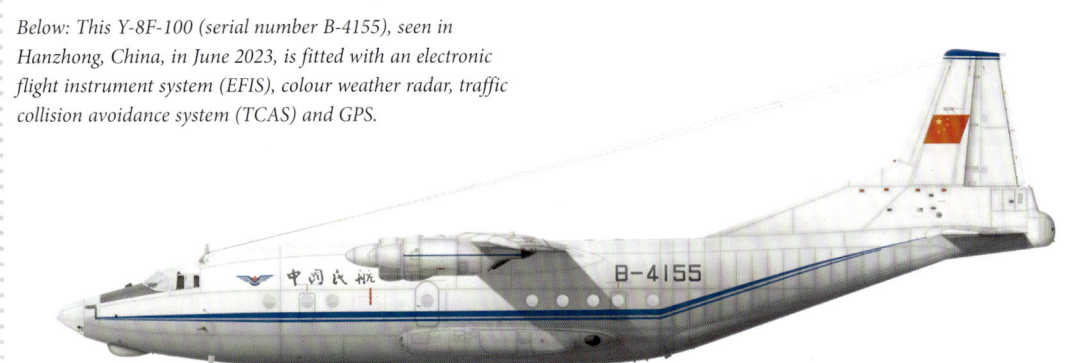

SHAANXI Y-8C
This Y-8C (serial number 55415), the basic transport version of the type, was seen in Zhengzhou International Airport, China, in April 2023.

HOLD CAPACITY
The hold capacity of the aircraft is enormous: 20,000kg (44,092lb) of cargo; 96 troops or 82 paratroops; or up to 60 medevac patients.

REAR CARGO RAMP
The Y-8C includes a rear cargo ramp, which was first introduced on the Y-8B model.

PRESSURIZATION
The early versions of the Y-8 – also known as the Category I Platform – were hampered by a lack of pressurization in the cargo hold. This meant they could only operate at higher altitudes with a maximum of 14 passengers, who were provided with a pressurized crew cabin in the forward part of the fuselage. A version with a fully pressurized cargo hold was developed, as the Y-8C, which made its maiden flight in December 1990. This revised aircraft is also designated as the Category II Platform.

Y-8 PARA DROP
Chinese People's Liberation Army (PLA) paratroopers jump from a Y-8 aircraft towards the end of a two week-long Pakistan–China military exercise in Jhelum in November 2011. Eighty-two paratroopers can be carried in the main compartment.

Mil Mi-17 (1981)

TYPE • Helicopter **COUNTRY** • Soviet Union

SPECIFICATIONS (Mi-171)

DIMENSIONS:	Length: 25.3m (83ft); Rotor diameter: 21.1m (69ft 3in); Height: 5.65m (18ft 6in)
WEIGHT:	Maximum take-off: 13,000kg (28,660lb)
POWERPLANT:	Two 1435kW (1924shp) Klimov TV3-117VM turboshaft engines
MAX SPEED:	250km/h (155mph)
RANGE:	610km (980 miles)
CEILING:	5000m (16,404ft)
CREW:	3
ARMAMENT:	24 troops or 12 stretchers or 4000kg (8818lb) internal payload or 5000kg (11023lb) externally slung

One of the world's most successful helicopters, the Soviet Mi-17 has been produced in vast numbers and remains ubiquitous in the air arms of many nations. China has even ordered new Mi-171s as recently as 2019.

Left: A Russian military Mil Mi-17 performs in the annual British air show at Fairford, England, on 14 July 2022.

ENGINES
By replacing the 1044kW (1400hp) powerplant of the Mi-8 with two 1434kW (1923hp) Isotov TV3-117MT engines, designers gave the Mi-17 an improved cruising speed and hovering ceiling and an increased maximum take-off weight.

TAIL ROTOR
One of the main recognition points distinguishing the Mi-17 from the Mi-8 is the repositioned tail rotor. This has been moved from the right side of the tailboom to the left.

GLOBAL EXPORT
The Mi-17, which first appeared in 1975, is the designation given to export variants of the venerable Mil Mi-8, which first flew in July 1961, the same aircraft being referred to as the Mi-8MT in Soviet/Russian service. Differing from earlier Mi-8 variants in that it is fitted with larger Klimov TV3-117MT engines, rotors and transmission, together with fuselage improvements for heavier loads, the Mi-17 was adopted by a huge number of nations across the world.

Mi-17M 'Hip H'
Czechoslovakian military aircraft were divided between the Czech Republic and Slovakia when these countries became independent nations. Most Czech Mi-17s are painted in this green/grey scheme.

Mi-17 ON TAKE-OFF
An all-new titanium rotor hub was designed for the Mi-17. The Mi-8 prototype had a four-bladed main rotor, as used on the Mi-4 'Hound', but all production aircraft, including Mi-17s, have been fitted with a five-bladed rotor.

Kamov Ka-27 (1982)

TYPE • Helicopter **COUNTRY** • Soviet Union

The Kamov Ka-27 (NATO reporting name 'Helix') is a ship- or shore-based anti-submarine and search-and-rescue helicopter with a normal crew of three, and Kamov's trademark coaxial, contra-rotating rotor configuration.

SPECIFICATIONS

DIMENSIONS:	Fuselage length: 11.3m (37ft 1in); Rotor diameter: 15.9m (52ft 2in); Height: 5.4m (17ft 8in)
WEIGHT:	Maximum take-off: 11,000kg (24,251lb)
POWERPLANT:	Two Klimov TV3-117KM turboshaft engines each rated at 1618kW (2200shp)
MAX SPEED:	290km/h (180mph)
RANGE:	700km (435 miles)
CEILING:	3500m (11,480ft)
CREW:	1-3 plus 1-3 specialists
ARMAMENT:	Up to 2000kg (4409lb) of weapons and stores

Right: Part of the Baltic Fleet, '41 Yellow' is a modernized Ka-27M of the Russian Navy, based at Chernyakhovsk air base in the Russian enclave of Kaliningrad during 2018.

KA-27, UKRAINE NAVY
The Ukrainian Navy has been a major operator of the Ka-27. During the recent conflict with Russia, some members of the international community have also given Ukraine Halos from their own imported stocks.

AVIONICS
The original Ka-27 (izdelye D2 or izdeliye 500), Helix-A, featured the Osminog automated search/attack system, consisting of an Initsiativa-2KM radar, VGS-3 Ros-V dipping sonar, tactical display and digital computer.

TWIN ROTORS
Using two rotors spinning in opposite directions the Soviet Kamov design bureau dispensed with the tail rotor normally used to give directional stability in helicopters. The design enabled shorter rotor blades to be used, which made it suitable for use aboard ships. The 'Helix' series of helicopters, which includes anti-submarine, assault transport and search-and-rescue versions, served aboard a variety of Soviet warships.

WEAPON OPTIONS
Weapons options include torpedo or 6–8 250kg (550lb) depth charges, including the Zagon series guided depth charges. Up to 36 sonobuoys can be carried.

NAVAL DEPLOYMENT
A Russian Navy Ka-27 assigned to the Russian destroyer *Admiral Vinogradov* (DDG 572) flies near the guided-missile cruiser USS *Vella Gulf* (CG 72) while conducting operations in the Gulf of Aden.

Sikorsky HH-60 (1982)

TYPE • Helicopter COUNTRY • United States

The HH-60 is an angel of mercy. This helicopter flies from ship decks and shore bases to bring salvation to those in peril. For its own special operations requirements, the USAF developed the advanced HH-60G Pave Hawk.

SPECIFICATIONS (HH-60 Rescue Hawk)

Dimensions:	Length: 19.76m (64ft 10in); Main rotor diameter: 16.36m (53ft 8in); Height: 5.18m (17ft)
Weight:	Empty 6114kg (13,480lb)
Powerplant:	Two General Electric T700-GE-401C turboshafts each rated at 1417kW (1900hp)
Max speed:	234km/h (145mph)
Range:	Operational radius 463km (287 miles)
Ceiling:	4206m (13,800ft)
Crew:	4
Armament:	Armament: two 7.62mm (0.3in) M134 Miniguns pintle-mounted each side of the cabin

Above: A HH-60H Rescue Hawk – the rescue role has made the HH-60 an extremely adaptable platform. Even when in service with the US Air Force, missions often take place over water.

Refuelling probe
The refuelling probe extends nearly to the point of clearing the rotor disc. The tanker aircraft are variants of the C-130 equipped with wingmounted drogues that trail behind the aircraft.

Cabin
The cabin area has room for stretcher cases and their attendants, or a rescue team of four who retrieve the downed aircrew if they are unable to reach the helicopter themselves due to injury.

HH-60G Pave Hawk
Pave Hawks give downed USAF aircrew the chance of rescue from behind enemy lines. They also extend the reach of Special Forces on covert operations.

HH-60G PAVE HAWK
The Pave Hawk derives from modifications made by the USAF to some standard UH-60s it acquired in the early 1980s. Fitted with an air refuelling probe and additional fuel tanks in the cabin and with machine guns changed from .30in to .50 calibre (12.7mm) XM218s, these helicopters were referred to as 'Credible Hawks' and entered service in 1987. Further developments resulted in the HH-60G and MH-60G Pave Hawk, both used for the infiltration, exfiltration and resupply of special forces by day or night.

PARARESCUE PRACTICE
Two US Air Force pararescuers from the 101st Expeditionary Rescue Squadron (ERS) practise a tandem live hoist onto an HH-60G Pave Hawk helicopter during a combat search and rescue (CSAR) exercise.

General Dynamics EF-111A Raven (1983)

TYPE • Electronic warfare **COUNTRY** • United States

SPECIFICATIONS

DIMENSIONS:	Length: 23.18m (76ft); Wingspan: 19.2m (63ft) spread; Height: 6.1m (20ft)
WEIGHT:	Maximum take-off 40,337kg (88,928lb)
POWERPLANT:	Two 8380kg (18,500lb) Pratt & Whitney TF-30-P3 turbofans
MAX SPEED:	2272km/h (1412mph); service ceiling
RANGE:	Combat radius 1495km (929 miles)
CEILING:	13,725m (45,000ft)
CREW:	2
ARMAMENT:	Two AIM-9L Sidewinders on pylons (optional)

Bitter experiences over North Vietnam led the USAF to identify a need for a dedicated electronic warfare aircraft. The EF-111A combined the proven airframe of a fast, sturdy, long-range strike aircraft with the electronic equipment to enable bombers to reach their targets.

This page: The EF-111A shown here, 66-0019, was assigned to the 430th ECS at Cannon AFB, New Mexico, in 1993. Delivered by General Dynamics in 1967 as an F-111A, the aircraft was assigned to the 4480th TFW at Nellis AFB, Nevada, and was involved in Operation Harvest Reaper, the work-up programme for the first Combat Lancer F-111A deployment to Vietnam.

Attack radar
The EF-111A's APQ-160 attack radar, used for accurate mapping, and its terrain-following radar (TFR) were retained, the latter to permit low-level penetration.

SIR pod
The system integrated receiver (SIR) pod and a pair of antenna fairings either side of the fin resembled those found on the EA-6B Prowler and contained receivers for the ALQ-99 jamming system.

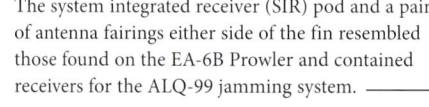

Sidewinders
As well as a defensive avionics suite, the Raven could be equipped with a pair of AIM-9s for self-defence. These were mounted on shoulder pylons attached to the outer pair of four wing pylons normally reserved for fuel tanks and datalink pods.

DEVELOPMENT
Development of the EF-111A Raven electronic support aircraft began in January 1975, when the USAF contracted with Grumman Aerospace to modify two F-111As to serve as electronic warfare platforms. The aircraft's high speed, long range and substantial payload made it ideally suited to this role. To accommodate the extra 2718kg (6000lb) of equipment Grumman added a 4.88m (16ft) canoe-type radome under the fuselage and a pod mounted on top of the tailfin.

EF-111A MOUNTAIN FLIGHT
The EF-111A Raven first entered the public eye in April 1986, when UK-based aircraft provided jamming support for F-111E fighter-bombers attacking targets in Libya in response to that nation's alleged support for terrorist activities. The last EF-111As were withdrawn in 1999.

Xi'an Y-7 (1984)

TYPE • *Transport* **COUNTRY •** *China*

SPECIFICATIONS (Y-7-100)

DIMENSIONS:	Length: 24.2m (79ft 5in); Wingspan: 29.67m (97ft 3in); Height: 8.6m (28ft 2in)
WEIGHT:	Maximum take-off 21,800kg (48,061lb)
POWERPLANT:	Two Dongan WJ-5E turboprop engines, 1800kW (2400shp) each equivalent
MAX SPEED:	505km/h (314mph)
RANGE:	910km ((565 miles)
CEILING:	8750m (28,707ft)
CREW:	3; plus 52 passengers
ARMAMENT:	None

Another aviation product of the Soviet Union that has been licence-built in China was the Antonov An-24 turboprop regional transport, produced by Xi'an as the Y-7, in a number of successively improved versions, and also manufactured for export as the MA-60.

Above: This Chinese Navy (PLANAF) Xi'an Y-7G, registration 9082, was photographed in Changchun Longjia International Airport in July 2022.

IMPROVED VERSION
Work on an improved version, the Y-7-100, was launched in the mid-1980s and this emerged essentially as an unlicenced variant of the An-26. It features a fully pressurized fuselage, a rear cargo ramp, more powerful WJ-5E turboprop engines, an auxiliary turbojet and a modernized cockpit.

XI'AN Y-7G
A PLAAF Xi'an Y-7G, registration 55017, dating from February 2022. The Y-7G is a military variant of the MA60, in turn a stretched version of the Xi'an Y7-200A.

POWERPLANT
Power for the Y-7 comes from two Dongan WJ-5E turboprop engines, each generating 1800kW (2400shp).

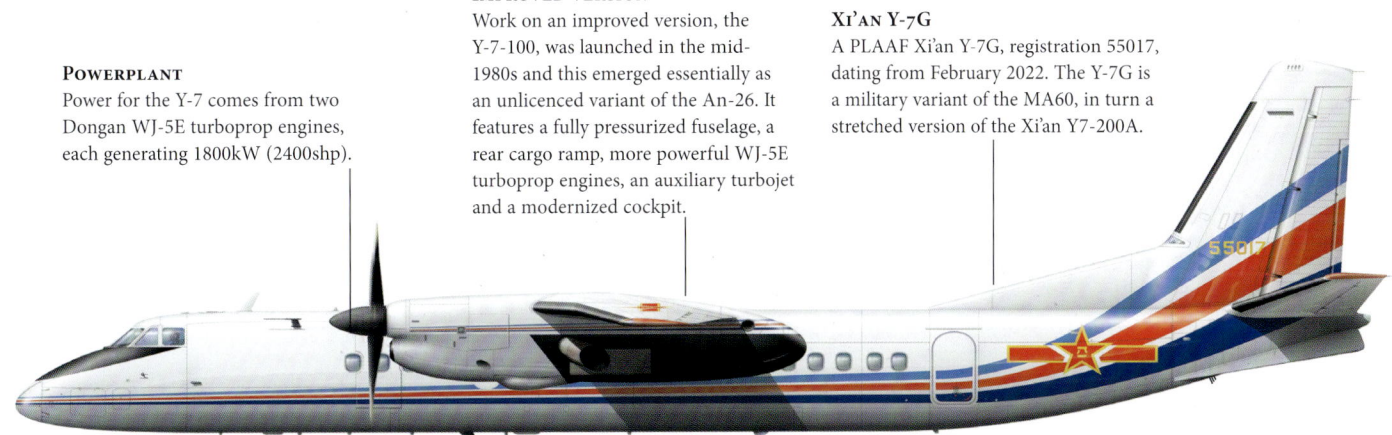

HYJ-7 BOMBER TRAINER
A unique version of the Y-7 employed by the PLAAF is the HYJ-7 (also designated Y-7LH), a training aircraft based on the Y-7-100C2 passenger transport. Introduced in the late 1990s, the aircraft is used to train navigators and bombardiers destined to operate the H-6 bomber. Changes compared to the transport version include a large observation gondola on the starboard side of the fuselage with glazing to represent the nose of the H-6. Equipment includes a bombsight, a bomb computer and a new navigation system.

XI'AN Y-7 TAXIING
Xi'an began work on the Y-7 in the mid-1960s and a first example of this aircraft – actually a Chinese-assembled An-24T cargo transport – flew on 25 December 1970. The upheavals of the Cultural Revolution in China meant that large-scale production of the Y-7 did not commence until around 1982. The first aircraft for the PLAAF began to be delivered in around 1984.

Vickers VC10 (1984)

TYPE • Transport COUNTRY • United Kingdom

SPECIFICATIONS (Type 1101)

DIMENSIONS:	Length: 48.36m (158ft 8in) Wingspan: 44.55m (146ft); Height: 12.4m (39ft 6in)
WEIGHT:	Maximum take-off 151,898kg (334,878lb)
POWERPLANT:	4 × Rolls-Royce Conway Mk 301 turbofans, 100kN (22,500lbf) thrust each
MAX SPEED:	935km/h (580mph)
RANGE:	7560km (4154 miles)
CEILING:	13,000m (43,000ft)
CREW:	4+
ARMAMENT:	None

The Vickers VC10 was a 'great' of the 1960s golden age, when jet-powered air commerce was new and airlines pampered travellers. It was chosen without hesitation by many carriers, with a military version joining the Royal Air Force.

Above: The VC10 K.2 was an in-flight refuelling tanker. Five of this type of aircraft were produced for the RAF.

PERFORMANCE
With a standard VC10 length fuselage, large forward loading door and strengthened cabin floor, the more powerful VC10 supplied was fast, with a maximum speed of 935km/h (580mph) and good range, up to 7560km (4154 miles).

REFUELLING
The RAF VC10 refuelling aircraft could extend three drogue-stabilized refuelling lines. The aircraft had its own refuelling probe in the nose, so it could also take on fuel while airborne.

VC10 K.3
This particular VC10 K.3 refuelling tanker, ZA148, first flew in 1962, having a long career as a civil airliner before its tanker conversion and entry in RAF service in 1984.

CIVIL AIRLINER
In the 1960s, the Vickers VC10 offered a high standard of transportation. The VC.10 was largely underwritten by BOAC, the airline since replaced by British Airways, but the aircraft appeared in many colours: Ghana Airways leased one to Middle East Airways, which stationed the VC10 in Beirut. The success of the VC.10 jetliner paralleled faithful service by its military counterpart.

RAF SERVICE
In RAF service, the VC10 was repurposed to several key roles, including aeromedical evacuation, VIP transportation, freight carrying and aerial refuelling tanker. The final British military flights of the VC10 took place in 2013.

Lockheed Martin F-16 Fighting Falcon C/D (1984)

TYPE • Multi-role fighter **COUNTRY** • United States

SPECIFICATIONS (F-16C)

DIMENSIONS: Length: 15.09m (49ft 6in); Wingspan: 9.45m (31ft); Height: 5.09m (16ft 8in)

WEIGHT: 16,057kg (35,400lb) loaded

POWERPLANT: Either one 10,800kg (23,770lb) thrust Pratt & Whitney F100-PW-200 or one 13,150kg (28,984lb) thrust General Electric F110-GE-100 turbofan

MAX SPEED: 2142km/h (1320mph)

RANGE: Combat radius 925km (525 miles)

CEILING: 15,240m (50,000ft)

CREW: 1

ARMAMENT: One General Electric M61A1 multi-barrelled cannon; seven external hardpoints for up to 9276kg (20,450lb) of ordnance

The main production version of Lockheed Martin's illustrious F-16 fighter has been the F-16C (and its two-seat counterpart the F-16D), which had structural and avionics improvements over the F-16A and B.

Below: NF-16D variable in-flight simulator aircraft (VISTA). Operated by the US Air Force Test Pilot School at Edwards Air Force Base, it was redesignated as the X-62A in June 2021. The NF-16 VISTA started life as a F-16D Block 30, which later received numerous upgrades and modifications.

F-16C FIGHTING FALCON
Less visible but important changes introduced in the F-16C/D included a more advanced cockpit with a GEC head-up display (HUD).

JAMMER
The F-16C/D aircraft had an enlarged base for the vertical tail. This was originally intended to accommodate the internal airborne self-protection jammer (ASPJ), although the US Air Force later abandoned this plan.

RADAR
The F-16C/D was fielded with an all-new radar, the Westinghouse AN/APG-68 multimode system replacing the previous AN/APG-66.

IMPROVED VARIANTS

First flown on 19 June 1984, the single-seat F-16C was an altogether more capable warplane than the F-16A/B. Together with the two-seat F-16D model, they introduced a range of improvements that considerably boosted their combat potential. In terms of identification, the F-16C/D aircraft could be distinguished by an enlarged base for the vertical tail, forming a distinctive 'island' at the bottom of the fin.

WEAPON OPTIONS
The F-16C/D's new radar provided increased range, enhanced resolution and a wider selection of operating modes. Importantly, the new radar was compatible with more capable weapons, including the latest AIM-120 advanced medium-range Air-to-air missile (AMRAAM) for beyond-visual-range aerial engagements and the AGM-65D Maverick with imaging infrared guidance for air-to-ground missions.

Beriev A-50 (1985)

TYPE • *AEW&C* COUNTRY • *Soviet Union*

SPECIFICATIONS

DIMENSIONS:	Length: 49.59m (162ft 8in); Wingspan: 50.5m (165ft 8in); Height: 14.76m (48ft 5in)
WEIGHT:	Maximum take-off: 190,000kg (418,878lb)
POWERPLANT:	Four Aviadvigatel/ Perm D-30KP series 2 turbofans, each rated at 117.68kN (26,455lbf)
MAX SPEED:	810km/h (503mph)
RANGE:	5100km (3169 miles)
CEILING:	12,200m (40,026ft)
CREW:	15
ARMAMENT:	None

Developed from the Il-86 by Beriev, the Beriev A-50 (NATO reporting name 'Mainstay') is a strategic airborne early warning and control (AEW&C) aircraft based on the Il-76MD (A-50) or Il-76MD-90A (A-100) platform.

Above: Shown as it appeared in circa 2005, '46 Red' wears the standard white and grey livery of the A-50 fleet. As well as the obvious rotodome, the A-50 differs from the Il-76 in featuring a nose refuelling probe and deletion of the tail turret.

REFUELLING PROBE
The nose-mounted in-flight refuelling probe was initially removed from the developmental 'Mainstay' because of problems encountered with airflow across the aircraft during the fuel transfer phase.

ROTODOME
Influenced by the design of America's E-3 Sentry, the 'Mainstay's primary radar is positioned in a rotodome mounted above the fuselage.

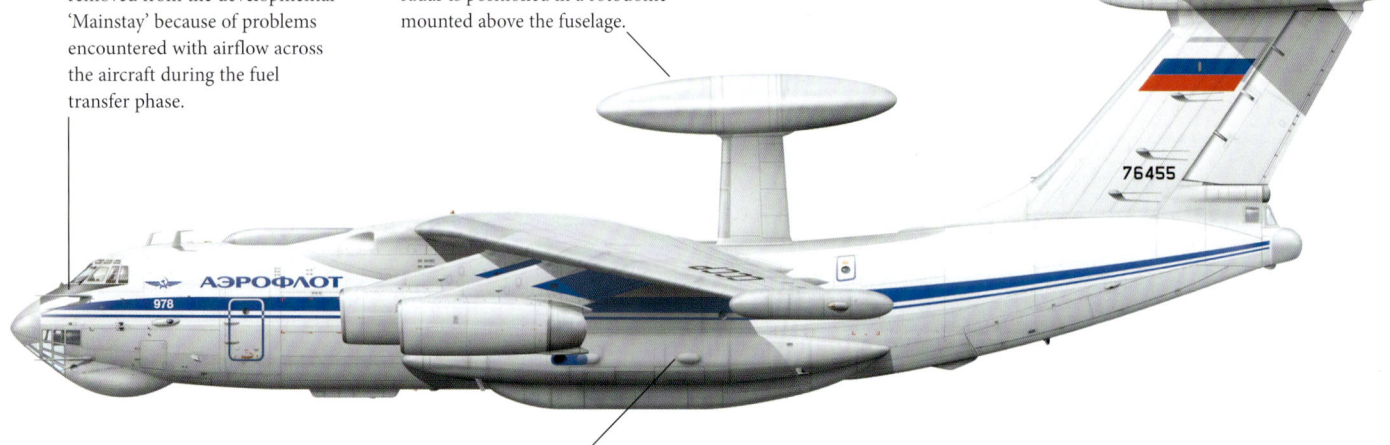

STABILITY
At high altitude the stability of the A-50 was found to be relatively poor. Additional finlets were added to the lower fuselage above the undercarriage bays to alleviate the problem.

RADAR ASSET
In 1978 work began on a replacement for the primitive Tu-126 'Moss' AEW&C platform in service with the Soviet forces. An all-new radar system, with its associated computer equipment and other sensors, was to be incorporated into the Il-76 airframe by Beriev. Production of the A-50 'Mainstay' began in 1983 and, although there were a number of early problems, the aircraft matured into an effective radar and command asset.

BERIEV A-100
Beriev was contracted in 2007 to develop the A-100 aircraft as a successor to the A-50, using the new Il-76MD-90A airframe. The Premier radar has active electronic scanning in elevation and mechanical scanning in azimuth. The aircraft also features new navigation and communication systems plus additional antennas likely for passive surveillance.

Rockwell B-1B Lancer (1985)

TYPE • Bomber COUNTRY • United States

Although its days in US Air Force service are now numbered with plans to retire the B-1B in the future in favour of the forthcoming B-21 Raider, the 'Bone' remains a prized asset in the inventory.

SPECIFICATIONS

DIMENSIONS:	Length: 47.8m (147ft); Wingspan: 41.8m (137ft) unswept or 24.1m (79ft) fully swept; Height: 10.24m (33ft 7.25in)
WEIGHT:	Maximum take-off 163,300kg (360,000lb)
POWERPLANT:	Four General Electric F101-GE-102 turbofan engines each developing 133.45kN (30,000lbf) of thrust with afterburning
MAX SPEED:	Mach 1.2 (1481.76m/h)
RANGE:	11,675km (7255 miles)
CEILING:	Around 10,668m (35,000ft)
CREW:	4
ARMAMENT:	Up to a maximum of 56,250kg (125,000lb) ordnance

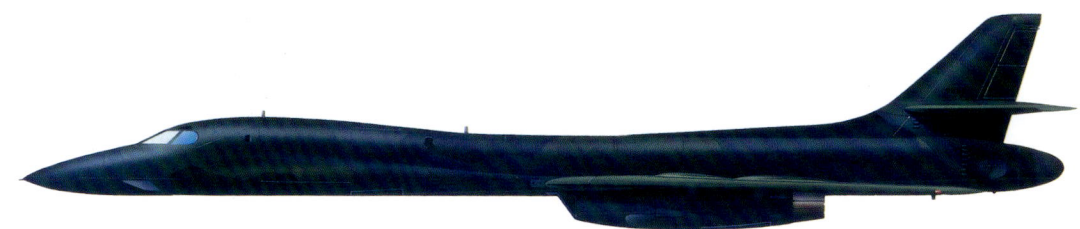

Above: The original B-1A had an ejection capsule like the F-111, but the B-1B has separate crew compartments and individual ejection seats for the pilots and systems operators.

ENGINE INTAKES
The engine intakes have been designed to shield the engine compressor fans from hostile radar beams. Since the compressor would otherwise give a strong radar return, this feature automatically reduces the bomber's signature.

CREW
Pilot and co-pilot sit side-by-side in a cockpit with both digital and analogue instruments. The B-1B is flown like a fighter, using a stick and rudder pedals. Crew members 3 and 4, known as the OSO and DSO (offensive and defensive systems operators), sit side-by-side behind the pilots.

COMBAT OPS
In its conventional guise, the B-1B first saw combat during Operation *Desert Fox*, a series of airstrikes mounted against Iraq in December 1998. The following year, the aircraft was employed during Operation Allied Force, the NATO campaign against Serbia. Since then, the B-1B has been a regular contributor to US military campaigns in Afghanistan, Iraq, Libya and Syria, successively adding new conventional weapons capabilities to improve its utility, initially as part of the Conventional Mission Upgrade Program (CMUP).

WING
The wing has seven-segment leading-edge flaps and six-segment trailing-edge flaps. There are no ailerons, and roll control is effected by spoilers.

LANCER AT TAKE-OFF
It may lack the glamour of the stealthy B-2 flying wing or the reverence accorded to the 50-year-old B-52 Stratofortress, but the Rockwell B-1B Lancer is a highly advanced supersonic bomber. With its tremendous capacity to carry immense loads of nuclear and conventional weaponry, the Lancer became America's primary low-level, supersonic, nuclear strike asset.

Bell AH-1W SuperCobra (1986)

TYPE • Helicopter   COUNTRY • United States

SPECIFICATIONS

DIMENSIONS:	Length: 17.68m (58ft); Rotor diameter: 14.63m (48ft); Height: 4.11m (13ft 6in)
WEIGHT:	Empty 4627kg (10,200lb); loaded 6690kg (14,749lb)
POWERPLANT:	2 x General Electric T-700 turboshafts, 1212kW (1625hp)
MAX SPEED:	352km/h (219mph)
RANGE:	590km (367 miles)
CEILING:	4495m (14,747ft)
CREW:	2
ARMAMENT:	1 x M197 20mm (0.79in) cannon in undernose turret, four underwing hardpoints for guided anti-armour or air-to-air missiles, Minigun pods, or unguided high explosive rockets

Derived from a project to supply the Shah of Iran with an upgraded version of the AH-1T, the SuperCobra prototype with its twin T-700 engines served as the basis for the development of the current Marine Corps AH-1W.

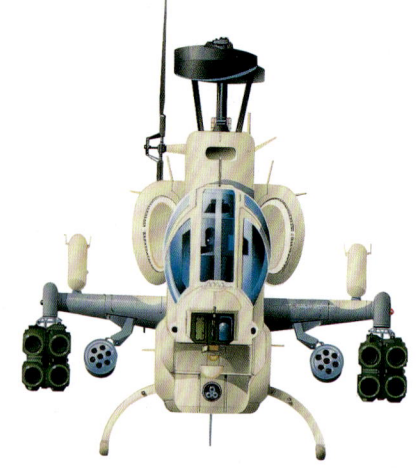

Right: The stub pylons provide not only the means to carry a large weapon load but also act as miniature wings, providing valuable extra lift when the Cobra is in forward flight.

POWERPLANT
Cobras have been powered by a variety of engines over the years. Marine aircraft generally have two engines, as an added safety factor for long over-water operations.

TAIL
The Cobra's tail rotors are made from an aluminium honeycomb with a stainless steel skin and leading edge.

CANNON TURRET
The General Electric turret houses the deadly 20mm (0.79in) M197 cannon. The turret can swing through 110 degrees either side of the nose.

MARINE CORPS COBRA

Until 2020, Marines used the AH-1W 'Whiskey Cobra', a warrior for the hi-tech battlefield that was as formidable in many situations as the Army's newer Apache, which came along years after the AH-1W first flew. The 'Whiskey Cobra' excelled at amphibious warfare, flying from ship decks or from land. Pilots of this thin, graceful ship praised its nimble flying qualities as well as its flexibility and fighting prowess.

BELL SUPER COBRA AH-1W
The Marine Corps retired its AH-1W SuperCobras in 2020, but maintained its fidelity to the Cobra family by switching to the advanced AH-1Z Viper.

Changhe Z-8 (1986)

TYPE • Helicopter COUNTRY • China

As a reverse-engineered variant of the amphibious Aérospatiale Super Frelon, the Changhe Z-8's basic design has continued to be developed in China. Updated versions remained in production at the end of 2023 for both civil and military use.

SPECIFICATIONS (Z-8KA)

DIMENSIONS:	Length: 19.4m (63ft 8in); Rotor diameter: 18.9m (62ft); Height: 6.76m (22ft 2in)
WEIGHT:	Maximum take-off 12,074kg (26,619lb)
POWERPLANT:	Three 1156kW (1550shp) Changzhou WZ-6 (Turbomeca Turmo) turboshaft engines
MAX SPEED:	315km/h (196mph)
RANGE:	820km (510 miles)
CEILING:	6000m (19,685ft)
CREW:	2–3
ARMAMENT:	None

Below: This Z-8KA, as seen in November 2008, wears an overall dark olive-green scheme and is coded in the 6x9x range which, at that time, identified it as belonging to the 15th Airborne Army, under direct command of the PLAAF in Beijing. It is Bort White 18 and has the serial number Z8KA-02.

POWERPLANT
Further development by Harbin saw the proven Z-8 airframe upgraded into a new amphibious utility variant as the Z-8F with Pratt & Whitney Canada PT6A-67B turboshaft engines of greater power than the original Turbomeca Turmo units.

CHANGHE Z-8K
This Z-8K, with a code in the 51x1x range, identifies it as belonging to the Eastern Theatre Command's Transportation and Search and Rescue Brigade, based at Nanjing. It is finished in a three-tone disruptive scheme with additional red cross markings.

INTERNAL CAPACITY
The Z-8 has a powerful heavy-lift capacity via its internal cabin, being able to transport 27–30 troops or 15 stretchers or up to 5000kg (11,023lb) internal payload.

REVERSE ENGINEERING
China took delivery of 12 SA 321 Super Frelon between 1975 and 1977 primarily for use in the ASR and SAR roles, designating the helicopter the Z-8. Impressed with the Z-8's performance, the decision was taken to reverse engineer the design and put it into local production. Work on the project began in 1975. Budgetary concerns saw work on Chinese Z-8 production stall for a time, but the first locally produced helicopter flew for the first time in 1985.

PLA NAVY Z-8
The late 2000s saw the Z-8 developed into an 18-passenger civil helicopter, the Avicopter AC313, utilizing digital avionics systems and with composites used for around 50 per cent of the airframe. Known variants of the new helicopter are the basic Z-18 naval transport with folding tail boom and rotor blades and nose-mounted weather radar and FLIR/TV turret.

Antonov An-124 (1986)

TYPE • Transport **COUNTRY** • Soviet Union

The world's heaviest military transport aircraft, the An-124 (NATO reporting name 'Condor') is a long-range military freighter with a double-deck fuselage loaded through an upward-hinging fuselage nose and rear ramp.

SPECIFICATIONS (An-124 Condor)

DIMENSIONS:	Length: 69.10m (226ft 8in); Wingspan: 73.30m (240ft 6in); Height: 21.08m (69ft 2in)
WEIGHT:	Maximum take-off 392,000kg (864,210lb)
POWERPLANT:	Four Ivchenko Progress D-18T series 3 turbofans, each rated at 229.8kN (51,654lbf)
MAX SPEED:	850km/h (528mph)
RANGE:	14,200km (8823 miles)
CEILING:	11,600m (38,058ft)
CREW:	6
ARMAMENT:	None

Above: An An-124-100. A truly enormous aircraft, for several years Russia struggled to keep many of its 26-strong An-124 fleet operational. RF-82041 had languished on the ground since 1993 until it was returned to flying status in November 2014.

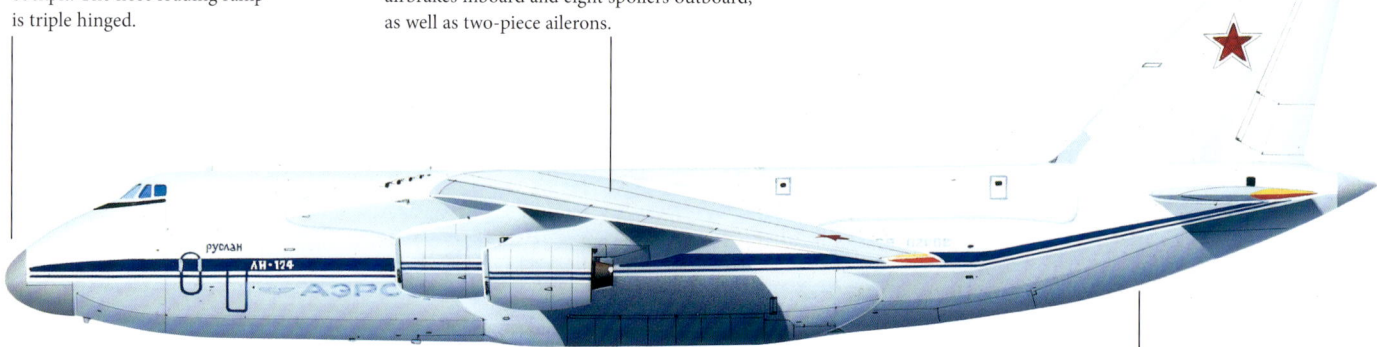

NOSE SECTION
Hinging at this point, the entire nose section of the aircraft swings upward and over the cockpit. The nose loading ramp is triple hinged.

WINGS
Antonov produced a wing of exceptional design for the An-124. It features full-span leading-edge slats, huge three-section flaps, four airbrakes inboard and eight spoilers outboard, as well as two-piece ailerons.

REAR LOADING
At the rear of the aircraft, a pair of clamshell doors open hydraulically, allowing the deployment of a rear loading ramp. This may be fixed at an intermediate position for loading from truck-bed height.

STRATEGIC LIFT
A true giant of the air, the Antonov An-124 Ruslan, known as the 'Condor' in the West, is the largest production aircraft in the world, dwarfing even the Lockheed C-5 Galaxy. In fact, this aircraft is so large that the cockpit is almost 9m (30ft) high when the aircraft is on the ground. The An-124 is a successful military airlifter of great strategic importance because of its ability to carry tanks, missiles and heavy equipment over long ranges.

AN-124 IN UNITED STATES
A Russian AN-124 is parked at Moffett Federal Airfield, California, during the War on Terror years. The contracted AN-124 transported 129th Rescue Wing deployment cargo to Afghanistan because the high operational tempos of Operations Iraqi Freedom and Enduring Freedom kept US C-17 Globemaster III and C-5 Galaxy aircraft fully engaged.

Boeing AH-64 Apache (1986)

TYPE • Helicopter COUNTRY • United States

Hughes developed the AH-64 Apache in response to the Warsaw Pact's massive armoured strength. Produced by McDonnell Douglas, the AH-64 can engage tanks, often at a safe distance, even at night and in bad weather.

SPECIFICATIONS (AH-64A)

DIMENSIONS:	Length: 14.97m (49ft 2in); Main rotor diameter: 14.63m (48ft); Height: 4.66m (15ft 4in)
WEIGHT:	Maximum take-off 9525kg (20,995lb)
POWERPLANT:	2 × General Electric T700-GE-701 turboshaft engines, 1265kW (1700hp)
MAX SPEED:	293km/h (182mph)
RANGE:	428km (265 miles)
CEILING:	6400m (21,000ft)
CREW:	2
ARMAMENT:	1 × 30mm (1.18in) M230 chain gun cannon, up to 16 AGM-114 Hellfire laser-guided missiles or up to 76 folding fin rockets; various other combinations of rocket projectiles, guns and missiles

Right: The Apache's main landing gear has shock struts to absorb impact and a kneeling facility to allow for air transportation. Each landing strut has a one-time, high-impact absorbing capability, so reducing injury to the crew in the event of a forced landing.

ROTOR BLADES
Constructed of fibreglass, stainless steel and composites, the main rotor blades are proof against hits by 23mm (0.9in) cannon shells. They have swept tips for increased performance.

ISRAELI AH-64A
Israel has been an AH-64 operator since September 1990, and paints its Apaches in an IR-suppressive olive drab finish. For operations in the southern Lebanon, aircraft carry an IR-reflective 'V' identification marking on the rear fuselage.

TANK HUNTING
One of the leading battlefield helicopters in the world, the tandem-seat AH-64, which has the gunner forward and pilot aft, uses high-tech sensors, a chain gun cannon and far-reaching Hellfire missiles to destroy tanks and other key targets. At night or in bad weather – or even in dust storms, as during Operation *Desert Storm* – the Apache crew can monitor enemy tank movements, using the PNVS (pilot's night-vision system) and TADS (target acquisition and designation system) to pinpoint and fire at targets.

WEAPONS PYLONS
The Apache's weapons pylons are articulated to provide the desired elevation for various fire control modes and for aerodynamic/handling purposes. When an Apache lands, the pylons automatically translate to ground stow mode, so that they are parallel with level terrain.

IRAQ CLOSE SUPPORT
A US Army Apache of the 1st Battalion, 101st Aviation Regiment, Fort Campbell, Kentucky, provides ground forces with air support from Forward Operating Base Speicher in Iraq, 21 October, 2005.

Atlas Cheetah (1987)

TYPE • Fighter **COUNTRY** • South Africa

SPECIFICATIONS

DIMENSIONS:	Length: 15.40m (50ft 6.5in); Wingspan: 8.22m (26ft 11.5in); Height: 4.25m (13ft 11.5in)
WEIGHT:	Maximum take-off c. 16,500kg (36,300lb)
POWERPLANT:	One 7200kg (15,876lb) thrust SNECMA Atar 9K-50 turbojet
MAX SPEED:	2337km (1452mph)
RANGE:	Combat radius 1200km (745 miles)
CEILING:	17,000m (55,774ft)
CREW:	1
ARMAMENT:	Two DEFA 30mm (1.18in) cannons (single-seat aircraft only); plus up to 4000kg (8800lb) of other weapons

For many years the South African Air Force relied on two principal international sources – France and Israel – for its combat aircraft. A UN arms embargo then forced South Africa to develop its own: the Cheetah.

Right: A flight-refuelling probe, here projecting visibly from the cockpit, was one of the first extra items of equipment added by Chile, Israel, Peru, Venezuela and South Africa to their upgraded Mirage aircraft. The South African Air Force used a number of converted Boeing 707s as tanker aircraft.

CAMOUFLAGE
The Cheetah C used a new two-tone grey camouflage scheme with a large, darker diamond-shaped panel to obscure the shape of the delta wing in flight. The SAAF markings on the wings were also toned down until they were barely visible.

LENGTHENED NOSE
The longer nose of the Cheetah C housed an Israeli-designed EL/M-2032 radar. This could pick up an air-to-air target at (32km (20 miles) and could track targets while scanning for others.

MODIFIED MIRAGE
In July 1968, the South African manufacturer Atlas unveiled a much-modified Dassault Mirage III, renamed Cheetah. Developed with Israel's help, the Cheetah resembled that country's Kfir fighter. South Africa converted about 38 single- and two-seat Mirage IIIs into Cheetahs and used them as the frontline cutting edge of its air combat force. In addition, to the two cannons found on single-seat versions, the Cheetah carried a remarkable variety of smart air-to-air and air-to-ground weapons guided by an indigenous designator pod.

AIR-TO-AIR MISSILES
This Cheetah C is armed with two V3C Darter short-range AAMs. The Darter entered SAAF service in 1990 and has an off-boresight capability of 20 degrees when used in conjunction with the Cheetah pilot's helmet-mounted sight.

HIGH PERFORMANCE
The pouring of a new wine into a familiar bottle produced a first-rate warplane with superb performance. The Cheetah was a delight to fly, and pilots revelled in being turned loose to fling this powerful ship around the sky.

Tupolev Tu-160 (1987)

TYPE • Bomber COUNTRY • Soviet Union

SPECIFICATIONS
(Tu-160 Blackjack-A)

DIMENSIONS:	Length: 54.10m (177ft 6in); Wingspan: 55.70m (182ft 9in) spread; Height: 13.10m (43ft)
WEIGHT:	Maximum take-off 275,000kg (606,261lb)
POWERPLANT:	Four 25,000kg (55,125lb) Kuznetsov NK-321 turbofans
MAX SPEED:	2000km/h (1243mph)
RANGE:	14,000km (8694 miles)
CEILING:	18,300m (60,040ft)
CREW:	4
ARMAMENT:	Up to 16,500kg (36,383lb) of munitions

The concept of the supersonic bomber dates back to the 1950s, but it was not until the 1980s that it became truly viable, first with America's Rockwell B-1 and then with Russia's Tu-160, code-named Blackjack by NATO.

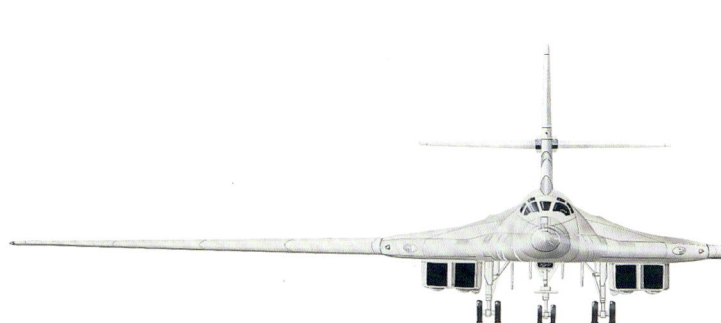

Left: Each main undercarriage strut holds a six-wheel bogie, the wheels arranged in three pairs. These retract backwards to lie in the wing centre section between the fuselage and engine nacelles.

CREW
The Tu-160 is flown by a crew of four, comprising two pilots seated side-by-side, and two navigators behind. One of the latter is known as the 'navigator-operator' and is responsible for aiming the weapons, while the other navigator is responsible for en-route navigation.

PAINT SCHEME
Wearing a white paint scheme, presumably as protection against nuclear flash (like Britain's V-bomber force of the early 1960s), this Tu-160 was part of the 184th Heavy Bomber Regiment, based at Pryluky in the Ukraine.

DEFENCES
Mounted under the tail of the Tu-160 is a battery of 72 chaff/flare dispensers. Built into the fuselage is a defensive avionics suite, with receiver antenna and jammers located in the tail bullet fairing, tailcone and under flush dielectric panels in the leading edge of the wing glove section.

UKRAINIAN TU-160s
Comparable to but much larger than the Rockwell B-1B, the TU-160 entered production in 1984 and the first operational aircraft were deployed in May 1987. Thirty-six were operational in 1989 (out of a planned total of 100) with the Soviet Air Force, divided between the 184th Air Regiment in the Ukraine and the 121st Air Regiment at Engels Air Base. The Ukraine-based aircraft were eventually returned to Russia, along with 600 air-launched missiles, as part of a deal that involved paying off a commercial debt.

TU-160 WITH EXTENDED WINGS
A Tupolev Tu-160 is put through its paces during an air show in Russia. The Tu-160 emerged from a multi-mission bomber competition in which the competitors included a Tupolev proposal (using elements from the Tu-144 supersonic airliner), a Sukhoi design based on the T-4 aircraft and the Myasishchev M-18. Tupolev won the competition, and the Tu-160 remains in Russian service today.

Modern Era

Many of the military aircraft in operation today are often increasingly digitized developments of late Cold War types. Modern aircraft are broadly categorized according to their technological generations, but the growth of mid-life upgrades and improvement programmes over the last couple of decades increasingly blur the boundaries between those generations. Today, there is no reason why a warplane designed in the 1950s can't carry a radar, for example, that is every bit as sophisticated as the types found on new-generation aircraft. Yet even as manned aircraft are truly reaching state-of-the-art, the greater transformation lies in the rise of unmanned aerial vehicles (UAVs). Combined with exponentially powerful artificial intelligence (AI), they augur a potentially autonomous and pilotless future of military aviation.

REFUELLING
An F-22A Raptor and two F-15C Eagles participate in a refueling mission with a KC-135 from the Mississippi Air National Guard over eastern Florida, September 2008.

Kamov Ka-29 (1987)

TYPE • *Helicopter* COUNTRY • *Soviet Union*

A ship-based assault helicopter able to transport assault troops from ship to shore, the Kamov Ka-29 (NATO reporting name 'Helix-B') is also capable of providing mobile fire support for naval infantry.

SPECIFICATIONS

DIMENSIONS:	Fuselage length: 11.3m (37ft 1in); Main rotor diameter: 15.9m (52ft 2in); Height: 5.4m (17ft 8in)
WEIGHT:	Maximum take-off 11,500kg (25,353lb)
POWERPLANT:	Two Klimov TV3-117V turboshaft engines, each rated at 1618kW (2200shp)
MAX SPEED:	290km/h (180mph)
RANGE:	460km (286 miles)
CEILING:	4300m (14,108ft)
CREW:	3
ARMAMENT:	One 7.62mm (0.3in) Glagolev-Shipunov-Gryazev GShG-7.62 four-barrelled MG, one optional 30mm (1.18in) Shipunov 2A42 autocannon fixed forward firing under fuselage; up to 4000kg (8818lb) of weapons and stores

Above: A Ka-29TB. One of the pre-production development airframes, '25 Blue' featured different instrumentation for flight test operations and lacked the fuselage-mounted 30mm (1.18in) cannon fitted as standard to production machines.

FUSELAGE
Compared to the Ka-27, the Ka-29 features a wider, more angular forward fuselage containing a cabin with space for 12–16 troops or four stretchers with eight passengers.

VARIANTS
A range of specialist helicopters developed from the Ka-27, including the Ka-29 for naval assault, the Ka-31 radar picket and Ka-32 civilian helicopter.

FIRE SUPPORT
In fire support configuration, the main armament of the Ka-29 is the Shturm-V anti-tank missile system as also used on the Mi-24. This adds electro-optical sight and radio-command guidance antenna under the nose plus up to eight 9M114 (AT-6 Spiral) tube-launched anti-tank missiles. Unguided weapons options include up to four pods for 80mm (3.1in) unguided rockets, UPK-23-250 gun pods, bombs or incendiary tanks. An optional 2A42 30mm (1.2in) single-barrel cannon can be fitted on the port weapons rack.

ASSAULT TRANSPORT
In assault transport configuration the Shturm-V is removed, but a GShG-7.62 4-barrel Gatling-type 7.62mm (0.3in) machine gun and 1800 rounds of ammunition is fitted in the front of the cockpit on the starboard side.

KA-29, RUSSIA
The same aircraft as depicted in the profile above, '39 Yellow' was photographed in July 2021 over St Petersburg. Note the rocket pod fitted to the port weapons wing.

Boeing/Grumman E-8 J-Stars (1989)

TYPE • *Reconnaissance* COUNTRY • *United States*

SPECIFICATIONS (E-8A)

DIMENSIONS:	Length: 46.6m (152ft 10in); Wingspan: 44.4m (145ft 8in); Height: 13m (42ft 8in)
WEIGHT:	Maximum take-off 150,142kg (331,000lb)
POWERPLANT:	Four 8698kg (19,200lb) Pratt & Whitney TF33-102C turbofans
MAX SPEED:	945km/h (590mph) cruising
RANGE:	9266km (5745 miles)
CEILING:	12,802m (42,000ft)
CREW:	21 or 34
ARMAMENT:	None

During the Gulf War, the USAF introduced a new aircraft to track down Iraqi missiles in the famous 'Scud Hunt'. This was the Boeing/Grumman E-8A Joint Surveillance Target Attack Radar System, or J-Stars.

This page: Built as a Boeing 707-323C for American Airlines, N8411, seen here, was the second of two development E-8As. It made its first flight as an E-8A on 31 August 1989, following the first aircraft, which took to the air on 22 December 1988.

CREW
On a standard mission the E-8 carries a crew of 21, comprising three flight crew and 18 systems operators. On a long endurance mission this is increased to 34, with six flight crew and 28 systems operators.

ANTENNA FAIRING
The large 'teardrop' fairing under the E-8A's centre section housed the antenna for the Flight Test Data Link. Fitted to test aircraft only, it was used during Operation *Desert Storm* for transmitting data over long distances back to central commands in Saudi Arabia.

RADOME
The most prominent external feature is a 12m (40ft) long canoe-shaped radome under the forward fuselage, housing a 7.3m (24ft) long side-looking phase array antenna.

MISSION PURPOSE
The E-8 joint surveillance target attack radar system, or Joint STARS, is an airborne battle management, command and control, intelligence, surveillance and reconnaissance platform. Its primary mission is to provide theatre ground and air commanders with ground surveillance to support attack operations and targeting that contributes to the delay, disruption and destruction of enemy forces.

AERIAL REFUELLING
An E-8C J-Stars from the 116th Air Control Wing, pulls away after refuelling from a KC-135 Stratotanker with the 459th Air Refueling Wing. As the E-8As carried so much test equipment, the number of operator consoles within the cabin was reduced to ten. The E-8C has 17 consoles, plus one for the management of defensive electronics.

Northrop B-2 Spirit (1989)

TYPE • Bomber **COUNTRY** • United States

SPECIFICATIONS	
Dimensions:	Length: 21.03m (69ft); Wingspan: 52.43m (172ft); Height: 5.18m (17ft)
Weight:	152,635kg (336,500lb) typical take-off
Powerplant:	4 x General Electric F118-GE-100 turbofans, 84.52kN (19,000lbf)
Max speed:	About 764km/h (475mph) at high altitude
Range:	8334km (5178 miles)
Ceiling:	15,240m (50,000ft)
Crew:	2
Armament:	Up to 18,144kg (40,000lb) of disposable stores carried in two weapons bays in underside of centre section

The bat-winged B-2 stealth bomber is the most expensive warplane ever built, with a total costs price tag of around $2 billion million per aircraft. Tricky to maintain, the small fleet of Spirits comprise some of the most potent weapons in the US arsenal.

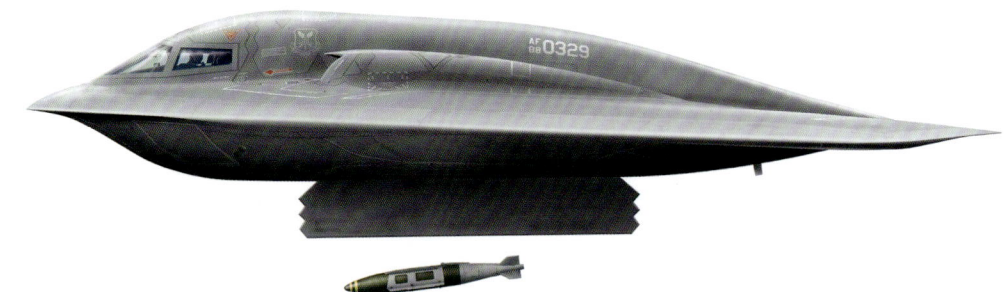

Above: B-2A Spirit serial number 88-0329 'Spirit of Missouri', also now flown by the 509th Bomb Wing's 13th Bomb Squadron. This aircraft is shown dropping a single 907kg (2000lb) JBU-32 JDAM precision-guided weapon, one of the type's most important conventional stores.

Wing planform
The leading edge is swept at 33 degrees, while the trailing edge features an unusual sawtooth configuration designed to trap radar energy. The engine intakes feature S-shaped curves.

Engines
The four General Electric F118-GE-100 turbofans exhaust through V-shaped outlets set back and above the trailing edges to shield these from ground-based sensors.

RAM coating
As well as its radar-defeating shape, the B-2 uses radar-absorbing coatings, which require special maintenance and a considerable support infrastructure.

Destructive Force
In addition to B61 and B83 free-fall nuclear bombs, the Spirit can carry basic and penetrator versions of the 907kg (2000lb) GBU-31 JDAM, and up to 80 of the 227kg (500lb) GBU-38 JDAMs. The Spirit can also accommodate 16 907kg (2000lb) Mk 84 general-purpose bombs, the 2041kg (4500lb) GBU-28 laser-guided bomb, the AGM-154 joint stand-off weapon, 16 joint air-to-surface stand-off missiles (JASSMs) and, most recently, the enormous 13,154kg (29,000lb) GBU-57 massive ordnance penetrator (MOP). If required, the B-2 can also carry Mk 62 Quickstrike sea mines.

B-2 Landing
The 'Spirit of Kitty Hawk, a B-2 Spirit from Whiteman Air Force Base, Missouri, lands at Fairchild Air Force Base, Washington. Among the roles performed by the B-2 is as part of the USAF's continuous bomber presence, maintained as a deterrent force in the Asia-Pacific region.

British Aerospace Harrier GR.7 (1989)

TYPE • *VTOL fighter* COUNTRY • *United Kingdom*

SPECIFICATIONS

DIMENSIONS:	Length: 14.36m (47ft 1.5in); Wingspan: 9.25m (30ft 4in); Height: 3.55m (11ft 7.75in)
WEIGHT:	Maximum take-off 14,061kg (31,000lb)
POWERPLANT:	One 96.7kN (21,750lbf) Rolls-Royce Pegasus vectored-thrust turbofan
MAX SPEED:	1065km/h (661mph)
RANGE:	Combat radius 277km (172 miles)
CEILING:	15,240m (50,000ft)
CREW:	1
ARMAMENT:	Provision for up to 4082kg (9000lb) (short take-off) or 3175kg (7000lb) (vertical take-off) of stores

The GR.7 became the definitive Harrier II for the RAF. They have been used over the Balkans, Iraq and Afghanistan, and fly from RN carriers as part of Joint Force Harrier. Conversion with new avionics and structural improvements created the GR.9.

Above: This heavily armed GR.7 includes a cannon pod in its weapon stores. RAF Harrier GR.7s saw their first active combat during the NATO bombing of former Yugoslavia in 1999.

CHAFF/FLARE LAUNCHERS
Usually for a Western tactical aircraft, the AV-8B(NA) is equipped with chaff/flare launchers on the upper surfaces of the rear fuselage, each side of the ram air inlet.

ZG746
The aircraft seen here, ZG476, is a Harrier GR.Mk.7 of No. 4 Squadron, one of two Harrier GR.7 units formerly based at Gütersloh, Germany. After a spell at Laarbruch, the squadron returned to the UK (RAF Cottesmore) in 1999.

CANNON POD
The AV-8B can be fitted with a single GE GAU-12A 'Equalizer' five-barrel Gatling-type cannon. This occupies the port under-fuselage pod, with 300 rounds of ammunition contained in the starboard pod. The GR.7, however, is armed with a 25mm (1in) version of the Royal Ordnance Factory's Aden gun, two of which can be fitted under the fuselage.

MANOEUVRABILITY
Leading-edge root extensions (LERX) are fitted to enhance the Harrier's air combat agility by improving the turn rate, while longitudinal fences (LIDS, or lift improvement devices) are incorporated beneath the fuselage and on the gun pods to capture ground-reflected jets in vertical take-off and landing, to give a much bigger ground cushion and to reduce hot gas recirculation.

HARRIER GR.9
A Royal Air Force British Aerospace Harrier GR.9 conducts a combat patrol over Afghanistan on 12 December 2008. An upgrade of the GR.7, the GR.9 included improved avionics and weapons.

Boeing F-15E Strike Eagle (1989)

TYPE • Multi-role fighter **COUNTRY** • United States

While the McDonnell Douglas (later Boeing) F-15 Eagle established itself as the West's premier air superiority fighter, further development led to the F-15E Strike Eagle, an all-weather ground-attack aircraft that retains all the air-combat capabilities of the original Eagle.

SPECIFICATIONS

DIMENSIONS:	Length: 19.43m (63ft 9in); Wingspan: 13.05m (42ft 10in); Height: 5.63m (18ft 5in)
WEIGHT:	36,741kg (81,000lb) maximum take-off
POWERPLANT:	2 × Pratt & Whitney F100-PW-220 afterburning turbofans, 106kN (23,830lbf)
MAX SPEED:	2655km/h (1650mph)
RANGE:	1271km (790 miles) radius
CEILING:	18,290m (60,000ft)
CREW:	2
ARMAMENT:	1 × 20mm (0.79in) rotary cannon plus up to 11,000kg (24,250lb) of external ordnance

Below: The only US Air Force Eagles based permanently outside the United States are assigned to the 48th Fighter Wing at RAF Lakenheath, England. Its squadrons comprise one of F-15C/Ds and two of F-15Es. An example of the latter is serial 98-0133, operated by the 492nd Fighter Squadron 'Bolars'; note the blue tail flash.

EPAWSS
Officially known as the Eagle II, the F-15EX is based on the F-15QA and deliveries to the US Air Force began in 2021. One significant difference compared to the F-15QA is the F-15EX's Eagle passive active warning survivability system (EPAWSS) that provided enhanced self-protection capabilities.

AIM-120 AMRAAM
Carried on the shoulder launch rails, the AMRAAM confers a powerful air-to-air capability, this being a 'fire and forget' missile with active guidance.

F-15EX Advanced Eagle
Originally known as the Advanced Eagle, and now officially named Eagle II, this is the first example of the F-15EX to be handed over to the US Air Force. The jet wears the tail codes of Eglin Air Force Base, Florida, from where it is flown by the 40th Flight Test Squadron.

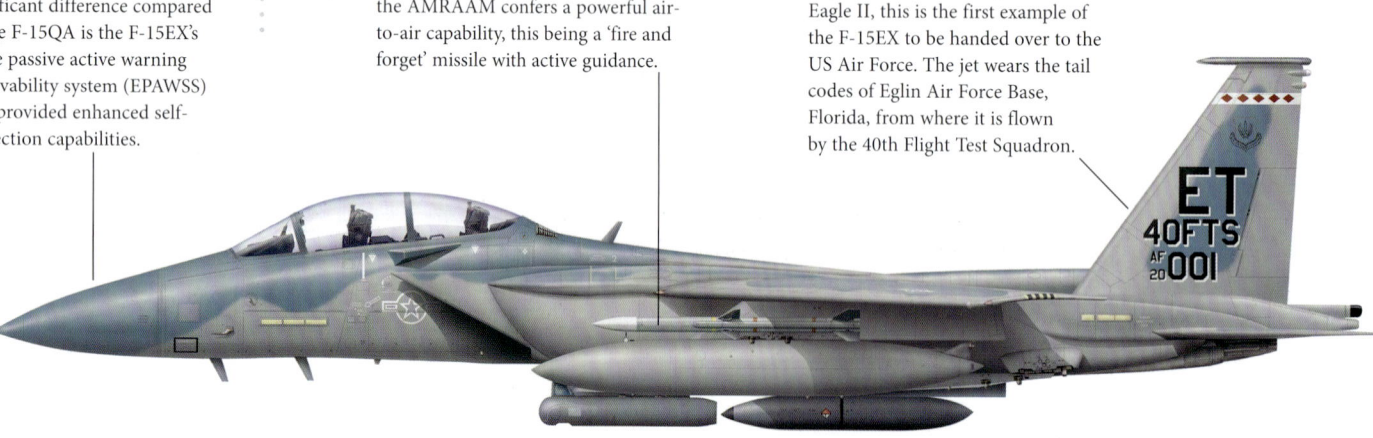

STRIKE EAGLE SERVICE
The Strike Eagle entered service with the 40th Tactical Training Wing at Luke AFB in 1988. Initial operational deliveries then followed to the 4th TFW at Seymour Johnson AFB, North Carolina. The F-15E made its combat debut during Operation *Desert Storm* in 1991, when its primary mission was to seek and destroy Iraqi 'Scud' mobile ballistic missiles. Since then, the F-15E has been at the forefront of every major US military campaign.

F-15E AFGHANISTAN
An F-15E from the USAF's 4th Fighter Wing flies over Afghanistan in April 2006. The USAF rotated detachments of F-15Es to south west Asia, from where they conducted close air support missions for troops on the ground engaged in rooting out insurgent sanctuaries and support networks.

Mikoyan MiG-29K (1989)

TYPE • *Fighter* COUNTRY • *Soviet Union*

A medium-weight multi-role carrier fighter, the MiG-29K (NATO reporting name 'Fulcrum') is superficially similar to the original MiG-29, but uses an all-new airframe and has a host of advanced performance and combat features.

SPECIFICATIONS

DIMENSIONS:	Length: 17.3m (56ft 9in); Wingspan: 11.99m (39ft 4in); Height: 4.4m (14ft 5in)
WEIGHT:	Maximum take-off 22,400kg (49,384lb)
POWERPLANT:	Two Klimov RD-33MK afterburning turbofan engines, each rated at 88.3kN (19,840lbf) with afterburning
MAX SPEED:	2200km/h (1370mph)
RANGE:	1850km (1150 miles)
CEILING:	17,500m (57,415ft)
CREW:	1
ARMAMENT:	One 30mm (1.2in) autocannon; up to 4500kg (8800lb) of weapons and stores

Above: An aircraft of the 100th Independent Shipborne Fighter Aviation Regiment is '42 Blue', shown here with external tanks. Note the long canopy used for both the twin- and single-seat versions.

MiG-29K
'34 Blue' is a MiG-29K of the 100th Independent Shipborne Fighter Aviation Regiment based at Severomorsk-3 air base near Murmansk in the north of Russia for operations with the Northern Fleet.

Cannon
Like its land-based MiG-29 predecessor, the aircraft is fitted with a GSh-30-1 single-barrel 30mm (1.2in) cannon in the port leading-edge root extension.

CARRIER-CAPABLE
The project to develop a carrier-capable Fulcrum dates back to the 1980s, with plans to equip an air wing for the Soviet Navy's planned carrier fleet. Compared to the MiG-29, the MiG-29K possesses a larger wing with double-slotted flaps, leading-edge vortex controllers, arrester hook and other naval features. It was developed alongside the land-based MiG-29M in the mid-1980s. Both aircraft were intended to have a new N010 radar, now known as the Zhuk. The first of two MiG-29K prototypes took to the air on 23 June 1988.

Stores
The MiG-29K can carry 4500kg (9921lb) of weapons and stores on nine hardpoints (eight under the wing and one under the fuselage).

MiG-29K TEST FLIGHT
A Russian Navy MiG-29K undergoes a test flight. Appearing in late 1999, the modernized MiG-29K was the izdeliye 9.41. Compared to the original 9.31, the new version sports increased-area flaps, a new digital fly-by-wire control system and upgraded engines and systems. The two-seat version is the MiG-29KUB or izdeliye 9.47.

AMX International AMX (1990)

TYPE • Fighter **COUNTRY** • Brazil/Italy

A lightweight fighter-bomber and reconnaissance aircraft, the programme to develop the subsonic AMX was launched by Brazil and Italy in the late 1970s with production lines established in both countries.

SPECIFICATIONS

DIMENSIONS:	Length: 13.23m (43ft 5in); Wingspan: 8.87m (29ft 1in); Height: 4.55m (14ft 11in)
WEIGHT:	Maximum take-off 13,000kg (28,660lb)
POWERPLANT:	One Rolls-Royce Spey 807 turbofan rated at 49.1kN (11,000lbf) of thrust
MAX SPEED:	1053km/h (654mph)
RANGE:	3336km (2073 miles)
CEILING:	13,000m (43,000ft)
CREW:	1
ARMAMENT:	One 20mm (0.787in) M61A1 Vulcan six-barrel Gatling cannon (Italy) or two 30mm (1.1in) cannons (Brazil), plus up to 3800kg (8378lb) of ordnance and fuel tanks carried on five external hardpoints

Right: The AMX tactical support aircraft, developed jointly by Italy and Brazil, follows the trend of international co-operation established by the European nations.

COCKPIT
The AMX has an advanced cockpit designed to reduce pilot workload. The OMI/Selenia head-up display is complemented by an Aeritalia multifunction head-down display, which can present TV/IR and synthetic map displays.

WEAPONRY
The AMX is cleared to use a wide variety of weapons, including freefall and retarded Mk 82, Mk 83, Mk 84 bombs and the Skyshark munitions dispenser system. Sidewinder or Piranha AAMs can also be carried for self-defence.

NOSE CODE
The 51 nose code and the cat and mouse fin badge identify this aircraft as one of those delivered to the first operational AMX unit: Gruppo 103, Stormo 51, based at Istrana. The unit was previously equipped with the Fiat G-91R.

INHERENT RESOLVE
A total of 187 examples were built for the Italian Air Force and 79 for the Brazilian Air Force in both single-seat and two-seat (AMX-T) forms. The AMX features good short-field performance and advanced nav/attack systems and has been used in combat by both operators. Italian examples have been employed on reconnaissance missions in the Middle East as part of Operation Inherent Resolve, while the Brazilian jets have been used during counter-narcotics operations.

AIR SHOW AMX
An Italian Air Force AMX prepares to land during an air show performance. The Italian jets that were previously designated as the A-11A and TA-11A have undergone an upgrade, resulting in the A-11B and TA-11B.

Ilyushin Il-80 & Ilyushin Il-82 (1990)

TYPE • Transport **COUNTRY** • Soviet Union

A strategic airborne command post, the Il-80 is intended to provide survivable nuclear command and control functions for the country's highest authorities in wartime. The Il-82 is a strategic radio-relay aircraft that is designed to be used in conjunction with the Il-80.

SPECIFICATIONS (Il-80 Maxdome)

Dimensions:	Length: 59.54m (195ft 4in); Wingspan: 48.06m (157ft 8in); Height: 15.81m (51ft 10in)
Weight:	Maximum take-off 215,000kg (473,993lb)
Powerplant:	Four Kuznetsov/Samara NK-86 turbofans, each rated at 127.48kN (28,660lbf)
Max speed:	850km/h (528mph)
Range:	7000km (4350 miles)
Ceiling:	11,000m (36,089ft)
Crew:	Unknown
Armament:	None

Above: Developed and converted from an Il-76MD transport by Beriev in Taganrog, the maiden flight of the first Il-82 took place on 29 April 1987. In the event of a major war, the aircraft is designed to serve the General Staff of the Russian Armed Forces, operating in concert with the Il-80 airborne command post as part of wartime nuclear command and control.

AEROFLOT SCHEME
Four Il-86 airliners were converted into Il-80 airborne command posts and originally operated with an Aeroflot colour scheme and titles.

COMMS SUITE
The aircraft possesses a comprehensive communication suite including a large canoe-shaped fairing on the forward fuselage for satellite communication antennas.

SUBMARINE COMMS
Fairings are mounted in front of the tailfin and under the rear part of the fuselage and the Fregat very low frequency (VLF) communication system is carried to communicate with submerged submarines.

AIRLINER CONVERSION
Designed by Ilyushin in Moscow, the Il-80 was converted from the Il-86 wide-body airliner and operated by a flight crew of four plus an undisclosed number of operational crew. Hardened against the effects of nuclear explosion, the Il-80 also features additional fuselage fuel tanks and provision for in-flight refuelling. Series manufacture of the Il-86 and conversion to Il-80 was undertaken by Voronezh Aircraft Production Association (VASO) in Voronezh, Russia.

IL-80 MAXDOME
An Il-80 'Maxdome', clearly showing its huge communications equipment bulge, flies during the celebration of the centenary of the Russian Air Force on 12 August 2012 in Zhukovsky, Russia.

MQ-1 Predator (1991)

TYPE • UAV   COUNTRY • United States

SPECIFICATIONS

Dimensions:	Length: 8.22m (27ft); Wingspan: 14.8m (48ft 6in); Height: 2.1m (6ft 10in)
Weight:	1020kg (2250lb)
Powerplant:	Rotax 914F turbocharged four-cylinder engine, 86kW (115hp)
Max speed:	217km/h (135mph)
Range:	1100km (675 miles)
Ceiling:	7620m (25,000ft)
Crew:	0
Armament:	2 hardpoints and provisions to carry combinations of: 2 × AGM-114 Hellfire (MQ-1B), 4 × AIM-92 Stinger (MQ-1B), 6 × AGM-176 Griffin air-to-surface missiles

The Predator is one of the most famous military drones. It has a strange-looking downward-pointing tail and is driven by a propeller at the rear. When it was developed in the early 1990s the Predator was not armed, but today's drones can carry two laser-guided Hellfire missiles.

Right: The Hellfire missile of this US Predator can be seen on its port wing. The first Hellfire tests with the Predator were conducted in February 2001.

Disassembly
The Predator can be dismantled into six main component parts, for easier transportation to a distant theatre.

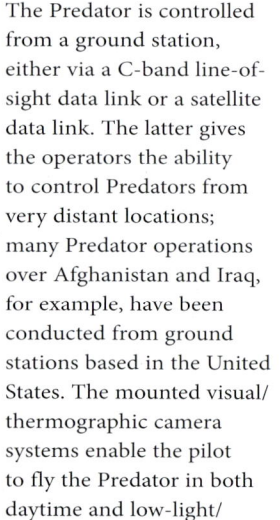

Drone carrier
The Predator can carry a smaller drone to the target area, then carry out its own mission after launching it.

Male drone
Predator is a MALE drone (medium-altitude, long-endurance), meaning it can fly fairly high and stay on station (in the target area) for a long time.

GROUND CONTROL
The Predator is controlled from a ground station, either via a C-band line-of-sight data link or a satellite data link. The latter gives the operators the ability to control Predators from very distant locations; many Predator operations over Afghanistan and Iraq, for example, have been conducted from ground stations based in the United States. The mounted visual/thermographic camera systems enable the pilot to fly the Predator in both daytime and low-light/night conditions.

PREDATOR PATROL
Predator drones are used by the armed forces of several countries and also by the CIA, principally for aerial reconnaissance and forward observation roles. Recently, the US hunter-killer UAV role has been largely taken by the General Atomics MQ-9A Reaper, developed from the Predator.

Sukhoi Su-30 (1992)

TYPE • Multi-role fighter COUNTRY • Russian Federation

SPECIFICATIONS (Su-30MKI)

DIMENSIONS:	Length: 21.94m (71ft 11in); Wingspan: 14.7m (48ft 2in); Height: 6.4m (20ft 11in)
WEIGHT:	Maximum take-off 34,000kg (74,957lb)
POWERPLANT:	Two AL-31FP (izdeliye 96) thrust-vectoring turbofans each providing 122.6kN (27,550lbf) thrust at maximum afterburning
MAX SPEED:	2120km/h (1320mph)
RANGE:	3000km (1864 miles)
CEILING:	17,300m (56,758ft)
CREW:	2
ARMAMENT:	One 30mm (1.2in) Gryazev-Shipunov GSh-30-1 Gast autocannon; up to 8130kg (17,920lb) of weapons and stores

The concept behind the Su-30 (NATO reporting name 'Flanker') began with the two-seat Su-27UB Flanker-C combat trainer, which retained operational capability but possessed the same fire control system and weapons of the single-seat Su-27.

Above: Russian Su-30SMs have been involved in combat operations over Syria since 2015, initially providing fighter escort for Russian attack jets and strategic bombers; the aircraft have also conducted their own air-to-ground strikes.

PERFORMANCE
The Su-30 is a fast aircraft, delivering maximum speeds of more than twice the speed of sound in some variants.

COCKPIT
The original Su-30 (also known as the Su-27PU) simply added a tactical situation display in the rear cockpit and an in-flight refuelling probe.

HARDPOINTS
All Su-30s can carry up to 8000kg (17,637lb) of stores on 12 hardpoints. Air-to-air loads include a wide variety of radar-guided and infrared-guided air-to-air missiles.

SU-30 LINES
Developed by the Sukhoi Design Bureau in Moscow, series production of the Su-30 is undertaken at two separate facilities in Irkutsk and Komsomolsk-on-Amur. The Irkutsk Su-30 is built by the Irkutsk aircraft plant, which is part of the Irkut Corporation. This line began with the Su-30MKI for India, developed and built in cooperation with Hindustan Aircraft Limited (HAL). The Irkutsk line is more advanced: these aircraft are characterized by their canard foreplanes, thrust-vectoring engines and an open-architecture fire control system based around the electronically scanned Bars radar.

RED FLAG SU-30
Indian Air Force maintainers prepare a Su-30MKI aircraft during a 'Red Flag' exercise (a multinational advanced aerial combat training exercise) at Nellis Air Force Base, Nevada in 2008.

AIDC F-CK-1 Ching-Kuo (1992)

TYPE • *Fighter* COUNTRY • *Taiwan*

Work began on Taiwan's first home-grown fighter aircraft in the early 1980s, when the decision was taken to field a replacement for the Northrop F-5 and Lockheed F-104 then in service with the Republic of China Air Force (ROCAF).

SPECIFICATIONS (F-CK-1A)

DIMENSIONS:	Length: 14.21m (46ft 7.5in) including probe; Wingspan: 8.53m (28ft); Height: 4.65m (15ft 3in)
WEIGHT:	9072kg (20,000lb) normal take-off
POWERPLANT:	Two ITEC (Garrett/AIDC) TFE-1042070 (F125) turbofans each rated at 42.08kN (9460lbf) thrust with afterburning
MAX SPEED:	1275km/h (792mph)
RANGE:	1100km (680 miles)
CEILING:	16,760m (55,000ft)
CREW:	1
ARMAMENT:	One M61A1 Vulcan 20mm (0.787in) rotary cannon, plus a typical air-to-air armament of two Tien Chien 2 and two TC-1 AAMs, or alternative air-to-ground stores

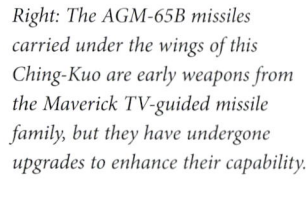

Right: The AGM-65B missiles carried under the wings of this Ching-Kuo are early weapons from the Maverick TV-guided missile family, but they have undergone upgrades to enhance their capability.

COCKPIT
The two-seat F-CK-1B Ching-Kuo has a separate windscreen and canopy, the latter opening to port for cockpit access. Pre-production and production single-seat aircraft have an F-16-style single-piece windscreen and canopy that hinges aft.

SIDEWINDERS
The Ching-Kuo is armed with a pair of Tien Chien 1 (AIM-9L Sidewinder) missiles at its wingtips, and a pair of semi-recessed TC-2 AAMs may also be carried in a tandem arrangement in the underside of the fuselage.

GOLDEN DRAGON RADAR
AIDC's Golden Dragon radar allows the Ching-Kuo to detect targets at a distance of up to 150km (93 miles). This radar, combined with Taiwanese missiles and Ching-Kuo's impressive flight performance, make the aircraft a formidable interceptor.

DESIGN

The Ching-Kuo was developed in Taiwan to help that country overcome the considerable restrictions placed on foreign imports. The country had intended replacing its ageing fleet of F-104 Starfighters with the Northrop F-20 Tigershark, but this proved impossible when the US government placed an embargo on this and any other comparable advanced fighter. American expertise was therefore bought in from General Dynamics, Garrett, Westinghouse, Bendix/King and Lear who helped to finalize a design in 1985.

CNAF FCK-1 CHING-KUO
The first Ching-Kuo aircraft were delivered to the Chinese Nationalist Air Force in 1994, although sales of the F-16 Fighting Falcon to Taiwan in 1992 reduced its production to a mere 130 aircraft.

Xi'an JH-7 (1992)

TYPE • Bomber COUNTRY • China

SPECIFICATIONS (JH-7A)

DIMENSIONS: Length: 22.32m (73ft 3in); Wingspan: 12.8m (42ft); Height: 6.22m (20ft 5in)

WEIGHT: Maximum take-off 28,475kg (62,777lb)

POWERPLANT: Two WS-9 turbofan engines each rated at 91.26kN (20,520lbf) thrust with afterburning

MAX SPEED: Mach 1.52

RANGE: Around 1760km (1090 miles) with one in-flight refuelling

CEILING: 16,000m (52,000ft)

CREW: 2

ARMAMENT: One 23mm (0.9in) twin-barrel GSh-23 cannon plus a maximum of 9000kg (20,000lb) of disposable stores carried on nine hardpoints

Known by the Western reporting name 'Flounder' and named 'Flying Leopard' in China, the JH-7 is a twin-engine supersonic tactical strike and maritime attack aircraft, development of which began in the mid-1970s. The definitive version is the JH-7A.

Left: A Xian JH-7A FBC-1 Flying Leopard frontline bomber of the Chinese People's Liberation Army Air Force, seen here at Dyagilevo airfield in 2018.

JH-7A
This JH-7A, serial number 73270, serves with the People's Liberation Army Air Force's 126th Air Brigade, part of the Nanning Base within Southern Theatre Command, and stationed at Liuzhou.

STRUCTURE
In terms of structure, the JH-7A benefitted from an increased use of composite materials, including in the redesigned wing and the tail.

MISSION FOCUS
The JH-7A's long-range capability and two-person crew lend themselves to the suppression of enemy air defences (SEAD), when fitted with a variety of different jamming or electronic intelligence (ELINT) pods. It appears that PLAAF tactics for the SEAD version involve the aircraft working in two-ship 'hunter-killer' teams, with one carrying pods to locate and/or jam hostile radio-frequency emitters and the other being armed with Russian-made Kh-31P or Chinese YJ-91 anti-radiation missiles to target them.

HARDPOINTS
For the carriage of its expanded range of weapons, the JH-7A's new wing was fitted with a pair of additional hardpoints and two more pylons were also added below the engine air intakes, to allow for the carriage of navigation/targeting pods.

JH-7 AFTERBURNERS
Chinese Air Force JH-7 taking off at Dyagilevo air base in Russia. The design of the JH-7A had been finalized by early 2001, and the flight-test programme was launched in July 2002, when the first prototype of this version made its maiden flight.

Saab JAS 39 Gripen (1993)

TYPE • Fighter COUNTRY • Sweden

In the early 1980s, Sweden began development of the JAS 39 Gripen as its new fighter aircraft to replace the Draken and Viggen in Swedish Air Force service and to take on fighter, attack and reconnaissance roles, as emphasized by its JAS (Jakt, Attack, Spaning) designation.

SPECIFICATIONS (JAS 39C)

- **DIMENSIONS:** Length: 14.8m (49ft); Wingspan: 8.4m (27ft 7in); Height: 4.5m (14ft 9in)
- **WEIGHT:** Maximum take-off 14,000kg (30,865lb)
- **POWERPLANT:** One Volvo RM12 turbofan engine rated at 80.5kN (18,100lbf) thrust with afterburning
- **MAX SPEED:** Mach 2 (2469km/h)
- **RANGE:** 3200km (2000 miles), ferry
- **CEILING:** 15,240m (50,000ft)
- **CREW:** 1
- **ARMAMENT:** One 27mm (1.063in) Mauser BK-27 cannon, plus up to 5300kg (11,700lb) of disposable stores on eight hardpoints

Left: This Gripen belongs to the Flygvapnet's F7 Wing at Såtenäs in the country's Southern Air Command.

CLIMB RATE
Carrying a full warload, a Gripen can reach 10,000m (33,000ft) in less than two minutes from starting its take-off roll.

CANARD FOREPLANES
In designing the JAS-39, Saab retained the tried and tested aft-mounted delta wing configuration, with swept canard foreplanes, an arrangement that makes for excellent manoeuvrability at all speeds and altitudes.

MIXED WEAPONS
This Gripen is seen in a mixed attack/defence load with two BK90 (DWS 39) glide weapons on the inboard wing pylons and Rb 99 (AIM-120) AMRAAMs on the outboard pylons. AIM-9 Sidewinders are mounted on the wingtip pylons.

JAS 39E/F
Following the upgrade to C/D standard, Saab began work on a third-generation Gripen, with the vastly more capable Gripen JAS 39E/F. The JAS 39E/F boasts a new General Electric F414-GE-39E engine for improved performance as well as the ability to carry additional fuel and weapons, including on two new fuselage stations. The new fighter is also equipped with a Selex ES-05 Raven active electronically scanned array (AESA) radar and a Skyward-G infrared search and track system, and can carry the Litening laser designator pod and RecceLite reconnaissance pod.

THE GRIPEN AS DOGFIGHTER
Subdued markings, a very low-visibility colour scheme and the Gripen's small size all combine to make it a difficult opponent in close-in dogfighting. Some pilots, however, have noted that the aircraft's holographic head-up display is so large that it can produce distinctive green flashes of sun 'glint' that are sometimes bright enough to betray the JAS-39's position.

McDonnell Douglas/Boeing C-17 Globemaster (1993)

TYPE • *Transport* COUNTRY • *United States*

SPECIFICATIONS
(C-17 Globemaster III)

DIMENSIONS:	Length: 53m (174ft); Wingspan: 51.76m (169ft 10in); Height: 16.79m (55ft 1in)
WEIGHT:	Maximum take-off 265,352kg (585,000lb)
POWERPLANT:	Four Pratt & Whitney F117-PW-100 turbofan engines each rated at 179.9kN (40,440lbf) thrust
MAX SPEED:	Mach 0.875 (1080km/h)
RANGE:	4480km (2780 miles) with 71,214kg (157,000lb) payload
CEILING:	14,000m (45,000ft)
CREW:	3
ARMAMENT:	None

A superlative transport aircraft, the McDonnell Douglas C-17 was developed for the US Air Force as a long-range military transport able to deliver large equipment, supplies and troops directly to small airfields.

Left: Viewed from the outside, the C-17A is deceptive. Although roughly the same size as the earlier C-141B Starlifter, its cargo capacity is three times greater. The C-17A handles 'outsize' cargo, although its external dimensions are such that it occupies reasonable ramp space at busy airfields.

CREW
An advanced flight deck means the C-17 can be operated by a crew of just three: the pilot and co-pilot in the cockpit, plus a loadmaster.

C-17 GLOBEMASTER III
In Royal Canadian Air Force service, the C-17A is designated as the CC-177. In December 2014, Canada announced it was acquiring a fifth Globemaster III, delivered in March 2015.

VERSATILITY
As well as the rapid strategic delivery of troops and cargo to main operating hubs or directly to forward bases in the deployment area, the C-17 can undertake tactical airlift and airdrop missions as well as aeromedical evacuation. Typical loads include main battle tanks, armoured vehicles, trucks or trailers loaded and unloaded via a large aft ramp, or up to 102 fully equipped paratroopers. The overall design is optimized for operations from small, austere airfields.

SHORT TAKE-OFF/LANDING
The aircraft can land and take off from runways as short as 1067m (3500ft), aided by the thrust reversers on all four engines. These provide enough thrust to reverse the aircraft and can also be operated in-flight drag for maximum-rate descents.

C-17 FLIGHT DECK
The flight crew of a C-17 Globemaster III, manning their highly digitized flight controls, approach the rotation point for lift-off during a US Air Force Weapon School's mission employment phase at Nellis Air Force Base, Nevada, in 2006.

Mikoyan MiG-29M and MiG-35 (1994)

TYPE • Fighter **COUNTRY** • Russian Federation

Superficially similar to the original Mikoyan MiG-29, the MiG-29M and MiG-35 (both with the NATO reporting name 'Fulcrum') feature an all-new airframe to enhance the multi-role capability of this medium-weight fighter.

SPECIFICATIONS (MiG-29M)

DIMENSIONS:	Length: 17.3m (56ft 9in); Wingspan: 11.99m (39ft 4in); Height: 4.4m (14ft 5in)
WEIGHT:	Maximum take-off 24,500kg (54,013lb)
POWERPLANT:	Two Klimov RD-33MK afterburning turbofan engines, each rated at 52.96kN (11,905lbf) thrust dry, and 88.3kN (19,840lbf) with afterburning
MAX SPEED:	2100km/h (1300mph)
RANGE:	2000km (1240 miles)
CEILING:	16,000m (52,493ft)
CREW:	1 or 2
ARMAMENT:	One 30mm (1.2in) Gryazev-Shipunov GSh-30-1 autocannon in the port leading-edge root extension; up to 5500kg (12,125lb) of weapons and stores

Above: A pre-production MiG-35 single seater '702 Blue' was constructed in October 2016 and is depicted as it appeared during tests with Russian Aerospace Forces in January 2017.

TWO-SEATER
The two-seat version is the MiG-29M2 (izdeliye 9.47S), the first of which flew on 24 December 2011, and was followed by the initial single-seat MiG-29M (izdeliye 9.41S) that took to the air on 3 February 2012.

MiG-29M2
'747 Blue' was one of two prototypes intended for Syria, although the aircraft never went into production due to the deterioration of the political situation in that country. '747 Blue' was retained by RAC MiG and demonstrated at several airshows during 2012 and 2013.

CANNON
The MiG-29M retained the GSh-30-1 single-barrel 30mm (1.8in) cannon in the port leading-edge root extension of its MiG-29 forebear.

ADVANCED WEAPONS

Advanced ordnance of the MiG-29M comprise up to six medium-range R-77 (AA-12 Adder) air-to-air missiles with active radar guidance, or up to six short-range R-73 (AA-11 Archer) air-to-air missiles with infrared guidance. Air-to-surface options include four subsonic Kh-35 (AS-20 Kayak) or supersonic Kh-31A (AS-17 Krypton) anti-ship missiles, four Kh-31P (AS-17 Krypton) anti-radiation missiles, four Kh-38M, four Kh-59M (AS-18 Kazoo) air-to-ground missiles or four KAB-500Kr TV-guided bombs.

MiG-29M2
MiG-29M2 '747 Blue' (the aircraft depicted in the artwork above) here conducts a demonstration flight at the International Aviation and Space salon MAKS-2013. The MiG-29M utilizes the N041 (or Zhuk-M1SE for export) radar, OLS-UM infrared search and track (IRST) sensor and an optional T-220 targeting pod.

Harbin Z-9 (1994)

TYPE • Helicopter COUNTRY • China

Aérospatiale flew its Dauphin 2 for the first time in 1975 and the helicopter proved highly successful. Developed from the Dauphin 2, more than 200 examples of the Chinese Harbin Z-9 have been built to date and the type has been widely exported.

SPECIFICATIONS (Z-9C)

DIMENSIONS:	Length: 12.11m (39ft 9in); Main rotor diameter: 11.94m (39ft 2in); Height: 4.01m (13ft 2in)
WEIGHT:	Maximum take-off 4100kg (9039lb)
POWERPLANT:	Two 632kW (848shp) Zhuzhou Aeroengine Factory WZ-8A (Turbomeca Arriel) turboshaft engines
MAX SPEED:	305km/h (190mph)
RANGE:	1000km (620 miles) with auxiliary fuel tank
CEILING:	4500m (14,800ft)
CREW:	1–2
ARMAMENT:	Two lightweight torpedoes

Left: A People's Liberation Army Harbin Z-9 helicopter prepares to land during the PLA Barracks Open Day at Shek Kong Barracks in Hong Kong, China, 2 May 2011. The Z-9A and Z-9B were followed by the armed Z-9W with optional weapons pylons, gyro stabilization and a roof-mounted optical sight that first flew in 1987, with the first weapons tests taking place in 1989.

NAVAL VARIANT
A naval variant, the Z-9C, is a licenced version of the Eurocopter AS565 Panther utilized for SAR and ASW duties.

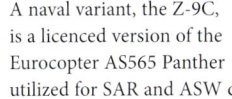

POWERPLANT
The Z-9C runs from two 632kW (848shp) Zhuzhou Aeroengine Factory WZ-8A (licensed Turbomeca Arriel) turboshaft engines.

HARBIN Z-9C
This PLA Navy (PLAN) Z-9 served aboard the Type 054A missile frigate *Zhoushan* (529). It was recorded leaving the flight deck of British frigate HMS *Cornwall* in August 2009, following joint anti-piracy operations in the Gulf of Aden.

LICENSED PRODUCTION
China acquired a licence to produce the Dauphin 2 in July 1980 and Harbin Aircraft Manufacturing Company (HAMC) flew the first locally assembled helicopter the following year which was designated Z-9. After constructing an initial batch of 28 helicopters from French components, HAMC produced the largely indigenous multi-role Z-9B with the more powerful Zhuzhou Aeroengine Factory WZ-8A turboshaft engine.

NAVAL Z-9C
A Z-9C flies off the deck of a Singapore Navy missile frigate RSS *Intrepid* during an exercise in 2015. The Pakistan Naval Air Arm also operates the type from its Zulfiquar frigates as the Z-9EC.

Kamov Ka-31 (1995)

TYPE • *Helicopter* COUNTRY • *Russian Federation*

The Ka-31 NATO reporting name 'Helix') is a radar picket helicopter developed for the Soviet (later Russian) Navy for detection of airborne and sea targets. Subsequently, a version of the Ka-31 was developed for battlefield surveillance.

SPECIFICATIONS

DIMENSIONS:	Fuselage length: 11.3m (37ft 1in); Main rotor diameter: 15.9m (52ft 2in); Height: 5.64m (17ft 8in)
WEIGHT:	Maximum take-off 12,500kg (27,558lb)
POWERPLANT:	Klimov TV3-117VMAR turboshaft each rated at 1618kW (2200shp)
MAX SPEED:	250km/h (155mph)
RANGE:	600km (373 miles)
CEILING:	3500m (11,483ft)
CREW:	2
ARMAMENT:	None

Left: An Indian Navy Kamov Ka-31 lands on the flight deck of USS Bunker Hill. *India is not the only export customer for this helicopter. In September 2007, China ordered nine Ka-31s, delivered between 2010 and 2012.*

VARIANT
The Ka-31 has been upgraded in the Ka-35 variant. This retains the Ka-31's airframe, powerplant, transmission and rotors and is intended as part of a battlefield reconnaissance and target acquisition system.

FUSELAGE
The Ka-31 features a modified fuselage with a widened nose and forward cabin, as seen on the Ka-29. An all-new landing gear is fitted with four legs that are raised hydraulically when radar is deployed.

INDIAN CONTRACT
The first export order for the new helicopter came from India in August 1999, covering four Ka-31s to serve aboard aircraft carrier *Viraat* and three *Krivak III*-class frigates. A contract for five further helicopters was signed in February 2001. Manufactured by Kumertau Aviation Production Enterprise, the first series-production helicopter was flown on 16 May 2001 and all nine were delivered to India between 2003 and 2004. Another Indian order followed in August 2009 covering five more Ka-31s.

RADAR
Primary sensor is the E-801 pulse-Doppler radar, stowed horizontally below the fuselage for cruise flight.

KAMOV K-31 ON DISPLAY
An order for the Russian Navy was placed in November 2008 for the Ka-31R version. Likely features of the Ka-31R include software improvements and Russian-standard communications and friend or foe identification.

Kamov Ka-52 (1995)

TYPE • Helicopter **COUNTRY** • Russian Federation

SPECIFICATIONS

DIMENSIONS:	Fuselage length: 13.87m (45ft 6in); Main rotor diameter: 14.5m (47ft 6in); Height: 5.05m (16ft 6in)
WEIGHT:	Maximum take-off 11,300kg (24,912lb)
POWERPLANT:	Two Klimov/St Petersburg VK-2500 turboshafts each rated at 1790kW (2400shp)
MAX SPEED:	300km/h (186mph)
RANGE:	460km (286 miles)
CEILING:	5500m (18,045ft)
CREW:	2
ARMAMENT:	One 30mm (1.2in) Shipunov 2A42 autocannon semi-rigidly mounted on starboard fuselage side; up to 2000kg (4409lb) of weapons and stores

A side-by-side two-seat reconnaissance/combat helicopter with the primary role of battlefield reconnaissance, the Ka-52 (NATO reporting name 'Hokum') can also coordinate helicopter strike groups and attack ground and aerial targets.

Above: '22 Yellow' of the 319th Independent Helicopter Regiment (319 OVP) based at Chernigova. This unit was the first to operate the Mi-24 in 1971. 319 OVP currently consists of a squadron of Mi-8AMTSh transport helicopters, a squadron of Ka-52s and a handful of Mi-26s.

ROTOR CONFIGURATION
Unlike other helicopters in its class, the Ka-52 utilizes Kamov's trademark coaxial, contra-rotating rotor configuration and features a crew escape system using ejection seats.

MISSION SYSTEM
The BREO-52 mission system is based around the Myech-1U radar and GOES-451.24 electro-optical turret and Vitebsk-52 self-defence suite.

UNDERWING PYLONS
Stores are carried on six underwing pylons, of which the outer two can only carry 9M39 Igla-V (SA-18 Grouse) air-to-air missiles (two per pylon). The primary weapon system is the Shturm-VU anti-tank guided missile system.

IN COMBAT

The Ka-52 has been extensively used in combat by Russian Aerospace Forces in support for the Syrian Civil War. Ukrainian sources suggest that a Ka-52 (and a Su-30SM) took part in an attack on Ukrainian Navy warships in the Black Sea in November 2018, launching unguided rockets. Ka-52s have seen action during the 2022 Russian invasion of Ukraine with four helicopters lost within the first two weeks of the conflict.

KA-52 'O61 YELLOW'
Photographed in 2009, this Ka-52 is caught with the undercarriage just beginning to retract. The tall co-axial rotors cancel out torque and remove the need for a vulnerable tail rotor, in the event of the rudder becoming inoperable the helicopter remains airworthy.

Mil Mi-171 (1995)

TYPE • Helicopter **COUNTRY** • Russian Federation

SPECIFICATIONS	
Dimensions:	Length: 25.3m (83ft); Main rotor diameter: 21.1m (69ft 3in); Height: 5.65m (18ft 6in)
Weight:	Maximum take-off 13,000kg (28,660lb)
Powerplant:	Two 1790kW (2400shp) Klimov VK-2500-03 turboshaft engines
Max speed:	250km/h (155mph)
Range:	610km (980 miles)
Ceiling:	5000m (16,404ft)
Crew:	3
Armament:	Provision for a pintle-mounted machine gun at either or both doors

One of the world's most successful helicopters, the Soviet Mi-17 has been produced in vast numbers and remains ubiquitous in the air arms of many nations. China is no exception and has ordered new Mi-171s as recently as 2019.

Left: A Czech Air Force Mil Mi-171. Some Mi-171s had an IR jammer and flare dispenser to protect the helicopter from MANPADS attacks and a few Mi-171s have been photographed with a SATCOM antenna installed on top of the tail boom.

Radar/tracking turret
The helicopter carries both terrain-following radar and an infrared search and track (IRST) system turret.

Mil Mi-171E
The yellow code is in the 53×1×1 range, meaning it belongs to the Western Theatre Command's 5th Transportation and SAR Brigade, based at Chengdu/Qionglai.

Payload
The helicopter has a payload of 24 troops or 12 stretchers or 4000kg (8818lb) internal payload or 5000kg (11023lb) externally slung.

Mi-171 Variant
The excellent service delivered by the cheap, robust and versatile Mi-17 saw development of the improved Mi-171 variant in 1995 with a weather radar installed in the chin. Earlier Mi-17s were subsequently upgraded to Mi-171 standard. Several Mil-171s have been upgraded with a search light, IRST turret, flare/chaff launchers and a terrain following radar in the nose for SAR missions.

Czech Mi-171
A Czech Mi-171 helicopter seen at Leopoldsburg airfield in Belgium in 2021. In recent years, Mi-171s have reportedly been providing electronic jamming support for Z-10 attack helicopters utilizing a new ECM pod, first observed in late 2020.

Saab Erieye (1996)

TYPE • *AEW&C* **COUNTRY** • *Sweden*

Sweden has a relatively long history of airborne early warning and control (AEW&C) development, with the latest mission equipment being based around Saab's popular Erieye system, originally developed by Ericsson to meet a Swedish Air Force requirement.

SPECIFICATIONS (Saab 340 AEW&C)

DIMENSIONS:	Length: 20.57m (67ft 6in); Wingspan: 21.44m (70ft 4in); Height: 6.97m (22ft 10in)
WEIGHT:	Maximum take-off 13,155kg (29,000lb)
POWERPLANT:	Two General Electric CT7-9B turboprops each rated at 1390kW (1870hp)
MAX SPEED:	487km/h (303mph)
RANGE:	2858km (1776 miles)
CEILING:	7620m (25,000ft)
CREW:	Two plus four mission personnel
ARMAMENT:	None

Right: The latest iteration of the Erieye is the Erieye ER (Extended Range) that is the cornerstone of the Saab GlobalEye multi-sensor AEW&C platform. The GlobalEye is based on a Bombardier Global 6000/6500 airframe.

AIR AND SEA SURVEILLANCE
The Erieye offers a multi-mission capability, allowing the aircraft to undertake surveillance of air and sea, across an area of over 500,000km² (193,051 sq miles) horizontally and over 18,288m (60,000ft) vertically.

PAKISTAN SAAB 2000 ERIEYE
The Pakistan Air Force was originally set to receive five Saab 2000 Erieye AEW&C aircraft, but this number was later reduced to four, the first two examples being delivered to Pakistan in December 2009 and April 2010, respectively.

ERIEYE
The Erieye is an active electronically scanned array (AESA) radar that is compact enough to be installed in a business jet-sized airframe. This system began to be tested in the mid-1980s and was then introduced by the Swedish Air Force onboard the Saab 340B twin turboprop, known in Swedish service as the S 100 Argus. Subsequently, the same Erieye radar was adapted for installation in the Saab 2000 twin turboprop and the Embraer EMB-145 regional jet.

POWERPLANT
Power for the Saab 340 Erieye comes from two General Electric CT7-9B turbofans each rated at 1390kW (1870hp).

EMBRAER E-99 ERIEYE
For Brazil, the Erieye system has been fitted to the airframe of an EMBRAER R-99, which is in turn based on an airframe based on the ERJ 145 civil regional jet. The resulting aircraft is the E-99.

AH-64D Longbow/AH-64E Guardian (1997)

TYPE • Helicopter **COUNTRY** • United States

The AH-64D Apache Longbow established itself as the leading combat helicopter of the modern age. Equipped with its prominent Longbow millimetre-wave fire control radar and Hellfire missiles, it was a formidable hunter in Afghanistan and Iraq.

SPECIFICATIONS (AH-64E)

Dimensions: Length: 14.68m (48ft 2in); Main rotor diameter: 14.63m (48ft); Height: 4.72m (15ft 6in)

Weight: Maximum takeoff 10,432kg (23,000lb)

Powerplant: 2 × General Electric T700-GE-701D turboshaft engines 1486kW (1994shp)

Max speed: 300km/h (186mph)

Range: 476km (295 miles)

Ceiling: 6400m (21,000ft)

Crew: 2

Armament: 30mm (1.18in) automatic Boeing M230 Chain Gun; Hellfire II semi-active laser or Hellfire RF missiles; Hydra-unguided rockets; multiple AAM and SAM options depending on mission parameters

Left: The AH-64E Apache is the latest iteration of the Apache family. It features more powerful engines, upgraded transmission and new rotor blades. Here an AH-64E conducts deck landing qualifications on the assault ship USS Peleliu *(LHA 5).*

Fire control radar
The Longbow millimetre wave fire control radar allows the aircraft to identify a target and launch an attack in 30 seconds.

Chain gun
The AH-64D is equipped with the 30mm (1.18in) automatic Boeing M230 chain gun, equipped with 1250 rounds of ammunition and firing at a cyclical rate of 625rpm.

Hellfire missiles
When working in the close-support role, the AH-64D can carry 16 Hellfire missiles on four four-rail launchers and four air-to-air missiles.

FIRE CONTROL

The heart of the Longbow variant is the Northrop Grumman millimeter-wave Longbow radar. This system delivers fast and accurate target designation information to the two-man crew. The advantage of the millimetre wave is that it gives accurate results even when visibility is poor or when there is a lot of ground clutter. It is also highly resistant to enemy countermeasures.

AH-64E GUARDIAN
A Task Force Lightning Horse AH-64E Apache Guardian from 1st Battalion, 25th Aviation Regiment, 25th Combat Aviation Brigade. Like the AH-64D, the AH-64E can also be fitted with the Longbow fire control system.

Bell V-22 Osprey (1997)

TYPE • VTOL helicopter **COUNTRY** • United States

SPECIFICATIONS

DIMENSIONS:	Length: 19.09m (62ft 8in); Rotor diameter: 11.58m (38ft); Wingspan: 14.36m (47ft 1in)
WEIGHT:	Maximum take-off 27,406kg
POWERPLANT:	Two 4586 kW (6150hp) Allison T406-AD-400 turboshaft engines
MAX SPEED:	556km/h (345mph)
RANGE:	3892km (2418 miles)
CEILING:	7930m (26,000ft)
CREW:	3
ARMAMENT:	None

From an interesting experiment the Bell V-22 Osprey tilt-rotor aircraft has evolved into a valuable operational asset, able to fly faster and for longer than most helicopters, while still retaining the ability to take off and land vertically.

Left: Each Osprey rotor has three high-twist tapered blades with elastomeric bearings and a power-folding mechanism. A transverse cross-shaft connects the two rotors and is unloaded during normal operations, but can drive both proprotors in the event of losing an engine.

CREW
The V-22 is operated by a pilot flying in the right-hand seat, as in a helicopter, a co-pilot in the left-hand seat and a crew chief.

WINGS
Slightly swept forward, the wing is fitted with two sections of single-slotted flaperons for roll control and extra lift, these being operated by the fly-by-wire control system. The wing centre-section houses the drive gearbox, rotor-phasing equipment and rotor brakes.

TILT-ROTOR DESIGN
The Bell V-22 owes its existence to the Bell XV-15 tilt-rotor testbed aircraft, developed in the mid-1970s. The XV-15 was powered by two Lycoming T-53 turboshaft engines driving three-bladed 7.62m (25ft) metal rotors; the engines were located in wingtip nacelles. For take-off, the engine/rotor assembly operated vertically, generating downward thrust and enabling the aircraft to lift off like a helicopter. Once off the ground, it could either fly like a helicopter or as a conventional aircraft, the engine/rotor assembly tilting horizontally in the latter case.

CABIN
The cabin can accommodate 24 troops, 12 litters or internal cargo, for which a 907kg (2000lb) hoist system is fitted.

MV-22 OSPREY
An MV-22 Osprey assigned to Marine Vertical Lift Squadron (VMM)161 approaches to land on the flight deck of the *San Antonio*-class amphibious transport dock ship Pre-Commissioning Unit (PCU) Anchorage (LPD 23) in 2013.

RQ-4 Global Hawk (1998)

TYPE • UAV COUNTRY • United States

SPECIFICATIONS

DIMENSIONS:	Length: 14.5m (47ft 8in); Wingspan: 39.9m (131ft); Height: 4.7m (15ft 3in)
WEIGHT:	Gross weight 14,628kg (32,250lb)
POWERPLANT:	Rolls-Royce F137-RR-100 turbofan engine, 34kN (7600lbf)
MAX SPEED:	629km/h (391mph)
RANGE:	22,779km (14,154 miles)
CEILING:	18,000m (59,000ft)
CREW:	Unmanned
ARMAMENT:	None

Global Hawk was developed to fly at extremely high altitudes and undertake long missions. It is not a combat platform designed to launch missiles at the enemy, but instead undertakes long-range reconnaissance missions over land or sea.

Left: An aerial view of the maiden flight of the second US Navy RQ-4A Global Hawk unmanned aerial vehicle (UAV) en route to Edwards Air Force Base, California from Palmdale, California on 7 June 2005.

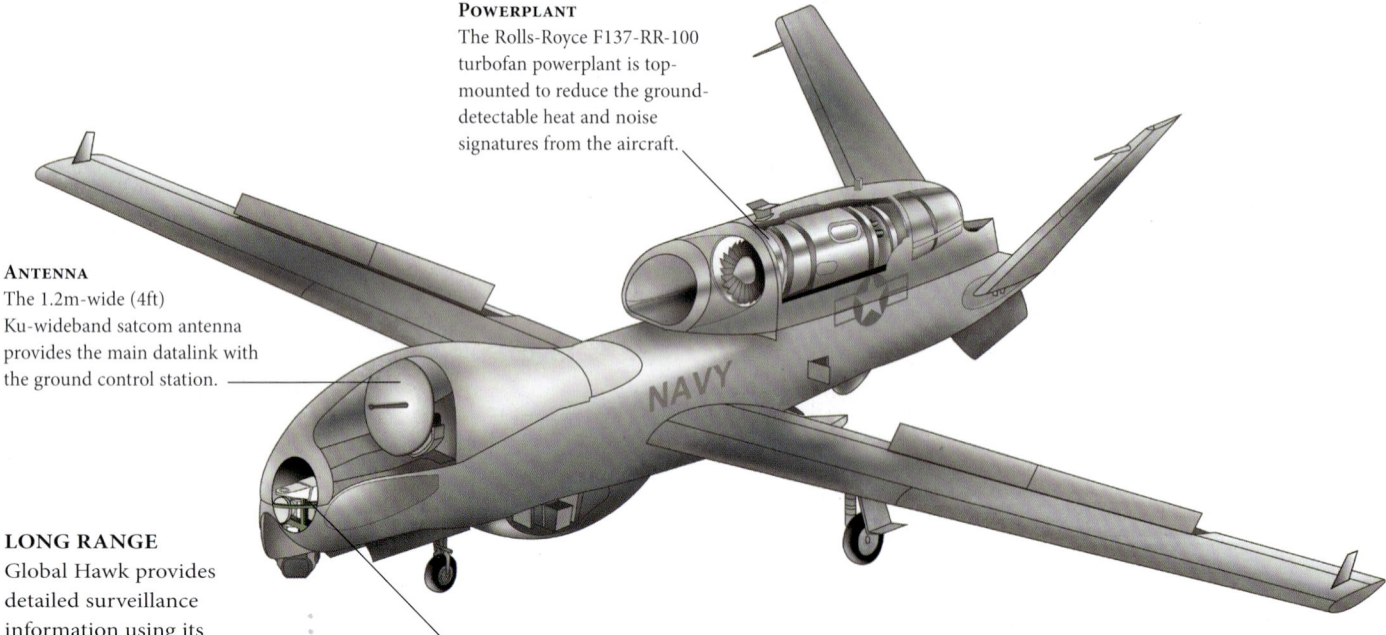

POWERPLANT
The Rolls-Royce F137-RR-100 turbofan powerplant is top-mounted to reduce the ground-detectable heat and noise signatures from the aircraft.

ANTENNA
The 1.2m-wide (4ft) Ku-wideband satcom antenna provides the main datalink with the ground control station.

LONG RANGE
Global Hawk provides detailed surveillance information using its cameras, thermal-imaging equipment and radar. It can stay aloft for more than 35 hours and operates at an altitude of almost 20km (12 miles). The UAV also offers considerable reach: it is designed to fly to a target area more than 1500km (930 miles) away, to remain there for a whole day or more, and then return home safely. To demonstrate its great range, a Global Hawk was flown nonstop from the United States to Australia, setting several records in the process.

INTEGRATED SENSOR SUITE
The integrated sensor suite (ISS) consists of a synthetic aperture radar (SAR), electro-optical (EO) and thermographic camera (IR) sensors.

RQ-4A GLOBAL HAWK
Global Hawk can taxi to the runway, take off, fly to its destination and return home again without direct human control. It can even land itself. If necessary, the drone can be retasked by its controllers and sent to another area or be told to carry out a new mission.

Shenyang J-11 and J-11A (1999)

TYPE • Fighter **COUNTRY** • China

After turning to the Soviet Union to fulfil its immediate fourth-generation fighter needs with the Sukhoi Su-27 'Flanker', China also secured a deal for the licence production of those same jets, with the local work to be undertaken by Shenyang.

SPECIFICATIONS (J-11A)

DIMENSIONS:	Length: 21.9m (69ft 6in) without probe; Wingspan: 14.7m (48ft 3in); Height: 5.7m (18ft 9in)
WEIGHT:	Maximum take-off 33,000kg (72,753lb)
POWERPLANT:	Two AL-31F series 3 turbofans, each rated at 122.5kN (27,558lbf) with afterburning
MAX SPEED:	Mach 2.35 (2,500 km/h)
RANGE:	3530km (2190 miles)
CEILING:	19,000m (62,336ft)
CREW:	1
ARMAMENT:	One GSh-301 30mm (1.2in) cannon, plus disposable stores carried on 10 hardpoints

Above: This J-11A, 'Yellow 10182', is believed to be from the 7th Division's 19th Regiment, based at Zhangjiakou. It is wearing special red stripe markings on the IR seeker, tail stinger and wing tip rails and a double stripe behind the canopy, as special markings for the 70th China Victory Day parade in August 2015.

SHENYANG J-11A
The Shenyang J-11A was a Chinese-produced multi-role variant of the J-11. Included was a modernized cockpit featuring two small colour multifunctional displays (MFDs).

AVIONICS
Some J-11As have also received a GPS display in the cockpit. Meanwhile, both J-11s and J-11As have been seen fitted with a new UV-band missile approach warning system (MAWS), with antennas behind the cockpit and on the sides of the tailfins.

CAMOUFLAGE
The camouflage is a variation of the darker grey and it has anti-collision strips on the nose, intakes and fins. It is thought to be from the 4th Regiment of the 2nd Air Division, based at Liuzhou.

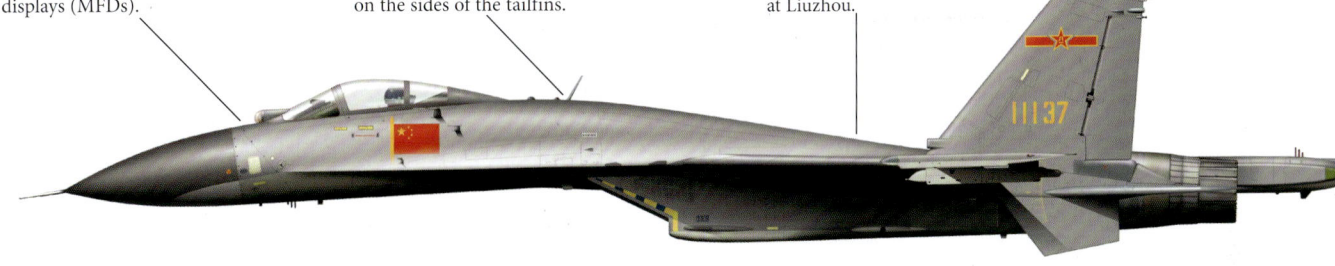

RUSSIAN DEAL

When finally signed in December 1995, the licence production deal covered the Chinese manufacture of 200 Su-27SKs. They would be supplied to China in kit form together with their Lyulka (later NPO Lyulka/Saturn) AL-31F turbofan engines and various items of weaponry. It seems that the deal also included the introduction of progressively more Chinese-furnished components, although a minimum of 30 per cent Russian content was stipulated. The Shenyang-produced Su-27SK is known as the J-11 but is otherwise similar to the Russian product.

J-11, CHINA, 2021
A J-11 fighter takes off from a PLA military airport in a training session in east China's Zhejiang province in late August 2021. It remains unclear exactly how many J-11As were produced, although reliable sources suggest that perhaps around 65 examples were completed before production was discontinued in 2006.

AgustaWestland EH101/AW101 Merlin (1999)

TYPE • Helicopter **COUNTRY** • United Kingdom/Italy

The trend towards international collaboration in the military aviation industry is by no means confined to fixed-wing aircraft. The EH101 Merlin is an Italian-British venture. Since 2000 the helicopter has taken the designation AW101, to reflect the merger between Westland and Agusta.

SPECIFICATIONS (EH101 Merlin HM.Mk.I)

DIMENSIONS:	Fuselage length: 19.63m (64ft 5in); Main rotor diameter: 18.6m (61ft); Height: 6.63m (21ft 9in)
WEIGHT:	Maximum take-off 14,600kg (32,188lb)
POWERPLANT:	Three Rolls-Royce/Turbomeca RTM-322 turbines
MAX SPEED:	280km/h (174mph)
RANGE:	1389km (863 miles)
CEILING:	4575m (15,000ft)
CREW:	3
ARMAMENT:	Variety of ASW and anti-shipping munitions, sonobuoys etc.

This page: The AW101 is a highly survivable helicopter. Westland drop-tested the EH101 to a survivable velocity of 10.6m (35ft) per second.

ROTOR
The EH101's rotor uses both fibre-reinforced and metal components. The system is resistant to hits from 23mm (0.906in) shells and will continue running for 45 minutes with no oil in the gearbox.

FUSELAGE
The Merlin's fuselage was designed in four sections, with the front, centre-section and tailcone common to all versions. The tailboom in the ramp-loading versions is slimmer and dispenses with the tail-loading option. The fuselage structure is mainly of honeycomb aluminium-lithium alloy with bonded composite panels.

JOINT PROGRAMME

The EH101 helicopter was developed jointly by Britain's GKN-Westland Helicopters and Italy's Agusta. The programme was given the go-ahead by the British and Italian governments on 25 January 1984. The EH101 subsequently entered full production in both Italy and the United Kingdom, with orders placed by the Royal Navy (Merlin HM Mk 1) and RAF (Merlin HC.Mk.3), the Italian Navy, the Canadian Armed Forces, the Danish armed forces and the Tokyo Police.

EMERGENCY FLOATS
In an emergency ditching, the Merlin relies on four Kevlar-reinforced polythene floats, inflated by bottled helium. One float is positioned on either side of the nose, and two on the undercarriage sponsons.

AW101 FROM USS *BATAAN*
British Royal Marines depart the *Wasp*-class amphibious assault ship USS *Bataan* in an AgustaWestland AW101 (EH101) Merlin in 2024. The Merlin Mk 3, pioneered into RAF service by No. 28 Squadron, was the first helicopter to enter service with an integrated Defensive Aids Suite (DAS), which gave the highest level of self-protection of any UK military helicopter.

Sukhoi Su-27K and Su-33 (1999)

TYPE • *Naval fighter* **COUNTRY** • *Russian Federation*

SPECIFICATIONS

DIMENSIONS:	Length: 21.2m (69ft 6in) without probe; Wingspan: 14.7m (48ft 3in); Height: 5.72m (18ft 9in)
WEIGHT:	Maximum take-off 24,500kg (54,013lb)
POWERPLANT:	Two modified AL-31F series 3 turbofans, each rated at 122.5kN (27,558lbf) with afterburning
MAX SPEED:	Mach 2.17 (2679km/h)
RANGE:	3000km (1864 miles)
CEILING:	17,000m (55,775ft)
CREW:	1
ARMAMENT:	One GSh-301 30mm (1.2in) cannon, plus up to 6500kg (14,330lb) of stores carried on 12 hardpoints

Developed during the Cold War to serve aboard an ambitious series of aircraft carriers planned for the Soviet Navy, the Su-33 eventually only went to sea aboard a single Russian Navy aircraft carrier, the conventionally powered *Admiral Kuznetsov*.

This page: The Su-33 began life under the designation Su-27K, reflecting the fact it was a minimum-change carrier-based version of the 'Flanker-B' air superiority and air defence fighter.

SU-27K
The changes required for carrier operations included canard foreplanes, revised wing flaps, reinforced undercarriage and an arrester hook.

ENGINES
The AL-31F engines feature some modifications for operations at sea and the aircraft is equipped with an in-flight refuelling probe (plus optional buddy refuelling pod), unlike the land-based Su-27.

CHINESE VARIANTS

After the demise of the Soviet Union, two T-10K test aircraft were left in Ukraine and one of these was sold to China in 2007. It subsequently became the pattern for the J-15, which incorporates Chinese avionics, a modern glass cockpit and has a multi-role capability using indigenous weapons. The People's Liberation Army Navy Air Force took the J-15 to sea for the first time aboard the *Kuznetsov*-class carrier *Liaoning* in 2012. The J-15 is now also deployed on board the carrier *Shandong*, which is a Chinese version of the *Kuznetsov* design.

SU-27 TAKE-OFF
A Sukhoi Su-33 (Flanker-D) takes off at Zhukovsky airport, Russia, after undergoing modernization with the "Gefest" satellite navigation system.

Boeing F/A-18E Super Hornet (1999)

TYPE • *Naval fighter* COUNTRY • *United States*

When the US Navy demanded a more capable, strike-oriented version of the Hornet to replace the A-6E Intruder and F-14 Tomcat, McDonnell Douglas (later part of Boeing) developed the F/A-18E/F Super Hornet.

SPECIFICATIONS

DIMENSIONS:	Length: 18.31m (60ft 1in); Wingspan: 13.62m (44ft 8in); Height: 4.88m (16ft)
WEIGHT:	Maximum take-off 29,937kg (66,000lb)
POWERPLANT:	Two General Electric F414-GE-400 turbofans each rated at 98kN (13,000lbf) thrust with afterburning
MAX SPEED:	Mach 1.6
RANGE:	3300km (2070 miles)
CEILING:	15,240m (50,000ft)
CREW:	1
ARMAMENT:	One 20mm (0.787in) M61A1 Vulcan six-barrel rotary cannon, plus maximum weapon load of 8050kg (17,750lb) on six underwing pylons, three fuselage pylons and two wingtip pylons

Left: An F/A-18E Super Hornet from the 'Pukin' Dogs' of Strike Fighter Squadron 143 launches from the flight deck of the Nimitz-class aircraft carrier USS Dwight D. Eisenhower *(CVN 69) in 2007.*

RADAR
The initial Block I Super Hornets retained the APG-73 but this was quickly replaced by the Raytheon APG-79 active electronically scanned array (AESA) radar, which offers greatly increased target detection ranges and which can operate simultaneously in air-to-air and air-to-ground modes.

LANDING CAPABILITY
The Super Hornet design prioritized its increased 'bring back' ability. The Super Hornet does not have to dump expensive weapons for a carrier landing, being able to trap with 4100kg (9000lb) of stores.

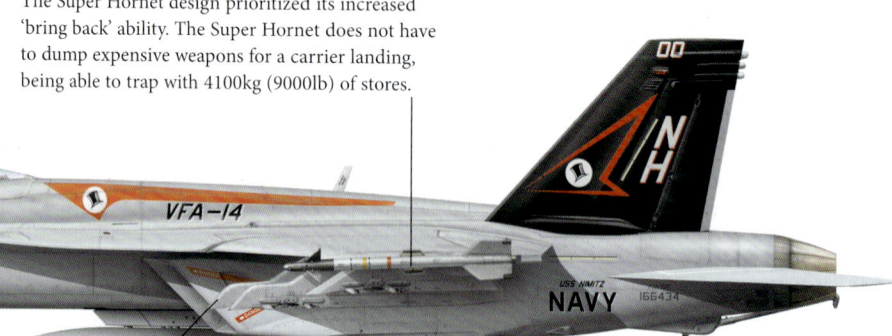

REPLACEMENTS
The Super Hornet was built in three versions. The F/A-18F replaced F-14 Tomcats in the USN's fighter (VF) squadrons, which became VFA (fighter-attack) units, and A-6E (VA) and F/A-18C (VFA) squadrons reformed on the F/A-18E. In practice, both types of Super Hornet are largely interchangeable and fly both ground-attack and air defence missions. The EA-18G Growler was a dedicated electronic warfare variant replacing the EA-6B Prowler.

WEAPONS PYLONS
The weapons pylons are canted or toed out because wind tunnel testing showed that underwing stores were at risk of colliding when dropped. Canting the pylons cured the problem, but causes extra drag.

SUPER HORNET UPGRADE
The Super Hornet is about 25 per cent larger than the F/A-18C and its F414 engines each have 35 per cent more power than the 'legacy' Hornet's F404s. Before settling on the Super Hornet, a modernized 'Tomcat 21' was proposed but abandoned, as was the two-seat Grumman A-12 Avenger stealth aircraft.

Lockheed Martin C-130J Hercules (1999)

TYPE • Transport   COUNTRY • United States

With a longer production run than any other military aircraft, the Hercules remains the Western airlifter of choice and its latest incarnation, the C-130J, continues the aircraft's impressive legacy with the US Air Force and a host of export operators.

SPECIFICATIONS

DIMENSIONS:	Length: 29.79m (97ft 9in); Wingspan: 40.41m (132ft 7in); Height: 11.84m (38ft 10in)
WEIGHT:	Maximum take-off 70,307kg (155,000lb)
POWERPLANT:	Four Rolls-Royce AE 2100D3 turboprop engines each rated at 3458kW (4637shp)
MAX SPEED:	670km/h (417mph)
RANGE:	3,00km (2100 miles) with 15,422kg (34,000lb) payload
CEILING:	12,310m (40,386ft)
CREW:	3 (2 pilots, 1 loadmaster)
ARMAMENT:	None

Left: A C-130J Super Hercules approaches Yokota Air Base, Japan on 29 March 2017. By 2020, the Super Hercules had secured orders from 24 operators in 20 different countries, and a global fleet of more than 450 aircraft had recorded more than 2 million flight hours.

COCKPIT
The C-130J has a fully digital cockpit with four colour multifunction displays and a separate head-up display (HUD) for each pilot. These avionics changes mean the J-model can be flown by a two-person crew, with no requirement for the previous flight engineer and navigator.

POWERPLANT
Thanks to its new engines, the C-130J provides a 40 per cent improvement in range over the 'legacy' Hercules, as well as an increase in speed by 21 per cent and a take-off distance reduced by as much as 41 per cent.

HOLD CAPACITY
The C-130J is capable of transporting heavy loads of equipment and personnel: 92 combat troops, 64 paratroopers, 54 passengers or an equivalent load of cargo.

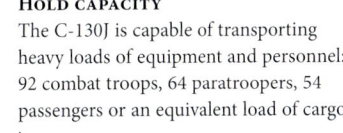
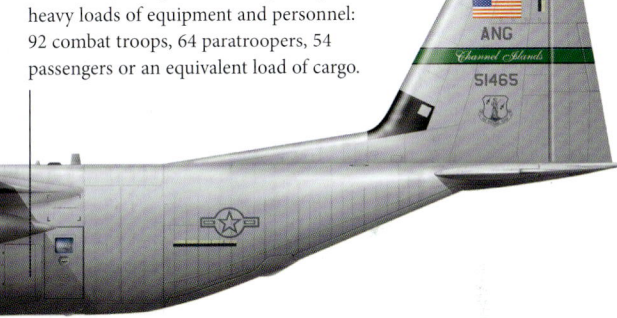

SPECIALIST VARIANTS
While the first examples handed over to the US Air Force were standard C-130J transports, delivered in January 1999, the service soon added new variants including the WC-130J 'Hurricane Hunter' for weather reconnaissance and the EC-130J Commando Solo electronic warfare aircraft, equipped to perform psychological operations. Further US Air Force include the HC-130J Combat King II that was developed to replace the HC-130N/P combat rescue tankers.

C-130J TAKE-OFF
The launch customer for the C-130J was the United Kingdom, which opted for a mix of 15 stretched C-130J-30 and 10 standard-length C-130J models, in 1994. The following year, the US Air Force placed its first orders for the type.

Mitsubishi F-2 (2000)

TYPE • Fighter COUNTRY • Japan

Looking very similar to the F-16, the Mitsubishi F-2 was developed for the Japan Air Self-Defense Force (JASDF) on the basis of that aircraft, with design work shared by Mitsubishi Heavy Industries and Lockheed Martin.

SPECIFICATIONS (F-2A)

Dimensions: Length: 15.52m (50ft 11in); Wingspan: 11.125m (36ft 6in) including wingtip pylons; Height: 16ft (4.9m)

Weight: Maximum take-off 22,100kg (48,722lb)

Powerplant: One General Electric F110-IHI-129 turbofan rated at 131kN (29,500lbf) thrust with afterburning

Max speed: Mach 1.7 (2099km/h)

Range: 833km (518 miles)

Ceiling: 18,000m (59,000ft)

Crew: 1

Armament: One 20mm (0.787in) JM61A1 six-barrel rotary cannon, plus maximum weapon load of 8085kg (17,824lb)

Left: Japan Air Self-Defense Force Mitsubishi F-2As on Guam. The JASDF F-2 force was struck a blow when Matsushima Air Base was hit by the powerful tsunami in March 2004, damaging 18 of these aircraft. Ultimately, however, all but a handful were restored to airworthy condition.

COCKPIT
One other subtle difference compared to the F-16 is the F-2's two-piece cockpit canopy, a more robust construction required to provide protection against bird strikes in the low-level maritime environment.

WING AREA
Compared to the F-16, the F-2 features a wing area increased by around 25 per cent as well as a fuselage 'stretched' by around 43cm (17in).

SUPPORT FIGHTER
Intended to meet the requirement for a so-called Support Fighter, or FS-X, replacing the Cold War-era Mitsubishi F-1, the multi-role F-2 was developed in the mid- to late-1980s and was originally optimized for air-to-surface missions, including anti-shipping strikes to protect Japan's sea lanes, although it is also equipped for air defence missions. The first prototype F-2 made its maiden flight on 7 October 1995 and production encompassed the single-seat F-2A and the two-seat F-2B with full combat capability, albeit with a somewhat reduced fuel load.

FLY-BY-WIRE
The aircraft's fly-by-wire flight control system and integrated electronic warfare system were developed in Japan and the aircraft is provided with a range of indigenously developed air-to-ground and air-to-air weapons.

JASDF F-2A
A Japan Air Self-Defense Force Mitsubishi F-2A taxis out at Andersen Air Force Base, Guam, in 2024. Thanks to the larger wing, there are two additional stores stations compared to the F-16, as well as additional fuel capacity for an extended range.

Aero Vodochody L-159 ALCA (2000)

TYPE • *Fighter/trainer* COUNTRY • *Czech Republic*

The L-159 ALCA traces its origins back to the enormously successful L-39 Albatros advanced jet trainer that was developed in the late 1960s and which became a standard aircraft of its type across much of the Warsaw Pact and for many of its allies.

SPECIFICATIONS (L-159A)

DIMENSIONS:	Length: 12.72m (41ft 9in); Wingspan: 9.54m (31ft 4in) with tip tanks; Height: 4.87m (16ft)
WEIGHT:	Maximum take-off 8000kg (17,637lb)
POWERPLANT:	One Honeywell/ITEC F124-GA-100 turbofan engine rated at 28.2kN (6300lbf) thrust
MAX SPEED:	936km/h (582mph)
RANGE:	1570km (980 miles)
CEILING:	13,198m (43,300ft)
CREW:	1
ARMAMENT:	Seven hardpoints under the wing and fuselage for a range of stores including 20mm (0.787in) gun pods, bombs, rockets and AIM-9 Sidewinder air-to-air missiles up to a total weight of 2340kg (5159lb)

Left: This underside view of an L-159 shows the stores- and ordnance-carrying capability offered by the seven underwing hardpoints, which can include AAMs, SAMs, rocket pods and cannon pods.

AVIONICS
The ALCA introduced NATO-standard avionics. These advanced systems are centred around an Italian-supplied Grifo multi-mode radar.

POWERPLANT
Initially intended to replace the Czech Air Force's ageing MiG-21 fleet, the ALCA introduced a US-made Honeywell F124 powerplant.

ISIS ATTACKS
The first export customer for the ALCA was Draken International, a US-based aggressor firm that used 21 of the jets for adversary training, flying against US and allied air arms and simulating advanced threat aircraft and missiles. In 2015, Iraq also placed orders for the ALCA with a contract for 15 examples, which were delivered in L-159A and L-159T variants. Since their delivery to Iraq in the same year, the ALCAs have seen considerable combat action against ISIS militants in the country.

L-159
An L-159 of the 21. základna taktického letectva – the 21st Tactical Aviation Base – at Čáslav. After transfers, conversions to two-seat standard and attrition, the Czech Air Force today operates 16 L-159As, with several more in storage.

L-159 WITH SIDEWINDERS
A Czech L-159, equipped with AIM-9 Sidewinders. Export Czech L-159T1s, delivered without radar, are now being updated to a combat-capable L159T+ standard, with Grifo radar, radar warning receivers and chaff/flare dispensers.

Boeing 737 (2001)

TYPE • Transport/Reconnaissance **COUNTRY** • China

The ubiquitous Boeing 737 airliner is one of the more unlikely aircraft in Chinese military service. The classic twinjet is used in two distinct roles by the PLAAF: as an airborne command post and as a VIP transport.

SPECIFICATIONS (737-300)

DIMENSIONS:	Length: 33.40m (109ft 7in); Wingspan: 28.88m (94ft 9in); Height: 11.13m (36ft 6in)
WEIGHT:	Maximum take-off 56,740kg (124,500lb)
POWERPLANT:	Two 89.0kN (20,000lbf) CFM International CFM563B1 turbofans, or optionally two 97.9kN (22,000lbf) CFM563B2
CRUISE SPEED:	908km/h (564mph)
RANGE:	4973km (3090 miles)
CEILING:	11,300m (37,000ft)
CREW:	Two
ARMAMENT:	None

Below: A PLAAF Boeing 737-300, as seen at Beijing Nanyuaairport in 2019.

CONVERSION UPGRADE
Work to transform two 737s as airborne command posts was reportedly conducted by Xi'an. The aircraft received a prominent fairing on top of the forward fuselage and two smaller fairings below the mid-section of the fuselage.

CAPACITY
In the VIP transport role, the Chinese Boeing 737-300 can carry 11 passengers in first-class accommodation, although seats can obviously be removed to create bespoke spaces.

CHINA'S BOEINGS
The PLAAF appears to have obtained a total of 14 Boeing 737s, in three different variants. These comprise eight 737-3Q8s, two 737-76Ds and four 737-85Ns. Initially, it appears that the entire 737 fleet was assigned to the 34th Transport Division for use as VIP transports. The aircraft were originally operated in a civilian-style China United Airlines paint scheme. This later gave way, at least on some of the aircraft, to full PLAAF insignia.

POWERPLANT
The Boeing 737-300 is powered by 89.0kN (20,000lbf) CFM International CFM563B1 turbofans, although it has the option of two 97.9kN (22,000lbf) CFM563B2 turbofans

PLAAF BOEING 737
PLAAF Boeing 737 (registration B-4021) at Poznan Lawica Airport, Poland, in July 2013. The aircraft was first delivered in September 1995. The Boeing 737 is familar as a commercial airliner flying short-haul and medium-haul routes. By February 2023, more than 11,700 of these aircraft were in service worldwide.

Dassault Rafale (2001)

TYPE • *Multi-role fighter* **COUNTRY** • *France*

SPECIFICATIONS (Rafale C)

DIMENSIONS:	Length: 15.27m (50ft 1in); Wingspan: 10.9m (35ft 9in); Height: 5.34m (17ft 6in)
WEIGHT:	Maximum take-off 24,500kg (54,013lb)
POWERPLANT:	Two Snecma M88-2 turbofans each rated at 75kN (17,000lbf) thrust with afterburning
MAX SPEED:	Mach 1.8 (2222km/h)
RANGE:	3700km (2300 miles), ferry, with three drop tanks
CEILING:	15,835m (51,952ft)
CREW:	1
ARMAMENT:	One 30mm (1.2in) GIAT 30M791 cannon, plus up to 9500kg (20,900lb) of disposable stores on 14 hardpoints

Described by the manufacturer as an omni-role fighter, the aircraft was designed to undertake a wide range of combat missions, which it now does on behalf of the French Air Force and French Navy, as well as an increasing number of export operators.

Above: The Aéronautique Navale, or French Fleet Air Arm, was the first operator to put the Rafale into frontline service, initially only in an air-to-air capacity, beginning in 2000. This Rafale M is assigned to Flottille 12F, one of three French Navy squadrons.

MULTI-SENSOR SUITE
As well as the Thales RBE2 multimode radar, the Rafale is fitted with the front sector optronics (FSO) system mounted just ahead of the windscreen. This combines an infrared search and track system with a TV sensor and laser rangefinder.

STEALTH TECHNOLOGY
Airframe radar cross-section is minimized by using appropriate materials and mould line, including serrated edges to the trailing edge of the wings and canards.

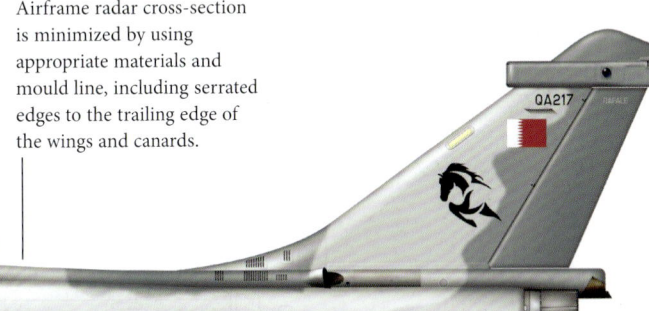

VARIANTS

The first variant was the Rafale F1, which offered air defence capabilities only. The next step was the Rafale F2, which provided a degree of precision ground-attack capability with the SCALP conventionally armed cruise missile, as well as different laser-guided bombs. The Rafale F3 was the first of the 'full-spectrum' jets, able to fly missions including reconnaissance, anti-shipping strike and nuclear deterrence. The Rafale family is also divided between twin-seat Rafale B and single-seat Rafale C and Rafale M, the latter for the French Navy.

TURBOFANS
The Rafale is powered by a pair of Snecma M88-2 turbofans specifically designed for the aircraft and intended to offer considerable thrust within a relatively small volume.

LIBYA, 2011
A French Rafale jet fighter overflies Libya on a security operation in 2011. In 2023 the F4 standard, which makes improvements to the sensors and self-defence suite, was qualified for service.

Airbus C295 (2001)

TYPE • *Tactical transport* **COUNTRY** • *Spain*

SPECIFICATIONS	
DIMENSIONS:	Length: 24.46m (80ft 3in); Wingspan: 25.81m (84ft 8in): Height: 8.66m (28ft 5in)
WEIGHT:	Maximum take-off 21,000kg (46,297lb)
POWERPLANT:	Two Pratt & Whitney Canada PW127G turboprop engines each rated at 1972kW (2644hp)
CRUISE SPEED:	Cruising 482km/h (300mph)
RANGE:	1555km (966 miles), with normal payload
CEILING:	9145m (30,003ft)
CREW:	Two (plus optional loadmaster)
ARMAMENT:	None

The C295 is one of the world's most widely used new tactical airlifters in the light-to-medium category. Originally manufactured by Construcciones Aeronáuticas SA (CASA), the C295 is now an Airbus product but continues to be built in Spain.

Left: A C295 flown by the Philippine Air Force. For maritime patrol and other complex missions, the C295 can be fitted with the Fully Integrated Tactical System (FITS), which integrates, controls and displays various different mission sensors.

COCKPIT
All versions of the C295 feature a glass cockpit with digital avionics, including four liquid-crystal displays that are compatible with night-vision goggles.

VIETNAMESE C-295
The C295 is one of relatively few Western designs in service today with the Vietnam People's Air Force (VPAF). A contract for three C295s was announced by Airbus in June 2014 and the first example had been handed over to the VPAF before the end of that year.

MULTI-ROLE
As of 2023, total orders of the C295 had exceeded 280 and the aircraft has been acquired by more than 30 countries. A key to its success is its versatility, which allows it to perform a wide range of missions beyond tactical transport. These include intelligence, surveillance and reconnaissance (ISR), airborne early warning and control (AEW&C), gunship and maritime patrol, with a variety of optional self-protection equipment available for missions in more contested environments.

HOLD CAPACITY
In transport configuration the C295 can carry 73 troops, 48 paratroopers, 12 stretcher patients or 7050kg (15,543lb) of cargo.

POLISH C295, 2023
Poland has taken 17 C295 aircrafy into service since 2001. The latest C295W version is equipped with winglets, improving efficiency and allowing a larger payload to be carried over a longer distance, including in hot and high conditions.

MQ-9 Reaper (2001)

TYPE • UAV COUNTRY • United States

The Reaper is a versatile sensor platform, equipped with various cameras, thermal imaging equipment, a laser rangefinder and synthetic aperture radar. It is used in US Homeland Security operations, monitoring coastlines and long borders, as well as in combat operations.

SPECIFICATIONS

DIMENSIONS:	Length: 11m (36ft 1in); Wingspan: 20m (65ft 7in); Height: 3.81m (12ft 6in)
WEIGHT:	Maximum takeoff 4760kg (10,494lb)
POWERPLANT:	Honeywell TPE331-10 turboprop, 671kW (900hp)
MAX SPEED:	482km/h (300mph)
RANGE:	1850km (1150 miles)
CEILING:	15,000m (50,000ft)
CREW:	Unmanned
ARMAMENT:	7 hardpoints: up to 680kg (1500lb) on the two inboard weapons stations; up to 340kg (750lb) on the two middle stations; up to 68kg (150lb) on the outboard stations; centre station not used

Left: The USAF first deployed the Reaper in 2007, conducting missions over Iraq and Afghanistan. Since then, the Reaper has seen a colossal amount of operational service and has so far been adopted by eight nations.

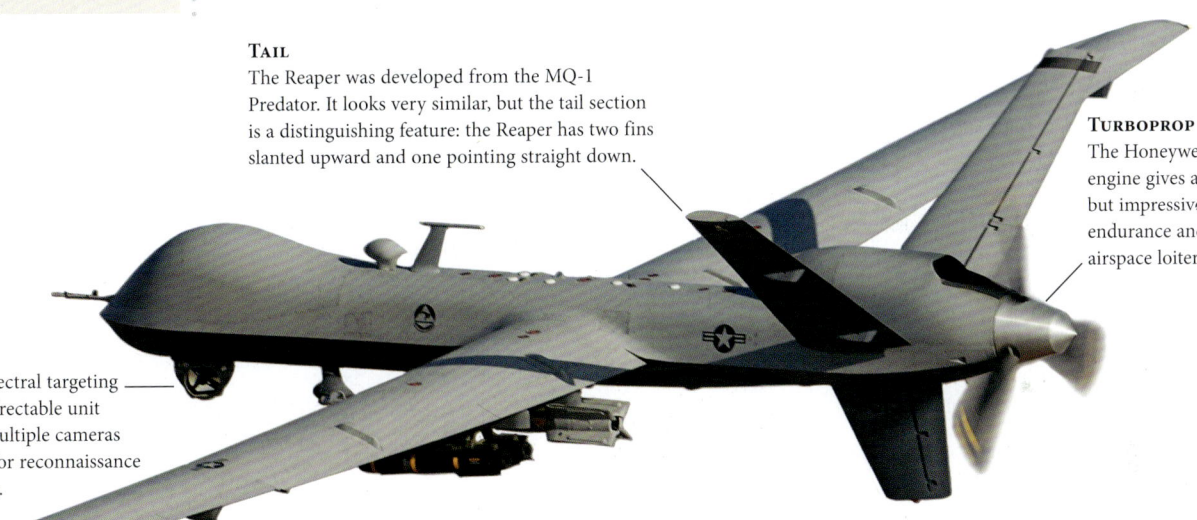

TAIL
The Reaper was developed from the MQ-1 Predator. It looks very similar, but the tail section is a distinguishing feature: the Reaper has two fins slanted upward and one pointing straight down.

TURBOPROP POWER
The Honeywell turboprop engine gives a low speed but impressive levels of endurance and battlefield airspace loitering.

MSTS
The multi-spectral targeting system is a directable unit containing multiple cameras and sensors for reconnaissance and targeting.

MARITIME SECURITY

The United States and a number of allied nations have explored the use of Reaper drones at sea, as part of anti-piracy and maritime security operations. The Reaper has also been approved for use in disaster situations, where it could search for survivors and assess damage to remote areas with its cameras and thermal imagers. The Reaper can remain on station for three times as long as a Predator and can carry out missions over a far greater distance.

ARMED REAPER
A fully armed MQ-9 Reaper taxis down an Afghanistan runway in 2007. Reapers have been heavily employed in counter-terrorist strikes using their Hellfire missiles, Paveway II laser-guided bombs or GPS-guided JDAM munitions.

Antonov An-140 (2002)

TYPE • Transport COUNTRY • Russian Federation

SPECIFICATIONS

DIMENSIONS:	Length: 22.61m (74ft 2in); Wingspan: 25.51m (83ft 8in); Height: 8.23m (27ft)
WEIGHT:	Maximum take-off 21,500kg (47,399lb)
POWERPLANT:	Two 1864kW (2500ehp) Motor Sich TV3-117VMA-SBM1 turboprops
MAX SPEED:	540km/h (336mph)
RANGE:	3050 km (1895 miles)
CEILING:	7200 m (23,622ft)
CREW:	2
ARMAMENT:	0

The An-140 was originally schemed as a replacement for the An-24 short-haul regional airliner with improved fuel efficiency, longer range, higher speed and improved comfort. This 52-passenger aircraft was repurposed by the Russian Ministry of Defence as a staff transport.

Left: An An-140 at the opening ceremony to celebrate the 100th anniversary of the Russian air force on 11 August 2012 at Zhukovsky, Russia.

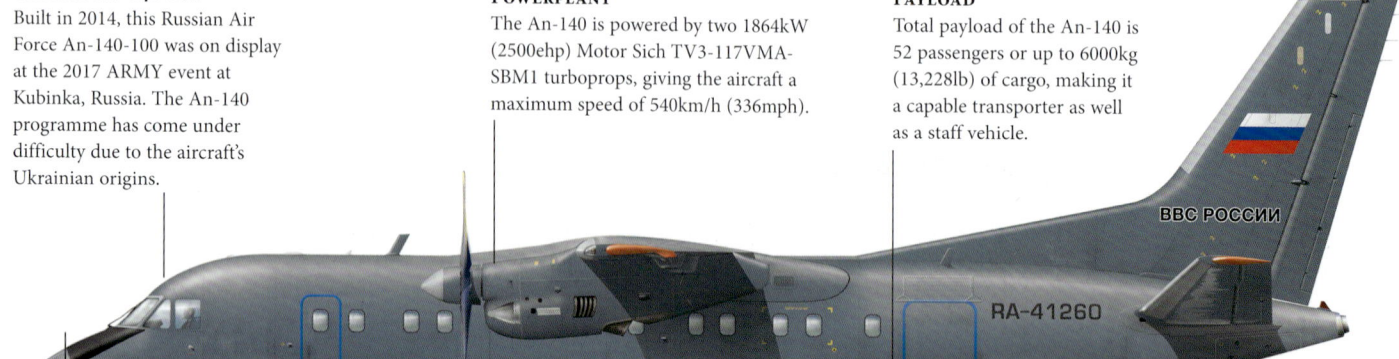

RUSSIAN AN-140-100
Built in 2014, this Russian Air Force An-140-100 was on display at the 2017 ARMY event at Kubinka, Russia. The An-140 programme has come under difficulty due to the aircraft's Ukrainian origins.

POWERPLANT
The An-140 is powered by two 1864kW (2500ehp) Motor Sich TV3-117VMA-SBM1 turboprops, giving the aircraft a maximum speed of 540km/h (336mph).

PAYLOAD
Total payload of the An-140 is 52 passengers or up to 6000kg (13,228lb) of cargo, making it a capable transporter as well as a staff vehicle.

VARIANTS

There are currently three variants of the An-140: the initial version, designed to deliver 52 passengers to 2200km (1367 miles); the An-140-100 is the standard production version with longer wing and increased weight to deliver 52 passengers to 2400km (1491 miles); and the IrAn-140 Faraz version assembled in Iran. Several planned An-140 variants have been proposed, including the stretched An-140-200, the An-140-330T military transport, the An-140C freighter and An-140T (An-142) military transport.

AN-140 NEAR MOSCOW
A Russian Air Force Antonov An-140 RA-41258 taxis out at Chkalovsky air base, located north-east of Moscow. Series production was launched in three locations, and production for Russian service is undertaken by Aviacor in Samara. Additional production takes place in Kharkiv, Ukraine and in Shahin Shahr, Iran.

Kamov Ka-226 (2002)

TYPE • *Helicopter* **COUNTRY** • *Russian Federation*

SPECIFICATIONS

DIMENSIONS: Fuselage length: 8.58m (28ft 2in); Main rotor diameter: 13m (42ft 8in); Height: 4.19m (13ft 9in)

WEIGHT: Maximum take-off 3400kg (7496lb)

POWERPLANT: Two Rolls-Royce Allison 250-C20R/2 turboshafts, each 346kW (465shp)

MAX SPEED: 220km/h (137mph)

RANGE: 600km (373 miles)

CEILING: 5000m (16,404ft)

CREW: 1 or 2

ARMAMENT: None

Developed as a general utility helicopter primarily for civilian use, the Kamov Ka-226 (NATO reporting name 'Hoodlum') was adopted by the Russian military in the early 2000s as a lightweight trainer aircraft.

Left: This Ka-226 Hoodlum was presented at the 12th MAKS-2015 (International Aviation and Space Show) in August 2015 in Moscow. At first the basic Ka-226 was powered by two Rolls-Royce Allison 250-C20R/2 turboshafts but from 2013, production aircraft were instead powered by two Turbomeca Arrius 2G1 turboshafts under the Ka-226T designation.

COCKPIT
The helicopter has a cockpit for a crew of two, with the option for a detachable module that can normally carry six passengers or a maximum of eight.

STUB WINGS
On the original Ka-26 the engines were mounted in prominent nacelles on the end of the stub wings. On the newer Ka-226, the stub wings are a mounting point for the undercarriage and the engines have moved atop the fuselage.

KA-266V
The Ka-226V (Ka-226.80) is the standard version for the Russian Ministry of Defence and features Allison engines, weather radar in the nose and dual controls.

IMPROVED KA-26

A much-improved development of the Cold War-era Ka-26, of which more than 800 were built, the Ka-226 was first flown in prototype form on 4 September 1997. Series manufacture began in 2002 at the Strela plant in Orenburg although only the first 10 were built at this factory before production switched to the Kumertau Aviation Production Enterprise (KumAPP). Most Ka-226 production has been for Russian state organizations.

KA-226, CHKALOVSKY
A Kamov Ka-226 helicopter, in full camouflage, lands at Chkalovsky. Users of the helicopter have included the Ministry of the Interior, Ministry of Emergencies and Federal Border Service, with others going to Gazprom.

Guizhou JL-9/FTC-2000 (2003)

TYPE • *Trainer* **COUNTRY** • *China*

A remarkable example of incremental development, the JL-9 (aka the FTC-2000, the export variant) mates a 21st-century wing-and-glass cockpit to the mid-1950s designed rear fuselage and tail surfaces of the MiG-21, versions of which have been produced in China since 1964.

SPECIFICATIONS (JL-9)

DIMENSIONS:	Length: 14.55m (47ft 9in); Wingspan: 8.32m (27ft 4in); Height: 4.1m (13ft 5in)
WEIGHT:	Maximum take-off 9850kg (21,716lb)
POWERPLANT:	One 53.89 kN (12,110lbf) thrust dry (76.53 kN (17,200lbf) with afterburner WP-14C Kunlun-3 turbojet engine
MAX SPEED:	1480km/h (920mph)
RANGE:	2500km (1553 miles) with maximum external fuel
CEILING:	16,000m (52,490ft)
CREW:	2
ARMAMENT:	One 23mm (0.9in) machine gun fixed forward-firing in forward fuselage; five hardpoints for a total of up to 2000kg (4409lb) external stores, including missiles, bombs or fuel tanks

Left: Building on the JL-9 platform, the FTC-2000 featured a new double-delta wing and a forward fuselage with side air intakes. Later variants include a dedicated Naval version for carrying out simulated carrier landings for training PLAN pilots, the JL-9G Swordfish, which made its maiden flight during 2009.

AVIONICS
The nose of the aircraft contains the JL-10GJ X-band pulse-Doppler fire-control radar with a 30km (18.6 mile) range.

COCKPIT
A new stepped cockpit and canopy design gave both student and instructor much better forward and downward views than the JJ-7 and the crew benefitted from much improved avionics.

HARDPOINTS
The JL-9 has five hardpoints for external stores. The FTC-2000G variant has seven hardpoints and is capable of carrying 3000kg (6614lb) of stores.

TRAINING PLATFORM
Revealed by Guizhou in 2001, at Zhuhai airshow, the FTC-2000 (Fighter Trainer China-2000) programme began as a private venture to deliver an inexpensive but effective training platform for fourth-generation fighter aircraft. The Chengdu JJ-7 trainer, a two-seat training version of the J-7 fighter (a Chinese-produced development of the MiG-21), was utilized as a basis for the new aircraft. To minimize costs the engine, empennage and mechanical controls of the JJ-7 were retained.

FTC-2000, 2018
An FTC-2000 performs at the Zhuhai airshow (2018) in Zhuhai, south China. The FTC-2000's cheap price and ready availability has seen it achieve modest export success with small numbers sold to Myanmar and Sudan. The FTC-2000G multi-role fighter trainer was the type acquired by Myanmar and at the time of delivery in late 2022 it was believed to be the cheapest light fighter available in the world.

Eurofighter Typhoon (2003)

TYPE • *Multi-role fighter* **COUNTRY** • *UK/Germany/France/Italy/Spain*

The pan-European Typhoon emerged out of a collaborative effort by Germany, Italy, Spain and the United Kingdom to produce a highly capable fighter aircraft that would be available for service from the late 1990s.

SPECIFICATIONS

DIMENSIONS:	Length: 15.96m (52ft 4in); Wingspan: 10.95m (35ft 11in); Height: 5.28m (17ft 4in)
WEIGHT:	23,500kg (51,809lb) maximum take-off
POWERPLANT:	2 × Eurojet EJ200 afterburning turbofans, 90kN (20,000lbf)
MAX SPEED:	2125km/h (1321mph)
RANGE:	1390km (864 miles) radius with 3 drop tanks
CEILING:	14,500m (47,570ft)
CREW:	1/2
ARMAMENT:	1 × 27mm (1in) cannon and up to 8000kg (17,637lb) of disposable stores carried on 13 under-fuselage and underwing stations

This page: DA.2, serial ZH588, was the first British prototype, and is seen here armed with AIM-9L Sidewinder missiles.

RADAR
The Captor-M radar's wide field of regard has significant benefits in both air-to-air and air-to-surface engagements, offering considerable power and aperture for enhanced angular coverage.

AIRFRAME
Only 15 per cent of the Typhoon's surface is metallic, thereby enhancing stealth and protection against radar-based systems. In all, 75 per cent of the aircraft is fabricated from carbon-fibre composites.

ENGINES
The two-spool EJ200 uses a single-stage turbine driving a three-stage fan and five-stage compressor with annular combustion with vaporizing burners. It can cruise at supersonic speeds without afterburning.

ORIGINS
The origins of the Typhoon lie in the European Fighter Aircraft (EFA) initiative that explored the characteristics of the future platform including deciding upon an unstable aerodynamic configuration with canard foreplanes, an active digital fly-by-wire control system, 'hands on throttle and stick' controls for the pilot and an advanced cockpit. In terms of structure, the aircraft would make extensive use of carbon-fibre composites and other advanced materials. The Typhoon was intended to offer a considerable edge over the advanced variants of the MiG-29 and Su-27 then under development in the Soviet Union.

WEAPONS HARDPOINTS
A Typhoon displays its underwing and underbelly hardpoints as it makes a roll. The end of the Cold War saw the Eurofighter programme lose momentum, but the aircraft is now well established in service, combat proven and has achieved some lucrative export orders.

Shenyang J-11B (2003)

TYPE • Fighter   COUNTRY • China

SPECIFICATIONS

DIMENSIONS:	Length: 21.9m (69ft 6in); Wingspan: 14.7m (48ft 3in); Height: 5.92m (19ft 5in)
WEIGHT:	Maximum take-off 33,000kg (72,753lb)
POWERPLANT:	Two Shenyang WS-10A 'Taihang' afterburning turbofans, 132kN (30,000lbf) thrust each (Block 02)
MAX SPEED:	2500km/h (1553mph)
RANGE:	3530km (2190 mies)
CEILING:	19,000m (62,000ft)
CREW:	1
ARMAMENT:	One 30mm (1.18in) Gryazev-Shipunov GSh-30-1 cannon with 150 rounds; 12 hardpoints: two under fuselage, two under air ducts, four under wings, four on wingtips, with provisions to carry PL-12 and PL-15 air-to-air missiles

Whereas the J-11 and J-11A involved Chinese licence production of the Su-27, with the J-11B the PLAAF finally achieved its ambition of fielding a more capable fourth-generation 'Flanker' over which it had full control.

Above: This Shenyang J-11B, coded 70005, is thought to be from the 89th Brigade, Northern Theatre Command, based at Pulandian.

RADAR/FIRE CONTROL
The J-11B not only provided the PLAAF with a 'Flanker' having domestically produced engines but also Chinese avionics, including a Type 1493 multi-mode pulse-Doppler radar and fire control system.

COCKPIT
The J-11B also employs a Chinese-developed modern cockpit with digital displays, including five colour multifunction displays (MFDs) and a new wide-angle holographic head-up display (HUD).

ENGINES
The first batch of J-11Bs were grounded due to problems with their WS-10s and were ultimately re-engined with Russian-made AL-31Fs. By 2009, engine issues appear to have been resolved and subsequent J-11Bs were powered by the WS-10A as standard.

DEVELOPMENT
The J-11B series, which has the NATO reporting name 'Flanker-L', represents the full Chinese indigenization of the original Soviet-era Su-27 design. The project appears to have been launched in 2002 and, after testing some systems on a pair of modified J-11s, the first true J-11B prototype reportedly took to the air in June 2004. Production likely began sometime in 2006, although there seem to have been delays relating to both the new flight control system and the domestically produced WS-10 Taihang engines.

AIRSHOW MANOEUVRES
A J-11B fighter aircraft from the People's Liberation Army Air Force (PLAAF) performs at Dafangshen airport in China's Jilin province, during the biannual Changchun airshow in 2015.

Beriev Be-200 (2003)

TYPE • *Flying boat*   **COUNTRY** • *Russian Federation*

SPECIFICATIONS (Be-200ChS)

DIMENSIONS:	Length: 32.05m (105ft 2in); Wingspan: 32.78m (107ft 7in); Height: 8.9m (29ft 3in)
WEIGHT:	Maximum take-off 41,000kg (90,390lb)
POWERPLANT:	Two Ivchenko Progress D-436TP turbofans, each rated at 73.55kN (16,535lbf)
MAX SPEED:	700km/h (430mph)
RANGE:	3150km (1957 miles)
CEILING:	8100m (26,575ft)
CREW:	2
ARMAMENT:	Up to 57 passengers or 5000kg (11023lb) cargo or 12,000kg (26,455lb) water or retardant

A multi-purpose turbofan-powered amphibian, initially developed for search and rescue, the Beriev Be-200 is probably best known today as a civilian firefighting aircraft, although further variants are planned.

Above: '20 Yellow' is the first Be-200 built for Russian Navy use and delivered in February 2020. Named Aleksander Mamkin, the aircraft was lost with all eight on board on 14 August 2021 when it crashed while fighting fires in Turkey.

SAR EQUIPMENT
This includes a weather radar in the nose, an electro-optical turret under the port wing, a display at the observer's station in the cabin and a searchlight.

PAYLOAD
With a flight crew of two pilots in basic versions, plus optional observer and flight technician, the Be-200 can carry a payload of up to 5000kg (11,023lb) in the cabin.

BE-200ES
Azerbaijan became the first foreign customer for the Be-200 by purchasing a single example of the multi-role Be-200ES in April 2008. FHN-10201 is operated by the Azerbaijan Ministry of Emergency Situations as a fire-fighting, cargo or 43-seat passenger aircraft.

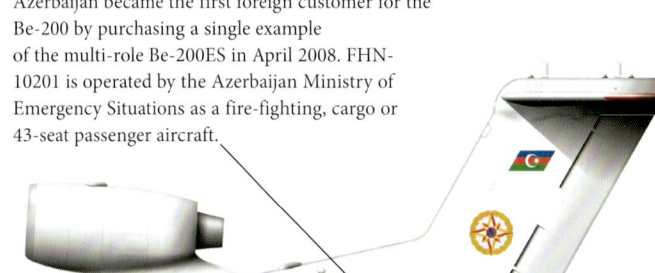

EMERGENCY ROLES
For SAR operations, the aircraft is capable of carrying up to 57 survivors/passengers in seats or up to 30 stretcher patients. Additionally, four droppable rescue capsules may be carried under the wing (containing dinghy or life raft, plus medical, radio and other equipment). As a firefighter, up to 12,000kg (26,455lb) of water is contained in eight tanks under the cabin floor. Water can be dropped in a single salvo within one second, via doors in the hull.

BERIEV BE-200, RUSSIA
A Beriev Be-200 amphibian plane lands. The Be-200 was developed by Beriev at Taganrog as a smaller version (roughly half the size) of the Beriev A-40 amphibian (NATO reporting name Mermaid) developed for the Soviet Navy during the 1980s.

Chengdu J-10 (2004)

TYPE • Fighter COUNTRY • China

Numerically the most important fully indigenous fighter in PLAAF use, the J-10 is nothing less than a milestone in China's military aviation, as the country's first truly fourth-generation multi-role fighter to enter production and service.

SPECIFICATIONS

Dimensions:	Length: 16.03m (52ft 7in); Wingspan: 9.25m (30ft 4in); Height: 5.43m (17ft 10in)
Weight:	Maximum take-off 20,500kg (45,195lb)
Powerplant:	One Saturn AL-31FN turbofan engine rated at 79.43kN (17,860lbf) of thrust with afterburning
Max speed:	Mach 2.1 (2593km/h)
Range:	2250km (1400 miles)
Ceiling:	17,000m (56,000ft)
Crew:	1
Armament:	One 23mm (0.9in) twin-barrel GSh-23 cannon plus a maximum of 6,800kg (15,400lb) of disposable stores carried on 11 hardpoints

Left: Minor modifications to the single-seater and two-seater versions of the J-10A produced the J-10AY and J-10SY subvariants that were operated by the PLAAF's aerial demonstration team, known as August 1 (or Ba Yi).

Avionics
As well as the Type 1473 pulse-Doppler fire-control radar, avionics includes GPS/inertial navigation system, air data computer, radar warning receiver, digital fuel management system, mission management system and an ARINC429 databus for stores management.

Cockpit
The J-10's advanced cockpit is equipped with a wide-angle head-up display (HUD), two monochrome multifunctional displays (MFDs) and one colour MFD. The pilot is provided with a helmet-mounted sight and 'hands on throttle and stick' (HOTAS) controls.

Chengdu J-10B
A J-10B, serial number 10537, of the PLAAF's 5th Regiment, 2nd Division, based at Guilin, China.

VARIANTS
Three batches of the initial-production J-10 were swiftly followed by the J-10A with various minor avionics improvements including the more capable Type 1473G fire-control radar and a revised cockpit. Original J-10s were later modified to the same standards. A derivative of the J-10A is the J-10AH for land-based naval service. There is also a tandem-seat trainer version of the J-10A, designated J-10AS, which features a prominent single canopy and a large dorsal spine to accommodate the electronics displaced by the rear cockpit. In naval use, the trainer version is the J-10ASH.

DEMONSTRATION TEAM J-10
In a surprise move, a further modified aircraft for the demonstration team appeared in mid-2022. This features a prominent enlarged fuselage spine, leading to early suggestions that it may have been a dedicated SEAD model, tentatively dubbed J-10D. In fact, the 'big spine' J-10 is thought to be designated J-10CY and is a new version of the J-10C variant intended specifically for service with the August 1 team.

Antonov An-148 (2004)

TYPE • *Tactical transport* COUNTRY • *Russian Federation*

The Antonov An-148 was developed as a commercial regional jet and used by the Russian Ministry of Defence as an executive transport. Ukrainian production terminated after the breakdown in Russian–Ukrainian relations after 2014.

SPECIFICATIONS

Dimensions: Length: 29.13m (95ft 7in); Wingspan: 28.91m (94ft 10in); Height: 8.19m (26ft 10in)

Weight: Maximum take-off 43,700kg (96,342lb)

Powerplant: Two Ivchenko D-436-148 turbofans, each rated at 66.97kN (15,058lbf)

Max speed: 870km/h (541mph)

Range: 4400km (2734 miles)

Ceiling: 12,200m (40,026ft)

Crew: 2

Armament: Up to 85 passengers or 9000kg (20,000lb) cargo

Left: The interior cockpit of an Antonov AN-148-100V twin-engine turbojet aircraft, photographed in winter in St Petersburg, Russia.

Capacity
The An-148-100E has a maximum weight of 43,700kg (96,342lb), allowing it to carry 75 passengers 4400km (2734 miles). The first aircraft built to this order was handed over to the Russian Ministry of Defence in December 2013.

Powerplant
Suspended from the two high-set wings are two Ivchenko D-436-148 turbofans, each rated at 66.97kN (15,058lbf).

An-148-100E
RA-61728 was delivered to the Russian Air Force in November 2016 and is on the strength of 800 AVB. The last Russian-built An-148 was completed in October 2018.

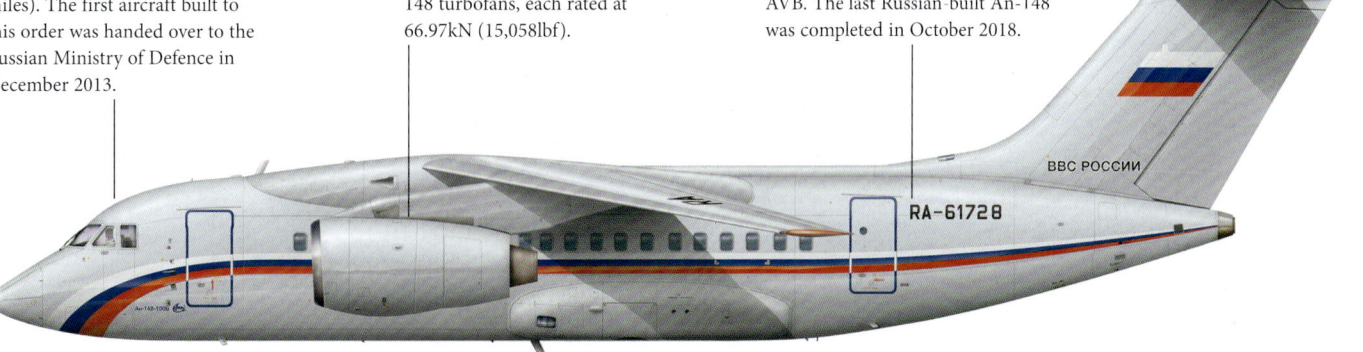

SERIES PRODUCTION
The An-148 programme was launched in 2001 to develop a successor to the Tu-134 regional jet, with the first flight taking place on 17 December 2004 in Kyiv. Series production has been undertaken by both Antonov in Kyiv, Ukraine and Voronezh Aircraft Production Association (VASO) in Voronezh, Russia. The factories co-operated in production with final assembly lines at both facilities. Initially planned as a commercial aircraft, the An-148 has mainly been ordered by Russian state operators.

AN-148 TEST FLIGHT
A brand-new An-148 is taken for a test flight in 2011. The initial variants were the An-148-100, the basic airliner, which was followed by the An-148-100A and 100B that both featured the same passenger capacity but featured a higher maximum weight allowing for greater range capability.

Shaanxi Y-8 and Y-9 Special Mission (2004)

TYPE • *Tactical transport* COUNTRY • *China*

SPECIFICATIONS (Shaanxi Y-8C)

DIMENSIONS: Length: 34m (111ft 6in); Wingspan: 38m (124ft 7in); Height: 11.16m (36ft 7in)

WEIGHT: Maximum take-off 61,000kg (134,100lb)

POWERPLANT: Four Zhuzhou WoJiang WJ-6 turboprops, developing 3170kW (4250hp)

MAX SPEED: 660km/h (410mph)

RANGE: 5615km (3489 miles)

CEILING: 10,400m (34,120ft)

CREW: 5

ARMAMENT: Two 23mm (0.9in) cannon

China embarked on domestic development of the Soviet-designed Antonov An-12, as the Shaanxi Y-8, which has had an illustrious career. During this time numerous improvements have been made, ultimately leading to the entirely reworked Y-9.

Above: The Y-9Q is the anti-submarine warfare (ASW) variant of the type.

FAIRINGS
The Y-8GX-3 has two large 'hamster cheek' fairings on the sides of the forward fuselage that likely house ECM antennas to provide long-range stand-off electronic jamming capabilities.

SHAANXI Y-8GX-3
The Y-8GX-3 is also known as the Y-8G and has the Western reporting name 'Mouse'. This Y-8GX-3 long-range electronic countermeasures (ECM) aircraft, serial number 30518, was seen in October 2020.

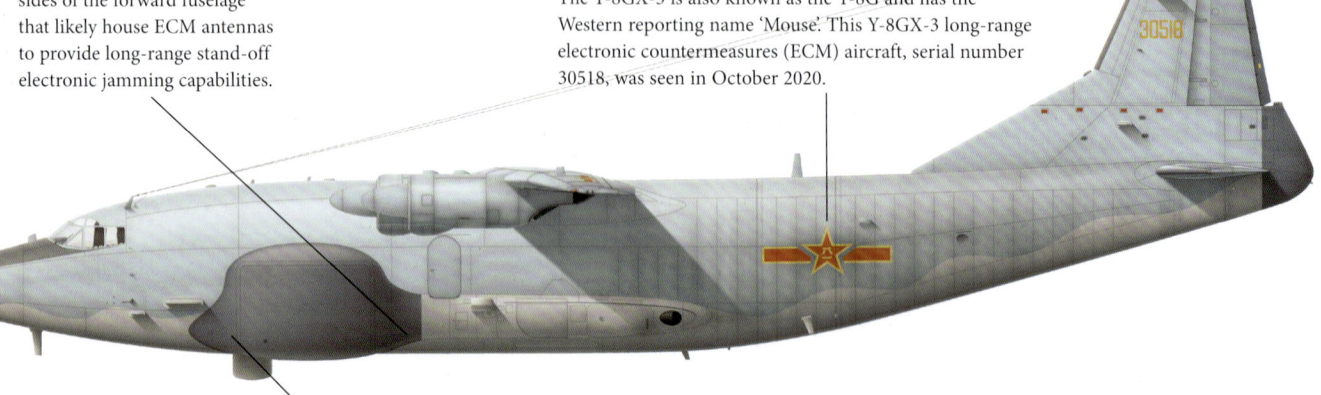

STAND-OFF JAMMER
The type includes SOJ (stand-off jammer) capabilities, which are designed to confuse and deceive enemy radar signals and communication systems.

SPECIAL MISSION VERSIONS

Somewhat confusingly, special mission versions based on both the Y-8 and the new-generation Y-9 transports use designations in the Y-8 Gaoxin, or High New programme series. Types under this category include: Y-8J 'Mask' airborne early warning variant; Y-8GX-7 psychological warfare version; and Y-8GX-8 electronic intelligence (ELINT) aircraft, among many others.

SHAANXI Y-8 LOADING
The basic Y-8 transport can carry up to 20,000kg (44,092lb) of cargo. Alternative loads include up to 96 soldiers or 82 paratroopers in the main cargo compartment. The aircraft can also be configured for medical evacuation (medevac), in which capacity it can carry 60 soldier casualties on stretchers, or 20 walking wounded plus three medical attendants.

Lockheed Martin F-22 Raptor (2005)

TYPE • *Stealth fighter* COUNTRY • *United States*

Widely regarded as the most capable air superiority fighter in service anywhere in the world, the F-22 is capable of both air-to-air and air-to-ground missions. It has been designed to combine stealth, performance, agility and integrated avionics in a single airframe.

SPECIFICATIONS (F-22A)

DIMENSIONS:	Length: 18.92m (62ft 1in); Wingspan: 13.56m (44ft 6in); Height: 5.08m (16ft 8in)
WEIGHT:	Maximum take-off 37,875kg (83,500lb)
POWERPLANT:	Two Pratt & Whitney F119-PW-100 turbofans, each rated at 156kN (35,000lbf) of thrust with afterburning
MAX SPEED:	Mach 2.25 (2778km/h)
RANGE:	2897km (1800 miles) with two external fuel tanks
CEILING:	19,812m (65,000ft)
CREW:	1
ARMAMENT:	One 20mm (0.787in) M61A2 Vulcan six-barrel Gatling cannon, plus up to eight air-to-air missiles carried in three internal bays, or an equivalent load of precision-guided air-to-ground munitions

Left: The two Pratt & Whitney F119 engines allow the Raptor to accelerate and cruise at speeds of about Mach 1.8 without using afterburners.

DATALINK
The pilot can call upon the Raptor's AN/ALR-94 passive receiver system to track an emitting target without having to reveal the F-22's presence, while targeting information can also be acquired from other platforms via a fighter datalink.

AIRFRAME
The low-observable characteristics of the F-22 are ensured through the use of a clean, angular airframe configuration, with jagged edges on any panels that may reflect electro-magnetic energy back to a hostile radar.

MANOEUVRABILITY
High manoeuvrability is achieved through a triplex fly-by-wire flight control system, two-dimensional thrust vectoring nozzles on the twin Pratt & Whitney F119 turbofan engines and an airframe that incorporates negative static stability.

ATTACK AIRCRAFT
Although originally planned as an air dominance fighter, the F-22 has latterly emerged as a powerful attack aircraft. In the air-to-ground configuration, the aircraft can carry two 907kg (2000lb) GBU-32 joint direct attack munitions (JDAM) internally and up to eight GBU-39 small diameter bombs (SDBs). Whether carrying JDAMs or SDBs, the Raptor can also carry two AIM-120s and two AIM-9s for self-defence. The F-22A made its combat debut in this role, striking targets in Syria in September 2014.

F-22 RAPTORS OVER HAWAII
Two F-22 Raptors fly over Joint Base Pearl Harbor-Hickam, Hawaii, in 2019. While the Raptor currently rules the roost in the US Air Force fighter community, the service is well into planning for its replacement under the multifaceted Next Generation Air Dominance (NGAD) programme, which is expected to provide both manned and unmanned platforms that will offer capabilities in excess of even the F-22.

T-50 Golden Eagle (2005)

TYPE • *Trainer* COUNTRY • *South Korea*

The Korea Aerospace Industries (KAI) T-50 family includes supersonic advanced jet trainers, combat-capable trainers and light combat aircraft. It bears some similarity to the F-16 with Lockheed Martin having provided input into the design of the South Korean jet.

SPECIFICATIONS

Dimensions: Length: 13.14m (43ft 1in); Wingspan: 9.45m (31ft) with wingtip missiles; Height: 4.94m (16ft 2in)

Weight: Maximum take-off 12,300kg (27,117lb)

Powerplant: Two General Electric F404 turbofans each providing 78.7kN (17,700lbf) of thrust with afterburning

Max speed: Mach 1.2

Range: 1851km (1150 miles)

Ceiling: 14,630m (48,000ft)

Crew: 2

Armament: One three-barrelled version of the M61 Vulcan 20mm (0.787in) rotary cannon, plus a range of ordnance including bombs, rockets and missiles on five external hardpoints and two wingtip missile rails

Left: A pair of Indonesian Air Force T-50i Golden Eagles taxi at Halim Perdanakusuma Air Force Base in East Jakarta.

T-50I
The Indonesian Air Force received 16 T-50i jets, the first of which entered service in September 2013. Although designated as T-50 trainers, all 16 Indonesian examples are fully combat capable, making them equivalent to TA-50s.

Tandem seats
The TA-150 has tandem seats for its two pilots beneath the high-mounted canopy, which provides excellent visibility.

FA-50 VARIANT
Very similar to the TA-50 is the FA-50, a dedicated light combat version, with increased internal fuel capacity, improved radar and electronic warfare systems, a Link 16 tactical datalink and compatibility with a wider range of weapons. South Korea placed orders for a reported 60 FA-50s, which are in the process of replacing a portion of its ageing F-5E/F Tiger IIs. Precision-guided weapons options include the AGM-65 Maverick air-to-ground missile and GBU-38/B joint direct attack munitions (JDAM).

Munitions
Precision-guided weapons options include the AGM-65 Maverick air-to-ground missile and GBU-38/B joint direct attack munitions (JDAM).

EMERGENCY LANDING
Demonstrating its flexible flight characteristics, a South Korean Air Force TA-50 participates in an emergency landing and take-off exercise on a controlled public road in 2013.

Bombardier CRJ (2005)

TYPE • Tactical transport COUNTRY • China

SPECIFICATIONS (Challenger 600)

DIMENSIONS:	Length: 26.77m (87ft 10in); Wingspan: 21.21m (69ft 7in); Height: 6.22m (20ft 5in)
WEIGHT:	Maximum take-off 23,133kg (51,000lb)
POWERPLANT:	Two General Electric CF34-3B1 turbofan engines, 38.84kN (8730lbf) thrust each
CRUISE SPEED:	819km/h (509mph)
RANGE:	5,206km (3,235 miles)
CEILING:	12,500m (41,000ft)
CREW:	2 + 1
ARMAMENT:	None

Several examples of the popular twin-engine Canadair Regional Jet (CRJ) series, produced by the Canadian Bombardier company, are operated by the 100th Air Regiment, which is the PLAAF's VIP transport arm.

Above: This People's Liberation Army Naval Air Force (PLANAF) Bombardier Challenger 600, registration B-4701, was seen at Shanghai Pudong Airport in July 2022.

BOMBARDIER CL-600-2C10 CHALLENGER 870
This Bombardier Challenger 870, registration B-4067, was first delivered to the PLAAF in September 2014.

PASSENGERS
The internal cabin seating can be configured for anywhere between 18 and 30 passengers. The CRJ700 variant has seating for a maximum of 78 passengers.

POWERPLANT
As with the conventional airliner, the military aircraft is powered by two General Electric CF34-3B1 turbofan engines of 38.84kN (8730lbf) thrust each, mounted either side of the rear fuselage.

VIP TRANSPORT

In Chinese military service, these smaller business jets complement the larger, longer-range A319 and Boeing 737 and are used to move high-ranking personnel and government officials over shorter distances, primarily within China. The PLAAF operates five CRJ200ER aircraft, which are also known as CL-600-2B19 Challenger 600s. The Chinese military also flies a dozen examples of the CRJ700, which are also known as CL-600-2C10 Challenger 870s.

CIVILIAN AIRLINER
A stretched derivative of the CRJ200, the CRJ700 offers seating for a maximum of 78 passengers, in its regional airliner version, although the configuration used by the PLAAF is unclear. Additionally, the CRJ700 features new wings with leading-edge slats and a slightly widened fuselage with a lowered cabin floor.

Shaanxi KJ-200 (2005)

TYPE • *AEW&C* COUNTRY • *China*

Fitted with a characteristic 'balance beam' radar above its fuselage, the KJ-200 (under the Western reporting name 'Moth') is one of a handful of different airborne early warning and control (AEW&C) aircraft types in the PLAAF inventory.

SPECIFICATIONS (KJ-200A)

DIMENSIONS:	Length: 34m (111ft 6in); Wingspan: 38m (124ft 8in); Height: 11m (36ft 1in)
WEIGHT:	Maximum take-off 61,000kg (134,482lb)
POWERPLANT:	Four Zhuzhou WoJiang-6 turboprop engines developing 3169kW (4250hp) each
MAX SPEED:	660km/h (410mph)
RANGE:	5500km (3418 miles)
CEILING:	10,400m (34,121ft)
CREW:	2 plus 5 mission crew
ARMAMENT:	None

Below: The KJ-200H is the People's Liberation Army Navy (PLAN) version of the model, with low-visibility finish and insignia.

COMMUNICATIONS ANTENNAS
A series of antennas arranged above the forward fuselage are thought to be used for VHF and UHF communications, while likely communications intelligence (COMINT) antennas are located below the forward fuselage.

KJ-200A
This KJ-200A, serial number 30672, is designed for airborne early warning and control (AEW&C) duties, and is part of the 26th Division, PLAAF.

RADAR ANTENNA
The JY-06 active electronically scanned array (AESA) radar is housed in a prominent fairing mounted on top of the fuselage, similar to the Swedish Saab Erieye.

CHINESE SERVICE
The loss of the second prototype in an accident in 2006 appears to have led to delays in the KJ-200 programme. The findings of the accident also appear to have led to some redesign work. But the KJ-200 eventually entered PLAAF service with the 76th Electric Warfare Regiment within the PLAAF's 26th Special Mission Division at Wuxi/Shuofang, serving alongside the larger KJ-2000. Examples are also flown by the People's Liberation Army Navy.

KJ-200 TAXIING
In late 2016 an improved version of the aircraft appeared, known as the KJ-200A. This is distinguished by a revised nose containing a new (reportedly weather) radar, rather than the previous 'Pinocchio'-type thimble radome. Existing KJ-200s are apparently being modified to KJ-200A standard.

Xi'an KJ-2000 (2005)

TYPE • *AEW&C* **COUNTRY** • *China*

SPECIFICATIONS

DIMENSIONS:	Length: 46.6m (152ft 8in); Wingspan: 50.5m (165ft 7in); Height: 14.8m (48ft 6in)
WEIGHT:	Maximum take-off 175,000kg (385,809lb)
POWERPLANT:	Four PS-90A turbofan engines developing 157kN (35,300lbf)
MAX SPEED:	900km/h (560mph)
RANGE:	5500km (3418 miles)
CEILING:	10,200m (6388 miles)
CREW:	5 (flight crew); up to 11 mission crew
ARMAMENT:	None

Known to Western intelligence as 'Mainring', the KJ-2000 is China's first and, to date, largest dedicated airborne early warning and control (AEW&C) aircraft. Its early production was disrupted by issues surrounding the development of its radar.

Below: The KJ-2000 is an early warning and control aircraft (AEW&C), made up of domestically designed electronics and radars installed on a modified Ilyushin Il-76 airframe.

PHASED ARRAY RADAR
The Type 88 comprises three active electronically scanned array (AESA) antennas arranged in a triangular configuration within a fixed rotodome.

KJ-2000
As well as the main radar, carried in a rotodome, the KJ-2000 has received a Chinese-made command, control, communications, computers, intelligence, surveillance and reconnaissance (C4ISR) system, including identification friend or foe (IFF) and datalink systems.

AVIONICS
The aircraft's sensors have a detection range against fighter-sized targets of 470km (290 miles); range against ballistic missiles: 1200km (750 miles); maximum simultaneous targets tracked are 60–100.

RADAR
In 1999, the first A-50I prototype was delivered to Israel where it was planned to receive its radar. However, apparently in response to US pressure, Israel withdrew from the deal, leaving the A-50I without radar and stranded temporarily in Israel. China then had to develop its own radar, known as the Type 88, and the A-50I then arrived in China in 2002 to have it installed.

KJ-2000
China's KJ-2000 airborne early warning and control system aircraft performs at the Airshow China 2014 in Zhuhai, south China's Guangdong province on 11 November 2014.

Leonardo C-27J Spartan (2006)

TYPE • *Tactical transport* COUNTRY • *Italy*

SPECIFICATIONS

DIMENSIONS:	Length: 22.7m (74ft 6in); Wingspan: 28.7m (94ft 2in); Height: 9.64m (31ft 8in)
WEIGHT:	Maximum take-off 21,000kg (46,297lb)
POWERPLANT:	Two Rolls-Royce AE2100-D2A turboprops each rated at 3458 kW (4637hp)
MAX SPEED:	602km/h (374mph)
RANGE:	5852km (3636 miles), ferry
CEILING:	9144m (30,000ft)
CREW:	2 (plus optional loadmaster)
ARMAMENT:	None

In service with 16 operators, the C-27J is a thoroughly modernized version of the Aeritalia G.222 transport designed in the 1970s that was optimized for short take-off and landing operations from rough airstrips.

Left: A Slovak Air Force Leonardo C-27J shows its capabilites during an air show. The manufacturer claims that the C-27J offers the largest cargo bay in its class as well as the best descent and climb rate.

COCKPIT
The aircraft features a glass cockpit that reduces crew workload, with five colour multifunction displays, a radar and a comprehensive communication suite.

CARRYING CAPACITY
The C-27J is a spacious transporter, with capacityfor 60 troops, 46 paratroopers, 36 stretcher patients or equivalent cargo load.

C-27J IMPROVEMENTS
The C-27J differs primarily from its predecessor in having new Rolls-Royce AE 2100 engines driving six-blade propellers and also features a fully digital avionics suite. Compared to the G.222, the new aircraft offers a 35 per cent increase in range and a 15 per cent faster cruising speed. First flown in prototype form on 24 September 1999, the C-27J is configured primarily for tactical airlift missions, but can be rapidly reconfigured to undertake other roles, including medical evacuation.

AUSTRALIAN C-27J SPARTAN
The Royal Australian Air Force's No. 35 Squadron operates 10 C-27Js. Acquired via US Foreign Military Sales channels, the first RAAF C-27J took to the air from Alenia Aermacchi's facility in Turin in December 2013.

SLOVAKIAN AIR FORCE C-27J
The C-27J has enjoyed solid export sales, largely on account of its flexible platform. Thanks to a range of roll-on/roll-off kits, the aircraft can also be adapted for firefighting, intelligence, surveillance and reconnaissance (ISR), maritime patrol and gunship missions.

Hongdu JL-10 (2006)

TYPE • *Trainer* COUNTRY • *China*

An advanced trainer and light combat aircraft, the JL-10 was developed with the assistance of Yakovlev OKB in Russia. Both subsonic and supersonic variants have been produced and the type is expected to replace both JL-8 and JL-9 in PLAAF service.

SPECIFICATIONS

DIMENSIONS:	Length: 2.4m (40ft 8in); Wingspan: 9.4m (30ft 10in); Height: 4.7m (15ft 5in)
WEIGHT:	Maximum take-off 1600kg (25,574lb)
POWERPLANT:	Two 24.7kN (5553lbf) static thrust dry (41.2kN/9262lbf) with afterburner) Ivchenko-Progress AI-222K-25F turbofan engines
MAX SPEED:	1730km/h (1074mph)
RANGE:	2600km (1600 miles) with maximum external fuel
CEILING:	16,000m (52,000ft)
CREW:	2
ARMAMENT:	Nine hardpoints for a total of up to 3000kg (6618lb) external stores including missiles, bombs or fuel tanks

Left: The JL-10 is undergoing continual modifications and upgrades. For example, further development has seen the development of a naval variant with strengthened twin-wheel nose gear plus launch bar and tail hook, which is believed to be designated JL-10J and will serve aboard the new carrier Fujian.

HARDPOINTS
This JL-10 can carry up to 3000kg (6618lb) of external stores. The newer L-15B variant is able to carry a total payload of 3500kg (7716lb), is capable of Mach 1.4 and boasts a shorter take-off and landing distance.

COCKPIT
The aircraft was designed from the outset with advanced features such as a glass cockpit, this being compatible with night-vision goggles so the pilot can conduct more effective operations in darkness conditions.

WING DESIGN
Like the Yak-130, the JL-10 is equipped with distinctive leading-edge extensions (LERX) and a large vertical fin and rudder, both features designed to impart excellent control at high angles of attack.

YAKOVLEV INFLUENCE
The Hongdu JL-10 was a clean-sheet design intended to feature such elements as fly-by-wire and HOTAS control for the student and instructor. In 2000, the Yakovlev Design Bureau was contracted as a technical and scientific consultancy for the aircraft, then known as the L-15, a designation it retains for export versions. Yakovlev's influence in the design is readily apparent as the aircraft bears a strong resemblance to the Yak-130, over 100 of which serve with Russian Aerospace Forces, and notably utilizes the same engines as the Russian aircraft.

FLARE COUNTERMEASURE
A Hongdu JL-10 trainer jet performs during the 2022 China Aviation Industry Conference and Nanchang Air Show on 26 November 2022, dispersing multiple flares as a countermeasure against infrared homing missiles.

General Dynamics F-16E/F (2007)

TYPE • *Multi-role fighter* **COUNTRY** • *United States*

The F-16E/F (or Block 60) iterations of the Fighting Falcon/Viper emerged from several studies aimed at winning orders from the United Arab Emirates (UAE). Collectively known as Desert Falcons, the F-16E is a single-seater while the F-16F is a two-seater.

SPECIFICATIONS
(F-16 Block 60)

DIMENSIONS:	Length: 15.04m (49ft 4in); Wingspan: 9.4488m (31ft); Height: 5.09m (16ft 8.5in)
WEIGHT:	Maximum takeoff 20,865kg (46,000lb)
POWERPLANT:	One General Electric F110-GE-132 turbofan, 144.57kN (32,500lbf)
MAX SPEED:	Mach 2.02 (2447k/mh)
RANGE:	Maximum ferry range with external fuel: 3700km (2300 miles)
CEILING:	18,000m (60,000ft)
CREW:	2
ARMAMENT:	One M61A1 20mm (0.787in) cannon; six under-wing hardpoints, three fuselage hardpoints, two wingtip pylons for various bombs, AAMs, ASMs and rocket pods

Below: A Lockheed Martin F-16E Block 60, 3056, serial number 00-6031, from the 1st Squadron, UAEAF.

AVIONICS
The avionics suite of the Desert Falcon includes a Northrop Grumman AN/APG-80 multi-mode active electronically scanned array (AESA) agile-beam radar, which is claimed to offer a range of around three times greater that of a US Air Force Block 50 F-16C.

ARMAMENT
Typical weapons of this type include: AIM-9 Sidewinder, AIM-7 Sparrow, AIM-120 AMRAAM, IRIS-T, Python 4 air-to-air missiles; ground attack: 70mm unguided rocket pods, AGM-65 Maverick missile, Mk82, Mk83, Mk84 guided or unguided bombs, BGU-39 small diameter bomb.

F-16F BLOCK 60
A Lockheed Martin F-16F Block 60 3023 (serial no. 00-6078), from the 2nd Squadron United Arab Emirates Air Force. The UAE opted for the F-16 but in an orthodox configuration, although the suite of sensors and weapons were among the most advanced of any of the Vipers.

INTO SERVICE
The first Block 60 F-16 took to the air at Fort Worth on 6 December 2003, with a phased development programme that ensured the first aircraft could be handed over to the UAE even before the completion of system integration and flight testing. The UAE ordered 80 Block 60 aircraft, comprising 55 single-seat F-16Es and 25 two-seat F-16Fs. In this way, the initial Standard 0 aircraft began to be handed over in September 2004 to begin training of UAE aircrews in the United States. In May 2005, the first Standard 1 aircraft with basic operational capabilities was delivered.

UAE F-16E, ARIZONA
A United Arab Emirates F-16E Desert Falcon, Block 60, flies over southern Arizona during a Red Flag event at Nellis Air Force Base, Nevada. Such events give the opportunity to test defensive technologies – the Desert Falcon is notably well protected against enemy air defence systems, using the Northrop Grumman Falcon Edge integrated electronic countermeasures system.

PAC JF-17 Thunder (2007)

TYPE • *Fighter*   COUNTRY • *China/Pakistan*

The JF-17 fighter was developed jointly by China's Chengdu and the Pakistan Aeronautical Complex (PAC), the aircraft being tailored to the Pakistan Air Force's requirement for a successor to its ageing Chengdu F-7, Dassault Mirage, Nanchang A-5 and Shenyang F-6 fleets.

SPECIFICATIONS (JF-17A Block 2)

DIMENSIONS:	Length: 14.93m (49ft); Wingspan: 9.44m (31ft); Height: 4.77m (15ft 8in)
WEIGHT:	Maximum take-off 12,384kg (27,302lb)
POWERPLANT:	One Klimov RD-93 turbofan engine rated at 84.4kN (19,000lbf) thrust with afterburner
MAX SPEED:	Mach 1.6 (1976km/h)
RANGE:	1352km (840 miles)
CEILING:	16,920m (55,510ft)
CREW:	1
ARMAMENT:	One 23mm (0.787in) twin-barrel GSh-23 cannon plus up to 1500kg (3300lb) of disposable stores carried on seven external hardpoints

Above: This Block I Thunder is operated by the Pakistan Air Force's Combat Commander School (CCS), which includes a JF-17 squadron at PAF Base Mushaf. The PAF had received eight pre-production by mid-March 2008, but the Thunder did not formally enter service until early 2010, when 26 Squadron was re-equipped with the type at PAF Base Kamra.

RADAR
The initial radar is a Chinese KLJ-7(V)2 radar of the mechanically steered type, which was also codeveloped and coproduced by PAC.

JF-17 THUNDER
The first export operator to receive the JF-17 was Myanmar, although few details about its fleet are available. The type was apparently introduced to serve with the Myanmar Air Force in late 2018.

POWERPLANT
The aircraft is powered by a Russian-designed RD-93 turbofan engine, an improved version of the RD-33 that is found in the MiG-29.

BLOCK 3
While the original Block 1 jets were all single-seaters, the Block 2s are provided in both single-seat JF-17A and two-seat JF-17B versions, the latter also being used for lead-in fighter training. Most recently, Chengdu and PAC have developed a Block 3 version of the single-seat JF-17A, which introduces a KLJ-7A active electronically scanned array (AESA) radar, as well as an improved fly-by-wire flight-control system, infrared search and track system, helmet-mounted display and a larger holographic wide-angle head-up display for the pilot.

PAKISTAN AIR FORCE JF-17
A Pakistan Air Force PAC JF-17 Thunder fighter jet aircraft taxis to the runway at the Paris Air Show in 2019. As well as steadily re-equipping the PAF fighter force, the JF-17 has found some export success with aircraft having been delivered to Myanmar and Nigeria.

Mil Mi-28 (2007)

TYPE • Helicopter **COUNTRY** • Russian Federation

Compared to the earlier Mi-24, the Mi-28 (NATO reporting name 'Havoc') combat helicopter is optimized for the anti-tank mission, possessing improved manoeuvrability, performance and battlefield survivability.

SPECIFICATIONS (Mil Mi-28N Havoc)

DIMENSIONS:	Fuselage length: 16.88m (55ft 5in); Rotor diameter: 17.2m (56ft 5in); Height without radar: 3.82m (12ft 7in)
WEIGHT:	Maximum take-off 11,500kg (25,300lb)
POWERPLANT:	Two Klimov/St Petersburg VK-2500 turboshafts, each rated at 1641kW (2200shp)
MAX SPEED:	280km/h (174mph)
RANGE:	200km (120 miles)
CEILING:	5000m (16,400ft)
CREW:	2
ARMAMENT:	One 30mm (1.2in) Shipunov 2A42 autocannon flexibly mounted under nose; up to 2350kg (5181lb) weapons and stores

Above: '04 White' features the overall grey finish applied to some Mi-28s, a finish that makes it look even more like the Boeing AH-64 Apache it superficially resembles. Compared to its US counterpart, the Mi-28 is slower but has a more powerful main gun and is better armoured.

CANNON
Designed for ground attack against personnel, positions and light vehicles, one single-barrel 30mm (1.8in) 2A42-2 cannon is mounted in an under-nose turret with 250 rounds.

MI-28 HAVOC-A
'022 Yellow' was the second prototype Mil-28, completed in 1983 and differed from later examples with its three-blade tail rotor. Three further prototypes were built to Mi-28A standard before development stalled in the early 1990s.

MODERNIZATION
The latest development intended for Russian service is the Mi-28NM, which features a new avionics suite, including new radar, electro-optical turret, laser missile homing system compatible with new versions of Ataka plus Khrizantema missiles and a new self-defence suite. Powered by VK-2500P engines with 'hot and high' capability, an initial test batch of eight much-modernized Mi-28NMs was ordered by Russia in December 2017 with plans to buy another 98.

WEAPON PYLONS
Mil-28s carry their stores on four underwing pylons. The standard anti-tank configuration comprises eight (or a maximum of 16) 9M120 Ataka-VN (AT-9 Spiral-2) anti-tank guided missiles carried in eight-tube clusters.

MiL MI-28N
Production Mi-28N helicopters began to be delivered by Rostvertol to the-then Russian Air Force in January 2008 when delivery was accepted of the first two production Mi-28Ns. These entered service with the evaluation and operational conversion centre in Torzhok.

MQ-8B Fire Scout (2009)

TYPE • UAV COUNTRY • United States

SPECIFICATIONS

DIMENSIONS: Length: 7.3m (24ft); Rotor diameter: 8.4m (27ft 6in); Height: 2.9m (9ft 8in)
WEIGHT: 1430kg (3150lb)
POWERPLANT: Rolls-Royce 250, 313kW (420hp)
MAX SPEED: 213km/h (132mph)
RANGE: 8 hours endurance
CEILING: 6100m (20,000ft)
CREW: N/A
ARMAMENT: Options for rocket pods, ASMs and precision-guided bombs

Fire Scout was developed for the US Navy to provide UAV functions such as surveillance and reconnaissance, targeting information (for both land and naval forces) and fire support, but it is mainly intended for non-combat duties.

Left: An MQ-8C Fire Scout, attached to the 'Wildcards' of Helicopter Sea Combat (HSC) Squadron 23, assigned to the Independence-variant littoral combat ship USS Jackson *(LCS 6) prepares to take off over the South China Sea in 2022.*

RADAR TRACKING
The Fire Scout drone's radar can track multiple contacts at once. It can also be fitted with a landmine/IED aerial detection system.

ROTOR
The original RQ-8A Fire Scout had a three-blade rotor, but performance was improved dramatically in the MQ-8B's four-blade rotor.

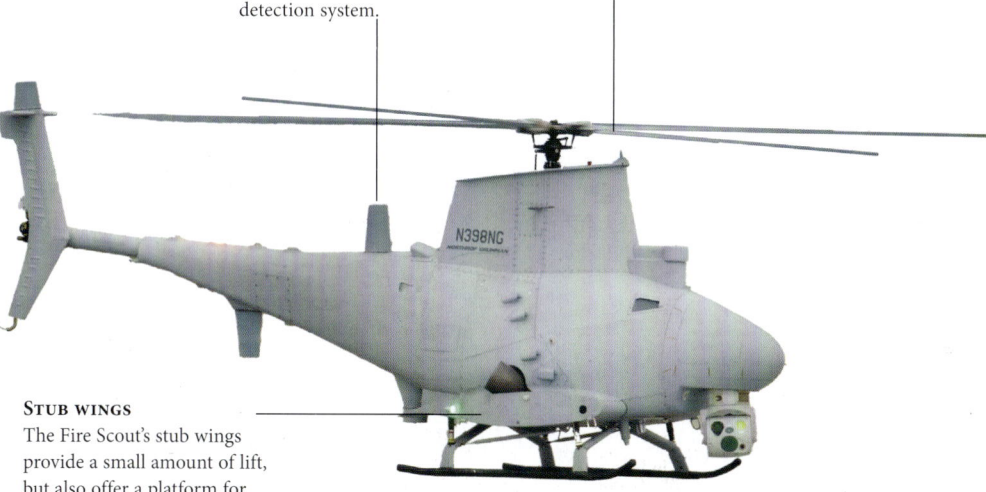

STUB WINGS
The Fire Scout's stub wings provide a small amount of lift, but also offer a platform for mounting weapon systems and electronics pods.

DEVELOPING RECORD

Fire Scout can fly at over 200km/h (124mph) and stay in the air for more than five hours with a full load. With a lighter load it can fly for up to eight hours. US Navy shipboard deployments of the Fire Scout began in 2008. The type subsequently saw operational service over Afghanistan, Libya (where one was shot down) and off the coast of Africa, the latter in the context of anti-piracy options. By 2013, the Fire Scout had clocked up more than 8000 hours of flight time, including 5000 hours in Afghanistan.

US NAVY RQ-8A
An RQ-8A Fire Scout prepares for the first autonomous landing aboard the amphibious transport dock ship USS *Nashville* (LPD 13) on 17 January 2006. With an on station endurance of over four hours, the Fire Scout system is capable of continuous operations.

Boeing 737 AEW&C (2010)

TYPE • *AEW&C* COUNTRY • *United States*

SPECIFICATIONS

DIMENSIONS:	Length: 33.6m (110ft 4in); Wingspan: 35.8m (117ft 2in); Height 12.5m (41ft 2in)
WEIGHT:	Maximum take-off 77,565kg (171,000lb)
POWERPLANT:	Two CFM International CFM56-7B27A turbofans each rated at 121kN (27,300lbf) thrust
CRUISE SPEED:	853km/h (530mph)
RANGE:	6500km (4000 miles)
CEILING:	12,500m (41,000ft)
CREW:	Two, plus 6–10 mission personnel
ARMAMENT:	None

Boeing's latest airborne early warning and control (AEW&C) aircraft is based on the popular Next Generation 737 airliner that is combined with the Northrop Grumman multi-role electronically scanned array (MESA) radar mounted above the fuselage.

Below: The Boeing E-7A Wedgetail was Australia's solution for an Airborne Early Warning and Control system.

UPGRADEABLE
Thanks to its open mission systems (OMS) design, the 737 AEW&C can be quickly upgraded with new capabilities, including being made compatible with the battle management command and control (BMC2) system, NATO's latest tactical situational awareness architecture.

MESA RADAR
Although non-rotating, the MESA radar uses active electronically scanned array (AESA) technology to provide 360-degree coverage and can operate in air and maritime modes out to a range of around 370km (230 miles).

E-7A WEDGETAIL/AEW&C
The first E-7A for the Royal Australian Air Force arrived in Australia in November 2009 and was officially accepted into service in May 2010. The sixth and final example was handed over to the RAAF in May 2012 and initial operational capability was announced in November 2012.

BOEING 737 AEW&C
Boeing's latest airborne early warning and control (AEW&C) aircraft is based on the popular Next Generation 737 airliner that's combined with the Northrop Grumman multi-role electronically scanned array (MESA) radar carried within a fixed antenna mounted above the fuselage. Compared to the older E-3 Sentry airborne warning and control system (AWACS), the 737 AEW&C offers increased capabilities coupled with much-reduced operating costs and manpower requirements.

TURKISH AIRFORCE BOEING 737 E-3 SENTRY
The Boeing 737-700 is one of four variants of the Boeing 737 Next Generation series. The Boeing 737 AEW&C is specifically a conversion of the 737-700IGW.

E-2D Advanced Hawkeye (2010)

TYPE • *AEW&C* COUNTRY • *United States*

SPECIFICATIONS

DIMENSIONS: Length: 17.6m (57ft 8.75in); Wingspan: 24.56m (80ft 7in); Height: 5.58m (18ft 4in)

WEIGHT: Maximum take-off 26,082kg (57,500lb)

POWERPLANT: Two Allison/Rolls-Royce T56-A-427A turboprops each rated at 3800kW (5100hp)

MAX SPEED: 650km/h (400mph)

RANGE: 2708km (1682 miles), ferry

CEILING: 10,600m (34,700ft)

CREW: 2 plus 3 mission personnel

ARMAMENT: None

Since the mid-1960s, the US Navy has relied upon successive iterations of the carrier-based E-2 Hawkeye to provide airborne early warning and control (AEW&C) capabilities to ensure the defence of the carrier battle group.

Left: The first example of the E-2D was delivered to the US Navy's VAW-120 'Greyhawks' at Naval Air Station Norfolk, Virginia, in July 2010. The type has also been acquired by several export customers, with Egypt, France and Japan all opting to replace their previous Hawkeyes with the latest D-model.

E-2D ADVANCED HAWKEYE
This US Navy E-2D, BuNo 168593, is shown in the markings of Carrier Airborne Early Warning Squadron 126 (VAW-126), the 'Seahawks'. Since then, the aircraft has been transferred to VAW-120, which serves as the Fleet Replacement Squadron responsible for training crews on both the E-2 Hawkeye and the C-2 Greyhound.

AVIONICS
The aircraft has a new mission computer and tactical workstations for the avionics operators, as well as an all-glass cockpit for the flight crew.

RADOME
The centrepiece of the E-2D's capabilities is the Lockheed Martin AN/APY-9 active electronically scanned array (AESA) radar housed in a rotating radome above the fuselage.

ADVANCED HAWKEYE
The E-2D Advanced Hawkeye not only serves as the 'eyes and ears' of the fleet, but has also taken on a 'digital quarterback' role, in particular managing complex missions including missile defence. When employed in the increasingly important missile defence role, the E-2D makes use of the Cooperative Engagement Capability (CEC), which allows it to work in conjunction with Standard Missile SM-6 surface-to-air missiles to tackle various different missile threats.

E-2D IN FLIGHT
Compared to earlier Hawkeyes, the E-2D also features an aerial refuelling capability, increasing its potential time on station to five hours and increasing total mission time from four to seven hours.

Yakovlev Yak-130 (2010)

TYPE • *Trainer* COUNTRY • *Russian Federation*

SPECIFICATIONS

DIMENSIONS:	Length: 11.49m (37 ft 9in); Wingspan: 9.84m (32ft 3in); Height: 4.76m (15ft 7in)
WEIGHT:	Maximum take-off 10,290kg (22,679lb)
POWERPLANT:	Two Ukrainian Progress/Zaporizhzhya AI-222-25 turbofans each rated at 24.5kN (5510lbf)
MAX SPEED:	1060km/h (659 miles)
RANGE:	1600km (994 miles)
CEILING:	12,500 m (41,013ft)
CREW:	2
ARMAMENT:	Up to 3000kg (6614lb) weapons and stores

An advanced trainer and lead-in fighter trainer (LIFT), the Yak-130 (NATO reporting name 'Mitten') supplements L-39C Albatros and two-seat versions of combat aircraft in service with Russian Aerospace Forces.

Left: A Yakovlev Yak-130 takes off during a Victory Day parade rehearsal at Kubinka air force base. It was planned to equip combat units with the Yak-130 as a cheaper trainer to 'download' hours from fighter aircraft and maintain pilot proficiency but has apparently proven so far too expensive for such additional orders.

PLANNED VARIANTS
Planned variants include the Yak-130M, a modernized version with a new mission computer, new head-up display (HUD) and laser rangefinder in the nose.

YAK-130
Appearing at the Kubinka Air display in August 2021, '30 Red' is finished in three-tone blue-grey camouflage and equipped with a pair of external tanks but no armament on its pylons.

TEST FLIGHTS
In March 2002, the Russian Air Force approved the Yak-130, formally selecting it over the MiG-AT. Three aircraft were completed for tests, the first of these flying in April 2004. The second flew in April 2005 and the third in March 2006. The third was lost due to a fly-by-wire system failure in July 2006. Factory trials were completed in 2005 but the 2006 crash led to more changes. In December 2009, the Yak-130 completed state evaluation, clearing it for military use.

HARDPOINTS
Production aircraft can carry up to 3000kg (6614lb) of weapons and stores on six underwing hardpoints, plus one centreline hardpoint and two wingtip stations.

YAK-130, AIR SHOW, 2012
'134 White' carries a pair of PTB-450 external tanks on the innermost pylons, along with two B-13L rocket pods carrying five 122mm (4.8in) unguided rockets each, and two R-73 air-to-air missiles on the outermost pylons.

Wing Loong II/Chengdu GJ-2 (2011)

TYPE • UAV COUNTRY • *China*

SPECIFICATIONS

DIMENSIONS:	Length: 11m (36ft 1in); Wingspan: 20.5m (67ft 3in); Height: 4.1m (13ft 5in)
WEIGHT:	4200kg (9259lb)
POWERPLANT:	One turboprop engine
MAX SPEED:	370km/h (230mph)
RANGE:	32 hours endurance
CEILING:	9900m (32,000ft)
CREW:	N/A
ARMAMENT:	480kg (1060lb) of air-to-surface weapon

China is the world's second-largest drone market, only the United States is larger. As one might therefore expect, China has produced an impressive array of military drones for a variety of applications, including the Wing Loong II/Chengdu GJ-2.

Left: A Chinese UAV pilot tests the operation and control system for a Wing Loong II drone at the assembly plant of AVIC Chengdu.

PERFORMANCE
Maximum speed is raised to 370km/h (230mph) and Wing Loong II possesses the same 20-hour endurance of its immediate forebear.

SAR RADAR
In addition to the EO turret, Wing Loong II also features an SAR radar in its nose for acquiring ground targets in poor weather. A datalink antenna for guiding missiles is installed under the starboard forward fuselage.

ANCESTRY
The predecessor of the GJ-2 is the GJ-1/WD-1K Wing Loong (Pterosaur) I developed by the Chengdu Aircraft Research & Design Institute (CADI) in concert with Chengdu Aircraft Corporation (CAC) and Guizhou Aircraft Industry Corporation (GAIC). Broadly similar in size, layout, mission and capability to the US MQ-1 Predator, the GJ-1 first flew in October 2007. Meanwhile the larger and more powerful UCAV was being developed by CADI and GAIC, designated the GJ-2 Wing Loong II. Both UAVs are intended for medium-altitude and long-endurance (MALE) UAV duties.

EO TURRET
The GJ-2 is fitted with an electro optical (EO) turret under the nose containing FLIR, TV and laser rangefinder/designator for tracking and locking on to ground targets in all-weather conditions.

WING LOONG II, DUBAI, 2017
It is believed that UAE acquired the first 15 Wing Loong IIs. These have subsequently been involved in the Libyan civil war armed with BA-7 ATGMs and have suffered several losses in action.

Shaanxi Y-9 (2012)

TYPE • Transport COUNTRY • China

SPECIFICATIONS

DIMENSIONS:	Length: 36.07m (118ft 4in); Wingspan: 38m (124ft 8in); Height: 11.3m (37ft 1in)
WEIGHT:	Maximum take-off 265,352kg (585,000lb)
POWERPLANT:	Four WoJiang WJ-6C turboprop engines each rated at 3805kW (5103hp)
MAX SPEED:	650km/h (400mph)
RANGE:	2200km (1400 miles) with 15,000kg (33,069lb) payload
CEILING:	10,400m (34,100ft)
CREW:	4
ARMAMENT:	None

In the early 2000s, China set about developing a reworked version of its Y-8 four-engine tactical transport, to create an aircraft much better suited to the requirements of the PLAAF in the 21st century. The result was the Y-9, which reportedly has the Western reporting name 'Claw'.

Left: In terms of the Y-9's airframe compared to that of the Y-8, the wing and the fuselage were subject to a redesign. Other improvements include small vertical stabilizers mounted on the horizontal tailplanes, likely to improve stability at low speeds. The aircraft is reportedly able to carry almost twice as much fuel as the Y-8, for a considerable improvement in range.

FLIGHT DECK
The flight deck is equipped for operations by a flight crew of four. These are provided with six colour multifunction displays, with an electronic flight instrument system (EFIS) to display primary flight data.

PAYLOAD
The total payload capacity of this spacious transporter is 98 paratroopers, 72 medevac patients or 20,000kg (44,092lb) of cargo. Loads of up to 13,200kg (29,101lb) can be airdropped via the rear ramp.

SHAANXI Y-9
Serial number 10255 is operated by the 4th Division's 10th Air Regiment, within the Western Theatre Command. The unit's new generation Y-9s serve alongside Y-8Cs based at Chengdu-Qionglai air base.

FOREIGN INVOLVEMENT
As well as Antonov, the engine supplier Pratt & Whitney Canada (P&WC) was also involved in the development of this new-generation airlifter, which was originally known as the Y-8F600, or, alternatively, the Category III Platform. Other Western companies also contributed at this early stage, especially in terms of the aircraft's modern avionics. Development was reportedly launched in 1999, with Antonov and P&WC having joined the programme by the early 2000s.

Y-9 PREPARES FOR TAKE-OFF
The new solid nose of the Y-9 (in contrast to the glazing of the Y-8) covers a navigation radar and the aircraft also features an advanced communication suite. Collision-avoidance systems are also included for safer operations, especially in poor weather.

Tupolev Tu-214 (2012)

TYPE • *Transport* COUNTRY • *Russian Federation*

SPECIFICATIONS

DIMENSIONS: Length: 46.14m (151ft 5in); Wingspan: 41.8m (137ft 2in); Height: 13.89m (45ft 7in)

WEIGHT: Maximum take-off 110,750kg (244,162lb)

POWERPLANT: Two Aviadvigatel PS-90A turbofans, each rated at 158.28kN (35,583lbf)

MAX SPEED: (estimated) 900km/h (559mph)

RANGE: (estimated) 8000km (4971 miles)

CEILING: (estimated) 12,100m (39,700ft)

CREW: 3 + unknown mission crew

ARMAMENT: None

The Tupolev Tu-214 airliner (NATO reporting name 'Cookpot') has been repurposed for a variety of military duties in the Russian Air Force, particularly as a reconnaissance, command post, radio relay, escort jamming and maritime patrol aircraft.

Left: Transaero Tu-214 RA-64549 lands at Domodedovo International Airport in 2012. As a civil airliner, the Tu-214 can carry more than 200 passengers.

TU-214 VARIANT
The Tu-214R variant utilizes a multi-band radar system with a side-looking radar with flat antennas on the forward fuselage sides and a 360-degree surveillance radar in a radome under the rear fuselage.

COMMUNICATION CENTRE
The Tu-214PU-SBUS is equipped with the SBUS-214 (Spetsyalnyi Bortovoy Uzel Svyazi) onboard communication centre.

TU-214PU-SBUS
This Tu-214PU-SBUS is another military Tu-214 variant, in this case acting as an airborne control post (PU, Punkt Upravleniya).

MILITARY PURPOSE
The Tu-204 passenger aircraft first flew on 2 January 1989. The Tu-214 version has increased take-off weight and first flew on 21 March 1996. Series production of the Tu-204 was launched at the Aviastar-SP plant in Ulyanovsk; 52 were built before production of the Tu-204 ceased and the Tu-214 began to be manufactured at the KAZ plant in Kazan. Limited commercial orders led to the plans to develop the Tu-214 for military use.

TU-124 DOMODEDOVO AIRPORT
In 2002, the Russian Armed Forces' Main Intelligence Directorate (GRU) ordered Tupolev and KAZ to begin research and development leading to two Tu-214R aircraft, or izdeliye 411. The first Tu-214R took to the air on 24 December 2009 and the type had begun operational missions by December 2012.

Changhe Z-10 (2012)

TYPE • Helicopter **COUNTRY** • China

Initially developed by Harbin in concert with the Kamov Design Bureau during the 1990s, the Z-10 would become China's standard medium-attack and anti-tank helicopter, with more than 200 aircraft in service.

SPECIFICATIONS (Z-10K)

Dimensions: Length: 14.15m (46ft 5in); Rotor diameter: 12m (39ft 4in); Height: 3.85m (12ft 8in)

Weight: Maximum take-off 7000kg (15,432lb)

Powerplant: Two 930–957kW (1247–1283shp) AVIC WZ-9 turboshaft engines

Max speed: 270km/h (170mph)

Range: 800km (500 miles)

Ceiling: 6400m (21,000ft)

Crew: 2

Armament: One 23mm (0.906in) revolver gun or 25mm (0.9in) PX-10A chain gun; four hardpoints with a total capacity of 1500kg (3307lb), able to carry a variety of gun pods, air-to-ground, and air-to-air guided missiles

Left: The extraordinarily slim profile of the Z-10 is evident in this front view of the helicopter. The Z-10 adopts the conventional stepped tandem cockpit design of most modern helicopter gunships.

Changhe Z-10K
This Z-10K belongs to the 4th (Rotary Wing) Brigade of the PLAAF, based at Huangpi. It wears a three-colour tactical camouflage

Anti-tank missiles
carries AKD-10 anti-tank missiles under its stub wings. These are carried in sets of four hung from each pylon, but only the nearest two of the outboard set are visible in the artwork.

UPGRADES
Production Z-10s entered service in 2012. Since then the helicopter has been subject to an upgrade programme that saw upward-pointing engine exhaust nozzles added to reduce the helicopter's infrared signature, with additional armour plates added to the cockpit and engines. A new IFF system and antenna for the BeiDou satellite navigation system were also added. An upgraded export version designated the Z-10ME was unveiled in 2018.

Gun armament
The Z-10 features a 23mm PX-10A chain gun mounted under the chin, aimed by the gunner in the forward cockpit by helmet mounted display.

Z-10 FORMATION FLYING
A formation of Z-10 helicopters conduct a demonstration at the Tianjin International Helicopter Expo in northern China's Tianjin Municipality on 9 September 2015.

Kazan Ansat (2013)

TYPE • Helicopter COUNTRY • Russian Federation

A lightweight general utility helicopter with two pilots seated side-by-side with dual controls, the Ansat-U (for training, or *Uchebnyi*) was developed for the Russian military under the *Pervokursnik* (first-year student) programme.

SPECIFICATIONS

DIMENSIONS:	Length: 13.54m (44ft 5in); Rotor diameter: 11.5m (37ft 9in); Height 3.56m (11ft 8in)
WEIGHT:	Maximum take-off 3300kg (7275lb)
POWERPLANT:	Two Pratt & Whitney Canada PW207K turboshaft engines with FADEC system, each rated at 477kW (640shp)
MAX SPEED:	275km/h (171mph)
RANGE:	508km (316 miles)
CEILING:	5700m (18,701ft)
CREW:	1 or 2
ARMAMENT:	None

Below: Virtually all Ansats built have been the Ansat-U, basic military training version, as seen here. However, during 2005 Kazan flight tested the Ansat-2RC (Razvedchik-Celeukazatel, reconnaissance-target indicator) featuring a reworked narrow fuselage with two tandem cockpits, forward-looking infrared turret, optional weapons and stores on four pylons, plus a fixed 12.7mm (0.5in) machine gun. Only one was built.

POWERPLANT
The Ansat is powered by two Pratt & Whitney Canada PW207K turboshaft engines fitted with the full authority digital engine (FADEC) system of aircraft engine management.

ANSAT-U
Pictured as it appeared during the Airshow to celebrate 100 years of the Russian Air force in 2012, '40 Yellow' is typical of the Ansat fleet which are currently finished in overall dark grey.

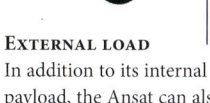

CAPACITY
Aside from a single pilot, the Ansay is able to carry up to 10 passengers; with the passenger seats removed, it can transport 1000kg (2200lb) of cargo internally.

EXTERNAL LOAD
In addition to its internal payload, the Ansat can also carry an underslung load of 1300kg (2860lb).

DESIGN MODIFICATIONS
Designed and manufactured by Kazan Helicopters in Tatarstan, development of the Ansat was launched in 1993 as a new lightweight twin-engine helicopter. The maiden flight of the first prototype took place on 17 August 1999. Series production began in 2004 for civilian customers but an early crash contributed to the fly-by-wire control system of the original version not being certified for passenger flights. As a result, in 2012 work began on Ansat-1M with conventional mechanical flight controls.

ANSAT-U TRAINER
In September 2001, the Ansat-U was selected for the Russian armed forces, for basic and advanced training, to replace the elderly Mil Mi-2. State acceptance trials were completed in October 2009 and the first eight helicopters in the initial order were delivered during 2009 and 2010.

Shenyang J-15 (2013)

TYPE • *Naval fighter* **COUNTRY** • *China*

After the demise of the Soviet Union, two T-10K test aircraft (prototypes of the Sukhoi Su-27K) were left in Ukraine and one of these was sold to China in 2007. It subsequently became the pattern for China's J-15 fifth-generation fighter.

SPECIFICATIONS (estimated)

Dimensions:	Length: 22.28m (73ft 1in); Wingspan: 15m (49ft 3in); Height: 5.92m (19ft 5in)
Weight:	Maximum take-off 32,500kg (71,650lb)
Powerplant:	2 × WS-10 afterburning turbofans, 89.17kN (20,050lbf) thrust dry
Max speed:	Mach 2.4 (2963km/h)
Range:	3500km (2200 miles)
Ceiling:	20,000m (66,000ft)
Crew:	1 or 2
Armament:	1 × 30mm (1.18in) GSh-30-1 cannon; 12 × external hardpoints with a capacity of 6500kg (14,300lb)

Left: A Shenyang J-15 prepares for launch from a Chinese carrier. With China testing an electromagnetic aircraft launch system and a steam-powered catapult, Fujian will permit catapult assisted take-offs, and an appropriately equipped J-15 (sometimes known as the J-15T) has been noted under test.

Shenyang J-15
A J-15 of the PLANAF, operating from the aircraft carrier *Liaoning* in around 2017. Equipped for short take-off but arrested recovery (STOBAR) operations from China's first carriers, at least 36 J-15 production aircraft have been completed, with as many as 60–70 examples likely planned.

Electronic warfare
In addition to the fighter variant, there is also the J-15D electronic warfare version, equivalent to the US Navy's EA-18G Growler, with electronic warfare jamming pods on the wingtips.

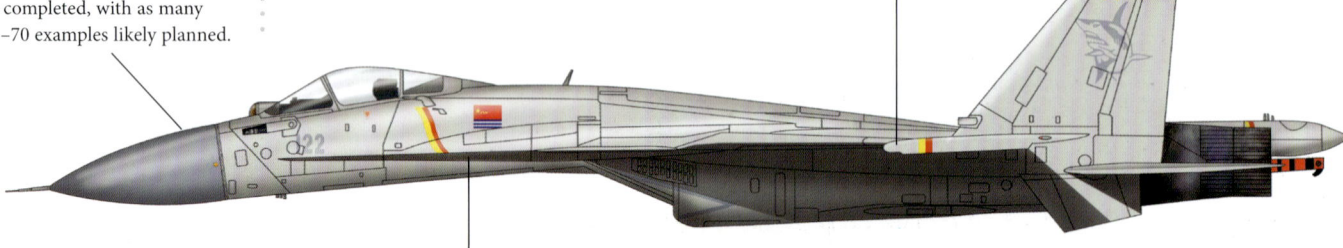

Armament
The J-15 has one GSh-301 30mm (1.18in) cannon, plus up to 6500kg (14,330lb) of stores carried on 12 hardpoints, including: PL-8 and PL-12 air-to-air missiles, YJ-83KH land attack missile, YJ-83K anti-ship missile.

CARRIER DEPLOYMENT
The People's Liberation Army Navy Air Force took the J-15 to sea for the first time aboard the *Kuznetsov*-class aircraft carrier *Liaoning* in 2012. The J-15 is now also deployed on board the carrier *Shandong*, which is a Chinese version of the *Kuznetsov* design and is likely also to form the main part of the air wing of China's third carrier, named *Fujian*, which was launched on 17 June 2022.

J-15 COMING INTO LAND
The T-10K aircraft sold to China in 2007 subsequently became the pattern for the J-15, which incorporates Chinese avionics, a modern glass cockpit and a multi-role capability using indigenous weapons.

HAL LCA Tejas (2013)

TYPE • Fighter COUNTRY • India

The Hindustan Aeronautics Limited (HAL) Light Combat Aircraft (or LCA) known to the Indian Air Force as the Tejas, is the first fully indigenous Indian fighter aircraft since the HAL HF-24 Marut that was designed in the 1950s.

SPECIFICATIONS (Tejas Mk 1)

DIMENSIONS:	Length: 13.2m (43ft 4in); Wingspan: 8.2m (26ft 11in); Height: 4.4m (14ft 5in)
WEIGHT:	Maximum take-off 13,500kg (29,762lb)
POWERPLANT:	One General Electric F404-GE-IN20 turbofan rated at 90kN (20,200lbf) thrust with afterburning
MAX SPEED:	Mach 1.6 (1975km/h)
RANGE:	3200km (1986 miles) with two external drop tanks
CEILING:	16,500m (50,000ft)
CREW:	1
ARMAMENT:	One 23mm (0.9in) twin-barrel GSh-23 cannon plus up to 5300kg (11,685lb) of disposable stores carried on eight external hardpoints

Left: A HAL LAC Tejas of the Indian Air Force performs aerobatic manoeuvres with accompanying smoke trails at Aero India 2015 (a biennial air show and aviation exhibition).

MULTI-MODE RADAR
The Indian defence industry has focused on developing its own multi-mode radar for tactical engagements, rejecting initial reliance of imported options.

POWERPLANT
The aircraft is powered by a single General Electric F404-GE-IN20 turbofan, rated at 90kN (20,200lb) thrust with afterburning

TEJAS MK 1
Serial number LA-5018 was the Indian Air Force's first Tejas Mk1 to be completed to the Final Operational Clearance (FOC) standard. It was delivered to No 18 Squadron 'Flying Bullets' at Sulur Air Force Station in May 2020.

DELAYED PROGRAMME
The origins of the Tejas date back to the early 1980s, when the Indian Air Force (IAF) began planning for a successor to the MiG-21, as well as Gnat and Ajeet light fighters then in service. The project received the formal go-ahead in 1983, with plans to locally develop the turbofan engine as well as fly-by-wire technology, composite aero-structures, electronic warfare and radar systems. After many delays, the first of two development aircraft took to the air on 4 January 2001.

TEJAS TAXIING
The first series-production aircraft for the IAF was handed over in January 2015 and the service received 20 Tejas Mk 1 aircraft in an initial operating capability (IOC) version before switching to the full operating capability (FOC) model, deliveries of these 20 jets beginning in late 2019.

Sukhoi Su-34 (2014)

TYPE • *Strike aircraft* COUNTRY • *Russian Federation*

Originally intended to fulfil air interdiction, reconnaissance and escort-jamming missions, the Su-34 (NATO reporting name 'Fullback') is currently primarily used as a long-range strike aircraft, or a 'tactical bomber' in Russian parlance.

SPECIFICATIONS

DIMENSIONS:	Length: 24.8m (81ft 4in); Wingspan: 14.7m (48ft 2in); Height: 6.08m (19ft 11in)
WEIGHT:	Maximum take-off 45,100kg (99,428lb)
POWERPLANT:	Two Lyulka-Saturn/Moscow AL-31F (izdeliye 99V) turbofans each rated at 75.22kN (16,909lbf) dry and 122.58kN (27,558lbf) with afterburning
MAX SPEED:	Mach 1.6 (1975km/h)
RANGE:	4000km (2485 miles) ferry range
CEILING:	15,700m (51,509ft)
CREW:	2
ARMAMENT:	One 30mm (1.2in) Gryazev-Shipunov GSh-30-1 autocannon; up to 8000kg (17,637lb) of weapons and stores

This page: '42 Blue' was the first flying prototype Su-27I and first flew on 13 April 1990. It was subsequently redesignated Su-34.

AVIONICS
Mission avionics include the K-102 targeting and navigation complex that brings together a central computer, data display system, Platan laser/TV sight and navigation system.

NOSE AND COCKPIT
The Su-33 features a widened 'platypus' nose seating the pilot and the navigator/weapons system operator side-by-side in a titanium-alloy armoured box.

WEAPON STATIONS
The Su-34 can carry a total of up to 8000kg (17,637lb) of weapons and stores on 12 pylons: six under the wing, two under the engine intake ducts and two in tandem between the engines.

COMBAT EVALUATION
In October 2006, the Su-34 was cleared for initial operational service, although during its trials phase test aircraft were involved in the conflict in Chechnya, being used to jam enemy air defence radars in 2000 and 2002. Two prototypes also took part in the war with Georgia in August 2008, again in an escort-jamming role. The 'second stage' of state evaluation lasted between 2006 and 2011 with the first operational regiment fully equipped with 24 new Su-34s by October 2013 at Voronezh.

DEMO FLIGHT
A Russian Sukhoi Su-34 '38 red' perfoms a demonstration flight in Zhukovsky during the MAKS 2015 airshow. In its basic form, the Su-34 is a bomber intended primarily for tactical interdiction, although reconnaissance duties can be performed with the addition of mission-specific pods.

Sukhoi Su-35S (2014)

TYPE • *Fighter* **COUNTRY** • *Russian Federation*

SPECIFICATIONS

DIMENSIONS:	Length: 21.9m (71ft 10in); Wingspan: 14.7m (48ft 2in); Height: 5.9m (19ft 4in)
WEIGHT:	Maximum take-off 34,500kg (76,059lb)
POWERPLANT:	Two thrust-vectoring Saturn AL-41F-1S (izdeliye 117S) turbofans each rated at 86.29kN (19,400lbf) dry and 137.29kN (30,865lbf) with afterburning
MAX SPEED:	2400km/h (1500mph)
RANGE:	3600km (2200 miles)
CEILING:	18,000m (59,055ft)
CREW:	1
ARMAMENT:	One 30mm (1.2in) Gryazev-Shipunov GSh-30-1 autocannon; Up to 8000kg (17,637lb) of weapons and stores

A formidable single-seat, heavy long-range, multi-role fighter developed from the Su-27, the Su-35S (NATO reporting name 'Flanker') is for the Russian Air Force a lower-cost supplement to the new-generation Su-57.

Above: An Su-35S. The Su-35 is a late fourth-generation (4 + +) air superiority fighter. It can undertake ground-attack missions, but has been used in its primary role over Ukraine, where it outclasses all likely opponents.

RADAR
A modern Irbis passive electronically scanned array radar, reportedly capable of detecting hostile aircraft at a range of 400km (882lb) was also introduced as part of a revised fire-control system.

WEAPONS LOAD
The Su-35S can carry up to 8000kg (17,637lb) of weapons and stores on its 12 hardpoints. All types of air-launched tactical missile currently in the Russian inventory can be employed.

Su-35BM
The Su-35BM (*Bolshaya Modernisatsiya*, or 'big modernisation') was launched in the early 2000s for export, using Sukhoi funds, and the prototype, '901', first flew on 19 February 2008. This aircraft became the Su-35S.

SU-35S
The most obvious external change of the Su35S from the earlier Su-35 airframe is the deletion of the canard foreplanes, Sukhoi having concluded that the manoeuvrability lost with the removal of these control surfaces could be more than made up for with the thrust vectoring nozzles while dispensing the weight penalty imposed by the canards. As a result the Su-35S more closely resembles the Su-27 from which it was originally developed rather than the earlier iteration of the Su-35 design.

CHINESE SU-35S
After the successful service introduction of the Su-35S in Russia, orders for the export Su-35 version were placed by China and Egypt. China received its first four aircraft in December 2016 with a further two batches of 10 each being delivered during 2017 and 2018.

Airbus A400M Atlas (2014)

TYPE • Refueller **COUNTRY** • France

SPECIFICATIONS

DIMENSIONS:	Length: 45.1m (148ft); Wingspan: 42.4m (139ft 1in; Height: 14.7m (48ft 3in)
WEIGHT:	Maximum take-off 141,000kg (310,852lb)
POWERPLANT:	Four Europrop TP400-D6 turboprops each rated at 8200kW (11,000hp)
MAX SPEED:	Mach 0.72
RANGE:	8700km (5400 miles), ferry
CEILING:	12,200m (40,000ft)
CREW:	3 or 4 (2 or 3 pilots, 1 loadmaster)
ARMAMENT:	None

A multinational effort, the A400M Atlas programme was launched in 2003 to provide a new airlifter for seven European nations: Belgium, France, Germany, Luxemburg, Spain, Turkey and the United Kingdom.

Left: The A400M is the only tactical tanker offering a third refuelling point for large aircraft refuelling and as an alternative to pods. To monitor day and night air-to-air refuelling operations, the A400M can be fitted with three cameras controlled from the cockpit by the co-pilot, suppressing the need for visual observers.

LUFTWAFFE A400M
Germany initially planned to acquire 75 A400Ms, before reducing its order to 60 aircraft, then to 53. At one point, Berlin also planned to try and sell off 13 of its allocated aircraft, but plans were subsequently revised, and all 53 examples are operated by the Luftwaffe.

PROPELLERS
The aircraft is fitted with eight-bladed scimitar-type propellers. Each pair of propellers on each wing is contra-rotating, producing a lot more lift that propellers running in the same direction.

PAYLOAD
The A440M has a carrying capacity of 116 combat troops or paratroopers, 66 stretcher patients or an equivalent load of cargo

The A400M was designed from the outset to combine the qualities of tactical and strategic military transports, as well as being able to operate as an aerial refuelling tanker. As a result, the aircraft has the long-range performance and internal capacity of a strategic airlifter, but is able to operate from short, semi-prepared airstrips in austere locations. The A400M recorded its first flight on 11 December 2009 and an initial production aircraft was delivered to the French Air Force in August 2013, entering service the following year.

A400M STEEP TAKE-OFF
The A400M can carry a greater payload than a C-130, but do so under more challenging conditions. For example, it can deliver a 25,000kg (55,116lb) cargo load to an unprepared airstrip less than 750m (2461ft) long while carrying enough fuel to complete a 930km (578 mile) return trip.

Shaanxi KJ-500 (2014)

TYPE • *AEW&C* **COUNTRY** • *China*

SPECIFICATIONS

DIMENSIONS:	Length: 11.50m (37ft 8in); Wingspan: 40m (131ft 3in); Height: 11m (36ft 1in)
WEIGHT:	Maximum take-off 70,000kg (154,324lb)
POWERPLANT:	Four Zhuzhou WoJiang-6C (FWJ-6C) turboprop engines developing 3803kW (5100hp) each
MAX SPEED:	550km/h (324mph)
RANGE:	5700km (3542 miles)
CEILING:	10,500m (34,449ft)
CREW:	5 (flight crew); up to 19 mission crew
ARMAMENT:	None

Following on from the KJ-200, with its 'balance beam' radar above the fuselage, China developed the KJ-500, as another mid-size AEW&C aircraft type. Its main characteristic is the primary surveillance radar housed in a more traditional fixed rotodome.

Left: A KJ-500 seen above the skies of Guangdong Province, China, in 2021. Outside China, KJ-500s are regularly encountered in Taiwan's Air Defence Identification Zone, or ADIZ, which covers the Taiwan Strait, but which also extends over territory in the People's Republic.

AESA RADAR
Compared to the ZDK-03, the KJ-500 has an active electronically scanned array (AESA) radar, instead of a PESA type. This is not installed in a rotating rotodome, but a fixed one, containing three individual antennas arranged in a triangular configuration.

ELINT SYSTEMS
The aircraft also has an electronic intelligence (ELINT) capability, for which it likely relies upon rectangular antennas on each side of the rear fuselage – these fairings are very similar to those found on the Y-9GX-8, a dedicated ELINT platform.

PLAAF SERVICE
The KJ-500 has proven to be a popular type with the PLAAF and its advanced AESA radar likely offers a very similar capability to the much more expensive and complex KJ-2000. At the same time, the Y-9 turboprop platform makes the KJ-500 much better suited to deployed operations, including flying to and from smaller airstrips on islands in the South China Sea, for example.

COUNTERMEASURES
For self-protection, the KJ-500 is fitted with countermeasures dispensers below the rear fuselage. Its detection range against fighter-sized targets is 470km (290 miles).

KJ-500 RADOME
This front view of a KJ-500 illustrates the elevated prominence of the fixed dorsal radome. Atop the main radome is a smaller fairing, apparently concealing a satellite communications (satcom) antenna.

Bayraktar TB2 (2015)

TYPE • UAV COUNTRY • Turkey

SPECIFICATIONS	
Dimensions:	Length: 8.5m (27ft 10in); Wingspan: 12m (39ft 4in) Height: 2.2m (7ft 2.4in)
Weight:	Maximum take-off 650kg (1433lb)
Powerplant:	7.5kW (10hp) internal combustion injection engine
Max speed:	222km/h (138mph)
Range:	27 hours endurance
Ceiling:	7600m (25,000ft)
Crew:	N/A
Armament:	4 laser-guided smart munitions

First flying in 2009, the TB2 is a medium-altitude combat drone manufactured in Turkey. It was used with success in the Syrian conflict, where it is reported to have destroyed Russian-made Pantsir air defence systems.

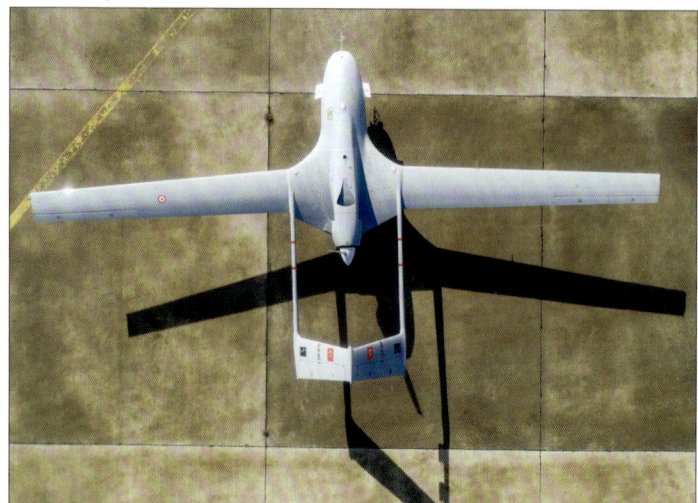

Left: A TB2 lands at Gecitkale Airport in northern Cyprus on 16 December 2019. The TB2 first flew in 2014 and has since acquired a proven combat record in multiple war zones.

Autonomous
The UAV has fully autonomous taxiing, take-off, cruise and landing modes, plus an automatic navigation and tracking system that can keep the aircraft flying purposefully even with the loss of GPS.

Communication range
The TB2 has a communication range of 300km (186 miles), giving the operator the ability to range deep behind enemy lines.

Endurance
Using its pusher configuration The TB2 can stay in the air for an astonishing 27 hours, the UAV platform giving an endurance far beyond the capabilities of a human pilot.

USE IN UKRAINE
In Ukrainian service, the drones were first used against separatist forces in 2020 and saw extensive use after the Russian invasion. The drone has a maximum weapons load of 150kg (330.7lb) and uses the lightweight MAM family of missiles. The MAM missile family is laser-guided. The MAM-C has a range of 8km (5 miles), MAM-L 15km (9.3 miles) and MAM-T 30–80km (18.6–49.7 miles). HE-Fragmentation, thermobaric and anti-armour warheads can be carried, depending on the mission.

BAYRAKTAR AKINCI
Another UAV from the Bayraktar stable is the twin-engine Akinci platform, which is enhanced with dual artificial intelligence avionics and can stay aloft for 24 hours carrying a wide range of smart munitions.

Lockheed Martin F-35 Lightning II (2015)

TYPE • *Stealth fighter* COUNTRY • *United States*

SPECIFICATIONS (F-35A)

DIMENSIONS: Length: 15.7m (51ft 5in); Wingspan: 11m (35ft); Height: 4.4m (14ft 5in)

WEIGHT: Maximum take-off 31,751kg (70,000lb)

POWERPLANT: One Pratt & Whitney F135-PW-100 turbofan rated at 190kN (43,000lbf) of thrust with afterburning

MAX SPEED: Mach 1.6 (1975km/h)

RANGE: 2736km (1700 miles)

CEILING: 15,240m (50,000ft)

CREW: 1

ARMAMENT: One 25mm (0.98in) GAU-22/A four-barrel rotary cannon, plus maximum weapon load of 8050kg (18,000lb) on four internal stations, plus six underwing stations

The result of the single most expensive programme in US Department of Defense history, the Lockheed Martin F-35 joint strike fighter was intended to combine high-end capabilities with a relatively low price tag.

Left: The F-35 has full STOVL capability. Two primary components provide vertical lift for hover: the LiftFan and 3-Bearing Swivel Module (3BSM). The LiftFan is mounted behind the cockpit. As the aircraft transitions to hover, doors open on top of the aircraft and two counter-rotating fans blow unheated air straight down, producing around half the downward thrust needed. The majority of the remaining vertical thrust is provided by the 3BSM at the rear of the aircraft.

HELMET SIGHT
Real-time imagery is streamed to the helmet, allowing pilots to 'look through' the aircraft. Pilots thus can see the entire environment surrounding them. The helmet also provides pilots with night vision using an integrated camera.

EODAS
The Electro-Optical Distributed Aperture System (EODAS) is the only 360-degree, spherical situational awareness system, sending high-resolution real-time imagery to the pilot's helmet from six infrared cameras mounted around the airframe.

EOTS
The low-drag, stealthy Electro-Optical Targeting System (EOTS) is behind a durable sapphire window. It is linked to the aircraft's integrated central computer through a high-speed fibre-optic interface.

VARIANTS

The F-35A is the only version to carry an internal cannon – the 25mm (0.98in) GAU-22/A – and will be the most prolific model. The F-35B is capable of STOVL operation and has a smaller internal weapon bay and reduced internal fuel capacity compared to the F-35A. It is also equipped for the probe and drogue method of aerial refuelling. The F-35C features larger wings and a strengthened undercarriage in order to cope with catapult launches and arrested landings.

F-35 TRAINING FLIGHT
All three JSF variants incorporate a high degree of stealth, intended to ensure the fighter could evade enemy detection and fly combat missions in contested airspace. As a result, it primarily carries its weapons internally, but external stores can be added when flying in more permissive environments.

Shenyang J-16 (2015)

TYPE • *Strike aircraft* **COUNTRY** • *China*

SPECIFICATIONS

DIMENSIONS:	Length: 21.9m (69ft 6in); Wingspan: 14.7m (48ft 3in); Height: 6.36m (20ft 9in)
WEIGHT:	Maximum take-off 35,000kg (77,162lb)
POWERPLANT:	Two Shenyang WS-10B afterburning turbofans, 135kN (30,000lbf) with afterburner
MAX SPEED:	Mach 2 (2470km/h)
RANGE:	3530km (2190 miles)
CEILING:	19,000m (62,000ft)
CREW:	2
ARMAMENT:	One 30mm (1.18in) Gryazev-Shipunov GSh-30-1 cannon with 150 rounds; munitions on 12 external hardpoints

Notably, the J-16 not only represents an apparent break with any previous 'Flanker' licence production agreements that may have existed with Russia, but also fully embodies the advances and true multi-role capability that China had long sought to incorporate in its J-11 fleet.

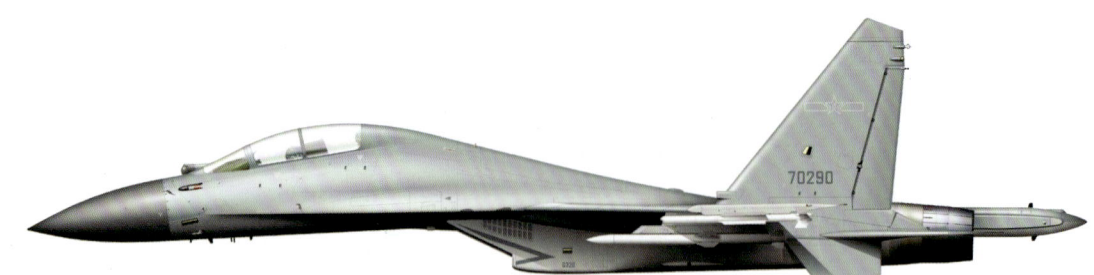

Above: This J-16 is from the Eastern Command's 98th Air Brigade based at Chongqing. She is coded 70290, making her 'grey 20' and carrying the serial 0320 on her intakes. This multi-role fighter is armed with a PL-10 imaging infrared air-to-air missile on the wingtip and a PL-15 extended-range, active radar-guided air-to-air missile beneath its wings.

RADOME
A reconfigured radome is thought to accommodate a different AESA radar optimized for electronic warfare. At the same time, the usual cannon and IRST/laser rangefinder housing have apparently been removed from the aircraft's nose.

J-16D
The J-16D is a dedicated electronic warfare version of the J-16. This J-16D was displayed at the Zhuhai International Airshow in September 2021.

ICM/JAMMING PODS
The wingtips carry the diagnostic ICM pods of the type and, on pylons beneath the wings and intake cheeks, are various jamming pods from the RKZ930 family.

FUSION
The J-16 seems to have originated in Chinese plans to develop an indigenous two-seat 'Strike Flanker' in preference to buying further batches of Su-30MKKs or Su-30MK2s from Russia. To achieve this goal, the J-16 ultimately emerged as a combination of the airframe of the two-seat J-11BS (itself a reverse-engineered Su-27UBK) and the domestically developed avionics and WS-10 series engines from the J-11B.

J-16 PRODUCING WING VORTICES
Outwardly, the J-16 may look similar to the Russian Su-30MKK, but it is understood to be very different under the skin, including a much greater use of modern composite materials in its construction.

Chengdu J-20 (2016)

TYPE • *Stealth fighter* COUNTRY • *China*

Apparently assigned the Western reporting name 'Firefang', and known in China as Mighty Dragon, the Chengdu J-20 is only the third stealth fighter to have entered service anywhere in the world, a major accomplishment for China's aerospace industry.

SPECIFICATIONS (J-20A)

DIMENSIONS:	Length: 20.4m (66ft 10in); Wingspan: 13.5m (44ft 4in); Height: 4.45m (14ft 7in)
WEIGHT:	Maximum take-off 37,013kg (81,600lb)
POWERPLANT:	Two Saturn AL-31FN turbofans each rated at 145kN (33,000lbf) thrust with afterburning
MAX SPEED:	Mach 2 (2470km/h)
RANGE:	3400km (2113 miles)
CEILING:	20,000m (65,617ft)
CREW:	1
ARMAMENT:	Disposable ordnance carried in one large weapon bay in the lower fuselage; optional additional ordnance on four underwing pylons

Left: *The J-20 is a highly manoeuvrable aircraft. There have been reports that the J-20 might receive engines with thrust-vectoring controls, for enhanced manoeuvrability, and a J-10B testbed fitted with a thrust-vectoring WS-10 began to be tested around 2018.*

AVIONICS SUITE
The J-20 incorporates an advanced avionics suite, with an active electronically scanned array (AESA) Type 1475 radar at its core. The mission avionics also include an apparent distributed situational awareness system, similar to the EODAS on the F-35 Lightning II.

CHENGDU J-20A
This J-20A serves with the PLAAF's 172nd Air Brigade, part of the Flight Test and Training Base, which operates from Cangzhou. This is one of two test and training units operating the J-20.

CHENGDU J-20A
Production of the J-20A continues today, with refinements being added to the aircraft. In late 2022, for example, a modified version of the J-20A under test at Chengdu. featured a slightly raised cockpit and a deeper spine, which likely contains additional avionics and possibly also more fuel. At the same time, the nose of the aircraft has been slightly reprofiled, probably to reduce drag. Other changes have been made to the engine intakes, likely in order to accommodate the WS-15s. By late 2022 it was assessed that total J-20 production had reached around 150 examples.

STEALTH DESIGN
In common with other stealth fighters, the J-20's design is based, to a significant degree, around the internal carriage of its weapons. Altogether, the J-20 has three internal weapons bays

J-20S IN FORMATION
The J-20 is a critical element of China's air defence capability. It is expected that the J-20 will eventually begin to incorporate new weapons, likely including a new long-range AAM. This missile will almost certainly be designed specifically for internal carriage, allowing the aircraft to be armed with six medium-range missiles in its main weapons bay, as opposed to the current four.

435

Kawasaki C-2 (2016)

TYPE • Transport COUNTRY • Japan

Japan's Kawasaki corporation developed the C-2 transporter as a successor to the Cold War-era C-1 transport, the redesign placing a greater emphasis on airlift missions in international peacekeeping contexts.

SPECIFICATIONS

DIMENSIONS: Length: 43.9m (144ft); Wingspan: 44.4m (145ft 8in); Height: 14.2m (46ft 7in)

WEIGHT: Maximum take-off 141,400kg (311,734lb)

POWERPLANT: Two General Electric CF6-80C2K1F turbofan engines each rated at 265.7kN (59,740lbf) thrust

MAX SPEED: Mach 0.82 (1012km/h)

RANGE: 9800km (6100 miles) ferry range

CEILING: 12,200m (40,000ft)

CREW: 3 (2 pilots, 1 loadmaster)

ARMAMENT: None

Below: The first prototype C-2 took to the air on 26 January 2010 and the JASDF introduced the type to service in early 2017.

KAWASAKI C-2
Following a pair of XC-2 prototypes, orders for 13 production C-2s had been placed on behalf of the Japan Air Self-Defense Force (JASDF) through to Fiscal Year 2019.

PAYLOAD
The hold of the C-2 is able to transport 120 fully equipped soldiers and a maximum payload of 37,600kg (82,720lb) of cargo. The fuselage can accommodate large cargoes such as the Patriot surface-to-air missile system or Mitsubishi H-60 series helicopters.

FLIGHT DECK
The modern flight deck is equipped with head-up displays and the flight crew can call upon the tactical flight control system that is combined with fly-by-wire flight controls to ensure safe low-level flying, including in adverse weather.

AIRFRAME
The C-2 airframe has also been adapted for electronic intelligence gathering, with a single RC-2 version having been converted from the second prototype.

KAWASAKI C-2, 403RD TACTICAL AIRLIFT SQN
A JASDF Kawasaki C-2 sits on the flightline prior to take off on Joint Base Elmendorf-Richardson, Alaska, during an international exercise in 2023. In a break from previous tradition in Japan, the C-2 has been offered for export, with the aim of challenging competitors such as the A400M and the Il-76.

PURPOSES
The C-2 is intended to allow the Japan Air Self-Defense Force (JASDF) to support rapid deployment of troops and cargo, as well as to assist in the response to natural disasters and other contingencies. Compared with the C-1, the C-2 has an improved range, increased speed and a larger payload. It also features a range of modern technologies including a tactical flight control system, auto-airdrop system, self-protection equipment and an air-to-air refuelling capability.

Xi'an Y-20 (2016)

TYPE • Transport COUNTRY • China

While the Y-20 has been overshadowed by its new-generation, near-contemporary, the J-20 stealth fighter, the four-engine, long-range transport is arguably as important in terms of the PLAAF's modernization and especially to Beijing's wider strategic ambitions.

SPECIFICATIONS (Y-20A)

Dimensions:	Length: 47m (154ft 2in); Wingspan: 50m (164ft 1in); Height: 15m (49ft 3in)
Weight:	Maximum take-off 180,000kg (396,832lb)
Powerplant:	Four Soloviev D-30KP-2 turbofan engines each rated at 117.68kN (26,460lbf)
Max speed:	925km/h (576mph)
Range:	7800km (4847 miles) with payload of two main battle tanks
Ceiling:	13,000m (42,651ft)
Crew:	3
Armament:	None

Left: The J-20 is a massive aircraft, with the conventional overall layout of a heavy military transport. However, it makes extensive use of composite materials in its construction, helping to reduce overall airframe weight.

Avionics
The Y-20 has a digital flight deck for the flight crew of three. Avionics otherwise include a digital fly-by-wire flight control system, satellite communications and a small forward-looking infrared (FLIR) sensor for take-off and landing in poor weather conditions.

Wings
The Y-20 has a broadly similar wingspan to the Il-76 but the wing itself is of a much more modern design, with high-lift devices and an efficient supercritical aerofoil section.

Xi'an Y-20
An early-production Y-20, serial number 11057 is on strength with the 12th Air Regiment of the 4th Division. The aircraft is based at Chengdu-Qionglai.

Y-20 COMPARISONS
The Y-20 is frequently compared to both the Il-76 and the US-designed C-17 Globemaster III. Based on its payload attributes, the Y-20 is smaller than the C-17 and much closer to the Il-76. However, even though it is somewhat shorter than the Il-76 overall, it importantly features a wider and taller cargo compartment than the Soviet design. This allows a wider variety of larger items to be carried, including main battle tanks.

Y-20 IN THE PLAAF
Overall, the Y-20 is set to play a very important role in helping the Chinese military meet its wider ambitions, including deploying and sustaining forces over much greater distances. The same basic airframe has also been further developed as an in-flight refuelling tanker, with rumours that an airborne early warning and control (AEW&C) derivative may also be planned.

CH-5 Rainbow (2017)

TYPE • UAV COUNTRY • *China*

SPECIFICATIONS	
DIMENSIONS:	Length: 11.2m (36ft 8in); Wingspan: 21m (68ft 10in); Height: 3.8m (12ft 6in)
WEIGHT:	2225kg (4905lb)
POWERPLANT:	Turbocharged piston engine
MAX SPEED:	220km/h (137mph)
RANGE:	10,600km (6214 miles)
CEILING:	9000m (29,500ft)
CREW:	N/A
ARMAMENT:	Anti-tank missiles; smart weapons

Another UCAV intended for the medium-altitude long-endurance (MALE) role is the China Aerospace Science and Technology Corporation's (CASC) CH-5 Rainbow, one of a series of UAV and UCAVs named 'Rainbow' that began with the twin-boom CH-1 of 2000.

Left: A CH-5 Rainbow on display. The CH-5 has perhaps unsurprisingly proved popular on the export market with Myanmar, Pakistan (illustrated below), Egypt, Saudi Arabia, Algeria and Iraq all known to operate the type.

LAYOUT
The aircraft has a straight-wing configuration with a pusher-type powerplant at the rear of the aircraft; by placing the engine here, electromagnetic interference of the sensors at the front is kept to a minimum.

PAYLOAD
The aircraft has a 1000kg (2200lb) payload capacity; it can carry two AR-1 lightweight air to surface missiles for ground attack and an SAR pod for reconnaissance.

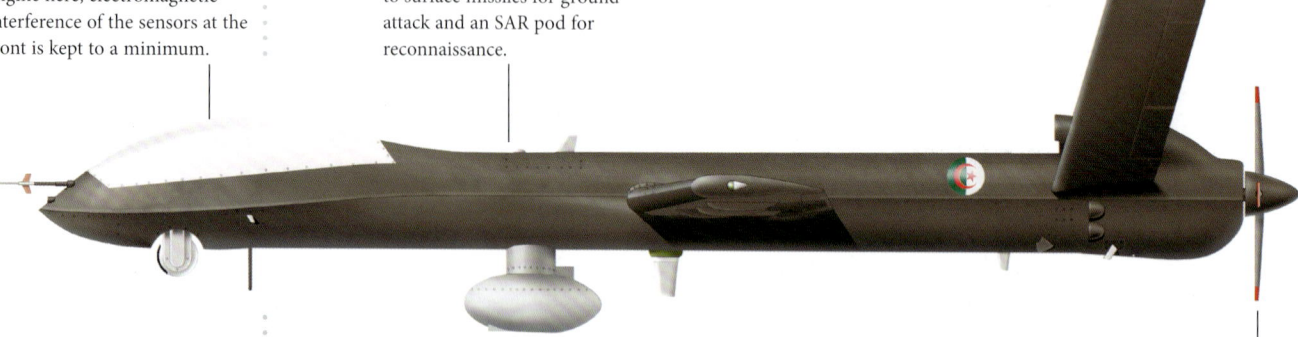

PISTON ENGINE
The piston engine of the CH-5, which is of an unidentified types, produces around half the power of the MQ-9 Reaper's turboprop, although the Chinese aircraft still has formidable endurance of about 60 hours.

REAPER ALTERNATIVE
Most of the Rainbow series were developed primarily for export with the CH-2, another twin-boom design, and the canard CH-3 following the initial CH-1. The CH-4 however is externally virtually identical to the General Atomics MQ-9 Reaper and has been further developed into the CH-5, the most recent of the Rainbow series to achieve production. The CH-5 is touted as offering the same capability as the MQ-9 at roughly half the cost, though this claim may be difficult to justify.

CH-5 WITH ANTI-TANK MISSILES
A CH-5 medium-altitude long-endurance (MALE) UAV system on display at the China International Aviation and Aerospace Exhibition, in Zhuhai, China, in 2016. It carries AR-1 or AR-2 compact supersonic anti-tank missiles.

Harbin Z-20 (2017)

TYPE • Helicopter COUNTRY • China

SPECIFICATIONS

DIMENSIONS:	Length: 19.54m (64ft 1in); Main rotor diameter: 16.2m (53ft 2in); Height: 4.98m (16ft 4in)
WEIGHT:	Maximum take-off approximately 10,000kg (22,000lb)
POWERPLANT:	Two 2000kW (2682shp) WZ-10 turboshaft engines
MAX SPEED:	320km/h (199mph)
RANGE:	560km (350 miles)
CEILING:	5400m (17,717ft)
CREW:	2
ARMAMENT:	Up to 1000kg (2200lb) of cargo internally, including 12–15 troops, or up to 4000kg (8800lb) externally slung; may be armed with up to eight KD-10 anti-tank IR guided missiles

Derived from the highly successful Sikorsky UH-60, the Harbin Z-20 is comparable to the latest US Blackhawk variants in capability and performance. It is intended to replace the Mil Mi-17 in the People's Liberation Army Air Force (PLAAF) service.

Left: Introduced into service in 2019 and in production ever since, an unknown number of Z-20s has been delivered to the PLAAF and the aircraft is also in service with the Chinese Army and Navy as well as with the civil Chinese People's Armed Police Force.

MAIN ROTOR
Although externally very similar to the UH-60, the Z-20 differs in certain key regards, most obviously in its use of a five blade main rotor as opposed to the Blackhawk's four blade unit, as well as in its utilisation of fly-by-wire controls.

EXTERNAL PAYLOAD
Up to 4000kg (8,800lb) can be slung externally under the helicopter; it can also be armed with up to eight KD-10 anti-tank IR guided missiles.

'COPYHAWK'
After acquiring 24 Sikorsky UH-60 Blackhawk helicopters in the 1980s, the US embargoed further deliveries following China's response to the 1989 Tiananmen Square protests. As a result an indigenous '10-tonne helicopter project' was initiated and the resulting aircraft bore an obvious resemblance to the Blackhawk. Derided as the 'Copyhawk' by some Western commentators, development of the Z-20 was delayed by the greater priority afforded to the Z-10 helicopter gunship and flew for the first time only in December 2013.

CABIN
The main cabin of the Z-20 is believed to be more capacious than that of the UH-60. It can take up to 1000kg (2200lb) of cargo internally

Z-20 DEPLOYING FLARES
A Z-20 puts on a pyrotechnic show with countermeasure flares at the 14th China International Aviation and Aerospace Exhibition in Zhuhai on 8 November 2022.

Embraer C-390 Millennium (2019)

TYPE • Transport **COUNTRY** • Brazil

Brazil's Embraer called upon its extensive experience in developing regional commercial aircraft, such as the popular E-Series, to produce a new-generation, multi-mission airlifter designed to stress mobility and operational flexibility with low operating costs.

SPECIFICATIONS

DIMENSIONS:	Length: 35.2m (115ft 6in); Wingspan: 35.05m (115ft); Height: 11.84m (38ft 10in)
WEIGHT:	Maximum take-off 86,999kg (191,800lb)
POWERPLANT:	Two IAE V2500-E5 turbofans each rated at 139.4kN (31,330lbf) thrust
MAX SPEED:	988km/h (614mph)
RANGE:	5820km (3610 miles), with 14,000kg (30,865lb) payload
CEILING:	11,000m (36,000ft)
CREW:	2
ARMAMENT:	None

Left: A Brazilian Air Force KC-390 Millennium at Le Bourget Airport, France, in 2023. When operating as a tanker (with the KC-390 designation), the aircraft is fitted with a pair of wing-mounted probe and drogue pods.

COCKPIT
The cockpit is equipped with head-up displays as part of an enhanced vision system and includes commercial-standard Rockwell Collins avionics.

WING
The design of the C-390 is based around a generously proportioned hold and a high-mounted wing with various high-lift devices to aid short-field performance.

HOLD CAPACITY
The hold can accommodate two M113 armoured personnel carriers, a Sikorsky H-60 series helicopter, 74 litters with life-support equipment, up to 80 soldiers or 66 fully equipped paratroopers.

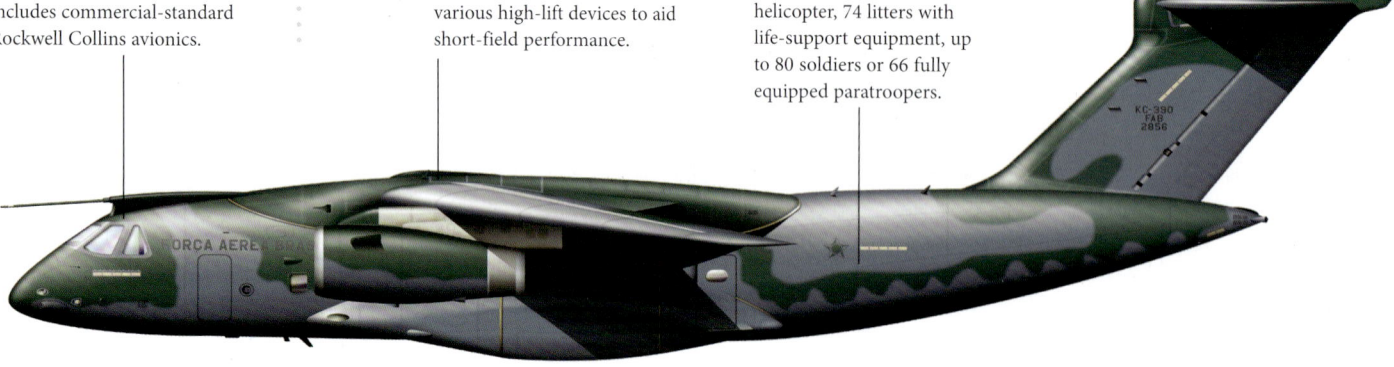

MISSION SPECTRUM
The C-390 is primarily intended to transport troops and cargo but its mission spectrum also includes medical evacuation, search and rescue, humanitarian relief and air-to-air refuelling of both fixed-wing aircraft and helicopters. First flown on 3 February 2015, the C-390 has been ordered by the Brazilian Air Force, which received the first of a planned 28 examples in September 2019. The C-390 aircraft has also been selected by Hungary and Portugal.

KC-390 AT FAIRFORD
An Embraer KC-390 Millennium demonstrates a steep angle of climb after taking off from RAF Fairford, England, during the Royal International Air Tattoo.

Sukhoi Su-57 (2021)

TYPE • *Stealth fighter* **COUNTRY** • *Russian Federation*

SPECIFICATIONS

DIMENSIONS:	Length: 20.1m (66ft); Wingspan: 14.1m (46ft 3in); Height: 4.6m (15ft 1in)
WEIGHT:	Maximum take-off 35,000kg (77,162lb)
POWERPLANT:	Two Saturn/Rybinsk AL-41F1 (izdeliye 117) thrust-vectoring turbofan engines each rated at approximately 9000kg (19,841lb) of dry thrust and 14,500kg (31967lb) with afterburning.
MAX SPEED:	2135km/h (1327mph)
RANGE:	3500km (2175 miles)
CEILING:	20,000m (66,000ft)
CREW:	1
ARMAMENT:	One 30mm (1.2in) Gryazev-Shipunov GSh-30-1 autocannon; up to 10,000kg (22,046lb) of weapons and stores in internal bays and external hardpoints

A fifth-generation, single-seat, heavy long-range multi-role fighter, the Sukhoi Su-57 (NATO reporting name 'Felon') features low observability, supersonic cruising speed (supercruise) and supersonic manoeuvrability.

Below: Despite bearing the serial number 01, this is actually the second full production machine, which was renumbered following the delivery flight crash of the first example in December 2019. The aircraft wears the badge of the 929th Chkalov State Flight Test Centre with which it serves.

CONFIGURATION
The Su-57's aerodynamic configuration is optimized for high lift-to-drag ratio at supersonic speed. The airframe includes fuselage side extensions that terminate in large moving flaps on the leading edge, to improve manoeuvrability.

T-50-9
The ninth prototype, and eighth flyable airframe of the Su-57, also known as the T-50-9. First flown on 24 April 2017, the aircraft wears a pixelated camouflage pattern.

MISSILE LOAD
New air-to-air missiles have been developed for internal carriage by the Su-57, including the beyond-visual-range R-77M and the very-long-range *izdeliye* 810. Two additional missiles are contained in 'quick launch' bays in underwing fairings; each of these bays carries one R-74M2 infrared-guided air-to-air missile. Air-to-surface weapons designed specifically for the Su-57 include the Kh-69 stand-off missile, with a range of 300km (186 miles) or more. The Kh-58UShK anti-radiation missile possesses folding wings and fins to fit in the internal bays.

101KS SUITE
The 101KS consists of an IR search-and-track (IRST) sensor, four UV missile-approach warning sensors, two directional infrared countermeasures turrets and one imaging IR sensor for low-level flying.

PRODUCTION SU-57
By December 2019, the Komsomolsk-on-Amur plant had completed the first pre-production Su-57 (T-50S-1) but the aircraft crashed during its handover flight. The second aircraft, T-50S-2, was completed in October 2020. This was the first to be delivered to the Russian Aerospace Forces in December 2021. In May 2019, Russian President Putin called for the purchase of 76 Su-57 fighters by 2028.

Glossary

AAM: Air-to-Air Missile

ADP: Automatic Data Processing

ADV: Air Defence Variant (of the Tornado)

Aeronautics: the science of travel through the Earth's atmosphere

AEW: Airborne Early Warning

AEW&C: Airborne Early Warning and Control radar

Afterburning (reheat): method of increasing the thrust of a gas turbine aircraft engine by injecting additional fuel into the hot exhaust duct between the engine and the tailpipe, where it ignites to provide a short-term increase of power

Aileron: an aerofoil used for causing an aircraft to roll around its longitudinal axis, usually fitted near the wingtips. Ailerons are controlled by use of the pilot's control column

ALARM: Air-Launched Anti-Radiation Missile

All-Up Weight: the total weight of an aircraft in operating condition. Normal maximum AUW is the maximum at which an aircraft is permitted to fly within normal design restrictions, while overload weight is the maximum AUW at which an aircraft is permitted to fly subject to ultimate flying restrictions

Altimeter: instrument that measures altitude, or height above sea level

AMRAAM: Advanced Medium-Range Air-to-Air Missile

Angle of Attack: the angle between the wing (airfoil) and the airflow relative to it

Aspect Ratio: the ratio of wingspan to chord

ASV: Air to Surface Vessel – airborne detection radar for locating ships and submarines

ASW: Anti-Submarine Warfare

ATF: Advanced Tactical Fighter

Autogiro: heavier-than-air craft which supports itself in the air by means of a rotary wing (rotor), forward propulsion being provided by a conventional engine

Automatic pilot (Autopilot): automatic device that keeps an aircraft flying on a set course at a set speed and altitude

AWACS: Airborne Warning and Control System

Bf: abbreviation for Bayerische Flugzeugwerke (Bavariant Aircraft Factories)

CAP: Combat Air Patrol

Chord: cross-section of a wing from leading edge to trailing edge

Convertiplane: vertical take-off and landing craft with wingmounted rotors that act as helicopter rotors for take-off, then tilt to act as conventional propellers for forward flight

Delta Wing: aircraft shaped like the Greek letter delta

Disposable Load: the weight of crew and consumable load (fuel, missiles etc.)

Electronic Combat Reconnaissance (ECR): a variant of the Panavia Tornado optimized for electronic warfare

Electronic Countermeasures (ECM): systems designed to confuse and disrupt enemy radar equipment

Electronic Counter-Countermeasures (ECCM): measures taken to reduce the effectiveness of ECM by improving the resistance of radar equipment to jamming

Elevator: a horizontal control surface used to control the upward or downward inclination of an aircraft in flight. Elevators are usually hinged to the trailing edge of the tailplane

ELF: Extremely Low Frequency. A radio frequency used for communication with submarines

ELINT: Electronic Intelligence. Information gathered through monitoring enemy electronic transmissions by specially equipped aircraft, ships or satellites

Empty Equipped (also known as Tare Weight): the weight of an aircraft equipped to a minimum scale, i.e. with all equipment plus the weight of coolant in the engines, radiators and associated systems, and residual fuel in tanks, engines and associated systems

EW: Electronic Warfare

FAC: Forward Air Controller. A battlefront observer who directs strike aircraft on to their targets near the front line

FAE: Fuel-Air Explosive. A weapon that disperses fuel into the atmosphere in the form of an aerosol cloud. The cloud is ignited to produce intense heat and heat effects

FLIR: Forward-Looking Infra-Red. Heat-sensing equipment fitted in an aircraft that scans the path ahead to detect heat from objects such as vehicle engines

FRS: Fighter Reconnaissance Strike

Gas turbine: engine in which burning fuel supplies hot gas to spin a turbine

Geodetic construction: a 'basket weave' system of aircraft construction producing a self-stabilizing framework in which loads in any direction are automatically equalized by forces in the intersecting set of frames, producing high strength for low weight

Helicopter: powered aircraft that achieves both lift and propulsion by means of a rotary wing (rotor)

HOTAS: Hands on Throttle and Stick. A system whereby the pilot exercises full control over his aircraft in combat without the need to remove his hands from the throttle and control column to operate weapons selection switches or other controls

HUD: Head-Up Display. A system in which essential information is projected onto a cockpit windscreen so that the pilot has no need to look down at his instrument panel

IFF: Identification Friend or Foe. An electronic pulse emitted by an aircraft to identify it as friendly on a radar screen

INS: Inertial Navigation System. An on-board guidance system that steers an aircraft or missile over a predetermined course by measuring factors such as the distance travelled and reference to 'waypoints' (landmarks) en route

IR: Infra-Red

Jet propulsion: method of propulsion in which an object is propelled in one direction by a jet, or stream of gases, moving in the other

JSTARS: Joint Surveillance and Target Attack Radar System. An airborne command and control system that directs air and ground forces in battle

Jumo: abbreviation for Junkers Motorenwerke (Junkers Motor Works)

Kiloton: Nuclear weapon yield, one kiloton (kT) being roughly equivalent to 1000 tons of TNT

Laminar Flow: airflow passes over an aircraft's wing in layers, the first of which, the boundary layer, remains stationary while successive layers progressively accelerate; this is known as laminar flow. The smoother the wing surface, and the more efficient its design, the smoother the airflow

LAMPS: Light Airborne Multi-Purpose System. Antisubmarine helicopter equipment, comprising search radar, sonobuoys and other detection equipment

Landing Weight: the AUW of an aircraft at the moment of landing

Lantirn: Low-Altitude Navigation and Targeting Infra-Red for Night. An infra-red system fitted to the F-15E Strike Eagle that combines heat sensing with terrain-following radar to enable the pilot to view the ground ahead of the aircraft during low-level night operations. The information is projected on the pilot's head-up display

LWR: Laser Warning Radar. Equipment fitted to an aircraft that warns the pilot if they are being tracked by a missile guiding radar beam

Mach: named after the Austrian professor Ernst Mach, a Mach number is the ratio of the speed of an aircraft or missile to the local speed of sound. At sea level, Mach One (1.0M) is approximately 1226km/h (762mph), decreasing to about 1062km/h (660mph) at 9140m (30,000ft). An aircraft or missile travelling faster than Mach One is said to be supersonic.

MANPADS: Man-Portable Air Defense Systems – surface-to-air missiles that can be carried and fired by a single individual

Maximum Take-Off Weight: the maximum AUW, due to design or operational limitations, at which an aircraft is permitted to take off

Megaton: Thermonuclear weapon yield, one megaton (mT) being roughly equal to 1,000,000 tons of TNT

MG: Machine gun (Maschinengewehr in German, hence MG 15)

Microlight: very light aircraft with a small engine; a powered hang-glider

Mk: mark (of aircraft)

MK: *Maschinenkanone* (automatic cannon, e.g. MK.108)

NATO: North Atlantic Treaty Organization

NBC: Nuclear, Chemical and Biological (warfare)

NVG: Night Vision Goggles. Specially designed goggles that enhance a pilot's ability to see at night

OBOGS: On-Board Oxygen Generating System. A system that generates oxygen, avoiding the need to rely on pre-charged oxygen bottles and extending the time a pilot can stay airborne during long transit flights over the ocean, for example

Operational load: The weight of equipment necessarily carried by an aircraft for a particular role

Payload: the weight of passengers and/or cargo

Phased-Array Radar: A warning radar system using many small aerials spread over a large flat area, rather than a rotating scanner. The advantage of this system is that it can track hundreds of targets simultaneously, electronically directing its beam from target to target in microseconds (millionths of a second)

PLSS: Precision Location Strike System. A battlefield surveillance system installed in the Lockheed TR-1 that detects the movement of enemy forces and directs air and ground attacks against them

Pulse-Doppler Radar: a type of airborne interception radar that picks out fast-moving targets from background clutter by measuring the change in frequency of a series of pulses bounced off the targets. This is based on the well-known Doppler Effect, an apparent change in the frequency of waves when the source emitting them has a relative velocity towards or away from an observer.

Ramjet: simple form of jet engine that is accelerated to high speed causing air to be forced into the combustion chamber, into which fuel is sprayed and then ignited

Rudder: movable vertical surface or surfaces forming part of the tail unit, by which the yawing of an aircraft is controlled

RWR: Radar Warning Receiver. A device mounted on an aircraft that warns the pilot if they are being tracked by an enemy missile guidance or intercept radar

SAM: Surface-to-Air Missile

SHF: Super High Frequency (radio waves)

SIGINT: Signals Intelligence. Information on enemy intentions gathered by monitoring electronic transmissions from his command, control and communications network

SLAM: Stand-off Land Attack Missile – a missile that can be air-launched many miles from its target

SLAR: Side-Looking Airborne Radar. A type of radar that provides a continuous radar map of the ground on either side of the aircraft carrying the equipment

Sound Barrier: popular name for the concept that the speed of sound (see Mach) constitutes a limit to to flight through the atmosphere to all aircraft except those specially designed to penetrate it. The cone-shaped shock wave created by an aircraft breaking the 'barrier' produces a 'sonic boom' when it passes over the ground

Spin: a spin is the result of yawing or rolling an aeroplane at the point of a stall

SRAM: Short-range Attack Missile

Stall: condition that occurs when the smooth flow of the air over an aircraft's wing changes to a turbulent flow and the lift decreases to the point where control is lost

Stealth technology: technology applied to aircraft or fighting vehicles to reduce their radar signatures. Examples of stealth aircraft are the Lockheed F-117 and the Northrop B-2

STOVL: Short Take-off, Vertical Landing

Stuka: abbreviation of *Sturzkampfflugzeug* (literally 'diving battle aircraft')

TADS: Target Acquisition/Designation System. A laser sighting system fitted to the AH-64 Apache attack helicopter

Take-Off Weight: the AUW of an aircraft at the moment of take-off

Thermal imager: Equipment fitted to an aircraft or fighting vehicle which typically comprises a telescope to collect and focus infra-red energy emitted by objects on a battlefield, a mechanism to scan the scene across an array of heatsensitive detectors, and a processor to turn the signals from these detectors into a 'thermal image' displayed on a TV screen

TIALD: Thermal Imaging/Airborne Laser Designator. Equipment fitted to the Panavia Tornado IDS enabling it to locate and attack precision targets at night

Turbofan engine: type of jet engine fitted with a very large front fan that not only sends air into the engine for combustion but also around the engine to produce additional thrust. This results in faster and more fuel-efficient propulsion

Turbojet engine: jet engine that derives its thrust from a stream of hot exhaust gases

Turboprop engine: jet engine that derives its thrust partly from a jet of exhaust gases, but mainly from a propeller powered by a turbine in the jet exhaust

Variable-Geometry Wing: a type of wing whose angle of sweep can be altered to suit a particular flight profile. Popularly called a Swing Wing

UAV/UCAV: Unmanned Aerial Vehicle/Unmanned Combat Aerial Vehicle – battlefield drones

VHF: Very High Frequency

VLF: Very Low Frequency

V/STOL: Vertical/Short Take-off and Landing

Wild Weasel: code name applied to specialized combat aircraft tasked with defence suppression

Window: strips of tinfoil cut to the wavelengths of enemy radars and scattered from attacking aircraft to confuse enemy defences. Also known as 'chaff'

Yaw: the action of turning an aircraft in the air around its normal (vertical) axis by use of the rudder. An aircraft is said to yaw when the fore-and-aft axis turns to port or starboard, out of the line of flight

Index

Admiral Kuznetsov (carrier) 389
Admiral Vinogradov (destroyer) 342
Aermacchi M.B.339A 339
   MB.326 339
   MB.339C 339
   MB.339K Veltro 2 339
Aero L-39 Albatros 308
   L-38 308
   L-39C 308
   L-39MS/L-59 308
   L-59 308
   L-159 308
Aero Vodochody L-159 ALCA 393
   L-159A 393
   L-159T 393
   L-159T1 393
   L159T+ 393
Aérospatiale Puma 287
   SA 330 Puma HC.1 287
   SA 330E (Puma HC.Mk 1) 287
   SA 330J 287
   SA 330L 287
Afghanistan 236, 264, 290, 295, 304, 322, 384, 397, 332, 334, 355, 358, 367–8, 372, 417
Agusta-Bell 212 286
   AB.212ASW 286
AgustaWestland EH101/AW101 Merlin 388
AH-64D Longbow/AH-64E Guardian 384
Aichi D3A 123
   D3A1 123
AIDC F-CK-1 Ching-Kuo 374
   F-CK-1A 374
   F-CK-1B 374
Airbus A400M Atlas 430
Airbus C295 396
   C295W 396
Airco DH.2 31
Airco DH.4 51
Airco DH.9 53
   DH.9A 53

Albatros B.II 15
   B.II (PK) 15
Albatros C.I & C.III 22
Albatros D.I & D.II 34
Albatros D.III 35
Albatros D.V 50
   D.Va 50
AMX International AMX 370
   A-11A 370
   A-11B 370
   AMX-T 370
   TA-11A 370
   TA-11B 370
Antietam, USS 243
Antonov An-22 281
Antonov An-26 299
Antonov An-30 293
   An-24RT 292
Antonov An-124 358
Antonov An-140 398
   An-140-330T 398
   An-140C 398
   An-140T 398
Antonov An-148 405
   An-148-100 405
   An-148-100A 405
   An-148-100B 405
   An-148-100E 405
Antrim, HMS 259
Arado Ar 196 121
   Ar 196A-5 121
Arado Ar 234 182
   Ar 234B-1 182
   Ar 234B-2 182
Arctic 88, 103, 108
Ardennes 182
Arizona, USS 113
Ark Royal, HMS 94, 148
Armstrong Whitworth Siskin 63
   Siskin IIIA 63
Atlantic Conveyor (container ship) 336
Atlantic Ocean 103, 106
Atlas Cheetah 360
   Cheetah C 360

Austro–Hungarian front (WWI) 56
Avdyeyev, M.V. 147
Avia B.534 72
   B.534-I 72
   B.534-IV 72
Aviatik C.I 21
   B series 21
Avro Canada CF-100 Canuck 225
   Mk 5 225
Avro Lancaster 153
   B.I (Special) 153
   Mk I 153
Avro Manchester 140
   Mk I 140
   Mk IA 140
Avro Vulcan 222
   B.Mk 2 222
   K.Mk 2 222

BAC 167 Strikemaster 289
   Jet Provost 289
   Mk 82A 289
   Mk 88 289
Bachem Ba 349 Natter 197
Balbo, Italo 56
Balkans 88, 154, 367
Barkhorn, Gerhard 82
Barnwell, Frank 36
Bartels, Heinrich 82
Bartini, Roberto Ludvigovich 162
Bataan, USS 388
Battle of Britain 83, 87, 88, 89
Bay of Pigs 210
Bayraktar TB2 432
Béchereau, Louis 46
Bell AH-1 HueyCobra 290
   AH-1G 290
   AH-1T 290
Bell AH-1J SeaCobra 305
   AH-1T 'Improved SeaCobra' 305
   AH-1W SuperCobra 305
   AH-1Z Viper 305
Bell AH-1W SuperCobra 356
   AH-1W 'Whiskey Cobra' 356

AH-1Z Viper 284, 305
Bell OH-58 Kiowa 278
   OH-58D 278
Bell P-39 Airacobra 161
   P-39L 161
   P-39Q 161
Bell P-63 Kingcobra 177
   P-39Q 177
   P-63A 177
   P-63C 177
Bell UH-1 Iroquois (Huey) 255
   UH-1B 255
   UH-1D 255
   UH-1H 255
   UH-1N 255
Bell V-22 Osprey 385
Bellonte, Maurice 62
Beriev A-50 354
   A-100 354
Beriev Be-10 'Mallow' 248
Beriev Be-200 403
   Be-200ChS 403
   Be-200ES 403
Berlin, bombing of 100, 162
Berlin Crisis 231
Berthold, Rudolf 48
Birkigt, Marc 42, 46
Bishop, William 29
Bismarck (battleship) 94
Blackburn Shark 71
   Mk I 71
Blackburn Skua 114
   Mk II 114
Blériot XI 12
Blitzkrieg 84, 99
Blohm & Voss BV 138 105
   BV 138 MS 105
   BV 138A 105
   BV 138B 105
   BV 138B-1 105
   BV 138C 105
Boeing 737 394
   737-3Q8 394
   737-76D 394

443

737-85N 394
737-300 394
Boeing 737 AEW&C 418
   737-700 418
   737-700IGW 418
   E-7A Wedgetail/AEW&C 418
Boeing AH-64 Apache 359
   AH-64A 359
Boeing B-17 Flying Fortress 85
   B-17E 85
   B-17F 85
   B-17G 85
Boeing B-29 Superfortress 181
   B-29A-5-BN 181
   B-29B 181
Boeing B-52 Stratofortress 220
   B-52D 220
   B-52G 220
   B-52H 220
Boeing CH-47 Chinook 264
   CH-37C 264
   CH-47C 264
   CH-47F 264
   MH-47 264
   MH-47E 264
Boeing E-3 Sentry 331
   E-3A 331
Boeing F-15E Strike Eagle 368
   F-15EX 368
   F-15QA 368
Boeing F/A-18E Super Hornet 390
   EA-18G 390
   F/A-18C 390
   F/A-18F 390
Boeing/Grumman E-8 J-Stars 365
   E-8A 365
   E-8C 365
Boeing KC-135 Stratotanker 236, 250, 362–3
   KC-135A 236
   KC-135E 236
   KC-135Q 236
   KC-135R 236
Boeing P-26 Peashooter 70
Boeing RC-135 263
   EC-135 263
   RC-135V 263
Bombardier CRJ 409
   Challenger 600 409
   Challenger 870 409
   CRJ200ER 409
   CRJ700 409
Bon Homme Richard, USS 223
Borneo 103
Bosnia 304
Boulton Paul Defiant 130
   Mk I 130
   Mk II 130
Bréguet 14 54
   14A2 54
   14B2 54
Bréguet 19 62
   19A2 62
Brewster F2A Buffalo 109
   F2A-2 109
Bristol Beaufighter TF.Mk X 141
   Mk IC 141
   Mk XIC 141
Bristol Blenheim Mk IV 97
Bristol Bulldog 69
   Mk II 69
   Mk IIA 69
   Mk IV 69
   Mk IVA 69
Bristol F.2A & F.2B 36
British Aerospace Harrier GR.7 367
British Aerospace Hawk 326
   T.Mk 1 326
   T.Mk 1A 326
British Aerospace Sea Harrier 338
   FA.2 338

FRS.Mk 1 338
T.4 338
Brown, Lt Harry Winston 115
Brown, Russell J. 192
Bunker Hill, USS 380
Büttner, Lt Karl-Heinrich 34

Cadbury, Maj E. 51
Cameron, Flt Sgt D. 142
Camm, Sir Sydney 189
Caproni Ca.3 56
Caproni Ca.133 77
Carter, George 188
Caudron R.11 A.3 20
   R.4 20
Caudron, René and Gaston 20
Cessna A-37B Dragonfly 280
   A-27B 280
   A-37A 280
   OA-37 280
Cessna O-2 Skymaster 279
   O-2A 279
   O-2B 279
CH-5 Rainbow
   CH-1 438
   CH-2 438
   CH-3 438
   CH-4 438
Changhe Z-8 357
   Avicopter AC313 357
   Z-8F 357
   Z-8K 357
   Z-8KA 357
   Z-18 357
Changhe Z-10 424
   Z-10K 424
   Z-10ME 424
Charles de Gaulle (carrier) 336
Chengdu J-7/F-7 276
   F-7B 276
   F-7M 276
   F-7MG 276
   J-7I 276
   J-7II 276
   J-7L 276
Chengdu J-10 404
   J-10A 404
   J-10AH 404
   J-10AS 404
   J-10ASH 404
   J-10AY 404
   J-10B 404
   J-10C 404
   J-10CY 404
   J-10D 404
   J-10SY 404
Chengdu J-20 435
   J-20A 435
Christmas Island 234
Collishaw, Lt Col Raymond 27
Consolidated B-24 Liberator 154
   B-24D 154
Consolidated PBY Catalina 90
   Mk I 90
   PBY-4 90
   PBY-5 90
   PBY-5A 90
Constellation, USS 205
Convair B-36 Peacemaker 213
   B-36J 213
   RB-36D 213
Convair B-58 Hustler 258
   B-58A 258
Convair F-102 Delta Dagger 240
Convair F-106 Delta Dart 252
   F-106A 252
   YF-106 252
Coral Sea, Battle of (1942) 101
Coral Sea, USS 333
Cornwall, HMS 123, 379
Costes, Dieudonné 62

Courageous, HMS 71
Crailsheim, Kurt von 19
Cuban Missile Crisis (1962) 232
Curtiss P-6 Hawk 73
   P-6E 73
   XP-6 73
   XP-6A 73
Curtiss P-36 & Hawk Model 75 115
   P-36A 115
Curtiss P-40 Kittyhawk 126
   P-40D 126
   P-40E 126
Curtiss SB2C Helldiver 169
   SB2C-1C 169
Curtiss SBC Helldiver 112
   Curtiss Model 73 112
   SBC-3 112
   SBC-4 112

Dalrymple, Capt Sydney 36
Dambusters raid 153
Dassault/Dornier Alpha Jet 329
   Alpha Jet A 329
   Alpha Jet E 329
Dassault Mirage 2000 332
   2000C 332
   2000N 332
Dassault Mirage F.1 314
   F.1A 314
   F.1AZ 314
Dassault Mirage III 261
   IIICJ 261
   IIIEP 261
   IIIRP 261
Dassault Rafale 395
   Rafale B 395
   Rafale C 395
   Rafale F1 395
   Rafale F2 395
   Rafale F3 395
   Rafale F4 395
   Rafale M 395
Dassault Super Etendard 336
Davies, Cdr Thomas P. 209
de Havilland, Geoffrey 31
De Havilland Hornet 206
   F.MK III 206
De Havilland Mosquito 152
   FB.Mk VI 152
   NF.Mk 30 152
   NF.Mk II 152
De Havilland Sea Vixen 254
   FAW.1 254
   FAW.2 254
De Havilland Vampire 204
   FB.5 204
   FB.9 204
   FB.50 204
Dean, Fg Off 188
Délage, Gustave 23, 30
Delville, Paul 20
Destainville, Lt 44
Dewoitine D. 520 120
DFW C.V 40
Dieterle, Hans 191
'Doolittle raid' 119
Dornier, Claudius 191
Dornier Do 17 122
   Do 17E-1 122
   Do 17E-2 122
   Do 17M 122
   Do 17P 122
   Do 17Z-2 122
Dornier Do 24 103
   Do 24K-2 103
   Do 24T-1 103
Dornier Do 217 171
   Do 217E 171
   Do 217K 171
   Do 217N-1 171
   Do 217N-2 171

Dornier Do 335 Pfeil 191
   Do 335 V1 191
   Do 335A-0 191
   Do 335A-1 191
   Do 335A-4 191
   Do 335A-6 191
   Do 335A-10 and A-12 191
Dorsetshire, HMS 123
Douglas A-1 Skyraider 205
   A-1H 205
   AD-1 205
   AD-4N 205
Douglas A-4 Skyhawk 238
   A-4B 238
   A4D-1 238
   A4D-3 238
   A4D-5 238
   A-4E 238
   A-4S 238
Douglas A-20 Havoc 131
   A-20B 131
   A-20C 131
   A-20G 131
Douglas A-26 Invader 194
   A-26A 194
   A-26B 194
   A-26C 194
   B-26 194
   B-26C 194
   B-26K 194
Douglas C-47 Skytrain 78
   C-53 Skytrooper 78
Douglas C-124C Globemaster II 218
   C-124A 218
Douglas C-133 Cargomaster 242
   C-133A 242
   C-133B 242
Douglas DB-7/A-20 Havoc/Boston 144
   A-20G-20 144
   A-20G-35-DO Havoc 144
   A-20G Havoc 144
Douglas TBD Devastator 101
   TBD-1 101
Dundas, Flt Lt John 87
Dwight D. Eisenhower, USS 390

E-2D Advanced Hawkeye 419
East China Sea blockade (1939) 86
Eastern Front (WWII) 84, 88, 129, 155–6, 161–2, 173, 175, 186
Eisenhower, Gen Dwight D. 78
Embraer C-390 Millennium 440
   KC-390 440
English Electric Canberra 219
   B-57 219
   B. Mk 2 219
   B(I)Mk 58 219
   B.Mk 6 219
English Electric Lightning 257
   F.Mk.1A 257
   F.Mk.2A 257
   F.Mk.3 257
   F.Mk.6 257
   T.5 257
Enterprise, USS 262, 298
Esnault-Pelterie, Robert 56
Essex, USS 212
Etrich, Igo 13
Eurocopter AS332 Super Puma 337
   AS 332L-1 337
Eurofighter Typhoon 401
   DA.2 401

Fairchild Republic A-10 Thunderbolt II
   A-10A 307
Fairey Albacore 132
Fairey Barracuda 174
   Mk II 174
Fairey Firefly 179
   Mk I 179
   Mk 4 179

Mk 5 179
Mk 6 179
Fairey Fulmar 148
  Mk I 148
  Mk II 148
Fairey Swordfish 94
  Mk I 94
  Mk II 94
Falklands War (1982) 222, 259, 287, 296, 323, 336, 338
Felixstowe F.2A 55
Fiat/Aeritalia G.91 247
  G-91R 247
  G-91T 247
  G-91Y 247
Fiat CR.32 76
  CR.32quater 76
  CR.32ter 76
Fiat G.55 Centauro 176
Fieseler Fi 156 Storch 99
FMA Pucará 323
  IA-58 323
Focke-Wulf Fw 189 Uhu 129
  Fw 189F-2 129
Focke-Wulf Fw 190 150
  Fw 190A 150
  Fw 190A-1 150
  Fw 190A-2 150
  Fw 190A-3 150
  Fw 190A-4 150
  Fw 190A-5/U2 150
  Fw 190A-8 150
Focke-Wulf Fw 200 Condor 137
  Fw 200C-3/U2 137
  Fw 200C-6 137
Focke-Wulf Ta 152 199
  Ta 152C 199
  Ta 152C-1 199
  Ta 152E 199
  Ta 152H 199
  Ta 152H-1 199
  Ta 152S-1 199
  Ta 153C-1 199
  Ta 153C-2 199
  Ta 153C-3 199
Fokker Dr.1 45
Fokker D.VII 57
Fokker Eindecker 19
  E.I 19
  E.II 19
  E.III 19
'Fokker Scourge' 14
Fonck, René 28, 46
Formidable, HMS 132
Formosa 168, 184
France, Battles for/of 97, 106, 120
Fujian (carrier) 426
Furious, HMS 39

Garros, Roland 19
General Dynamics EF-111A Raven 350
  F-111A 350
General Dynamics F-16E/F 414
General Dynamics F-111 270
  EF-111A Raven 270
  F-111A 270
  F-111B 270
  F-111F 270
Georgia 428
Gibson, Wg Cdr Guy 153
Gloster Gladiator 98
  Mk I 98
  Mk II 98
Gloster Javelin 237
  FAW.Mk 1 237
  FAW.Mk 2 237
  FAW.Mk 4 237
  FAW.Mk 6 237
  FAW.Mk 9 237
  GA.5 237
Gloster Meteor F.4 193

F.8 193
F.Mk 8 193
T7 193
Gloster Meteor III 188
  F.8 188
  F.I 188
  F.Mk I 188
  F.Mk II
  F.Mk III 188
  F.Mk IV 188
  PR.10 188
  T.7 188
Gloster Meteor NF.14 228
  NF.11 228
  NF.12 228
  NF.13 228
  T.7 228
Gontermann, Heinrich 35
Göring, Hermann 57
Gotha G.IV 49
  G.V 49
Gray, Lt Cdr John 195
Great Patriotic War (1941–45) 128
Greim, Gen Ritter von 99
Grenada 273, 283, 290, 304, 340
Grumman A-6 Intruder 269
  A-6A 269
  KA-6D 269
Grumman E-2 Hawkeye 271
  E-2D 271
  Lockheed E-2C 271
Grumman EA-6B Prowler 304
Grumman F-14 Tomcat 298
  F-14A 298
Grumman F4F Wildcat 145
  F4F-3 145
  F4F-3A 145
  F4F-4 145
Grumman F6F Hellcat 165
  F6F-5 165
  F6F-5N 165
Grumman F7F Tigercat 196
  F7F-1 196
  F7F-2N 196
  F7F-3 196
  F7F-3N 196
Grumman F8F Bearcat 201
Grumman F9F Panther 223
  F9F-2 223
Grumman FF and F2F 68
  F2F-1 68
  FF-1 68
  SF-1 68
  XF2F-1 68
Grumman OV-1 Mohawk 241
  JOV-1A 241
  OV-1A 241
  OV-1D 241
Grumman TBF Avenger 170
  TBF-1B 170
  TBF-1C 170
  TBF-1D 170
  TBF-1L 170
  TBM-1 170
  TBM-3H 170
  TBM-3P 170
Guam, USS 283
Guizhou JL-9/FTC-2000 400
  FTC-2000G 400
  JL-9G 400
Gulf of Aden 342
Gulf War (1990–91) 271, 287, 290, 304, 319, 332, 340
  Operation Desert Shield 273
  Operation Desert Storm 266, 280, 298, 307, 309, 313, 315, 331, 345, 359, 365, 368
Guynemer, Georges 23, 28, 46

HAL LCA Tejas 427
  Mk 1 427
Halberstadt D.V 33

D.I & D.II 33
D.III 33
Hancock, USS 169
Handley Page Halifax 142
  Mk III 142
  Mk VI 142
  Mk VII 142
Handley Page Hampden 108
  Hampden TB.1 108
Handley Page O/400 43
  O/100 43
Handley Page Victor 251
  B.1 251
  B.2 251
Hanriot HD.1 41
Hansa-Brandenburg W.12 58
Harbin Z-9 379
  Z-9A 379
  Z-9B 379
  Z-9C 379
  Z-9EC 379
  Z-9W 379
Harbin Z-20 439
Hartmann, Erich 82
Harvey, Lt 34
Hawaiian Islands 113
Hawker Fury 60–1, 66
  Mk I 66
  Mk II 66
Hawker Hunter 229
  F.Mk 6 229
  F.Mk 56 229
  F.Mk 56A 229
  F.Mk.1 229
  P.1067 229
  T.Mk.8 229
Hawker Hurricane 83
  Mk IIC 83
  Mk IID 83
  Mk IV 83
Hawker Sea Fury 210
  FB.50 210
  FB.Mk 11 210
Hawker Siddeley/BAe Harrier 300
  GR.1 300
  GR.3 300
  GR.7 300
Hawker Siddeley/BAE Nimrod 296
  MRA4 296
  MR.Mk 2P 296
Hawker Tempest 189
  Mk V 189
Hawker Typhoon 166
  Mk 1A 166
  Mk IB 166
Heinkel, Ernst 58
Heinkel He 51 80
  He 49A 80
  He 51 B-1 80
  He 51A 80
  He5 1A-0 80
  He 51A-1 80
Heinkel He 111 79
  He 111F 79
  He 111H 79
  He 111H-6 79
  He 111P 79
  He 111P-2 79
Heinkel He 162 198
  He 162A 198
  He 162A-2 198
Heinkel He 177 172
  He 177A-3 172
  He 177A-5 172
Henschel Hs 123 91
  Hs 123A 91
Henschel Hs 126 104
  Hs 126A-0 104
  Hs 126B-1 104
Hermes, HMS 123
Higley, Radioman 2nd Class E.L. 113

Hongdu JL-10 413
  JL-10J 413
  L-15 413
  L-15B 413
Hongzhaji-5 (H-5) 215
Hunter, Sqn Ldr P.A. 130
Hütter, Ulrich 191

Illner, Karl 13
Illustrious, HMS 148, 157
Ilyushin DB-3 100
  DB-3T 100
  DB-3TP 100
Ilyushin Il-2 Shturmovik 118
  Il-2M3 118
Ilyushin Il-4 128
  DB-3F 128
Ilyushin Il-20 and Il-22 294
  Il-22M-11 294
Ilyushin Il-28 Beagle 215
  Il-28D 215
Ilyushin Il-38 291
Ilyushin Il-76 and Il-78 316
  Il-38 316
  IL-76M 316
  Il-76MD 316
  Il-78M 316
Ilyushin Il-80 & Ilyushin Il-82 371
Immelmann, Max 19
Independence, USS 238
India and Pakistan wars 229, 233
Indian Ocean 123
Insall, Lt G.S.M. 16
Intrepid, RSS 379
Invincible, HMS 338
Iran-Iraq War 267, 298
Iraq 219–20, 236, 270, 278, 288, 295, 334, 340, 355, 367, 372, 384, 397, 398
see also Gulf War (1990–91)
ISIS 393
Israel 216, 261, 301
Israeli Aircraft Industries (IAI) Kfir 324
  Kfir C.2 324
  Kfir-C7 324
Iwo Jima 134, 169

Jacobs, Lt Josef 34
Jeannin Taube 13
John C. Stennis, USS 336
John F. Kennedy, USS 271, 315
Junkers Ju 87 Stuka 84
  Ju 87B 84
  Ju 87D 84
  Ju 87G 84
Junkers Ju 88 88
  Ju 88A 88
  Ju 88A-4 88
  Ju 88A-17 88
  Ju 88G-7a 88
Junkers Ju 88 Mistel 200
  Mistel 1 200
  Mistel 2 200
  Mistel 3a & S-3a 200
  Mistel S-1 200
  Mistel S-2 200

Kamov Ka-27 348, 364
  Ka-27M 348
Kamov Ka-29 364
  Ka-29TB 364
  Ka-31 364
  Ka-32 364
Kamov Ka-31 380
  Ka-31R 380
  Ka-35 380
Kamov Ka-52 381
Kamov Ka-226 399
  Ka-226T 399
  Ka-226V 399
Kawasaki C-2 436
  RC-2 436

445

Kawasaki Ki.45 Toryu 167
   Ki.45 Kai-a 167
   Ki.45 Kai-b 167
   Ki.45 Kai-c 167
   Ki.45 Kai-d 167
Kawasaki Ki.61 Hien 168
   Ki.61-I-KAIc 168
   Ki.61-I 168
   Ki.61-II 168
   Ki.61 Kai 168
Kazan Ansat 425
   Ansat-1M 425
   Ansat-2RC 425
   Ansat-U 425
Kearsage, USS 334
Khe Sanh siege (1968) 239
Khrushchev, Nikita 303
Königsberg (cruiser) 114
Korean War 164, 192, 194, 196, 205, 210, 212, 214, 217–18, 223, 233, 253
Kursk, Battle of 139

Ladd, Flt Lt Fred 170
Lavochkin-Gorbunov-Gudkov LaGG-3 146
Lavochkin La-5 173
   La-5F 173
   La-5FN 173
Lavochkin La-7 186
   La-7UTI 186
Le Gloan, Pierre 120
Lebanon 301, 340, 359
Leckie, Capt R. 51
Leonardo C-27J Spartan 412
Lexington, USS 68
LFG Roland C.II 32
Liaoning (carrier) 389, 426
Libya 270, 295, 298, 304, 333, 351, 355, 395, 417, 421
Likholetov, Petr Yakovlevich 173
Linebacker offensive 220
Linke-Crawford, Frank 41
Lockheed AC-130 Spectre 284
   AC-130A 284
   AC-130H 284
   AC-130J Ghostrider 284
   AC-130U 284
Lockheed C-5 Galaxy 288
   C-5B 288
   C-5M Super Galaxy 288
Lockheed C-130 Hercules 226
   C-130E 226
   C-130K 226
   C.Mk 1 226
   C.Mk 1P 226
Lockheed C-141 StarLifter 273
   C-141A 273
   C-141B 273
Lockheed F-80 Shooting Star 192
   F-80C 192
   P-80A 192
   P-80A-1-LO 192
   XP-80 192
Lockheed F-104 Starfighter 249
   F-104A 249
   F-104G 249
   XF-104 249
Lockheed F-117 Nighthawk 343
   F-117A 343
Lockheed Hudson Mk.1 127
Lockheed Martin C-130J Hercules 391
   C-130J-30 391
   EC-130J 391
   HC-130J 391
   WC-130J 391
Lockheed Martin F-16 Fighting Falcon A/B 312
   F-16B Block 20 MLU 312
Lockheed Martin F-16 Fighting Falcon C/D 353
   F-16A/B 353

NF-16D 353
X-62A 353
Lockheed Martin F-22 Raptor 362–3, 407
   F-22A 407
Lockheed Martin F-35 Lightning II 433
   F-35A 433
   F-35B 433
   F-35C 433
Lockheed P-3 Orion 266
   P-3C 266
Lockheed P-38 Lightning 149
   P-38H 149
   P-38J 149
Lockheed P2V-1 Neptune 209
   P-2J 209
   XP2V-1 209
Lockheed S-3 Viking 315
   ES-3A 315
   S-3B 315
Lockheed SR-71 'Blackbird' 265
   SR-71A 265
Lockheed T-2V1 SeaStar 243
   T-2V 243
   T-33A 243
   T-33B 243
Lockheed U-2 232
   U-2A 232
   U-2R 232
London, bombing of 18, 49, 189
Lucy, William 114
Ludendorff Bridge 182
Lufbery, Raoul 30
Lugansky, Sergey 135
LVG C.I & C.II 18
   C.III 18
   C.IV 18

Maas, Lt Bruno 22
Malayan Emergency 87, 206, 219
Malta 92, 132
Mannert L. Abele (destroyer) 185
Mannock, Capt Edward 29
Marrett, 1st Lt Samuel H. 102
Marseille, Hans-Joachim 82
Martin B-26 Marauder 158
   B-26B 158
   B-26C 158
Martin PBM Mariner 133
   PBM-3 133
   PBM-3R 133
   PBM-3S 133
Martin T3M and T4M 64
   T3M-2 64
   T4M-1 64
McCudden, Maj James 29
McDonnell Douglas AV-8B Harrier II 334
   AV-8B(R) 334
   AV-8B(R) Harrier II+ 334
McDonnell Douglas/Boeing C-17 Globemaster 377
   C-17A 377
   CC-177 377
McDonnell Douglas F-4 Phantom (exports) 295
   F-4E 295
   F-4K 295
   F-4M 295
   FG.1 295
   FGR.2 295
   RF-4EJ 295
McDonnell Douglas F-4 Phantom II 246
   F-4E 246
   F-4EJ 246
   F-4EJ Kai 246
   F-4G 246
McDonnell Douglas F-15 Eagle 319, 362–3
   F-15A 319
   F-15B 319
   F-15C 319
McDonnell Douglas F/A-18 Hornet 333
   CF-18 333

CF-188A 333
CF-188B 333
EF-18A 333
EF-18B 333
F/A-18A 333
F/A-18B 333
F/A-18C 333
F/A-18D 333
McDonnell F-101 Voodoo 244
   CF-101B 244
   F-101B 244
McDonnell F2H Banshee 212
   F2H-2 212
   F2H-2P 212
McDonnell FH-1 Phantom 208
   F2H-1 208
Mediterranean 88, 92, 94, 103, 132, 148
Messerschmitt Bf 109 82
   Bf 109E 82
   Bf 109E-4 82
   Bf 109G-6/R6 82
   Bf 109K-4 82
Messerschmitt Bf 110 89
   Bf 110C 89
   Bf 110C-4 89
   Bf 110G-4b/R3 89
Messerschmitt Me 163 Komet 183
   Me 163B 183
   Me 163B-1a 183
Messerschmitt Me 210 160
   Me 210A-0 160
   Me 210A-1 160
Messerschmitt Me 262 180
   Me 262B-1a/U1 180
   Me 262A-1a 180
   Me 262B-1a/U1 180
Messerschmitt Me 323 Gigant 175
   Me 323D-1 175
   Me 323E-2 175
   Me 323F-1 175
Messerschmitt Me 410 Hornisse 178
   Me 410A-2 178
   Me 410A-3 178
Mi-24 'Hind' 310
   Hind-A 310
   Mi-24P 310
   Mi-35M 310
Midway, Battle of 101, 106, 109, 170
Mikoyan-Gurevich MiG-3 155
   MiG-1 155
Mikoyan-Gurevich MiG-9 207
Mikoyan-Gurevich MiG-15 211
Mikoyan-Gurevich MiG-17 216
   MiG-17F 216
   MiG-17PF 216
Mikoyan-Gurevich MiG-21 233
   MiG-21B 233
   MiG-21bis 233
   MiG-21F 233
   MiG-21M 233
Mikoyan-Gurevich MiG-23 301
   MiG-23BN 301
   MiG-23MF 'Flogger-B' 301
Mikoyan-Gurevich MiG-27 322
   MiG-27L 'Flogger-J' 322
Mikoyan MiG-29 'Fulcrum' 328
   Fulcrum-A 328
   MiG-29BM 328
   MiG-29SD 328
   MiG-29SE 328
   MiG-29SM 328
Mikoyan MiG-29K 369
   MiG-29KUB 369
   MiG-29M 369
Mikoyan MiG-29M and MiG-35 378
   MiG-29M2 378
Mikoyan MiG-31 344
   MiG-25 344
   MiG-25MP 344
   MiG-31B 344
   MiG-31B/BS Foxhound 344

MiG-31D 346
MiG-31DZ 346
MiG-31M 346
Mil Mi-4 221
   Mi-4A 221
Mil Mi-8 282
   Mi-8MTV-5-1 282
   Mi-8 'Hip- C' 282
   Mi-8T 282
   Mi-17 282
Mil Mi-14 320
   Mi-14PL 320
Mil Mi-17 347
   Mi-17M 'Hip H' 347
   Mi-171 347
Mil Mi-26 341
Mil Mi-28 416
   Mi-28A 416
   Mi-28N 416
   Mi-28NM 416
Mil Mi-171 382
   Mi-17 382
   Mi-171E 382
Mitchell, Reginald 81, 87, 190
Mitsubishi A5M4 'Claude' 86
Mitsubishi A6M Reisen 'Zeke' (Zero) 111
   A6M2 111
Mitsubishi F-1 330
   F-2 330
Mitsubishi F-2 392
   F-2A 392
   F-2B 392
Mitsubishi Ki.46 Dinah 138
   Ki.46- II 138
   Ki.46-III 138
   Ki.46-IVa 138
Mitsubishi Ki.67 Hiryu 184
   Ki.67-I 184
Morane-Saulnier MS.406 95
   MS.406C-1 95
Morane-Saulnier Type N 24
MQ-1 Predator 372
MQ-8B Fire Scout 417
   MQ-8B 417
   MQ-8C 417
   RQ-8A 417
MQ-9 Reaper 397
Muller, Hans 48
Mussolini, Benito 99, 104

Nakajima B5N Tenzan 96
   B5N1 96
   B5N2 96
Nakajima Ki.43 Hayabusa 124
   Army Type 1 Fighter Model 1A, 124
   Ki.43-I, K.43-Ia, Ki.43-Ib and Ki.43-Ic 124
   Ki.43 II, Ki.43-IIa and -IIb 124
   Ki.43-III 124
   Ki.43-Kai 124
Nanchang CJ-6 260
   CJ-6A 260
Nashville, USS 417
Navarre, Jean 30
Nieuport 10 23
Nieuport 11 28, 30
Nieuport 16 28
Nieuport 17 30, 44
Nieuport 24 44
Nieuport 25 44
Nieuport 27 44
North Africa (WWII) 83, 92, 119, 120, 126, 136, 148, 154, 175
North American A-5A Vigilante 262
   A3J-1 262
   A-5B 262
   YA3J-1 262
North American B-25 Mitchell 119
   B-25C 119
   B-25J 119
North American F-86 Sabre 207, 214

F-86D 214
F-86F 207
F-86H 214
P-86A 214
North American F-100 Super Sabre 231
F-100A 231
F-100C 231
F-100D 231
North American P-51 Mustang 134
A-36A 134
P-51B 134
P-51C 134
P-51D 134
North American Rockwell OV-10 Bronco 292
Northrop B-2 Spirit 366
B-2A 366
Northrop F-5E/F Tiger II 317
F-5N 317
RF-5E 317
Northrop P-61 Black Widow 163
P-61A 163
P-61B 163
Norway 97, 114, 174
Nungesser, Charles 28

Okinawa 134, 168, 184, 185, 196
Operation Allied Force 295, 355, 367
Operation Barbarossa 155
Operation Daguet 332
Operation Desert Fox 295, 355
Operation Desert Shield 273
Operation Desert Storm 266, 280, 298, 307, 309, 313, 315, 331, 345, 359, 365, 368
Operation Eiche 104
Operation Enduring Freedom 358
Operation Harvest Reaper 351
Operation Inherent Resolve 370
Operation Iraqi Freedom 219–20, 358
Operation Joint Forge 309
Operation Mountain Sweep 264
Operation Orator 108
Operation Torch 120, 136
Operation Tungsten 174
Osterkamp, Theo 48
Ozawa, Chief Engineer 184

PAC JF-17 Thunder 415
JF-17A 415
JF-17B 415
Panama 290
Panavia Tornado 313
F.Mk 2 313
Tornado ADV 313
Paris, bombing of 99
Parker, Capt 34
Pearl Harbor 96, 109, 111, 115, 123
Peleliu, USS 384
Petlyakov Pe-2 156
Pe-2FT 156
Petlyakov Pe-8 139
Pfalz D.III 47
D.IIIa 47
Philippines 168
Pinsard, Lt Armand 42
Piper L-4 Grasshopper 159
Ploesti oil refineries 154
Polikarpov I-16 74
Type 28 74
Polikarpov I-153 125
Polikarpov, Nikolai N. 125
Polish operations 91, 95
Popkov, Vitaly 173
Porte, Lt Cdr John Cyril 55
Powers, Gary 232
Price, Maj Gen James L. 252
PZL P.7 75
PZL P.11c 75
PZL P.24 75

Rabaul 136, 169
Rakov, Lt Col Vasili I. 156
Ranger, USS 136
Reitsch, Hanna 99
Republic F-84F Thunderstreak 227
F-84 Thunderjet 227
Republic F-105 Thunderchief 250
F-105D 250
F-105G 250
Republic P-47 Thunderbolt 151
P-47D 151
P-47D-15 151
Reshetov, Aleksey 135
Richthofen, Manfred von 45, 57
Rickenbacker, Eddie 10–11, 46
Rockwell B-1B Lancer 355
Rodschinka, Bruno 48
Ronald Reagan, USS 333
Rosatelli, Chief Engineer 76
Royal Aircraft Factory Engine B.E.2c 14
Royal Aircraft Factory F.E.2b 37
F.E.2a, 2c, 2d 37
Royal Aircraft Factory S.E.5a 38
Royal Aircraft Factory R.E.8 29
RQ-4 Global Hawk 386
RQ-4A 386
Rumpler C.I 25
Russian front (WWII) 79, 82, 91, 93
Ryan FR-1 Fireball 195

Saab Erieye 383
E-99 383
Saab J-35 Draken 256
J-35F 256
J-35J 256
Saab JAS 39 Gripen 376
JAS 39C 376
JAS 39E/F 376
Saab SF-37 Viggen 202–3, 302
AJ-37 302
Saipan, USS 208
Saratoga, USS 64, 106, 109
Savoia-Marchetti SM.79 Sparviero 92
SBD-3 Dauntless 136
SBD-1 136
SBD-2 136
SBD-5 136
Scharnhorst (battleship) 114
Schleich, Eduard Ritter von 32
Schneider, Franz 18
Schneider, Walter 150
SEPECAT Jaguar 309
GR.Mk.1A 309
Jaguar A/E 309
Serbia 295, 355
Seversky, Alexander 102
Seversky P-35 102
P-35A 102
Shaanxi KJ-200 410
KJ-200A 410
KJ-200H 410
Shaanxi KJ-500 431
Shaanxi Y-8 346
Y-8C 346
Y-8F-100 346
Shaanxi Y-8 and Y-9 Special Mission 406
Y-8C 406
Y-8G 406
Y-8GX-3 406
Y-8GX-7 406
Y-8GX-8 406
Y-8J 406
Y-9Q 406
Shaanxi Y-9 422
Y-8F600 422
Shandong (carrier) 389, 426
Sheffield, HMS 336
Shenyang J-6/F-6 268
J-6A 268
J-6C 268
J-6III 268

Shenyang J-8 342
J-8A 342
J-8C 342
J-8F 342
J-8I 342
JZ-8F 342
Shenyang J-11 and J-11A 387
Su-27SK 387
Shenyang J-11B 402
Shenyang J-15 426
J-15D 426
J-15T 426
Shenyang J-16 434
J-16D 434
Shikunov, Fedor 161
ShinMaywa SS-2 318
US-1 318
US-1A 318
US-2 318
Shinonome (destroyer) 103
Short Stirling 143
Short Sunderland Mk III 107
Mk I 107
Siddeley S.R.2 Siskin 63
Siemens-Schuckert D.III 48
Sikorsky H-19 Chickasaw/S-55/HRS-1 217
HSS-1 Seabat 217
Sikorsky HH-60 349
HH-60G Pave Hawk 349
HH-60H Rescue Hawk 349
MH-60G Pave Hawk 349
Sikorsky MH-53 Pave Low 290
MH-53J 290
Sikorsky S-65/CH-53 Sea Stallion 277
CH-53D 277
CH-53E Super Stallion 277
CH-53G 277
CH-53K 277
Sikorsky UH-60 Black Hawk/SH-60
Sea Hawk 340
SH-60F 340
UH-60A 340
Sino–Japanese war 96
Sino–Soviet incident (1939) 125
Six-Day War (1967) 261
Smith, Herbert 27
Somalia 340
Sopwith Camel 39
Sopwith Dolphin 59
Dolphin II 59
Sopwith Pup 26
Sopwith Triplane 27
SPAD S.VII 42
SPAD S.XIII 46
SPAD XIII 10–11
Spanish Civil War (1936-39) 76, 122
Spät, Maj Wolfgang 183
Steindl, Lt 82
Strähle, Paul 35
Strasser, Peter 51
Suez Crisis (1956) 193, 219, 227
Sukhoi Su-7 253
Su-7B 253
Su-7BM 253
Su-7BMK 'Fitter-A' 253
Sukhoi Su-15 274
Sukhoi Su-24 321
Fencer-C 321
Su-24M 321
Su-24M2 321
Sukhoi Su-25 345
Su-25SM 345
Su-25UB 345
Su-25UBK 345
Sukhoi Su-27 'Flanker' 327
Flanker-B 327
Flanker-C 327
Su-27P 327
Su-27S 327
Su-30M2 327
Su-30SM 327

Su-35S 327
Sukhoi Su-27K and Su-33 389
J-15 389
T-10K-1 389
Sukhoi Su-30 373
Su-27PU 373
Su-30MKI 373
Su-30SM 373
Sukhoi Su-34 428
Sukhoi Su-35 429
Su-35BM 429
Su-35S 429
Sukhoi Su-57 441
T-50-9 441
T-50S-1 441
T-50S-2 441
Supermarine Scimitar 245
F.1 245
Supermarine Seafire 157
F.Mk III 157
LF.Mk III 157
L.Mk IIC 157
Mk IB 157
Mk IIC 157
Mk III 157
Mk V 157
Seafire 47 157
Supermarine Spitfire 87, 116–17
Mk I 87
Mk V 87
Mk IX 87
Mk VB 87
Mk VC 87
Supermarine Spitfire FR. Mk XIVE 190
FR.Mk XIV 190
Mk XII 190
Mk XIV 190
Supermarine Walrus 81
Mk I 81
Mk II 81
Seagull V 81
Syria 120, 295, 301, 321, 355, 373, 381, 407, 432

T-50 Golden Eagle 408
FA-50 408
T-50i 408
TA-150 408
Taiwan 431
Tamotsu, Lieutenant 86
Tank, Kurt 129
Tarantella, Sgt 76
Taranto harbour 94
Tarawa, USS 164
Thelen, Robert 34
Theodore Roosevelt, USS 340
Tirpitz (battleship) 174
Tokyo, bombing of 181
Transall C-160 275
C-160A 275
C-160D 275
C-160F 275
C-160NG 275
C-160T 275
C-160Z 275
Tupolev, Andrei N. 93, 303
Tupolev ANT-9 65
Tupolev SB-2 93
SB-2bis 93
Tupolev TB-3 67
Tupolev Tu-2 187
ANT-68 187
Samolyet 103U 187
Tu-2S 187
Tu-10 187
Tupolev Tu-16 Badger 230
Tu 16PM Badger-L 230
Tupolev Tu-22 Blinder 267
Tupolev Tu-22M 306
Tu-22M-3 306
Tupolev Tu-95 235

447

'Bear-C' 235	VC10 K.3 352	Vullierme, Marcel 54	Y7-200A 356
'Bear-D' 235	Vickers Wellington 110		Xi'an Y-20 437
Tu-95K 235	Mk I 110	Wagner, Lt Boyd 102	
Tu-95M 235	Mk IA 110	*West Virginia*, USS 185	Yakovlev, Aleksandr 135, 164
Tu-95MS 235	Mk IC 110	Western Front (WWI) 16, 18, 20, 22,	Yakovlev fighters 135, 164
Tu-95RT 235	Mk XV 110	39, 51	Yak-7A 164
Tupolev Tu-128 272	*Victorious*, HMS 132	Westland Lynx 335	Yak-9 164
Tupolev Tu-134 303	Vietnam War 194, 205, 216, 218, 220,	AH.1 335	Yakovlev Yak-1 135
Tu-124A 303	226, 231, 233, 236, 238–41, 246, 250,	AH.7 335	Yak-1M 135
Tu-134A 303	265, 270, 273, 279, 283, 290, 304,	AW159 335	Yak-7A 135
Tu-134B 303	343, 351	Westland Wessex 259	Yak-7V 135
Tu-134Sh 303	*Viraat* (carrier) 380	HAS Mk 3 259	Yak-9 135
Tu-134UB-KM 303	Vogt, Dr 168	HC.Mk 2 259	Yakovlev Yak-38 325
Tu-134UB-L 303	Voisin III 17	Westland Whirlwind 224	Yak-38 Forger-A 325
Tupolev Tu-142 3111	LAS 17	HCC.Mk 12 224	Yakovlev Yak-130 420
Tupolev Tu-160 361	Volkert, George 43	Wing Loong II/Chengdu GJ-2 421	Yak-130M 420
Tupolev Tu-214 423	Vought A-7 Corsair II 285	Winter War (1939-40) 65, 69, 125	Yermolaev, Vladimir 162
Tu-214PU-SBUS 423	A-7D 285		Yermolaev Yer-2 162
Tu-214R 423	A-7E 285	Xi'an H-6 297	Yokosuka MXY7 Ohka 185
	Vought F-8 Crusader 239	H-6E/F 297	Model 11 185
U-boats 55, 127, 133	F-8E 239	H-6H 297	Model 22 185
Ukraine war 294, 299, 342, 381, 432	XF8U 239	H-6K 297	Model 33 185
	Vought F4U Corsair 147	H-6N 297	
Vella Gulf, USS 342	F4U-1 147	Xi'an JH-7 375	Zeppelin-Staaken R.IV and R.VI 52
Verdun, Battle of (1916) 25, 30	F4U-5 147	JH-7A 375	*Zhoushan* (frigate) 379
Vickers F.B.5 16	F4U-5N 147	Xi'an KJ-2000 411	Ziegesar, Joachim von 48
Vickers Valiant 234	Vought OS2U Kingfisher 113	Xi'an Y-7 351	
B. Mk 1 234	OS2U-3 113	HYJ-7 351	
B.2 234	Vought SB2U Vindicator 106	MA-60 351	
Vickers VC10 352	SB2U-1 106	Y-7-100 351	
Type 1101 352	SB2U-3 106	Y-7G 351	
VC10 K.2 352	V-156-F 106	Y-7LH 356	

Picture Credits

AgustaWestland: 335

Airbus: 430 top

AirSeaLand.images: 12 bottom, 13, 14 bottom, 15 both, 17 top, 18 both, 19, 21 both, 22, 23 bottom, 24 bottom, 25, 27–43 all, 45–51 top all, 52 both, 53 top, 54 top, 55, 56 top, 57–58 all, 62–63 all, 67 bottom, 70–71 all, 80, 81, 89, 103–105 all, 109, 116, 128 both, 129, 132 top, 134, 138, 139, 146–147 all, 150, 172 both, 174 bottom, 175, 178, 179 bottom, 185, 191, 194, 196 both, 200, 202, 205–210 all, 213, 215, 218–219 all, 227–229 all, 234, 242, 243, 245 both, 254 both, 257, 258, 262, 283, 287, 300, 301, 303 top, 305, 323, 342, 344, 356, 360

Alamy: 9 top, 10 (Vernon Lewis Gallery/Stocktrek Images), 24 top (Chronicle), 44 bottom (Imago History Collection), 51 bottom (Old-Time Images), 53 bottom & 60 (Chronicle), 77 top (Interfoto), 114 top (Chronicle), 241 (Ivan Cholakov), 351 (Robbie Shaw), 354 (ITAR–TASS News), 374 (Dean West), 375 bottom (Stocktrek Images), 379 bottom (Then Chih Wey/Xinhua), 380 top (Stocktrek Images), 395 (Abaca Press), 400 both (Zuma Press), 406 (Imago), 408 bottom (Sipa US), 416 (Stocktrek Images), 421 top (Imaginechina), 421 bottom (Xinhua), 424 top (Imaginechina), 424 bottom (Associated Press), 426 both (Nurdiansyah Putra), 432 bottom (Kris Christiaens), 438 top (Imaginechina), 439 top (Associated Press), 439 bottom (Xinhua), 440 top (VDWI Aviation), 440 bottom (Avpics)

Amber Books: 75, 76, 78, 79, 82–88 all, 90–92 all, 94–99 all, 107, 108, 110, 111, 118–127 all, 135–137 all, 140–145 all, 149, 151–161 all, 163–166 all, 168, 170, 171, 173, 174 top, 176, 177, 179 top, 180–184 all, 187–190 all, 192, 193, 197–199 all, 216, 225 both, 233, 237, 239, 251, 253, 256, 267, 308, 313, 329, 352

Copyright Saab AB: 376, 383 both

Creative Commons/Alert5: 435 top

Dreamstime: 204 (Andrew Harker), 211 (Gary Blakeley), 224 (Andyperiam), 230 (Meoita), 260 top (Icholakov), 269 (Laxramper), 274 top (Id1974), 279 (Richair), 281 (Starper), 286 (Fotofritz), 289 (Clive117), 293 (Nimdamer), 295 (Dpimborough), 296 top (Laxramper), 296 bottom (Andrewharker402), 299 (Brutusman), 302 (Vander Wolf Images), 303 bottom (Mikefuchslocher), 310 (Peter Lovas), 314 (Viledevil), 316 (Aarrows), 320 top (Vander Wolf Images), 326 (Andrewharker402), 327 (Upadek), 328 (Caspar107), 337 (Igor Groshev), 339 (Ryan Fletcher), 341 (Theflightvideo), 347 top (Andrew Oxley), 347 bottom (Sapilous), 361 (Coprid), 370 (Chiaretz82), 371 (Meoita), 381 (Kison1979), 393 both (Info8234), 396 top (Gordzam), 396 bottom (Jozsef Soos), 398 top (Id1974), 399 top (Eagle2308), 403 (Dragunov1981), 404 top (Mayinxi), 404 bottom (Zhuanghua), 405 top (Igor Akimov), 405 bottom (Dragunov1981), 408 top (Gitraa31), 409 (blurf), 410 (Henryike), 412 bottom (Jozsef Soos), 415 (Vander Wolf Images), 418 (Bogacerkan), 423 bottom (Artzzz), 427 bottom (Vishwakiran), 430 bottom (Nfx702)

Dreamstime/Artyomanikeev: 294, 369, 380 bottom, 398 bottom, 420 both, 423 top, 425, 428

Eurofighter: 401

Getty Images: 14 top (Science & Society Picture Library), 56 bottom (De Agostini), 66 (Royal Air Force Museum), 69 (Museum of Flight Foundation), 77 bottom (De Agostini), 100 (SovFoto), 102 (Corbis), 114 bottom (Royal Air Force Museum), 115 (Albert Harlingue), 132 bottom (Print Collector), 276 bottom (Thomas Klar), 346 (Aamir Qureshi/AFP), 379 top (Bloomberg), 387 (Feature China/Future Publishing), 402 (AFP), 411 (Johannes Eisel/AFP), 413 both, 421 middle, 422 both & 431 both (VCG), 432 top (Anadolu), 434 & 435 bottom (VCG), 438 bottom (Power Sport Images)

GNU Free Documentation License 1.2/Sergey Krivchikov: 311 top

Library of Congress: 8

MoD/Crown Copyright (Open Government Licence): 26, 338 (LA (PHOT) Bunting)

NASA: 315

National Archives & Records Administration: 20

Naval History & Heritage Command: 59, 64, 68, 101 bottom, 106, 112, 113 both, 148, 169, 201, 271

Northrop Grumman: 417 middle, 419 both

Public Domain: 12 top, 16, 17 bottom, 23 top, 44 top, 54 bottom, 65 both, 67 top, 72, 73, 101 top, 130, 131, 133, 162 both, 195, 221, 248 top, 250 top, 272, 276 top

Shutterstock: 247 (Daniele Raffanti), 248 bottom (alexsol), 259 (Kev Gregory), 260 bottom (Ryan Fletcher), 268 (Rehan Waheed), 291 (JetKat), 318 (viper-zero), 320 bottom (Popsuievych), 322 (Alexandr Makedonskiy), 330 (i_moppet), 357 (Earnest Tse), 364 (JetKat), 378 (Media Works), 382 top (Jozsef Soos), 382 bottom (Thierry Weber), 385 (A Periam Photography), 394 (Konwicki Marcin), 427 top (Joe Ravi)

Shutterstock/Fasttailwind: 297, 306, 321, 345, 375 top, 389, 399 bottom, 429, 437 both, 441

Stavka: 74, 93 both, 167, 186

U.S. Air Force: 226 (TSgt Howard Blair), 263 top (MSgt Lance Cheung), 275 (TSgt Larry E. Reid Jr.), 277 (SSgt Hector Garcia), 288 (SA Julianne Showalter), 307 (MSgt Robert Wieland), 311 (SSgt Michael B. Keller), 319 (MSgt Thomas Meneguin), 324 (SSgt Angela Ruiz), 331 (TSgt James E. Lotz), 355 (SSgt Joshua Strang), 358 (SMSgt Christopher Hartman), 359 (TSgt Andy Dunaway), 362 (MSgt Scott Reed), 366 (SA Christina M. Rumsey), 367 (SSgt Aaron Allmon), 368 (MSgt Lance Cheung), 372 top (432nd Air Expeditionary Wing), 372 middle (SSgt John Bainter), 372 bottom (TSgt Effrain Lopez), 373 (SSgt Kenny Kennemer), 377 (MSgt Kevin J. Gruenwald), 386 both (Jim Shryne), 391 top (Yasuo Osakabe), 391 bottom (SSgt Alex Koening), 392 both (SA Tylir Meyer), 397 top, 397 bottom (SSgt Brian Ferguson), 407 (A1 Erin Baxter), 433 top (R. Nial Bradshaw), 436 (A1 Erin V. Currie)

U.S. Air National Guard: 280 both (SSgt Joseph Morgan)

U.S. Department of Defence: 212, 214, 217, 220, 222, 223, 231, 232, 235, 238, 240, 244, 246, 249, 250 bottom, 252, 255, 261, 263 bottom–266 all, 270, 273, 274 bottom, 278, 282, 284, 285, 290, 292, 298, 304, 309, 311 bottom, 312, 317, 325 both, 334, 343, 349, 350, 353, 365, 397 middle, 414, 433 bottom

U.S. Marine Corps: 9 bottom (SSgt Stacy-Ann Viner), 388 (Cpl Michele Clarke)

U.S. Navy: 236 (Lt Peter Scheu), 333 (MC3 Charles D. Gaddis IV), 336 (MC1 Denny Cantrell), 348 (MC2 Jason R. Zalask), 384 both (CWO3 Mark Leung), 390 top (MCS David Danals), 390 bottom (MC3 Class Wilyanna Harper), 417 top (MC3 Charles DeParlier), 417 bottom (Kurt Lengfield)

All artworks Amber Books, except for the following:

Edward Jackson (artbyedo.com): 65, 67, 99 top), 100, 128, 162, 346, 406, 410, 431, 439

Pavel Matviyenko: 260, 292, 400, 413, 420, 438

Rolando Ugolini: 265 bottom, 276, 284 bottom, 291 bottom, 293, 294, 297, 299, 303, 312 bottom, 316, 321 bottom, 327, 334, 342, 345, 351, 353 top, 354, 358 top, 368, 371, 373, 375, 379, 390, 391, 394, 398, 403, 405, 409, 411, 414, 416, 418, 423, 429, 432, 434

Ronny Bar: 12–13 all, 15–20 top all, 21, 22 top, 23–25 top all, 27–28 all, 30, 32–36 top all, 40, 42 bottom, 44 top, 46 top, 47–50 all, 55, 57–59 all

Teasel Studios: 70, 81, 115, 179, 195, 201, 328 top, 348, 357, 364, 369, 377 bottom, 378, 380–382 all, 387, 392, 393, 395, 396, 399, 402, 404, 407 bottom, 408, 412, 415, 419, 422, 424, 427, 430, 433, 435–437 all, 440, 441